THOMAS ALLNUTT BRASSEY
JOHN LEYLAND

THE

NAVAL ANNUAL
1900

Elibron Classics
www.elibron.com

Elibron Classics series.

ISBN 1-4212-4176-5 (paperback)
ISBN 1-4212-4175-7 (hardcover)

This Elibron Classics Replica Edition is an unabridged facsimile of the edition published in 1900 by J. Griffin and Co., Portsmouth.

"FORMIDABLE,"

BRITISH BATTLESHIP.

THE

NAVAL ANNUAL,

1900.

EDITED BY

JOHN LEYLAND.

PART I.—Commander C. N. ROBINSON, R.N.; Commander R. H. S. BACON, R.N., D.S.O.; Messrs. J. R. THURSFIELD, G. R. DUNELL, and DAVID HANNAY; an Anonymous Contributor; and the EDITOR.

PART II.—Lists of Ships: Commander C. N. ROBINSON, R.N., and the EDITOR; Plates: F. K. BARNES, M.I.N.A.

PART III.—Captain ORDE BROWNE, late R.A., Lecturer on Armour to the R.A. College.

PART IV.—FIRST LORD'S MEMORANDUM, NAVY ESTIMATES, BRITISH AND FOREIGN, and the GERMAN NAVY BILL.

1900.

PORTSMOUTH:

J. GRIFFIN AND CO., 2, THE HARD.

(BOOKSELLERS BY APPOINTMENT TO HER MAJESTY AND H.R.H. THE DUKE OF EDINBURGH.)

London Agents: SIMPKIN, MARSHALL & CO.

Foreign Agents:

PARIS: BOYVEAU & CHEVILLET, 22, RUE DE LA BANQUE.

NEW YORK: D. VAN NOSTRAND COMPANY. BERLIN: W. H. KÜHL.

HONG KONG, SHANGHAI, AND YOKOHAMA: KELLY, WALSH & CO.

TOKIO: Z. P. MARUYA & CO.

LONDON:
PRINTED BY WILLIAM CLOWES AND SONS, LIMITED,
STAMFORD STREET AND CHARING CROSS.

PREFACE

Owing to the fact that I have been appointed to raise and take command of the Sussex Company of Imperial Yeomanry for service in South Africa, it has been necessary for me to place in other hands the work of editing the *Naval Annual*, a work which it has been my pride and pleasure to perform for the last ten years.

Mr. John Leyland has kindly undertaken the duties of editor for the present year, and I trust that he will receive the same sympathetic consideration from our readers as his predecessors.

T. A. Brassey.

R.M.S. "Carisbrook Castle,"
Southampton,
March 31*st*, 1900.

The war in South Africa could not leave the *Naval Annual* unchanged. It has deprived the new volume of the editor who has conducted the long series of volumes so capably. Mr. Brassey having honoured me by placing the editorship of the issue of 1900 in my hands—an honour I greatly prize—my single object has been to give to it the character possessed by its predecessors. Neither Lord Brassey nor his son has been able to contribute. Commander C. N. Robinson, R.N., who has often written for the *Annual*, is responsible for the chapter on the Progress of the British Navy, usually from the pen of Mr. Brassey. He has also shared with me the work of revising the lists of British and Foreign Ships, and I am indebted to him for help in other ways. M. Weyl, who for many years reviewed the Progress of Foreign Navies, but was unable to take up the chapter last year, has since died, and the *Annual* has lost in him a valued contributor. A strong effort has been made to give the chapter its accustomed character of completeness and accuracy.

After an interval of a year, Mr. Thursfield contributes a chapter on the Naval Manœuvres, dealing with them descriptively and critically with exhaustive fulness, and Commander Bacon has written a most suggestive and useful chapter on the Tactics of Fast Craft.

A very important chapter is that on Naval Training, in relation to the abolition of the masted training squadron, which is from the pen of a writer of the highest competence, who remains anonymous. A contributor new to the *Annual* is Mr. David Hannay, who writes upon the Employment of Naval Brigades, treating the subject historically. Mr. Dunell again reviews the Progress of Marine Engineering.

In Part II., owing to the absence from England of Mr. Barnaby, much of the work of preparing the plans of ships has fallen upon Mr. F. K. Barnes, who has been associated with the *Naval Annual* from the beginning, and to whom I am indebted for much help in passing the present volume through the press. It will be seen that the number of plates has again been increased, and that the ships represented are much more numerous, owing to many of the diagrams having been redrawn. Plans will be found of our latest battleships, including the Formidable, Canopus, and Duncan classes, and of our most recent cruisers. Foreign ships are equally well represented, the Henri IV., and the latest cruisers, for example, being among the French plates, and the new armoured class among the Italian.

Part III. is again in the capable hands of Captain Orde Browne, who has replied to the erroneous strictures upon our guns offered by M. Claudinon in the French Chamber. Some notes will be found added to the chapter on Ordnance on the wheel and other mountings provided for the naval guns which have done such good work in South Africa, and certain of the mountings are illustrated.

Among the papers in Part IV. will be found a translation of the German Navy Bill, which is a most instructive document, both from the naval and political points of view.

Finally I have to express indebtedness to many correspondents in all parts of the world, and particularly to Mr. Charles de Grave Sells, of Sampierdarena, Italy, for valuable information, as also to the public press, which has contributed much to these pages.

Forest Hill, JOHN LEYLAND.
May, 1900.

CONTENTS.

PART I.

PART II.

PART III.

Captain Orde Browne, late R.A.

PART IV.

LIST OF ILLUSTRATIONS.

CHARTS AND DIAGRAMS.

PART I.

CHAPTER I.

Progress of the British Navy.

The standard of strength.

In preparing this statement of the progress of British warship construction it seems natural and appropriate to preface the chronicle of work done at home by a glance around the naval horizon. Several reasons incline me to this course, and not the least that we have it on the best authority that what is occurring abroad in the field of naval exertion has a most important bearing upon the promise and fulfilment of the various programmes of shipbuilding projected from year to year by our own Admiralty. The British standard of strength by sea is admittedly relative to something which depends on the action, actual or proposed, of others; and although it may be modified or limited by factors over which those others have no direct control, these factors again, be they political or material, must be always of secondary importance. It is not so much the multiplicity or the value of the interests to be protected that determines the strength of the defensive force as the number and activity of possible depredators. It appears therefore that this thought should be in one's mind when reviewing our naval progress, that whether we find the aims and ends authoritatively set forth as essential to have been achieved, or if, on the contrary, we discover retardation and postponement, the result should be measured against what is going on elsewhere, and no kind of persuasion or argument based upon other conditions should content or satisfy us.

Naval expansion of the Powers.

Regarding the naval horizon from this standpoint, it has to be confessed that the prospect is not altogether so clear as could be wished. There are signs to which we cannot shut our eyes, signs which may presage a coming storm, may herald passing squalls, or may be merely clouds which will disappear with the dawn. That there is unusual activity and movement in the naval establishments of several of the foreign Powers is obvious, but there is no reason to doubt the knowledge by our own naval authorities of all that is occurring in foreign dockyards and arsenals. The significance of the circumstances just referred to rests rather with the future than with the present. It is quite possible to agree with the Secretary of the Admiralty when he asserts that the Navy of this country is perfectly competent to deal with any attack upon us that could now be meditated by any possible combination of Powers, and yet to feel a certain amount of apprehension lest our preparations should not be altogether sufficient or adequate to enable us to continue in this entirely satisfactory condition. The indications which it appears to be most necessary to watch carefully are those which virtually denote the introduction of new naval elements in the direction of world policy.

Germany. In our naval survey it is Germany that looms largest because she exhibits a desire and a determination to raise herself from the position of a comparatively weak sea Power to that of one among the more potent. Elsewhere in the *Annual* will be found a translation of the material points in the far-reaching programme of shipbuilding for which the Emperor and his ministers are still seeking the authority of the Reichstag. That the necessary sanction will be obtained need not be questioned, and providing the work be carried out promptly, the extension of her fleet must become all that the most ardent naval enthusiasts among her people can wish. Of the Emperor's intentions with regard to his Navy there is no concealment. In his New Year's speech to his generals he is reported to have said:—

"At the beginning of the century the army of Frederick the Great had fallen asleep on its laurels. It was guided by senile generals, and its officers were ruined by luxury and stupid arrogance. Our punishment was severe; the army was thrown into the dust. Frederick's glory faded; his banners were broken. During seven years of hard slavery God taught us to recover ourselves under the pressure of an overbearing conqueror Our nation established general military service, which gained the greatest importance under my grandfather, who reorganised the army in spite of all stupid opposition. His spirit revived the army, and his confidence in God carried the army away to unforeseen victories. So he united the German nationalities. By our army Germany regained her position in the council of nations. You, my generals, must preserve and prove the old qualities in the new century. Simplicity and modesty in life and daily sacrifice to the royal service must be your rule. *As my grandfather did for the army so will I for the navy, carry out the work of reorganisation. The navy must be equal to the army Then I shall be enabled to procure for Germany the place among foreign nations which she has not yet obtained.*"

Practically the result of the German programme, if fully carried out, will be to double the Fleet during the next sixteen years. We should not hastily deem such an increase to be a menace to our supremacy at sea, and individually it cannot be so. At the same time the conditions are conceivable in which if such a new force were thrown into the balance against us it might make our position one of great jeopardy, for assuming that the German Fleet remained intact during the continuance of a war in which this country was involved with other sea Powers it might be used to impose terms of peace, or partition, inimical to our interests and the continued maintenance of our Colonial Empire.

Japan. Two other Powers are adding or projecting additions to their naval strength in a manner which demands notice. They are Powers which happily may be considered as actuated by motives towards this country which are entirely friendly. Both the United States and Japan have recently been engaged in naval war, and in each case the truth of Captain Mahan's teachings as to the influence of sea power has been borne home to their peoples. It is not surprising therefore to find that they are making great efforts to be strong on the ocean. When her present programme of shipbuilding is completed—and this will be almost immediately—Japan will have quadrupled her naval strength as compared with what it was when she drove China from the sea. As a result, there will be forged in the Far East a weapon of immense power, and one that we may be sure will be used skilfully and ruthlessly to enforce the wishes of the enlightened rulers of the Land of the Rising Sun. We have to look back over a somewhat longer period to obtain the contrast, but the rehabilitation of the Navy of the United States presents a picture quite as striking as that of the Eastern Power. The natural and characteristic aspiration of our American cousins that their Fleet shall be home-grown seems to be alone responsible for the fact that its increase has not advanced so rapidly as has that of Japan. The progressive strides, however, have been continuous, and are consistent with a continuity of policy which exhibits no signs of abatement or change. **United States.**

Turning from the newer naval elements to the old, symptoms of relaxed efforts, navally speaking, have yet to be discovered in either France or Russia. The keynote of the latest French programme is that the new ships they will build shall be individually superior in all respects to any possessed by this nation. It is unnecessary here to refer to the programme in detail, but it entails a total expenditure of over thirty-five millions sterling, proposed to be spread over seven years, and includes an outlay of something like four millions sterling annually upon new construction. There should be some consolation for those who are most pessimistic in this country about our naval superiority in the fact that in France continuity of policy in regard to warship construction is not a marked feature. There the "naval expert" has been able to work his will in a way which has been granted to him nowhere else. It may be that the convictions of General de la Roque and M. Claudinon are shared by few other French naval critics, or that the policy recommended by Captain Chevalier has little chance just now of being accepted, but successive changes at the French Ministry of Marine have had results so important and peculiar that we are justified in regarding with a certain amount of suspicion, not unmingled with satisfaction, the accomplishment of any project hatched at the Rue Royale. **France.**

As it is in France, so also is it in Russia, but with a difference. There all the ships of the old programme are now in hand, while **Russia.**

there are rumours of a fresh one. But if it would not be wise to altogether ignore the statements about further naval construction, it would be weak also to shut our eyes to the many circumstances which seem to be unfavourable to any immediate effort to launch out into increased expenditure in this direction. It is not likely that Russia will place more orders for ships abroad, and her own building facilities can hardly be able to stand a much heavier strain than that imposed upon them at the present time. An interesting light is thrown upon the subject in Mr. Jane's book "The Imperial Russian Navy," but the trial which has recently taken place at Sebastopol indicates internal difficulties in the way of insuring efficiency in the fleet which deserve equal attention. The system of providing stores seems to have been completely rotten. Coal, oil, wood and iron have been paid for which have either never been delivered, or have been proved to be of inferior quality. Out of some forty or fifty persons accused of defrauding the Government, the greater number have been found guilty, the least crime for which these naval officers and officials were condemned being the receipt of commission from merchants and contractors. Too much might easily be made of these circumstances, especially as the frauds here exposed seem to have been confined to those persons connected with the victualling and supply of the Black Sea Fleet. Our own naval annals, even at the time when we were gaining our greatest victories afloat, are too full of similar occurrences to justify us in laying much stress upon them as evidence of general naval weakness.

Italy, Austria.

Turning to the remaining Naval Powers, it cannot be said that any one of them has recently shown special cause for mention in a rapid survey such as this, although it may be noted that everywhere almost there is a similar inclination to maintain, if not to increase, strength at sea. Both Italy and Austria are adding to their Navies, although in nothing like the same proportion as the other Powers that have been mentioned. While in regard to the Fleets of the minor nations, perhaps the most significant feature of recent years has been the readiness which some of them have shown to part with their ships when conditions arose which created a profitable market. This is a matter it would be well to bear in mind when making our estimates of relative strength.

On the whole, if the naval outlook is one not altogether conducive to a state of contented and apathetic self-confidence, it cannot be said that it presents features quite so alarming as some would have us believe. This is evidently no time to slacken our preparations, and if our position at the present time is one as satisfactory as it appears to be, it is only because we are reaping the benefit of the large investments which were made, and the sacrifices to that end which were willingly suffered by the people of this country in recent years. That the nation realises that it has got good value for its money is beyond a doubt, and that it is quite ready to spend more in order to insure similar results is equally undeniable. If anything should shake the confidence which the people have consistently reposed in the ability and determination of their advisers to maintain the standard of naval strength essential to conserving Imperial interests, it would be, not a demand for further sacrifices, but a relapse into unwarrantable optimism and unwise parsimony.

The Navy and Transport to South Africa

There is one other subject to which reference may be made before passing to the summary of naval progress. In 1896 attention was called in this chapter to the circumstances that the clouds which had passed over our relations with foreign Powers about that time had compelled British citizens, not only in the Mother country, but in the Colonies, to recognise the vital importance of the British Navy to their security. This year it may be said that the war in South Africa, and the circumstances in connection with the transmission of our armies thousands of miles across the seas with as much safety as if they had but traversed our own territory, has provided an object-lesson of a similar nature to the world at large. It is certain that but for the knowledge abroad that our naval forces were ready to effectually deal with and determine any issue, we should long ere this have heard something more than mere rumours of interference on the part of some of our European friends. The feeling of maritime security has been a great source of confidence at home, and although popular misconceptions as to the use and application of naval power have led in some cases to apprehension and talk of invasion and the like, there is reason to believe that more intelligent views will in the end prevail. It will be surprising indeed if the nation does not see in the manifold illustration of what Captain Mahan has called the "silent" influence of sea power, which the circumstance of the South African struggle has called forth, renewed reasons for maintaining in the highest state of efficiency, and at any necessary standard, that force which is an essential element in the continuance of our national prosperity and progress.

Recent Expenditure.

In summarising the Admiralty proposals for 1900–1901, it will be useful to refer briefly to the Estimates for some few years past. Those presented to Parliament in March, 1895, proposed an expenditure of £5,700,000 on new construction, a sum considerably in excess of what had been spent in the three previous years. As a result, the year 1895–96 was remarkable for the unprecedented activity in ship-building displayed both in the dockyards and in the private establishments. At Portsmouth and Chatham two battleships of 15,000 tons were completed within two years from the date of their commencement. In the following year (1896–97) the Estimates again showed a large increase, the shipbuilding vote being raised to over seven millions sterling; also the number of ships completed exceeded the number completed in any previous year since 1893–94. In 1897–98, although the Estimates were nearly the same, there was a decrease in the shipbuilding vote, but the sum devoted to new construction still stood at something like seven and a half millions sterling. During the twelve months ending in March, 1898, however, in consequence of the dispute in the engineering trade, the sum left unspent on new construction amounted to £2,139,000. Each year in succession since that date there has been under expenditure of the moneys voted for shipbuilding; the total sum thus unexpended in the three years ending March, 1900, amounting to £4,343,000. These figures are undoubtedly large, and have been made the basis for hostile criticism of the Admiralty policy, particularly by the Navy League. Mr. Goschen has explained the cause of this under expenditure in his memorandum thus:—

"The abnormal activity in shipbuilding and engineering which was described in the statement for last year has continued during 1899–1900 and has seriously affected progress and expenditure on ships, machinery and armour. Delays in delivery of material, difficulties in securing adequate numbers of workmen, and other circumstances have caused the adequate earnings on contract work to fall short of the estimated amount by £1,400,000, though the estimate was carefully calculated on the basis of actual earnings in past years on ships of similar character, and on very close investigation of the possible output of armour . . . . . The fact that so large a number of ships now in construction are designed for exceptionally high speeds, and will therefore be equipped with propelling machinery of great power, also tends to affect the rate of progress. Machinery of this kind can only be produced by firms of the first rank, who are limited in number, and who, in many cases, have other important contracts in hand. Longer periods are required for the manufacture and erection of the machinery, with the natural result of more time being necessary for the completion of the ships."

The Navy Estimates.

The Estimates for 1900–01 amount to a net total of £27,522,600, being an increase of £928,100 beyond the sum voted for the previous year. The estimated expenditure on new construction is, however, less by £395,335 than the sum voted for the same purpose in 1899–1900, the figures being £8,460,146 against £8,855,481 in the previous year, but it is larger by £1,131,179 than the actual expenditure for that year, and if it should really be spent it will represent an expenditure larger by more than a million than has ever yet been reached. In connection with "the failure on the part of the contractors for armour, hulls, and machinery to earn the money," which he hoped would have been spent during the year, and in explanation of his not laying down more ships, Mr. Goschen, in a speech on the Estimates, on Feb. 26, said: "Our programme is limited to what we believe to be the output of the country in armour, hulls, machinery, and the vast number of accessories to be provided." The information upon which this statement was based appears to have been somewhat misleading, since, as was shown in the *Annual* for 1894, the private shipbuilding establishments are certainly capable of a very much larger output of warships than has been demanded of them in the past year. Lord Hopetoun, in his address as President of the Institution of Naval Architects, also ventures to doubt if "we have reached the limit of our producing power," and his Lordship is further of the opinion that "the difficulty of procuring armour-plate seems to be the more acute of the two questions." In regard to this question, too, it has been shown in the *Annual* that our armour-plate manufacturers declare themselves to be perfectly capable of meeting any demand that might be made upon them, and Sir Alexander Wilson, the chairman of Messrs. Cammell, recently stated that "there need be no alarm on the score of supplying armour to the Government so far as Sheffield was concerned." The probable explanation of the seeming discrepancy between the assertions of the First Lord, and those of the manufacturers and shipbuilders, is that Mr. Goschen was referring, not to such a demand as might be made on our resources in an emergency, but to the result of his enquiries and experiences when the Admiralty is competing in ordinary times with other employers in the same field. He says in his memorandum on the Estimates :—

Mr. Goschen's explanation of delay in construction.

"The experience gained in recent years that after the most careful calculations as to the probable earnings of contractors for hulls of ships, machinery, and armour, the expenditure for new construction has continually failed to reach the sum voted, has been taken into account in framing the Estimate for 1900–1901. If the contractors should earn more instalments than are estimated for in the proposed vote, a Supplementary Estimate would, of course, be necessary."

It is a commonplace that in any estimate of progress in naval construction the basis of calculation must be completed ships. Yet, although this is the case, it is far from infrequent to find writers on naval matters drawing up their estimates in terms of ships laid down and ships launched. The result cannot fail to be misleading to the public, while its tendency is also to make authority anxious to fill the yards with vessels in various stages of construction. It was, unless I am mistaken, Sir Edward Reed who first pointed out that capital in uncompleted ships was entirely unremunerative, since no return for the investment could be possible until the vessels were able to take their places in the fighting line. There was a time, indeed, when it was seriously argued in favour of keeping ships unfinished, that thereby the authorities were able to introduce into their construction or equipment the improvements and alterations suggested by experience which were constantly being made. The process, however, was extremely expensive and extravagant, and for many years past it has been the avowed policy of the Admiralty, when laying down new ships, to complete them as rapidly as possible. That this policy is the best possible can hardly be in question, but the point of these reflections is that in this survey of the year's naval progress there appear to be certain signs of a retrograde character exhibited by an extension in the period during which ships are under construction, and by the increase in the number of the new vessels in hand.

Programme of 1896–97.

In the *Annual* for last year it was pointed out that, although no new battleships had been completed in the twelve months under review, no less than sixteen were under construction or projected. Since that date only three have been completed; there are yet fifteen battleships under construction, and two more are to be laid down. The three completed battleships are the Canopus, Goliath, and Ocean, of the programme of 1896–97. The following tables, which have been taken in part from an article in *Engineering* and in part from the *Times* reports, give particulars of the steam trials of these three vessels during 1899.

Canopus, Ocean, Goliath Trials.

	30 HOURS' COAL CONSUMPTION.			30 HOURS' COAL CONSUMPTION AT HIGHER POWER.			FULL POWER.		
	I H.P.	Speed.	Coal per H P.	I.H.P.	Speed.	Coal per H.P.	I.H P.	Speed.	Coal per H.P.
		Knots	Lbs.		Knots	Lbs		Knots.	Lbs.
Canopus . . .	2813	11·3	1·82	10,457	17·2	1·68	13,780	18·5	1·72
Ocean . . .	2767	11·4	1·84	10,314	16·2	1·63	13,828	18·74	1·76
Goliath . . .	2807	11·7	1 73	10,413	17·1	1·54	13,918	18·4	1 91

THIRTY HOURS' COAL CONSUMPTION AT ONE-FIFTH POWER

	Mean Draught.	Total I H P.	Mean Revolutions	Steam in Boilers.	Speed.	Coal per I.H.P. per Hour.
	Feet.			Lbs.		Lbs.
Canopus . .	26	2812	64·2	230	11·3	1·82
Ocean . . .	—	2769	66·8	210	11·4	1·84
Goliath . .	—	2807	65·7	236	11·7	1·73

THIRTY HOURS' COAL CONSUMPTION AT FOUR-FIFTHS POWER.

	Mean Draught.	Total I.H.P.	Mean Revolutions.	Steam in Boilers.	Speed.	Coal per I.H.P. per Hour.
	Feet			Lbs.		Lbs.
Canopus . .	26	10,454	99·7	255	17 2	1·68
Ocean . . . {	—	10,314	—	259	16·2	1·6
	—	10,303	94·4	264	15·5	—
Goliath . .	—	10,413	100·25	273	17·3	1·54

EIGHT HOURS' FULL POWER.

	Mean Draught	Total I.H.P.	Mean Revolutions.	Steam in Boilers.	Speed.	Coal per I H.P. per Hour.
	Feet			Lbs.		Lbs.
Canopus . .	26	13,763	108·5	289	18 5	1·72
Ocean . . .	—	13,828	113	272	18 5	1·7
Goliath . .	26	13,918	108·2	290	18·4	1·91

The Canopus hoisted the pennant for the first time on December 5th, 1899, and has proceeded to the Mediterranean for a commission. The Ocean was commissioned on February 20th for the same station, and on March 27th the Goliath followed her sisters into active service. She is to become a flagship on the China station.

The remaining battleships of the same programme and type are the Glory, Albion, and Vengeance. Of these the two first-named would, it was hoped, be completed before the close of the financial year 1899–1900, but this anticipation has not been realised. The Glory was floated out of dock at Birkenhead on March 11th, 1899; she has been delivered by the contractors, and has completed her official trials.

Glory Trials.

The results of these are thus given in the *Times* reports. On the thirty hours' trial, at a nominal power of 10,250 I.H.P., the draught was 26 ft. fore and aft, and the trial was run with 240 lb. of steam in boilers. The vacuum was 27·4 in. starboard and 26·4 in. port. The revolutions were 99·4 starboard and 99·2 port. The mean H.P. of the thirty hours was 10,587, with a vacuum in smoke boxes of 0·34 in. The ship made four runs over the deep sea course, and the recorded mean speed was 16·78 knots, which is less than the actual speed, as the ship on the third run over the measured distance had to go out of her course to avoid a sailing vessel. The eight hours' full power trial took place on February 23rd. At 5.30 in the morning the vessel was got under way at Portland with 70 revolutions, which gave her a speed of 12½ knots, but the speed was steadily increased to full power by the time the new twenty-five fathom course-mark off the Cornish coast was reached at 8 o'clock. She then made four runs over the 23-mile course, and finished her trial on the return run up Channel. She drew 26 ft. fore and aft, and had 265 lb. of steam in her boilers. The vacuum was 27 in. starboard and 25·8 in. port, and the revolutions were 108·5 starboard and 106·7 port, with a total I.H.P. of 13,745. There was no air pressure, and the mean speed of four runs over the course was 18·124 knots. The mean vacuum in the smoke-boxes was 0·4 in., and the coal consumption worked out at 1·58 lb. per unit of power per hour. The principal dimensions of the Glory are:—Length, 390 ft.; beam, extreme, 74 ft.; displacement at load draught, 12,950 tons. Her propelling machinery, designed by Messrs. Laird, consists of two sets of triple-expansion engines, each having three vertical cylinders of 30 in., 49 in., and 80 in. in diameter respectively, with a piston stroke of 51 in. They each drive a four-bladed gun metal screw propeller. The engines are designed to develop 13,500 I.H.P. at full power. Steam is supplied by twenty water-tube boilers of the Belleville type, consisting of fifteen generators of nine elements, and five of eight elements, with an economiser to each boiler. The total heating surface of generators and economisers is 33,700 sq. ft., and the grate surface is 1,055 sq. ft. The vessel is to be completed for sea at Portsmouth Dockyard.

Albion delays.

The Albion having been launched at Blackwall, on June 21st, 1898, has been delayed by financial difficulties of the contractors for the engines, but special arrangements have been made for finishing their work, and it is hoped that she may be delivered this year. The Vengeance, which was not laid down until August, 1897, at Barrow, will not, it is anticipated, be delivered before July, 1900. She was launched at Messrs. Vickers, Sons, & Maxim's yard, at Barrow, on July 25th, 1899.

Vengeance.

The Vengeance closely resembles the Canopus type, though, having been ordered a year later, she embodies some changes in detail. The following description is from the *Times:* Length 390 ft., beam 74 ft., load draught 26 ft., displacement 12,950 tons. The Vengeance is built on the double bottom system, the inner as well as the outer skin being carried up the side of the ship to form the armour shelf 6 ft. below the load waterline. This armour extends for nearly two-thirds of the length of the vessel, forming with the armoured bulkheads across, at the fore and after end, a citadel 230 ft. in length and the full width of the ship. Within the length of this citadel are placed not only all the guns, but also the magazines, etc., necessary for the fighting of the ship. The side armour forming this citadel is 6 in. thick, while the end bulkheads have an average thickness of 10 in., all specially hardened. The ends of the ship are not left unprotected. The main belt is continued forward in the form of 2-in. nickel steel, which widens out so that the whole of the sides of the ship at the ram are coated with this thickness of metal. The Vengeance may thus ram an adversary's ship without her skin plating being ruptured. Again, at the stern the usual skin plating is doubled for a considerable part of the depth of the ship from the point where the ordinary armour ceases sternwards. There is a protective deck 2 in. thick from end to end, enclosing machinery, boilers, magazines, etc., while extensive coal bunkers are arranged along either side to assist in the protection of the ship from the fire of an enemy's guns.

The big guns are of the 12-in. type, each weighing about 50 tons. There will be four of these weapons, mounted in pairs in heavy barbettes situated at the forward and after end of the citadel and covered in with a specially large armour hood. These barbettes are 37 ft. in diameter and are built up of armour plates 12 in. thick, with teak planking within the walls. The guns fire shots of 850 lb. in weight, capable of penetrating 36 in. of wrought iron placed at the muzzle. Each shot needs more than 150 lb. of cordite. The ammunition is sent up from the magazine below through armoured hoists. In addition to the big guns there are twelve 6 in. Q.F. guns in casemates formed of 6-in. armour. Four of these guns fire right ahead and four astern, as well as on the broadside. The Vengeance will have thirty smaller guns. There will be a military top on each of the two masts, which are made specially lofty for signalling purposes.

The Vengeance has two screws, each driven by an independent set of triple-expansion engines, with three vertical cylinders of the collective power of 6,750 I.H.P., the aggregate being 13,500. This is attained with the engines making 108 revolutions and with a steam boiler pressure of 300 lb. per square inch, reduced to 250 lb. at the engines. There are twenty boilers of the Belleville type, with economisers and all recent improvements. Each boiler can be used independently of the others and works at 300 lb. per square inch. The boilers are arranged in three compartments, eight in each of the forward and middle boiler rooms, and four in the after room. There is no middle line bulkhead in the boiler rooms. The heating surface is 21,760 square feet in the main tubes, and 12,010 square feet in the economisers, the total being 33,770 square feet. The boiler tubes are all of British manufacture, as is also the material from which they are drawn. The tubes are all solid drawn, finished cold, carefully annealed after manufacture, and subjected to severe tests. The distilling machinery consists of two evaporators capable of evaporating from sea water 68 tons per 24 hours. The two distillers produce 40 tons of fresh aerated water per day for drinking, at 15 deg. Fahr. above that of the circulating water. The electric light machinery consists of three sets of combined engines and dynamos. There are four sets of engines and pumps for air compressing, two boat hoists, two refrigerating machines, two coal hoists, five blowing engines, etc., while for ventilating the ship there are eight electrically driven fans. There are also two steam fans. The ship will have two masts and two funnels, the latter being 11 ft. in diameter and rising to 90 ft. from the grates of the boilers. The upper deck is flush from bow to stern. This is the first battleship built by Messrs. Vickers, Sons, and Maxim.

Programme of 1897–98.; Formidable class. The three ships of the Formidable class laid down in 1898, and launched in a very incomplete state in order to make room on the slips for the building of three sister vessels, have made fairly good progress; their completion will depend upon the delivery of certain portions of their armour. It is unnecessary to repeat here a description of the class, as it was fully described in last year's *Annual* in this chapter, and in the First Lord's Memorandum (*Naval Annual*, 1898, p. 425).

Programme of 1898–99 The London, laid down at Portsmouth on December 7th, 1898, was launched on September 21st, 1899; the Bulwark, laid down at Devonport on March 20th, 1899, was launched on October 18th, 1899; and the Venerable, laid down at Chatham on January 2nd, 1899, was launched on November 2nd, 1899. The time occupied at the respective dockyards from the laying down of the keels to the launching of these vessels was thus nine and a half months at Portsmouth, seven months at Devonport, and 10 months at Chatham, their launching weights being: Venerable, 5,200 tons; Bulwark, 5,450 tons, and London 5,200 tons.

Venerable class. The Venerable, London, and Bulwark, are identical in form, displacement, and dimensions, with the Formidable class, but differ slightly in the distribution of armour protection. In the Venerable class the belt of side-armour extends much nearer the bow than in the Formidable class, described in the *Naval Annual* for 1899, while the fore armoured bulkhead has been dispensed with.

One of the features of the new ships is their comparative lightness of draught, which will enable them not only to pass through the Suez Canal without reduction of stores, but to manœuvre in many parts of the world where vessels of the Majestic class would be at a disadvantage.

The following description of the Venerable was given in the *Times* of November 11th, 1899:—Length between perpendiculars, 400 ft.; extreme breadth, 75 ft.; draught of water—forward, 25 ft. 3 in., aft, 27 ft. 3 in.; displacement, 15,000 tons. The armour, which will be treated by the improved Harveyed process, is 9 in. thick and 15 ft. deep amidships, while forward it varies from 3 in. to 7 in. in thickness. A rounded armour bulkhead is fitted at the after end of the belt of Harveyed steel, and varies in thickness from 9 in. to 12 in. The whole of this armour is being supplied by

Messrs. Brown. The two circular barbettes are also of Harveyed steel, the upper tier of plates being 12 in. thick both fore and aft, and the lower tier 8 in. and 6 in. thick forward and 6 in. thick aft. The main deck is protected by plating 2 in. thick on the fore side of the armour bulkhead amidships and plating 1½ in. thick forward. On the middle deck the plating is 1 in. thick from the armour bulkhead to the fore side of the forward barbette, where plating 2 in. thick slopes down to the lower deck. Messrs. Maudslay, Sons, and Field are the contractors for the machinery, the engines being of the inverted triple-expansion type. The indicated horse-power of the engines is 15,000, and this will give a speed of 18 knots. Steam will be supplied by twenty separate water-tube boilers of the Belleville type. The safety valves are loaded to a pressure of 300 lb. per square inch, the steam pressure being reduced to 250 lb. per square inch at the engines. The steering gear is Harfield's patent compensating type, and there is a steering engine in each engine-room, having separate steam and exhaust pipes. Each engine is of sufficient power to put the rudder from hard over to hard over (or through 40 degrees) in thirty seconds, when the ship is steaming at full speed. Provision is made for the stowage of 2,040 tons of coal, which is sufficient to enable the ship to steam for thirty days at a speed of ten knots.

The main armament of the Venerable consists of four 12-in. (46-ton) B.L. wire guns of improved pattern, mounted in pairs on turntables in the barbettes. The guns, which are fitted with all-round loading mountings by Messrs. Vickers, Sons, & Maxim, are protected by strong shields of 8 in. and 10 in. in thickness. The auxiliary armament mainly consists of twelve 6-in. Q.F. guns, eight being placed in casemates on the main deck and four in casemates on the upper deck. Besides these the ship will carry sixteen 12-pr. guns, equally divided between the upper and main decks, two 12-pr. boat and field guns, and eight 0·45 Maxim guns. Each of the two military tops will be fitted with three 3-pr. Hotchkiss guns. The vessel is pierced for four submerged torpedo-tubes, two on the broadsides forward, and two aft. She will carry fourteen 18-in. torpedoes, and five 14-in., the latter being fired with dropping gear from the ship's steamboats. The ship has two steel masts, 3 ft. in diameter, each mast carrying a fighting top. She will also have striking topmasts, standing 160 ft. above the water-line, with a truck semaphore at the main and a multiple fibre flashing-lamp at the fore. The boats are seventeen in number—namely, two 56 ft. steam pinnaces, one 40 ft. steam pinnace, one 40 ft. admiral's steam barge, one 42 ft. launch, one 36 ft. pinnace, two 34 ft. cutters, one 30 ft. cutter, one 32 ft. gig, one 28 ft. gig, one 24 ft. gig, one 27 ft. whaler, one 32 ft. admiral's galley, two 16 ft. skiff dinghies, and one 13½ ft. Balsa raft.

The ships of this class differ from those of the Canopus, Majestic, and Royal Sovereign classes in having the steering engines in the engine-rooms; and as the ship's fans and after capstan are driven by electric motors, there are no steam-pipes aft, which will improve the habitability of the ship. Instead of the ordinary controlling shafting, Brown's telemotor gear will be fitted for controlling the steering engines from five different positions. There will be loud-speaking telephones from the conning-tower to the starboard engine-room, instead of voice pipes. The helm signals will also be worked by Brown's telemotor gear. It may also be noted that the sternpost of the Venerable is of unusual form, the deadwood being cut away aft to increase the manœuvring power.

It is hoped that the Formidable, Irresistible and Implacable will be completed by March, 1901, and the London, Bulwark and Venerable during the autumn of that year.

Later Programmes. Duncan class.

The battleships still on the stocks are the Duncan and Cornwallis, laid down at the Thames Ironworks on July 19th, 1899, the Exmouth, at Messrs. Laird's, Birkenhead, and the Russell, at the Palmer Shipbuilding Company's yard, Jarrow, in the same year (these four belong to the supplemental programme of 1898–9), with the Montagu, at Devonport, on November 23rd, 1899, and the Albemarle at Chatham, on January 1st, 1900, of last year's programme.

These vessels are of what is known as the Duncan class (described in the First Lord's Memorandum, *Naval Annual*, 1899, p. 423), the principal features of which are to be high speed, with great stability and buoyancy. Protection is to be secured by vertical side-armour, 7 in. thick, of Harveyised steel, extending over 290 ft. of their length, and continued in gradually decreasing thickness forward to 3 in. at the bows, and by two protective steel decks, the turtle-backed deck being 2 in. and the main deck 1 in. in thickness. The principal dimensions are: length, 405 ft.; beam, 75½ ft.; mean water-draught, 26 ft. 7 in.; displacement, 14,000 tons. The armament will consist of four 12-in. breech-loading wire guns, placed in pairs in two barbettes, plated with 11-in. steel armour, one forward and the other aft; twelve 6-in. Q.F. guns in casemates with 6-in. armour, eight on the main deck and four on the upper deck; twelve 12-pr. Q.F. guns, similarly placed; six 3-pr. guns in the fighting tops; and four submerged torpedo-tubes. The vessels will be rigged like the Formidable, and will have two masts, with one fighting top on each.

The programme of 1900–1901 includes two battleships to be

built in the dockyards; these vessels will be of the same class as those which preceded them.

Armoured Cruisers building

Cressy class

The armoured cruisers in hand are of three classes, of 14,100 tons displacement, 12,000 tons, and 9,800 tons respectively, and are known as the Drake, Cressy, and Kent classes. Of the Cressy class, which was fully described in the *Naval Annual* of last year, six ships are under construction. The Sutlej was launched at the Clydebank on November 18th, 1899, and the Cressy at the Fairfield Company's Works, Govan, on December 4th. The remaining vessels of this class building are the Aboukir, at Fairfield, the Euryalus and Hogue, at Barrow, and the Bacchante, at Clydebank. In the First Lord's Memorandum for the current year a hope is expressed that the Sutlej and Cressy may be delivered by the contractors during 1900–1901. Two other vessels of the class are said to be well advanced.

Drake class.

The Drake class comprises the four armoured cruisers referred to in the First Lord's Memorandum of last year; they are the Drake, building at Pembroke, the Good Hope (late Africa), at Fairfield, the Leviathan, at Clydebank, and the King Alfred, at Barrow. These vessels are of 14,100 tons displacement, and 500 ft. long; they will have 6-in. broadside armour, carry two 9·2-in. and twelve 6-in. guns, while it is anticipated that with 30,000 I.H.P. they will steam 23 knots. Further particulars of these vessels were given in the *Naval Annual* of last year, and in the First Lord's Memorandum for 1900–1901 it is said to be "too soon to forecast their dates of completion with certainty."

Kent class.

The 23-knot cruisers of 9,800 tons displacement are four in number, two being those of the supplemental programme of 1898–99, briefly referred to in last year's *Naval Annual,* and two being the armoured first-class cruisers of last year's programme. They have been laid down—the Kent, at Portsmouth, the Essex, at Pembroke, and the Monmouth and Bedford, at Fairfield. The following description of the class is compiled from the *Times* reports and from other sources :—

These remarkable cruisers are to be 440 ft. in length and 66 ft. in breadth, with their displacement of 9,800 tons, and will be armed with fourteen 6-in. Q.F. guns, ten of which will be in casemate, and two forward and two aft on twin mountings. They will also have six 12-pr. Q.F. guns amidships on the main deck, and two forward and two aft on the upper deck. Thus, on a 1,200-ton less displacement they will be provided with two fewer 6-in. and four fewer 12-pr. guns than the Diadem class ; but will have many compensating advantages, as not only will they be armoured, but they will have a speed of 23 knots, against 20½ knots in the Diadems. Their 4⅛-in. armour will begin 106 ft. from the stern, and, forward of the vital parts, will taper off to 2 in. at the bows. The after end will have a 2-in. armoured deck, and there will be two protective decks, the thickness on the main deck being 1¼ in. and the lower deck ¾ in. The thickness of the casemate armour will be 4 in. Protection to the fore and after guns will be furnished by a 4-in. barbette surmounted by a shield of similar thickness, revolving with the guns, which are to be carried on Vickers mountings. The cruisers will carry about 1,500 tons of coal, or 500 more than the normal bunker capacity of the Diadems, and in order to provide for the extra weight of armour and coal have been designed throughout to produce a combination of lightness and strength. There will be no boat deck, for instance, the boats being carried on skid beams on the upper deck, and the protective decks will be thinner than in the Diadem class, while there will be one casemate less on each side. The complement will be about 600 officers and men. There will be thirty-one Belleville boilers, in three boiler rooms, with six stokeholds in three compartments, the longitudinal space beneath the protective deck occupied by the engines and boilers being 194 ft., of which one-third will be taken up by the engines. They will draw 24 ft. forward and 25 ft. aft in ordinary seagoing trim, though with all the stores on board they will be somewhat lower in the water, and they will have a freeboard of 19 ft. amidships. Ventilation will be afforded by means of electrical fans, as in the Formidable class, each compartment between the main bulkheads being separately ventilated by electrical motor fans. The Kent will carry two masts but no fighting tops. She is the longest ship ever laid down at Portsmouth, and the building slip has been lengthened 100 ft. in order to take the keel.

Protected Cruisers

Of first-class protected cruisers, the only vessel remaining uncompleted is the Spartiate, laid down at Pembroke on May 10th, 1897, and launched October 27th, 1898. The following reference to the construction of this ship, and to the other two cruisers building at Pembroke, is taken from the dockyard correspondence of the *Naval and Military Record* of February 15th, 1900, and appears to throw some light upon the causes which have led to this vessel being so long under construction:—

A rumour has been in circulation to the effect that overtime on the Spartiate is about to be increased to three-quarters of a day extra daily, but no order on the subject has yet been issued. Such a change would doubtless be very popular with workmen, but overtime at the best is not conducive to an economical output, and where, as on the Spartiate at Hobbs's Point Pier, darkness at an early hour of the evening and exposure to inclement weather have to be considered, the unwisdom of resorting to overtime on an increased scale is manifest According to the best information obtainable, it seems that the Admiralty have determined that a strenuous effort must be made to launch the Drake in October, and that the launch of the Essex will, if possible, not be delayed longer than March, 1901. This prospect seems, on the face of it, exceedingly flattering, but it does not require great discriminative power to perceive that its realisation will be a public misfortune. If the Drake is launched at the time stated it will be necessary

"GLADIATOR,"

BRITISH SECOND CLASS CRUISER.

for her to be afloat in the harbour at least eighteen months or two years. The Spartiate, which is a very much smaller vessel, will have lain seventeen months before she leaves here Had the vessel last named remained in the building-slip until the Drake, which is now being built there, was laid down, her present stage would undoubtedly have been much more advanced than it is, and at an identical outlay. At this yard, which has no dock or basin sufficiently large to accommodate vessels of the classes under consideration, the cost per ton of construction is enormously enhanced after they leave the slip, by reason of their isolation from the place where material is stored. and from the workshops in which material is prepared for working into the ship. The cost of the delays due to weather and other circumstances account for much of the increase, but the time necessarily sacrificed by workmen passing to and from the yard to a ship during working hours is a considerable item. These unfortunate drawbacks have been long realised by the local authorities, and from the cost of all ships built here a large percentage should, on this account, be deducted, when a comparison is made with similar ships built elsewhere. The Admiralty are aware of this, but they persist in directing vessels built here being launched many months before it is necessary to do so. In the case of the Spartiate the loss incurred must have reached £10,000 or £15,000 at the least, and on the Drake, which is a much larger vessel, with a more complicated construction, it will be considerably greater. With the Essex also afloat for twelve months the sacrifice will be prodigious. It will, therefore, be well for the reputation of the yard if the Admiralty can be prevailed upon to postpone both launchings until much later dates than those now contemplated

The first-class protected cruiser Amphitrite completed her contract steam trials during the year, as did also the second-class protected cruiser Hyacinth. A second-class cruiser of an improved Hermes class is to be laid down in one of the dockyards during the financial year. This is the only "protected" vessel in the new programme of construction.

Third-class Cruisers.

The Pioneer, laid down at Chatham, on December 16th, 1897, was floated on June 28th, 1899, and the Pandora, laid down at Portsmouth, on January 3rd, 1898, was floated on January 17th, 1900.

Nine similar third-class cruisers, of a slightly less displacement, have already been launched and completed.

Trials.

The following are the results of the trials of the recently completed vessels of this class. The trials of the earlier vessels were given last year:—

EIGHT HOURS' NATURAL DRAUGHT TRIALS.

	Mean Draught	Total I.H P.	Mean Revolutions.	Speed.
	Ft. In.			
Pioneer	—	5263	235·1	17·7
Prometheus	—	5183·6	196·8	19·8
Pyramus	13 8	5424	205 05	19·9

FOUR HOURS' FORCED DRAUGHT TRIALS.

	Mean Draught.	Total I.H.P.	Mean Revolutions.	Speed.
	Ft. In.			
Pioneer	—	7912	264½	20
Prometheus	—	7274·7	222·5	20·8
Pyramus	13 7½	7303	220·3	20·7
Perseus	13 2	7068·7	214·7	20

THIRTY HOURS' COAL CONSUMPTION TRIALS.

	Mean Draught.	Total I.H P	Mean Revolutions.	Speed.	Coal per I H P per Hour.
	Ft. In.				Lbs.
Pioneer . . .	—	3665	205·8	16·3	2·2
Prometheus . .	13 11	3556·9	177 2	17·5	2 01
Pyramus . . .	14 2½	3605	176·5	17·5	2·05

Last year provision was made in the Estimates for three new third-class cruisers of somewhat larger dimensions than the Pelorus class, and of higher speed. Mr. Goschen states in his Memorandum that after full consideration of all the circumstances, including the action taken by foreign Powers, and weighing the fact that the cost involved in building such vessels is out of proportion to their fighting value and sea-keeping qualities, it has been decided not to proceed with their construction. In their place the second-class cruiser of the Hermes type already mentioned is to be built. The design of this vessel is not yet complete, but she will be of about twenty-one knots speed. In his speech in the House of Commons Mr. Goschen stated that "there will be another opportunity for considering whether our programme should include any second-class cruisers," from which it appears that there is some uncertainty about the matter.

Sloops. In addition to the four sloops named last year as having been laid down (the Shearwater, Vestal, Mutine, and Rinaldo), two vessels of the same class, the Espiègle and Fantôme, have been laid down at Sheerness. These with the Condor and Rosario are the eight vessels of the old programme, a description of which was given in last year's *Annual*. The two last named will be completed this year, four next year, and the remainder in 1902. Two new twin-screw sloops are to be laid down this year.

Gunboats. Four gunboats were laid down in the latter part of 1897. The Thistle was launched from the London and Glasgow Shipbuilding Company's Yard, on June 22nd. She is a sister ship to the Dwarf. The Bramble and Britomart have undergone their contractors' speed-trials, and are completing for sea.

Destroyers. Trials. The following table, published in *Engineering* of the 22nd December, gives particulars of the trials of torpedo-boat destroyers in 1899. "These vessels," it is pointed out, "are required to make a three hours' trial at full power, for continuous speed, as well as a corresponding trial to insure that the coal consumption does not exceed the limit set of 2·5 lb. per unit of power. Fifteen destroyers are included in the list. One of these—the Fervent—belongs to the original 42 vessels of 26 to 27 knots speed. This vessel had at first locomotive boilers, which had to give place to water-tube generators. Her consort—the Zephyr—from the same works, will complete this fleet of 42 vessels. Of boats with speeds of 30 knots and over, 67 have been ordered, 12 of them within the year, 1899. Of these higher-speed craft, 43 have passed through their trials; so that even including 13 not yet floated, there are only 24 yet to pass the trying ordeal. Amongst these are the Viper, which with Parson's turbine is to steam 32 knots, but which at a preliminary trial has made 35·5 knots, and the Express, which is to steam 33 knots with Normand boilers, and reciprocating engines by Messrs. Laird. The Albatross, by Messrs. Thornycroft—also a 32-knot destroyer—has in her preliminary trials attained her speed easily."

Trials of Torpedo-Boat Destroyers in 1899

Firm.	Name of Vessel.	Indicated horse-power	Speed in Knots.	Pounds of Coal per I.H.P. per Hour.
Palmer	Spiteful	6596	29·901	2·32
	,,	6444	29·511	..
	Flirt	6720	30·039	..
Vickers	Leopard	6848	30·135	2·299
	,,	6415	30·139	..
	Otter	6265	30·274	2 490
	,,	6077	30·071	..
	Bittern	6366	30·354	2·450
	,,	6627	30·403	.
Thornycroft	Cygnet	462	13·040	1·820
	,,	6077	30·375	2·229
	,,	5729	30·305	..
	Coquette	5643	30·060	2·091
	,,	5917	30·211	..
	Cynthia	5857	30·127	2·381
	,,	5494	30·205	..
John Brown and Co. (late Clydebank Engineering Company)	Vulture	505	13·044	2·200
	,,	6222	30·172	2·340
	,,	6175	30·277	..
	Kestrel	456	13·089	2·060
	,,	6682	30·044	2·350
	,,	6600	30·030	..
Hawthorn	Mermaid	6468	30·149	2·670
	,,	6578	30·833	..
	Cheerful	497	13·100	1·720
	,,	5566	29·941	2·840
	,,	5912	30·152	..
Laird	Orwell	456	12·963	1·975
	,,	6445	30·282	2·670
	,,	6350	30·187	..
Fairfield	Leven	464	13·101	1·407
	,,	6201	30·201	2·095
	,,	6189	30·383	..
Hanna and Co.	Fervent	227	10·128	1·870
	,,	4085	26·730	..

Messrs. Palmer adopt the Reed boiler; Messrs. Vickers, Laird, and the Clydebank the Normand modified by the respective engineering managers; Messrs. Hawthorn, Thornycroft, and the Fairfield Company use the Thornycroft boiler.

New Destroyers.

The twelve new torpedo-boat destroyers ordered by the Admiralty, to be built by contract, are as follows:—

Racehorse, Roebuck, Greyhound	Messrs. Hawthorne, Leslie & Co., Newcastle-on-Tyne.
Lively, Sprightly	Messrs. Laird Bros., Birkenhead.
Myrmidon, Peterel, Syren	Palmer's Shipbuilding Co., Jarrow-on-Tyne.
Success	Messrs. Doxford, Sunderland.
Falcon, Ostrich	Fairfield Co., Glasgow.
Vixen	Vickers, Sons & Maxim, Barrow.

The Arab is also building at Clydebank. The Viper was launched on September 8th, by Messrs. Hawthorne, Leslie & Co., at Hebburn-on-Tyne.

The total number of vessels in this class is 108, of which forty-two have speeds from 26 to 27 knots, and all but two of this category have passed through their trials. There are sixty-two destroyers of the 30-knot class, and of these all but eighteen have passed through their trials. There are also four which have contract speeds ranging from 31 to 33 knots. One of these has obtained a speed of 32 knots on her preliminary trials and has been delivered. Another of equal speed is under construction, while a third is undergoing preliminary trials. The fourth vessel is the Viper mentioned above. On her preliminary trials for short periods the very high speed alluded to has been reached.

Two 25-knot torpedo-boats of the programme 1899–1900 are under construction, and two more are to be laid down in accordance with the new shipbuilding programme.

The Royal Yacht.

The following reference is made to the new Royal Yacht in the First Lord's Memorandum :—"The new Royal Yacht was ready for her steam trials at the beginning of January, but an accident, which occurred to her while undocking at Pembroke, besides damaging the vessel's bottom, revealed a serious miscalculation of weight which will make considerable alterations necessary before she can proceed with her trials." The yacht has been sent to Portsmouth, where she is undergoing alteration.

For several years past stress has been laid in these columns upon the necessity for adding to our resources in the way of what Mr.

Goschen has called "the Fleet's appendages." In his speech on the Estimates, Mr. Goschen referred at some length to this subject. He said that as regards telegraph ships, "we do not think we should be wise in buying or constructing such ships ourselves, but that we should do far better by utilising the services of private enterprise and the cable ships of the companies who, in time of peace, are engaged in that work, and are continually seeking under pressure of competition to improve their plant and to develop every new invention." The Admiralty have been in communication with the cable companies, and have been able to make arrangements for meeting any emergency which may arise. As regards repairing ships, Mr. Goschen explained that experience with the Vulcan and the Hecla led him to believe that it would be more economical and more conducive to efficiency to purchase ships of the Mercantile Marine for this purpose. It is hoped to buy some ships of this class now engaged in the transport service, which are reported to be very suitable and appropriate for the purpose. Mr. Goschen led the House to believe that no time would be lost in obtaining these ships, and fitting them up for repairing purposes. In reference to colliers, it is also thought that, on the whole, using private enterprise will give a better result than building government colliers. Experiments have been made in this direction, and colliers are now chartered for the year instead of by the voyage, in order to give their owners an opportunity of improving the ships and their plant. Four collier transports, with an ocean speed of not less than 10 knots, are to be engaged for the manœuvres and specially reported upon. Another of the Fleet appendages mentioned by Mr. Goschen is a distilling ship, and two vessels for this purpose will be engaged for about six weeks, from July 1st next, to deliver fresh water into the tanks of the ships employed in the manœuvres. Each of these vessels is to have a storage capacity of not less than 600 tons of water and to be capable of producing not less than 6,000 gallons of distilled water per day.

"Appendages" of the Fleet.

Personnel.

Engineers Corps.

The following regulations have been issued with regard to the promotion, status, and pay of the officers of the engineer corps:—

The list of Chief Inspectors of Machinery has been increased from five to eight, and that of Inspectors of Machinery from eight to thirteen.

The Engineer-in-Chief has been given the relative rank of a Rear-Admiral.

The rank of Staff Engineer has been abolished.

Chief Engineers will rank *with* Lieutenants of and above eight years' seniority, while Engineers on promotion will rank *with* Lieutenants of less than eight years' seniority, instead of as now, with but *after* Lieutenants.

In other respects the relative rank of Engineer officers remains unchanged.

Engineers will be given a new scale of pay, viz :—

On promotion	10*s* a day.
After four years	11*s*. „
After eight years	12*s*. „

And the allowance of 1*s*. a day at present paid to senior Engineers for all ships will be replaced by a scale varying, according to responsibility, from 1*s*. to 2*s*. 6*d*. a day.

Medical Corps

The number of inspectors-general and deputy-inspectors-general of hospitals and fleets has been increased, and the conditions required for promotion to these ranks have been modified. The period of the course of instruction at Haslar Hospital for surgeons on entry has been extended, and the award of prizes at the end of each session introduced. The number of medical officers allowed to undergo periods of study at medical schools has been considerably increased, and the privilege extended to the senior ranks. An additional professor has been appointed on the instructing staff at Haslar in connection with the study of diseases of foreign stations. Medical officers newly entered will in future be only required to provide themselves with a pocket-case of instruments, as all ships bearing medical officers, and naval and marine barracks, will be supplied with surgical instruments at the public cost within the next three years.

Numbers

The total number of officers and men, including both regulars and reserves, voted for 1900–1901 is 155,000, of which number about 40,000 belong to the latter category. The additions to the regular *personnel* are thus composed :—

220	Officers.
3,050	Petty Officers and Seamen.
150	Engine-room Staff.
200	Miscellaneous.
300	Marines.
320	Apprentices (Artisan rating).
4,240	

Mr. Goschen, referring to this increase, said :—" Can we get them ? We can get them. I asked for an increase last year of little more than 4000, and we have raised them. . . . Last year I said I hoped that when we had reached 110,000 we might stop, but events have developed fast. The action of other countries and the general needs of the service have been such that we have been compelled to advance to the present figure. I will not deny that there are drawbacks to this rapid increase, which in a few years has raised a *personnel* from 65,000 to 115,000."

Royal Marines

Two thousand five hundred and eighty-five recruits were raised for the Royal Marine Corps during the last year, the wastage of the corps during the twelve months amounting to 2078 men.

Naval Reserve.

Two new classes of reserves are to be created, one to be composed of seamen who have taken their discharge after twelve years' service, and the other of marines who have taken their discharge without pension. It is also proposed to introduce a short Bill enabling the

Admiralty to call up a certain proportion of the Royal Naval Reserve instead of the whole number as would be necessary as the law stands at present. The following are the resources from which the present reserves can be drawn:—28,000 Royal Naval Reserve, 9000 seamen-pensioners, of whom 4500 are in the seamen-pensioners' reserve, and 2800 marine pensioners, making a total of 40,000. These, with the 115,000 on the active list, give a total of 155,000, of whom, if the 6000 boys in training are deducted, there remain 149,000 who could be called upon for service.

Colonial Reserve.

In regard to the question of organising Colonial Naval Reserves some discussion has taken place in the press consequent on the publication of the result of a conference of Colonial Naval Commandants held at Melbourne. The following quotation from Mr. Goschen's speech on February 26th upon the estimates appears to be a concise summary of what has been done in this matter up to the present time:—

> "We are now inquiring as to how we can organise Naval Reserves in our colonies. The military instincts of the colonies have been so developed in connection with the war in South Africa as to inspire us with the hope that, if we can only find an adequate system of organisation, we shall be able to get a valuable contingent for our Navy from Australia, Canada, and elsewhere. But there are some difficulties in the way, partly on account of the differences in wages in the colonies and partly because we have not got the same appliances for training in the colonies that we have at home. The terms which would have to be given to the Colonial Reserve would be so much higher than the pay of the bluejackets that discontent would arise; and, on the other hand, if we did not give these high terms, and we required the men to undergo six months' training, we should not be able to get the colonials. That is a formidable difficulty, but it may be possible to solve it. There is another point. It would be difficult to get a sea-going Reserve for Australia at all, and, therefore, whether the Colonial Reserve will be a Naval Reserve or a Reserve for coast-defence is a matter for future consideration. But I thought it better to suspend negotiations on the subject until federation is established, as it would be better to organise with one authority for one Naval Reserve than to arrange for several Reserves with four or five distinct Governments in Australia. With Canada we are more advanced. There wages are not so high, and there the fishing industry is so managed that it may be agreeable for the fishermen to be employed for a period in a man-of-war during the time the fishing is closed. They have asked that we should reduce the time of training from six to four months in accordance with their local exigencies. No final decision has been taken in the matter, but I thought it my duty to the colonies to inform the House of the interest they have shown in this attempt to form a Colonial Naval Reserve."

In a letter in the *Times*, February 7th, 1900, Lord Brassey explained the steps taken in the Australian colonies towards the enrolment of a Colonial Reserve, and urged that, at a favourable opportunity, a conference should be held between an officer representing the Reserves Office at home and naval men appointed by the Federal Government.

CHAS. N. ROBINSON.

CHAPTER II.

The Progress of Foreign Navies.

The progress of naval expansion abroad is continuous, and the Great Powers are showing a tendency to rival one another in the preparation of shipbuilding programmes covering extended periods of time. The desire to attain a definite constitution for the fleets is strongly marked both in Germany and France, the purpose being to remove from the political sphere and from the wrangling of parliaments those operations which are related to National Defence. In the United States the progress is steady, and further important additions to the fleet are contemplated; in Russia the ships of the programme are being pushed forward in spite of many disadvantages, every slip being occupied at home, and those ships which are being built in foreign yards are advancing rapidly; and in Japan the completion of the scheme of Naval Development is well within sight. In regard generally to the types of ships, it may only be necessary to remark in this place that the improved methods of making armour, enabling greater protection to be given with thinner plates, are having a marked effect in ship construction. The question of type in the battleship, however, shows no sign of being settled, for while in France the tendency appears to be to large displacements, contrary to recent indications of policy, the new Italian ships are intended to be of moderate displacement, and to embody the elements of various classes of ships, uniting the general character of the battleship with some of the features of the cruiser.

France.

Naval Policy

In June, 1899, there was another change in the holder of the naval portfolio in Paris, when M. de Lanessan, a gentleman who had devoted much attention to naval subjects, succeeded M. Lockroy. His accession to office brought about a new direction of policy, and ultimately led to a definite and expanded ship building programme, which will presently be described, being adopted by the Government. Many changes were made in the central administration and in the constitution of the naval staff, the chief of the staff being given general authority and responsibility in all matters relating to pre-

"CHARLEMAGNE,"

FRENCH BATTLESHIP.

paration for war, both in regard to the *matériel* and *personnel*, and the officer now holding the important post is Vice-Admiral Bienaimé.

During the year a number of vessels have been completed for service. The Charlemagne and Gaulois were practically ready at the beginning of 1899, and the sister ship St Louis has entered upon her trials. The first trial early in March, 1900, was considered satisfactory, though some minor changes were necessary, and in another preliminary trial a speed of 17·61 knots was attained. The estimated speed is 18 knots, and particulars of the trials of the sister ships were given in the *Annual* last year. The third-class cruiser D'Estrées has been completed at Rochefort, and the gunboat Décidée (launched in 1899), with the destroyers Dunois and La Hire, at L'Orient. Some particulars of the trials of torpedo craft will be found below. In the private yards the vessels completed were the commerce destroyers Châteaurenault and Guichen, the destroyer Hallebarde, and a number of torpedo-boats. Ships completed.

The trials of the Guichen gave great satisfaction. The estimated horse-power of the cruiser is 23,000, but on June 27th, 1899, at the Îles d'Hyères, with 18,500 I.H.P. and 112 revolutions, the speed was 20 knots, the machinery working perfectly. Afterwards the cruiser steamed at 22 knots, and in the full speed trial, a slight modification of the screws having been made, the engines during a three hours' run worked up to 25,400 I.H.P. with 136 revolutions, the mean speed being 23·55 knots, with a coal consumption of 1·83 lb. per horse-power per hour. The boilers are on the Lagrafel and d'Allest principle, and the machinery was constructed at St. Denis by the Chantiers de la Loire. Everything worked well at the trials. The cruiser proceeded to Brest in March, 1900, to join the Northern Squadron. Trials Guichen.

The sister ship Châteaurenault, which was built by the Forges et Chantiers de la Méditerranée at La Seyne, was also under trial at Toulon in March, 1900. The machinery differs somewhat from that in the Guichen, the two engines which drive the lateral screws being side by side, and the third, for the middle line screw, further aft, while the boilers, which are upon the Normand-Sigaudy principle, are forward of the engines. On March 19th, with natural draught, the engines developed 19,000 I.H.P., giving a speed of 22·7 knots; later, with the engines working at 21,600 I.H.P., the speed was 23·123 knots. The contract speed is 23 knots with 23,000 I.H.P. Châteaurenault.

Two battleships have been launched—the Suffren and the Henri IV. The former took the water at Brest on July 25th, 1899, having been only 6½ months on the stocks. The ship, however, had only been advanced sufficiently to enable her to be launched, and the time Ships launched. Suffren.

occupied is no real indication of rapidity of construction. The Suffren has been constructed from the designs of M. Thibaudier, and belongs to the same type as most other recent French battleships which have been described in the *Naval Annual*. Generally she resembles the Iéna, but is somewhat larger, her displacement being 12,728 tons as compared with 12,052. She is thus the heaviest French battleship yet afloat, though the new programme provides for vessels of still greater dimensions. The hull is protected by a complete belt of Harveyed steel, with a maximum thickness of 11·8 in., which rises about 3 ft. 6 in. above the water-line, and above this extends a belt of 5·1 in. Harveyed steel, having a height of 6 ft. 6 in. At the level of the top of the principal belt is the armour-deck, 2¾ inches thick, with a splinter-proof deck at its lower edge, the intermediate space having cellular construction. The armament, which is something more powerful in secondary guns than in the Iéna, is entirely protected, including the ammunition hoists, while in the Iéna and Charlemagne there is an unprotected space below some of the guns. The ordnance of the Suffren is of the 1893–6 type. The cost of the ship will be £1,180,000. She has Niclausse boilers.

Henri IV. The Henri IV., which was launched at Cherbourg in August, after having been in hand twenty-five months, belongs to a special type.* She took the water with a displacement of 4000 tons towards her total of about 9000. She was designed by M. Bertin, and it is believed, according to the *Yacht*, that if she should prove satisfactory, she will mark the point of departure for the construction of a series of ships. The length is 354 ft. 4 in., beam 72 ft. 3 in., and maximum draught 22 ft. 11 in. The peculiarity of the ship is that she resembles the ordinary high freeboard type at the bows, but from about one-fourth of the length from the stem and abaft she resembles the monitor type. The proportion of length to beam is about 4·8 to 1. From the bows the superstructure follows the outline of the hull until a width of some 46 ft. is attained, and from that point, on both sides of the ship, the topsides, rising from the armour deck, become vertical and parallel to the middle line, and extend to the after gun-turret, turning inward as they approach it. On each side of the superstructure there is thus a low-freeboard space about 13 ft. wide, as well as the entire space abaft the after turret, this low-freeboard space being 3 ft. above the water-line. The citadel is amidships, with a quick-firer at each angle, and before and abaft it the upper works are narrower to admit of direct fire ahead or astern, and form another stage in the superstructure. The Henri IV. will offer a great contrast to most modern French vessels from the fact

* *Cf.* Pl. 44, Pt. II.

that her structure above water is of very reduced dimensions, and that the target she will present to an enemy will be smaller than in the case of any other vessel of like displacement. The ship has an almost complete armour belt of Harveyed steel, varying in thickness from 7 in. to 11 in. It terminates in a transverse bulkhead close to the after extremity of the ship, where the obliquity is excessive, and the form of the ship abaft it is of thin steel plates. The belt rises to the height of the armour deck, which has a maximum thickness of 2·3 in., on a plating of $\frac{3}{4}$ in. The superstructure and the redoubt amidships are protected by Harveyed steel, $3\frac{1}{4}$ in. to $4\frac{1}{2}$ in., including plating at the back. A notable feature of the ship is that no difficulty is anticipated to arise when her decks are flooded, special provision being made for the water passing off as she steams ahead. Exceptional stability is assigned to her, but for greater security there is a thin steel deck below the armour deck descending very low at the sides, and covering the machinery. The heaviest guns are two of 10·8-in., severally in hooded in turrets forward and abaft, which are protected by 9·4 in. of Harveyed steel at the base, and of 11·8 in. in the parts which revolve. The turrets are to be worked by electricity. The turret forward rises high above the water, and, though the base is protected, there would appear to be some danger of the turret falling over, owing to the pillar-like character of the structure if the ship should be seriously damaged below. Above the after turret and a little further forward is another turret for a 5·5-in. Q.-F. Four other guns of the latter calibre are to be placed at the angles of the redoubt, and apparently two others in positions not determined, while twelve 1·8-in. Q.-F. will be on the superstructure. Three engines, driving as many screws, will develop 12,000 horse-power, and be supplied by Niclausse boilers, and are intended to give a speed of 17 knots. The normal coal supply will be 725 tons, though 1100 tons can be stowed on board. The whole of the machinery has been made at the Indret works.

Jeanne d'Arc.

The armoured cruiser Jeanne d'Arc was launched at Toulon on June 8th, and is a vessel of important character, though of much debated qualities. Although on paper she may look better than the Powerful or the Diadem, there is no certainty that she presents a satisfactory union of qualities. The French have a high opinion of her, and a writer in the *Yacht* has expressed himself in the following terms: "A comparison between the Jeanne d'Arc and vessels of the Powerful and Diadem classes shows that the English are less ingenious than ourselves in the matter of naval construction. To know how to unite a powerful armament with sufficient protection and high speed upon a small displacement, is a veritable *tour de force*, and from this point of view it cannot be denied that the Jeanne d'Arc, notwithstanding her defects, is a *chef d'œuvre* of naval architecture." In comparison with armoured cruisers of the Drake class she is distinctly inferior in armament and protection. She was laid down in April, 1896. Her dimensions are: displacement, 11,329 tons; length, 477 ft. 2 in.; beam, 63 ft. 8 in.; mean draught, 26 ft. 7 in. At the water-line is a narrow belt of steel 6 in. to 3 in. thick, running from end to end, and above this another belt from 3·2 in. to 2 in. thick. The whole bows of the ship to the height of the upper deck are encased in 3-in. steel, so that she is well adapted for end-on action. Right ahead she brings to bear eight 5·5-in. Q.-F. and one 7·6-in. It is open to question whether they could all be discharged. Her armament will comprise two 7·6-in. guns, singly in turrets forward and abaft, eight 5·5-in. Q.-F. (four on each side, two being in sponsons), ten 3·9-in. Q.-F., sixteen 3-pr. Q.-F., eight 1-pr. Q.-F., two Maxims, and two submerged torpedo-tubes. The coal capacity is calculated to give a range of 13,500 miles at 10 knots, and her complement will number 626. The ship is well advanced, and ought to be completed for sea in 1901.

Amiral de Gueydon and Montcalm.

Three other important cruisers, the Amiral de Gueydon and Montcalm, sister ships of 9517 tons, and the Dupleix, an armoured cruiser of a smaller class, have also been launched. The Amiral de Gueydon took the water at Lorient on September 20th, 1899, and the Montcalm on March 28th, 1900, at the yard of the Forges et Chantiers de la Méditerranée at La Seyne. Both ships were designed by M. Bertin, and may be described as smaller Jeanne d'Arcs. They are 459 ft. long, with 63 ft. 8 in. beam, and 24 ft. 7 in. draught. Protection is given by a 6-in. Harveyed steel belt, surmounted by thinner plating, with a maximum thickness of 3¾ in., and armoured and splinter proof decks. Two 7·6-in. guns are in turrets singly forward and abaft, and eight 6·4-in. Q.-F. in protected casemates on each side, four 3·9-in. Q.-F. with semi-circular shields on the spar deck, and twenty-two smaller guns on the superstructure. The speed is to be 21 knots, and the range of action to exceed 10,000 miles at 10 knots. The Amiral de Gueydon is provided with Niclausse boilers, and the Montcalm with boilers of the Normand-Sigaudy type. The Dupetit-Thouars, building at Toulon, is a sister ship.

Dupleix.

The Dupleix, which was launched at Rochefort on March 28th, 1900, is an armoured cruiser also designed by M. Bertin, and, generally speaking, of the same class, though smaller, the displacement being 7700 tons. The cruiser has twenty-four Belleville boilers, intended to give a speed of 21 knots. Two cruisers of the same class, Desaix and Kléber, are building respectively at St. Nazaire and Bordeaux.

Jurien de la Gravière. Infernet.

Two other cruisers have been launched since the last appearance of the *Annual*—the Jurien de la Gravière at Lorient, and the Infernet at the Chantiers de la Gironde, Bordeaux. The former is a second-class sheathed cruiser of 5605 tons, and is a reproduction on a smaller scale of the commerce-destroyer Guichen, with a speed of 23 knots, at a smaller range of action. She will be more heavily armed, having eight 6·4-in. Q.-F. as compared with two 6·4-in. and six 5·5-in. in the other cruiser. The Infernet is a third-class sheathed cruiser of 2452 tons, intended to steam at 20·5 knots, and having an armament of two 5·5-in., four 3·9-in., and eight 1·8-in. quick-firers. She has Normand water-tube boilers, and was launched on September 7th, with all her machinery on board.

Types of New Cruisers.

It may be useful at this point to tabulate the displacement and armament of the new French armoured cruisers subsequent to the Jeanne d'Arc, in order to show the development and variation of type.

Vessel.	Displacement.	Armament.
Montcalm, Gueydon, Dupetit-Thouars	9,517	2 7·6-in.; 8 6·4-in.; 4 3·9-in.; 22 smaller.
Desaix, Kléber, Dupleix	7,700	10 6·4-in.; 16 smaller.
Condé, Sully, Gloire	10,000	2 7·6-in.; 8 6·4-in.; 6 3·9-in.; 22 smaller.
Marseillaise, Amiral Aube	10,014	2 7·6-in.; 8 6·4-in; 6 3·9-in.; 26 smaller.
C 11, C 12, C 13	12,416	4 7·6-in.; 16 6·4-in.; 24 smaller.

The Sully, which has been put in hand at La Seyne, has the advantage of much better protection than the Montcalm type, four of the 6·4-in. guns being in turrets and the others in armoured casemates all united to the side plates. The Marseillaise and Amiral Aube (C 9 and C 10), in an early stage of construction at Brest, are improvements on the same type, adding two 2·5-in. guns to the secondary armament. The three other cruisers belong to the new programme, and the *Yacht* gives the following description: displacement, 12,416 tons; length, 474 ft.; beam, 71 ft.; total horse-power of the three engines, 24,000; speed, 21 knots; effective range, 10,000 miles at 10 knots. Armament, four guns of 7·6-in., sixteen of 6·4-in., twenty of 1·8-in., four of 1·4-in., and five torpedo-tubes, two to be submerged. The cruisers will cost about £1,170,000 each, which exceeds the cost of recent French battleships.

Gunboats launched.

Two gunboats were also launched in 1899 in French yards—the Décidée at Lorient, since completed, and the Zélée at Rochefort. Both are of the Surprise type, displacement about 645 tons, and have an armament of ten quick-firers, the largest guns being two 3·9-in. These are slow-speed vessels (13 knots), with Niclausse boilers and engines of 1000 I.H.P.

Argus and Vigilante.

Much attention was directed in February, 1900 to the launch of the shallow-draught river gunboats Argus and Vigilante, built for the French Government by Messrs. Thornycroft at Chiswick. The dimensions are: displacement, 122 tons; length, 145 ft.; beam, 24 ft.; draught, 2 ft.; the engines are of 550 I.H.P., and the contract speed is 13 knots, with 80 tons coal capacity. The boats have been compared with the Woodcock and Woodlark, which were also built at Chiswick, but the French boats are better armed—one 12-pr. and five 6-prs., as compared with two 6-prs. and four rifle-calibre Maxims. The Vigilante, on an official trial of two hours' duration with a full load on board, attained a speed of 13·25 knots, and the Argus of 13·4 knots. The boats are intended for China.

Torpedo craft.

There has been much activity in the matter of torpedo craft. The Durandal, destroyer, attained a speed of 27·42 knots at Cherbourg in May, 1899, 26 knots being the contract, while the Hallebarde of the same class steamed at 27·2 on her official trials in July. The Dunois and La Hire torpedo gunboats have given great satisfaction at their trials. The destroyers Fauconneau, Pique, and Framée have been launched, and four of the same class, the Pertuisane, Escopette, Flamberge, and Rapière, have been put in hand at Rochefort. A number of first-class torpedo-boats have been completed.

Submarine boats.

Narval

The great attention which is being paid in France to submarine navigation may render interesting the following account of the submarine or submersible boat, Narval, which has been undergoing trials at Cherbourg. She is the latest of the class, and great hopes have been entertained of her success. The chief characteristic of the boat is that she navigates by steam on the surface, and by means of electricity below water. She was designed by M. Laubeuf, who won a gold medal in a competition opened by the Minister of Marine in 1897. The following are her dimensions: displacement, 106 tons; length, 111 ft. 6 in.; extreme beam, 12 ft. 4 in. In the original project, the boat was to have been propelled on the surface by steam machinery of 300 I.H.P., the stoking being with compressed coal, but it was afterwards decided to supply liquid fuel, and finally an engine of 250 I.H.P., constructed by MM. Brûlé, was adopted, with multitubular boilers on the system of M. Adolphe Seigle, having five injectors for stoking with heavy petroleum. This machinery is placed near the centre of the boat, while the electric gear, in regard to which there is nothing special to note, is further aft. The accumulators are on the "Fulmen" system. The following are details of the speed and range of the boat: on the surface, 252 miles at a speed of 11 knots with 23 hours' duration, or 624 miles at eight knots with 78 hours' duration; submerged, 25 miles at eight knots, 72 miles at five knots. It deserves to be noted that when navigating upon the surface, the petroleum motor will drive dynamos and recharge the accumulators, thus extending the range. The hull of the boat has a particular character, being double. The inner plating is thicker than the outer, and in the intervening space sea water circulates freely, the purpose being to offer greater resistance to projectiles. In general form the boat resembles an ordinary torpedo-boat. During the early trials some difficulty occurred, either through the electric accumulators or through the want of sub-division between the two shells of the boat, and some changes were effected. The Narval has an armoured conning tower, from which a telescopic funnel projects, and when it is intended to plunge the funnel is withdrawn and the aperture made water-tight. The plunge is effected by means of two pairs of horizontal rudders operated by a hand mechanism. The armament consists of four Whitehead torpedoes, and there are two Drzewiecki torpedo-tubes on each side and towards the upper part of the boat, which launch the torpedoes in the direction of the beam.

Owing to the faults developed in the Narval, work on the Sirène and Triton, which are of the same class, was suspended at Cherbourg, and there will be no effort to multiply vessels of the type until entirely conclusive results have been obtained from the trials of the Narval. The plans of four boats which are to be built at Rochefort, and to be named Farfadet, Gnome, Korrigan, and Lutin, have been prepared by M. Maugras, first-class naval engineer, and the design differs from that of the Narval. The Morse is another submarine boat of older type, launched at Cherbourg.

The New Programme of Naval Construction.

The Naval Estimates prepared for 1900 and the shipbuilding programme presented with them caused considerable discussion, and the plans of the Minister were opposed by the Budget Committee. The Government, therefore, with the view of putting an end to the instability which had characterised the naval policy of France, presented to the Chamber in January, 1900, a programme of naval expansion covering a series of years, accompanied by a remarkable memorandum. The plan had been elaborated by the Superior Council of the Navy, adopted after long deliberation by the Cabinet, and included a shipbuilding programme extending to the year 1907.

The explanatory statement opens with the remark that however desirous France may be for peace, she cannot forget the lessons of history. The programme instituted after 1870, and shaped afresh in 1896 and 1898, has been repeatedly modified and has not yet been completed. "The government considers that it is indispensable to do away with such errors, and to lay down in definite form what should be the organisation of the navy, adapting it, like that of the army, to the needs of defence on the one hand and to the resources in men and money on the other." It is not to be denied, says the Minister, that owing to the fluctuations in the directive idea a number of ships are unequal to the duties for which they have been built. It is therefore highly important that the programme of naval construction should be voted once for all in order that the number and types of ships may not be continually discussed, and that the private yards and those of the State may not find the programme subjected to annual modification with disastrous results upon the progress of naval construction.

The necessity of State policy

Plans for national defence

It should be remarked that the new programme is only part of a large scheme dealing with national defence. The defences of the coasts and ports are to be reorganised and completed; the dockyards and naval establishments in France, Algeria, and Tunis are to be improved and developed; and provision is to be made for organising naval bases abroad and adequate defences for the colonies.

The strength of the fleet.

The official view as to the constitution of the fleet is as follows: twenty-eight battleships (four squadrons of six battleships each, and four others as reliefs); twenty-four armoured cruisers (forming eight divisions each of three vessels); fifty-two destroyers or torpedo gunboats; 263 torpedo-boats; and thirty-eight submarine or submersible vessels. Deducting from this list those which have been completed or were in hand on January 1st, 1900, there remain to be constructed under the programme:

Six battleships.	Twenty-eight destroyers.
Five armoured cruisers.	112 torpedo boats.

Twenty-six submarine boats.

All these vessels are intended for European waters, and the local defence of the colonies is to be entrusted to various protected cruisers, gradually to be replaced by more powerful vessels. The memorandum proceeds to discuss the financial aspect of the question, and points out that in 1905, of the ships now remaining to be constructed to complete the programme, two battleships, three cruisers, twenty destroyers, seventy-six torpedo-boats and twenty-six submarine boats would be ready.

The Minister then discusses at some length the question of types, and insists upon the impossibility of combining in any vessel in an adequate degree the requirements of armament, protection, range of action, and speed. The ship is a compromise, and "scientific minds bow before the necessity which is imposed upon all navies of possessing at once ships powerful in artillery and well protected against gun fire, but without great speed or the ability to cover great distances, and also other vessels in which the gun power and protection are less, but the speed and range of action very considerable." The Minister goes on to argue that even those officers who extol so loudly the advantages of the *guerre de course* will still find battleships necessary to facilitate the issue and return of the cruisers by diversions against the blockading line. "In a word, they will demand battleships if only to protect the cruisers and to make their action efficacious." Many arguments are offered against the dangerous idea that armoured cruisers can replace battleships. **Types of ships.**

The new type of battleship has been designed by M. Bertin and adopted by the Board of Construction and the staff after long deliberation, in order to give satisfaction to the many criticisms formulated by admirals in command relating to the insufficient armament and protection of the existing battleships compared with those possessed by foreign Powers; and the example of England, Russia, Japan, the United States, and Germany is cited in support of the design. The government has no hesitation in presenting to the legislature this new type of battleship of 14,865 tons, with a cruiser of 12,600 tons. "Owing to the increase of displacement the battleship of 14,865 tons unites qualities of speed, range of action, protection and armament which are not found in any of our existing battleships, nor in an equal degree in the most recent British ships. The speed is to be 18 knots, and to be an actual speed, but may be exceeded, as in the case of most of the vessels designed by M. Bertin. The new ship may defy every similar vessel of foreign navies, and the range of action (4000 miles) is sufficient for a battleship. The type is remarkable above all in the matter of protection and armament. The side armour of Harveyed steel 11·8 inches thick will rise 8 feet above the water, extend from stem to stern, and be completed by 4-in. protection enveloping the whole of the bows. The principal armament will consist of four 12-in. guns coupled in turrets forward and aft, the turrets protected by 12 inches of steel, while the secondary armament will comprise eighteen 6·4-in. guns, nine on each side, of which twelve will be coupled in armoured turrets and six in armoured casemates, their bases united to the side armour. With such arrangements this battleship will be the most formidable engine of war **The new battle-ships.**

which has been constructed for any navy. If, as reason and experience indicate, victory falls in battle to the ship which can continue her gun-fire the longest, our new battleship should be superior to all similar vessels in other navies, since none are better protected, better armed, or more rapid. The cruiser of about 12,600 tons which appears in the programme is also the work of M. Bertin. She deserves, as an armoured cruiser, the greatest praise, for she will be capable of fighting against the most powerful analogous vessels in foreign navies. The effective speed is 22 knots, that is to say, the indicated horse-power will probably be almost equal to the British cruisers of the Drake type, though the nominal speed of these is 23 knots. The armament is very powerful and very well protected, and the range of action is such that the vessel may undertake all the operations required of vessels of the class."

The new cruisers.

The organisation of the fleet.

The programme concludes with an account of the torpedo-boats, and with the remark that, in a fleet scientifically organised, five classes of vessels should be found—battleships for the offensive defence of the territory against the battleships and cruisers of the enemy, to give support to the vessels employed in commerce destroying, to make diversions against blockading squadrons, and to go into action when favourable circumstances occur; armoured cruisers to chase the enemy's protected and auxiliary cruisers, for raids against the enemy's coasts and ports, for scouting, and the destruction of the destroyers and scouts of the adversary; and the three classes of destroyers, torpedo-boats, and submarine boats for the defence of the ports and squadrons and for offensive operations.

The Parliamentary Naval Committee proposes to reduce the period covered by the programme by one year.

Germany.

The new programme

The opening of the year 1900 was made remarkable by the presentation to the Reichstag of a new shipbuilding programme of extended duration. The German Navy Law of April 10th, 1898, had been found insufficient, though when it was presented it was supposed to be invested with something of definitive character. The new scheme, which was adopted by the Federal Council, was laid before the Reichstag with an important explanatory memorandum (given almost in full in Part IV.), which begins by declaring that the protection of national interests and especially of foreign commerce is a vital question. "For this purpose the German Empire requires peace—not only upon land, but upon the sea—not, however, peace at

"HERTHA,"

GERMAN SECOND CLASS CRUISER.

any price but peace with honour." The memorandum expounds certain naval principles which are familiar in this country. A war touching commercial interests is likely to last long, and will last longer according to the object of the superior enemy. To that enemy such a war might cost little comparatively, but if it proved unfortunate for Germany it would result in the destruction of her maritime commerce and perhaps of her colonies, and a commerce destroyed requires long to recover. Again, the result of a naval engagement would be to disable many ships, but the stronger adversary would be possessed of other forces; and therefore, though the fleet now ready and in hand might render a blockade difficult, it would be powerless to prevent it.

It is unnecessary to describe the details of the measure here, because the essential portions and many details are given elsewhere.

The fleet necessary to Germany, we are told, must have the tactical formation of two double squadrons of efficient battleships, with the essential auxiliaries of cruisers and torpedo-boats, and the second double squadron or fleet is to have the same constitution as that adopted for the first under the law of 1898. The first fleet is to consist of the most modern vessels, in order that it may be a "school for tactical training," and be always ready for an outbreak of hostilities, while the second fleet may consist of the older vessels, not all of which will be kept continuously in commission.* Although the fleet in contemplation is for employment in home waters, it is evidently intended to increase the number of ships abroad, and reference is made to the occupation of Kiao-chau and to the increase of foreign commerce. The last matter is enforced in an appendix to the programme, wherein the development of German maritime interests and affairs is expounded under the headings of the increase of population at home and in the colonies, the development of commerce at sea and of German shipping, the expansion of the shipbuilding industry, the enlargement and increase of harbours, the magnitude of the German fisheries, the protection of cables, and the growth and development of the colonies generally.

The German Government certainly takes a long look ahead, for the financial provision extends to the year 1916, and the ships will not then have all been completed. When the Navy Law of 1898 was adopted, a plan was laid down for determining the obsolescence of warships, which was really at the base of the building programme,

* The following is the proposed organisation:—Active fleet, one flagship; two squadrons, each comprising eight battleships; four divisions, each of four cruisers; four torpedo flotillas (eight divisions), comprising forty boats in all. Reserve fleet—the same constitution, but in regular commission only four battleships of each squadron, two divisions of cruisers, and eight torpedo-boats (being one of each torpedo division).

and under this scheme it is estimated that by the year 1917 seventeen battleships and coast-defence vessels, ten first and second-class cruisers, twenty-nine third-class cruisers and gunboats, and twelve divisions of various torpedo-boats will have become obsolete. In the list of ships thus to be condemned are the four ships of the Brandenburg class, the new armoured cruiser Fürst Bismarck, the vessels of the Hertha class, and some small cruisers which have not yet been built, and are indicated by letters only. The four ships of the Sachsen class, with the König Wilhelm, Kaiser, and Deutschland, will disappear from the active list in 1901, but after that date there will only be three condemnations up to the year 1914, when the now modern ships will begin to disappear from the list. As these vessels become obsolete others are to be laid down to take their places, so that the fleet should always remain of the strength designed.

Battleships of the programme of 1898.

Of the battleships of the older programme, "A" is in hand at Schichau's yard, Danzig, Kaiser Karl der Grosse ("B") is completing afloat at Hamburg (Blohm and Voss), "C" is building at Wilhelmshaven, "D" at Kiel, "E" at the Germania yard, Kiel, "F" at the Vulcan yard, Bredow, near Stettin, and "G" has yet to be put in hand.

Battleships launched. Kaiser Wilhelm der Grosse, Kaiser Karl der Grosse.

Two battleships have been launched since the last publication of the *Annual*—the Kaiser Wilhelm der Grosse, on June 1st, 1899, at the Germania yard, Gaarden, near Kiel, and the Kaiser Karl der Grosse (B), at the yard of Messrs. Blohm and Voss, Hamburg. Both are sister ships of the Kaiser Friedrich III., and the following are particulars of the former, which has been built to replace the old König Wilhelm: Displacement, 11,000 tons; length, 377 ft. 4 in.; beam of, 65 ft. 8 in.; mean draught of, 25 ft. 8 in. The vessel has three triple-expansion engines in completely separated water-tight compartments, driving as many screws and developing 13,000 horse-power, calculated to give a speed of 18 knots. The water-tube boilers are on the Schulz system. The normal coal capacity is 650 tons, which can, however, be increased to 1000 tons. The protection consists of an armour-belt of Harveyed nickel steel, 6 ft. 8 in. in width, with a thickness of from 6 to 12 in. along the forward four-fifths of the vessel. The stern section is, for the sake of lightness, only protected by a 3-in. curved armour-deck. The two turrets for the large guns have 10-in. armour plating, and the turrets and casemates for the 5·9-in. guns have 6-in. armour. The conning tower has a shield with a thickness of from 4 in. to 10 in. The armament consists of four long 9·4-in. guns in revolving turrets, twelve 5·9-in. Q.-F. guns in armoured casemates, six of the same calibre in revolving turrets, twelve 3·3-in. Q.-F., twelve 1·4-in., and eight machine guns.

The torpedo armament is to consist of one 21-in. submerged bow-tube, four 18-in. submerged broadside tubes, and one 18-in. over-water stern-tube. The newer ships have somewhat larger dimensions, displacing 11,700 tons, as may be seen in the tables, where the particulars given are those of "F," building at Stettin.

Armoured cruisers. Prinz Heinrich. The armoured cruisers "A" and "B" are of a type new to the German Navy, being smaller than the Fürst Bismarck—8868 tons as compared with 10,650 tons, two 9·4-in. guns instead of four, and ten 5·9-in. instead of twelve. On the other hand, the smaller cruisers, are to have 20·5 knots as compared with 19. "A," which has received the name of Prinz Heinrich, has been launched at Kiel. Her length is 396 ft., and beam 64 ft. 3 in. The belt is of 4-in. nickel steel and runs the whole length of the ship, reaching also to the height of the battery. The deck protection varies from 2 in. in thickness to 2¾ in. There are to be three engines of vertical triple-expansion type, supplied by fourteen Dürr boilers, developing 15,000 horse-power and giving the ship a speed of 20·5 knots. Her bunkers will hold 950 tons. She will carry two 9·4-in. guns in turrets; ten 5·9-in. Q.-F. guns—four in turrets and six in casemates; ten 3·4-in. Q.-F. guns protected by shields; ten 1·4-in. Q.-F. guns; four machine guns; and four torpedo-tubes, forward, aft, and one submerged on each broadside. Her complement will be 528 men, with 43 additional when she is used as a flagship.

Protected cruisers. Nymphe, Niobe. The new class of armoured cruisers appears to have replaced the protected Freya class (5650 tons), of which no more are at present projected, while the Gazelle, launched at the Germania yard, Kiel, in March, 1898, was the first of a new and smaller class. The Nymphe has been launched at the Germania yard and the Niobe at the Weser yard, Bremen, and "C," "D," "E," and "F" are in hand. They are an improvement on the Gazelle in the matter of protection, and have two submerged broadside 18-in. torpedo-tubes, instead of two above water and one submerged bow tube. The following are the characteristics: Displacement, 2600 tons; length, 328 ft.; beam, 38 ft. 7 in.; draught, 15 ft. 1 in.; armament, ten 4·1-in. Q.-F., with ½-in. shields, fourteen 1·4-in. Q.-F., and four machine guns. There is a 2-in. deck in three layers, with a cofferdam filled with cork and cellulose, or gelatine, for the better protection of the machinery; also an armoured conning-tower. The speed of the Gazelle is 19 knots, that of the Nymphe and Niobe 21·5 knots.

Gunboat Tiger. The gunboat Tiger, built to replace the Wolf, has been launched at Danzig, where another of the same class is in hand. They are boats of 894 tons and 13·5 knots, carrying eight 3·4-in. and six 1·4-in. Q.-F. guns with two Maxims.

Refits.

The Württemberg, like her three sisters, has now been modernised. She has received Dürr water-tube boilers intended for 6000 I.H.P., and some small additions to her armament. Wood has been suppressed almost everywhere, and has been replaced by iron or steel, while the deck has been coated with xylolith. The Hagen, one of the unsatisfactory coast-defence armourclads, has been taken in hand at Kiel for complete reconstruction. She has been cut in two and is being lengthened by 25 ft. by the addition of a middle section. The idea is to secure greater coal capacity, and if the reconstruction should be satisfactory other ships of the same class will also be transformed.

ITALY.

Launches. Garibaldi. Varese.

Two armoured cruisers have been launched—the Garibaldi, at Messrs. Ansaldo's, Sestri Ponente, on June 29th, and the Varese, at Messrs. Orlando's, Leghorn, on August 6th. They are improved types of the original vessels built under these names. The arrangement made by the Italian Government to permit the sale of the earlier vessels has operated favourably in two ways. It has given a great stimulus to naval construction, and has enabled the experience gained to be utilised for the advantage of the Italian Navy. The work has been carried forward very rapidly, and Messrs. Ansaldo, who have displayed much enterprise, have built four Garibaldis within about four years.* Their establishment has been extended, and they secured permission to set back the main line of the railway from Genoa to Ventimiglia, thus enabling them to build five masonry slips on which ships larger than the Powerful can be constructed. The new Garibaldi was laid down on September 1st, 1898, and having been on the stocks fourteen months, she was launched with the shafting in place in the engine-room, the auxiliary engines fitted, and the double-bottom and bilge pipes and valves finished. Similar activity has been shown by Messrs. Orlando, at Leghorn, who have launched three Vareses,† of which the last was laid down on September 4th, 1898, was launched in the presence of Admiral Bettolo on August 6th, 1899, and was completed for sea in February, 1900, having thus been in hand about eighteen months. The engines were built by Orlando Brothers, and the Belleville boilers partly in their establishment and partly (to gain time) at St. Denis.

The new ships are about 16 feet longer than their predecessors, much

* Garibaldi I (Garibaldi, Argentine); II. (Cristobal Colon, Spanish, lost at Santiago); III (Pueyrredon, Argentine); IV. (Italian)

† Varese I. (San Martin, Argentine); II. (Belgrano, Argentine); III. (Italian).

of the space being given to the machinery. The Garibaldi, like her sister, the Francesco Ferrucio at Venice, has Niclausse boilers, while the new Varese is fitted with Belleville boilers, twenty-four in number, the maximum I.H.P. being 13,500. The great object has been to increase the power with natural draught. In the ships sold to Argentina the natural draught power was 8000, but the new vessels will attain from 10,000 to 11,000. The armour is of nickel-steel, upon the special system employed at the Terni Works, and has a maximum thickness of 6 in. Improvements have been introduced also in the arrangement of the ammunition hoists and in the disposition of the guns for firing ahead and astern. The heaviest guns are one of 10-in. calibre in a turret forward, protected by 6-in. steel, and two of 8-in. coupled to a turret aft. There are ten 6-in. Q.-F. guns in the battery, of which the end pairs are respectively for bow and stern fire, and the same is the case with four other 6-in. guns on the upper deck. Ten 2·9-in. guns (six on deck and four in the battery) have a similar arrangement; there are also six 1·8-in. Q.-F. and two Maxims. The vessels are provided with powerful electric apparatus for lighting the ship, supplying the current to the searchlights and driving the ventilators, working the ammunition hoists, and supplying power for the gun turrets. Another improvement is the almost total abolition (with the exception of a little teak) of wood, this being replaced mostly by thin steel.

Agordat, Coatit. The torpedo cruisers Agordat and Coatit were both launched at Castellamare toward the end of 1899. The length is 287 ft. 6 in.; beam, 30 ft. 6 in.; mean draught, 11 ft. 1 in.; displacement, 1313 tons. Two engines of triple-expansion type, developing 8000 horse-power, and supplied with steam by eight water-tube boilers of the Blechynden type, with a pressure of 15 atmospheres, are to give a speed of 23 knots. The internal fittings are of soft steel, wood being used as little as possible, and in many cases aluminium is substituted, and the deck is of hardened steel extending along the whole length. The armament consists of four 4·7-in., eight 2·2-in., and two 1·4-in. Q.-F. guns, with two torpedo-tubes. **Torpedo Craft.** The Lampo and some other torpedo-boat destroyers, built for the Italian Government by Messrs. Schichau, of Elbing, have been launched. The length is 196 ft. 8 in., and the nominal speed 30 knots. The 28-knot destroyer Fulmine, built by Odero, at Sestri Ponente, has been making satisfactory trials, and the first-class boat Condore on a three-hours' run attained a mean speed of 26 knots with 2400 I.H.P.

New programme. It is yet uncertain how far the financial resources of the country will enable the Italian Government to carry out the shipbuilding programme. Admiral Bettolo has laid great stress upon the importance to Italy of swift, heavily-armed vessels and of torpedo craft. It was intended to lay down four of the former, under the suggested names of Genova, Pisa, Venezia, and Amalfi; while, in addition to the four destroyers being built by Schichau at Elbing, two others would be put in hand at Chiswick and four at Naples. The Italia and Lepanto were also to be reconstructed, by making alterations in the machinery, to give a speed of 18 knots, the substitution of lighter and more numerous guns, and the disposition of a great layer of coal abreast of the engines and boilers for protection.

The new armoured type. It is probable that the Government will be able to put in hand at once only two ships of the new type—at Spezia and Castellamare—out of the four in the programme. The class promises to be extremely interesting, and a diagram of it will be found in Part II.* The idea is to combine in a ship with the moderate displacement of 8000 tons, great speed and range of action with sufficient protection and an armament of not less than twelve 8-in. quick-firers, in addition to a secondary armament. Admiral Bettolo's conception of the new ship arose from the great difficulty experienced in perforating plates of hardened steel with the projectiles of 6-inch guns, and the advantage of providing numerous guns of uniform character and more powerful than the 6-inch, and he charged Col. U. E. Cuniberti, the chief naval architect, to endeavour to combine the necessary qualities, taking as the basis of his design a displacement of 8000 tons, an armament of twelve 8-in. quick-firers, 6-in. armouring, and a speed of 22 knots. Colonel Cuniberti believes that he has succeeded in the task, and the following are the particulars of his design: displacement, 8000 tons; length, 400 ft. 3 in.; beam, 63 ft. 1 in.; mean draught, 22 ft.; belt, 6-in. special steel, manufactured at Terni, citadel, 6-in., with 8-in. bulkheads, conning-tower 6-in., deck 1½-in.; armament, twelve 8-in. quick-firers coupled in six turrets (one forward, one aft and two on each broadside, all with their bases protected by 6 in. of steel, and giving a direct fire of six guns ahead and astern and eight on each beam), and twelve 3-in. quick-firers (eight on the superstructure and two on each side at the bows and at the stern. The torpedo-tubes will be two above water and two submerged. With 15,000 I.H.P. the stipulated speed during a 24 hours' trial is 22 knots, but the ships are expected to attain over 23 knots with 600 tons of coal on board and all stores complete. The coal capacity will be 2100 tons, sufficing for a range of 15,000 miles at an economical speed.

* Plate 57.

Russia.

The best way to treat the Russian Navy in this place may be to indicate the shipbuilding work now in progress. Practically the whole of the larger shipbuilding programme is now in hand. The two battleships, Peresviet and Oslabya (12,674 tons), which were described in the *Annual* last year, have been completed, and the first of them has been under trial. Like their sister, the Pobieda, which is in course of construction at the Baltic Yard, St. Petersburg, they have been regarded as presenting a combination of the qualities of the battleship and the cruiser, combining heavy armament (the heaviest guns being of 10-inch calibre) and satisfactory protection with a high speed, which, however, it may be remembered, is only 18 knots. They differ from preceding Russian ships in having three independent engines, each driving a screw, and in the placing of the 6-in. guns in separate positions, protected by from 2 to 5 inches of steel. The arrangement of these guns is unusual.* One is in the bow and five are on each side—one in a sponsoned casemate amidships, and four in two double (superimposed) sponsons forward and aft. The ships have also a large electric installation, which provides for much mechanical work on board. The steering gear, for example, may be operated by the customary electric apparatus or by Shubin's special method, as well as by hand, by steam, or by hydraulic pressure. In addition to these ships the Kniaz Potemkine Tavritchesky, of 12,480 tons, is in hand at Nicolaieff, the Retwisan, 12,700 tons, well advanced at Philadelphia, and the Tsarevitch, 12,900 tons, at La Seyne. All have been described in previous issues of the *Annual* and are sister ships with minor differences. A very full account of the Retwisan was given last year.

Battleships completing. Peresviet, Oslabya.

Battleships building.

To complete the new programme three battleships have recently been ordered to be put in hand—the Imperator Alexander III., at the Baltic Yard, St. Petersburg, the Borodino, at the New Admiralty Yard, and the Orel, at Galerny Island. They belong to the same class as the Retwisan, and the following are their characteristics: Displacement, 13,600 tons; length, 397 ft.; beam, 72 ft.; draught, 26 ft.; protection, Harveyed steel belt 9 in. thick, 5 in. on the chief gun positions, and a steel deck, 4 in. maximum thickness; armament, four 12-in., twelve 6-in. Q.F., twenty 3-in. Q.F., and about twenty-six smaller Q.F.; torpedo armament, six discharges, of which five submerged; water-tube boilers and engines of 16,000 horse-power; speed, 18 knots.

New battleships. Alexander III., Borodino, Orel.

* See Pl. 79, Pt. II.

Coast-defence ship.

Admiral Boutakoff.

The coast-defence ship, Admiral Boutakoff, of the Apraxine type which has been announced as being in hand, has at length been laid down, and it is reported that she will be somewhat larger and more powerful than her sisters.

Two armoured cruisers are under construction—the Bayan, of 7800 tons, at La Seyne, and the Gromoboi, which was launched at the Baltic Yard on May 20th, 1899. She was described in the *Annual* of 1898, but some other particulars may be added. She belongs to the same type as the Rurik and the Rossia, but is larger than either and better protected.* She has four 8-in. guns forward and aft in 6-in. armoured sponsons, much as in the sister ships, but of the sixteen 6-in. quick-firers, six are on each side also in sponsons, those forward and aft being under the positions of the 8-in. guns. Two 6-in. guns are further aft in recessed positions, and the two others are in the bow and at the stern. Twenty 3-in. guns are on the upper deck above the sponsons, in the bows, and at the stern, and twenty machine guns of small calibre are in various positions in the ship. This armament differs little from that of the Rossia, which has four 8-in. guns, sixteen 6-in. quick-firers, twelve 3-in. quick-firers, and thirty-six smaller. The Gromoboi has four submerged torpedo-tubes. The side-armour is of Harveyed steel 350 ft. long, 6 ft. high, and 6 in. thick, and the arched armour-deck is 3 in. thick at the sides, at each end being a bulkhead with 9-in. steel forward and 8 in. aft. These bulkheads rise to the height of the upper deck and protect the battery. The ship has thirty-six water-tube boilers developing 18,000 I.H.P.; the coal capacity is 2500 tons, giving a range of 19,000 miles at 10 knots, and the maximum speed is 20 knots; there are three screws.

Protected cruisers.

Of protected cruisers still in hand four are afloat, of which the Waryag, built at Philadelphia, was fully described in the *Annual* last year. The Pallada and Diana—cruisers of 6630 tons and 20 knots—were launched at Galerny Island respectively on August 27th and October 12th, 1899, and the Askold, 6000 tons, at the Germania Yard, Kiel, in March, 1900. The following are the particulars of the last-named: Length between the perpendiculars, 426 ft. 6 in.; beam, 49 ft. 3 in.; normal draught, 20 ft. 4 in.; armament, wholly quick-firing, twelve 45-calibre 6-in., twelve 3-in., eight 1·8 in., two 1·4-in., two machine guns; two submerged and two above water broadside torpedo-tubes for 15-in. torpedoes, and tubes at the bows and the stern; I.H.P., 19,000; bunker capacity, 1100 tons. The ship is built upon fine lines in order to provide for her high speed, and she has two light masts for signalling. The conning-tower is heavily plated with 5·6-in. Krupp nickel steel, and

Askold.

* See Pl. 77, Pt. II.

"PERESVIET,"
RUSSIAN BATTLESHIP.

the ship has a $1\frac{1}{2}$-in. steel deck, thickening to 3 in. at the sides. The hull is divided into watertight compartments, and a remarkable feature is that by means of electric pumps any of the large compartments can be cleared of water within an hour. Three triple-expansion engines driving as many screws, and supplied by nine water tube boilers on the Schulz system, in five watertight compartments, are to give a speed of 23 knots as the mean of a twelve hours' trial, though the constructors hope that 24 knots will be attained. There will be a great number of auxiliary engines on board for working the steering, boat, and anchor gear, as well as six large dynamos, four of them under the armoured deck, to provide light and driving force for use in certain parts of the vessel. There will be six large search lights. The Askold was designed by Herr Rauchfuss, of the Germania Yard, and has been constructed of German materials.

The Aurora, a sister of the Diana and Pallada, is almost ready for launching at Galerny Island; the Boyarin, of the Askold class, has been put in hand at Copenhagen; and the *Kronstadtski Viestnik* states that the first vessel to be built at the new State yard at Windau will be a cruiser of the same class. The Novik, a cruiser of 3000 tons, of which some account was given last year, is in hand at Elbing, and other cruisers of the class are in hand or contemplated in Russia.

Destroyers.

Of torpedo craft there are at the present time building at the Nevsky Works, St. Petersburg, thirteen destroyers; at the Ishora Admiralty Works, five; and at the Creighton Works, St. Petersburg, four, making a total of twenty-two destroyers, all of the "Sokol" type, and reproducing the features of the boat built by Messrs. Yarrow, in 1895. The Russians are also building at the Nevsky Works ten destroyers of 350 tons, from designs furnished by Messrs. Yarrow, and very similar in character to the Japanese destroyers lately launched at Poplar.

Ice-breakers.

Russia is continuing to build ice-breaking ships. The Ermack, of which some account has been given in the *Annual*, was employed in the spring of 1899 in useful work. On the 4th of April she freed six vessels from the ice, and brought them to St. Petersburg, and three days later she found twenty-six steamboats behind an ice bank, which she cut through, enabling them to come out, while a third excursion enabled her to set ten other steamships free of ice. She had complete success in freeing ninety various vessels from the ice off the mouth of the Neva. Another ice-breaker, the Ledokol, has been built in Finland for the purpose of clearing the water between Kronstadt and the forts, and has undergone her trials. She is 100 ft. long, with 22 ft. beam, and 5 ft. draught, and has triple-expansion engines, which, with 150 revolutions, are to give a speed of 11 knots at sea.

A third ice-breaker, to be known as the Odessa, is to be built at Elswick, and will be 157 ft. 6 in. long, with 42 ft. beam. She is intended for the port of Odessa.

Mishaps.

The Russian Navy has lately been unfortunate with several of its ships. The Gangoot, which foundered near Viborg in 1897, has not yet been raised, notwithstanding the great efforts that have been made. The Gromoboi, on her way from St. Petersburg to Kronstadt to complete her armament, grounded on a sandbank in the maritime canal in November, 1899, suffering the same mishap as her sister, the Rossia; but she was floated off without assistance, owing to a sudden rise in the water. The coast defence ship, General Admiral Apraxine, in the same month, struck during a snow-storm a reef off the Island of Gothland, and was in a very critical situation, her bow being raised up sharply and her bottom stove in below the fore turret. Salvage work went on under great difficulty, owing to the intense cold, and part of the rock on which she struck was blown away to prevent further damage. Later it was reported that she had been driven by a mass of ice upon a rock, which had pierced a fresh hole in her hull, but she was floated off, with the assistance of the Ermack, on April 26th, 1900. The battleship Poltava ran ashore near Libau in January, 1900, but was floated from a dangerous situation without serious damage.

United States.

Trials. Alabama.

Since the last issue of the *Naval Annual* three of the new battleships have been under trial. At the builders' trial of the Alabama on August 29th, 1899, off Delaware Bay, over a distance of eleven and a half miles, a speed of 17·2 knots per hour was attained during the last run under forced draught, with 165 lb. of steam and 113½ revolutions of the propellers. The wind at the time favoured the vessel. The run immediately preceding was made at the speed of 15·43 knots an hour, weather conditions being reversed. This made the average speed for the two runs under forced draught 16·23 knots per hour.

Kearsarge, Kentucky.

The battleship Kearsarge, on her full-power trials between Cape Ann and Cape Porpoise on September 25th, 1899, steamed at a mean of 16·84 knots. (Natural draught trials, April 3rd, 1900, 14·99 knots, 8,483, I.H.P.) Accounts of the trials of the Kentucky at Cape Ann on November 24th state that she maintained an average speed of 16·878 knots for sixty-six nautical miles. It was estimated that tidal allowance would make her record 17 knots. The trial was made in a somewhat rough sea. The highest steam pressure developed was 172 lb. Boilers and machinery worked smoothly for four hours under forced draught on her return to Newport News.

Particular interest attaches to later trials of the Kearsarge, because, like her sister the Kentucky, she is built upon the superimposed turret system, which has already been described in the *Naval Annual*, there being four 8-in. guns in turrets rigidly fixed on the top of the housing of the 13-in. guns, and turning with them. After the additional trials in April, 1900, Captain Folger, commanding the ship, made the following report: "The double turret was thoroughly tested, and is an assured success, both from the military and structural standpoints. There is no interference between the planes of the guns or inconvenience from blast or smoke. The structure was tested with simultaneous discharges from three guns. It is quite strong enough to withstand the united shock of the four guns of either turret, but the absence of a suitable electric device for a simultaneous discharge of all the guns prevented this final test. Both pairs of 8-in. guns were tested in simultaneous firing." The trials, however, were not considered conclusive, and are to be continued. Rear-Admiral Philip Hichborn, Chief Constructor of the United States Navy, has recently stated his objection to the double turret in these words: "In the last war the vessels of our Navy fulfilled all that was required of them in every particular, despite the intricate mechanism of the modern warship. None of these vessels, however, were fitted with the double turret, and I am so firmly convinced of the disastrous consequences of such a design in actual warfare that I fear the result would have been otherwise had the two battleships fitted with the double or superimposed turrets been in service. The Kentucky and Kearsarge, so far as the double turrets are concerned, are as yet untried, but although the design was widely exploited several years ago, no other nation has deemed it desirable to incorporate it in the construction of its battleships."

The new battle-ships. None of the new American battleships have the double turret system. The Alabama, Illinois, and Wisconsin were very fully described in the *Annual* last year by Lieut.-Comm. W. H. Beehler, U.S.N., in his Chapter on the American Navy, as were also the characteristics of their successors, the Maine, Missouri, and Ohio, which are now in an early stage of construction, and of the Arkansas class of monitors. The following are additional particulars of the armament of the Maine class:—

Maine class. The main armament will consist of four 12-in. 40-calibre guns, the secondary battery being composed of sixteen 6-in. Q.-F. guns, and, like the 12-in. guns, designed to use smokeless powder. There will be an auxiliary armament of something like twenty 6-prs., four automatic 1-prs., four Gatlings, and a field piece. The 12-in. guns will be mounted in the two turrets, and will have arcs of training of 280 deg. each. Five of the 6-in. guns will be on each side in the main deck battery, and will have arcs of fire of 110 deg. The bow 6-in. guns will have each a total sweep of 138 deg. Four 6-in. guns mounted in the two citadels on the upper deck will each have an arc of fire of 138 deg. Two of the 6-prs. will be mounted on the main deck, one on each side, between the two foremost of the 6-in. guns. Four others will be placed aft, while the rest will be mounted on the superstructure. All of the 6-in. guns will carry heavy cylindrical shields, and the fighting stations between these guns will be separated by splinter bulkheads 1½-in. thick. There will be two under-water torpedo tubes placed well forward. The manipulating rooms will be behind the armour belt and below the water-line.

Georgia, New Jersey, Pennsylvania.

The three battleships of the 1899 programme, Georgia, New Jersey, and Pennsylvania, have been delayed by the decision of Congress that a greater price than 400 dols. per ton should not be paid for armour, it being found impossible to procure satisfactory contracts at this rate for armour that would satisfy the ballistic tests required for steel manufactured by the Krupp process. A majority of the committee appointed to consider the subject reported favourably to a greater outlay in April, 1900. The following are the dimensions of the vessels, which are larger than the Alabama class, and are to steam at 18·5 knots: displacement, 13,500 tons; length, 420 ft.; beam, 75 ft.; draught, 24 ft. Rear-Admiral O'Neil, President of the Board of Construction, gave the following particulars concerning the new ships on March 27th, 1900, which do not, however, seem to be fully settled, and differ from others previously announced:—

"The armament will consist of four 12-in. guns of 40 calibres, to be mounted in pairs in two electrically controlled, elliptical, balanced turrets having inclined port-plates, one forward and one aft, each having an arc of fire of 270 deg.; four 8-in. guns of 45 calibres, in two turrets, one on each side forward of the beam, and having an arc of fire of 145 deg., that is, from right ahead to 55 deg. abaft the beam; twelve 6-in. 50-calibre guns, two of which will be mounted on the upper deck at the after ends of the superstructure, having an arc of fire of 145 deg., that is, from right astern to 55 deg. forward of the beam. Ten of the 6-in. guns to be mounted in broadside on the main or gun deck, having an arc of fire of 110 deg., that is, 55 deg. forward of, and 55 deg. abaft the beam, except the two forward guns, which will fire from directly ahead to 55 deg. abaft the beam. The secondary battery will consist of twelve 14-prs., twelve 3-prs., eight 1-prs., eight machine guns, and two field guns, a total of sixty-four guns. The vessels will also be provided with two under-water torpedo tubes. The speed will be 19 knots, and the coal capacity 2,000 tons.

"The armour protection to the hull will consist of a complete water-line belt of Krupp armour, 8 ft. wide over the central portion of the ship. This belt will be 9 in. thick at its upper edge and will carry this thickness downward for a distance of 5 ft., and then taper to 6 in. at the lower edge. Forward and aft of the heavy belt the armour will taper to 4 in. at the bow and stern. Above the main belt the side will be protected with 6-in. armour, which will cover all the 6-in. gun positions. The turrets for the 12-in. guns will be 10 in. thick, except the front plate, which will be 11 in. The turrets for the 8-in. guns will be 6 in. thick, except the front plate, which will be 6½ in. The 14-pr. guns will be protected by armour of 2 in. in thickness. In addition to the above protection, the vessels will carry a complete belt of cellulose and a curved protective deck of 3½ in. thick on the slopes and 2½ in. thick on the flat. In the circular defining the chief characteristics of the vessels, which will be issued by the department in due time, there will probably be a proviso that the government reserves the right to change the number and calibre of the guns, or the arrangement of the battery, or the thickness and distribution of the armour, at any time within six months after the date of the contract, provided such change does not exceed the weight therein provided for the abovenamed purposes. This will allow time for a proper test of the superimposed system of turrets on the Kearsarge class and leave the department free to adopt them in the later ships, if it seems desirable. The new ships will have high freeboard, and the after body will be carried to the same height as forward, thus affording abundant space for a flag-officer and his staff, the vessels being designed for flag-ships. It is not yet fully determined whether these vessels shall be sheathed and coppered."

Armoured cruisers California, Nebraska, West Virginia.

There has been delay also in putting in hand the armoured cruisers California, Nebraska, and West Virginia, provided for in 1899. Admiral O'Neil gives the following particulars concerning them:—"The 12,000-ton cruisers are to have a speed of 22 knots and are to carry 2000 tons of coal. The battery will consist of four 8-in. 45-calibre guns, mounted in pairs in turrets, one turret forward and one aft, each having an arc of fire of 270 deg.; fourteen 6-in. 50-calibre guns, four of which will be carried on the upper deck, having an arc of fire of 145 deg., two firing ahead and two astern. On the gun-deck ten 6-in. guns will be mounted, the forward and after pair being sponsoned to obtain end fire and the others in broadside. The secondary battery will consist of eighteen 14-prs., twelve 3-prs., eight 1-prs., eight machine guns and two field guns, a total of sixty-six guns. The vessels will also be provided with two under-water torpedo tubes. The armoured protection will consist of a complete water-line belt, 7 ft. 6 in. wide, 6 in. thick at its upper edge and tapering to 5 in. at the lower edge. This heavy belt extends from the forward to the after turret, and is completed to the stem and stern by 3½-in. armour. Above the belt the side is protected by 5-in. armour, which covers all the 6-in. gun positions. The turrets for the 8-in. guns are of 6-in. armour, except the front plate, which is 6½ in. In addition to the above described protection, the vessels will have a complete cellulose belt and a heavy curved protective deck extending the entire length of the ship. This deck will be 4 in. thick on its slopes and 3 in. thick on the flat. The question of sheathing is not yet fully determined."

Second-class cruisers.

Six small second-class cruisers—the Chattanooga, Cleveland, Denver, Des Moines, Galveston, and Tacoma—have been given out to contract. They were provided for in the votes of 1899, and the leading particulars are as follows:—Length, 292 ft., extreme 308 ft. 2 in.; extreme beam, 43 ft.; mean draught at trial displacement with two-thirds coal, ammunition and stores, 15 ft. 6 in.; extreme draught, fully loaded, 16 ft. 8 in.; trial displacement, 3,100 tons; full load displacement, 3,400 tons; bunker capacity, 700 tons; 4,500 I.H.P.; speed, 16·5 knots; range at 10 knots, 9,800 miles. The armament will consist of ten 5-in., eight 6-pr., and two 1-pr. Q.-F. guns, and four Colt machine guns. Eight 6-in. guns will be mounted on the main deck in recessed ports, the four foremost ones having a range from right forward to 60 deg. abaft the beam, and the four after ones from right aft to 60 deg. before the beam. The two remaining 5-in. guns will be behind shields on the spar deck. Four 6-prs. will be on the main deck, two forward and two amidships, and four more on the spar deck. The wood used will be reduced to a minimum. All the bulkheads on the gun and berth decks will be of metal, and where it is necessary to use wood, it will be treated with the electric fire-proofing process. A watertight deck covered with ½-in. plate will run from stem to stern, the sides sloping down to 3 ft. below the water-line, and on the top of the deck at the sides a belt of obturating material will be placed, covering the water-line for the whole length of the ship.

Torpedo craft

A full account of the torpedo-boats in hand was given in Lieut.-Comm. Beehler's chapter last year. As will be seen by the torpedo tables in Part II., a number of destroyers and first-class boats have either been launched in 1899, or will take the water in 1900. After many trials the Navy Department, in April, 1900, signed a contract with the Holland Submarine Torpedo Boat Company for the purchase of the Holland at the price of about £30,000, and agreed to pay about £35,000 each for any boats of the Holland type which it may purchase subsequently, provided that the boats are of similar dimensions to the more recent Holland, which is larger than the old one.

The second-class cruiser Charleston was wrecked at half-past five on the morning of the 22nd November, 1899, on a coral reef not marked on the chart, three miles to the north-west of the Guinapak rocks, north-east of the Island of Luzon. All on board were safely landed on Kamiguin Island, armed with rifles and two Colts, and were rescued. **Loss of the Charleston.**

The Naval Appropriation Bill was reported to the House of Representatives from the Naval Committee on April 5, 1900. The Bill carries an expenditure of over £15,000,000, which is the largest amount ever reported in the House. It recommends the construction of two battleships of 13,500 tons, three armoured cruisers of 13,000 tons, and three protected cruisers of 8,000 tons. **New Programme.**

JAPAN.

In the financial year 1900–1901 it is intended to put in hand a torpedo depôt ship of the Vulcan type, two third-class cruisers, two gunboats, seven torpedo-boats to be built abroad, and twenty-two to be built in Japan. All these vessels should be practically complete by the year 1903-4, and the Japanese fleet will then have been created and constituted according to the naval expansion scheme, although the financial provision for that scheme extends until the year 1905–6.*

The battleship Shikishima, which has already been fully described in the *Naval Annual*, left Portsmouth on Jan. 27th, 1900, for Yokosuka, to be completed for commission. She went out by way of the Suez Canal, and in order to pass through was lightened to a draught of 26 ft. 6 in. The full-power speed trials in September, 1899, between Dartmouth and Torbay had given a first mean of 19·027 knots, with 14,667 I.H.P., in twenty to twenty-six fathoms, a second mean of 19·056 knots, with 15,188 I.H.P., and a third mean of 18·723 knots, with 15,621 I.H.P., in shallow water of about fourteen fathoms. **Battleships. Shikishima trials.**

The gunnery and torpedo trials were carried out off Portsmouth before the ship left for Japan.

* *Cf.*, *Naval Annual*, 1898, p. 51.

In the torpedo trials fifty per cent. of the shots passed fairly under the centre of the target, and the remainder were but little off it, although the torpedoes were not fitted for the gyroscopic steering gear. The Elswick submerged tubes performed all their functions. A torpedo was put into each of them before the ship got under way, no further attendance being required beyond the pressing of an electric key on deck as the sights of the directors came successively on the target. The automatic return of the guide or shield within the ship also rendered attention to the tubes after firing unnecessary. In the gunnery trials three rounds were fired from each 12-in. gun, at various positions of training and elevation. It was found that, with a crew which had not seen the guns two days previous to the trials, an interval of only forty-five seconds between the rounds was required. The whole of the hydraulic mechanism worked with the greatest ease. The 12-in. guns of the Shikishima are of forty calibres, and fire an 850-lb. projectile with a velocity of 2,300 ft.-secs., and with a charge of 145 lb. of cordite. If any accident should occur to the hydraulic machinery, electric gear can be brought into use, and should this in turn fail, the whole of the machinery can be worked by hand. The breech mechanism of the guns is easily worked by hand. Firing can be done either by electricity or percussion. The fourteen 6-in. and twenty 12-pr. guns carried by the Shikishima had been tried in November, two rounds from each 6-in. and four from each 12-pr. being fired without a hitch. The 6-in. and smaller guns are provided with hoists for the supply of ammunition by electricity or hand. The former system is on a new design, which has given favourable results. An endless chain, running over a pulley at the top and a winch at the bottom, is made to work very rapidly backwards and forwards; thus, a cartridge or a projectile in a bag is hooked on to the chain at the bottom, the coned friction gear is brought into action, and the ammunition very rapidly hoisted. Then, while it is being unhooked at the top, another round is being hooked on the other side of the chain at the bottom, the machinery is reversed and the second charge hoisted, and so on. The reversing is done by having two cones, one on each side of the driving wheel. If hoisting in one direction, one cone is brought into operation, or if in the other direction, the other. A simple cut-off insures the cone being thrown out of gear at the right time for each lift.

Launches.

Asahi and Hatsuse.

Two battleships have been launched—the Asahi, at Clydebank, on March 3rd, 1899 (chronicled in the *Annual* last year), and the Hatsuse, at Elswick, on June 27th, 1899. Like the Mikasa, building at Barrow, they are sisters, with minor differences, of the Shikishima.

The following particulars concerning the Asahi are from the *Times*. She is the heaviest battleship ever built on the Clyde. The weight on the blocks when she was launched was 7,500 tons, and the principal dimensions are :—Length between perpendiculars, 400 ft.; length over all, 425 ft. 6 in.; breadth, extreme, 75 ft. 2½ in.; depth, moulded, 43 ft. 7½ in.; normal mean draught of water, 27 ft. 3 in.; displacement, 15,200 tons. The armament is of Elswick design and manufacture. Each pair of 12-in. guns commands an uninterrupted arc of training of 240 degrees. The manipulation of the turntables and all the operations of loading and laying the guns are performed by hydraulic power, and loading is practicable with the guns in any position of training. The guns and gunners are sheltered by means of heavy armoured shields, which revolve with the turntables. The secondary armament will consist of fourteen 6-in. Q.F. guns, each mounted in a separate casemate, twenty 12-pr. Q.F. guns, eight 3-pr. Q.F. guns, four 2½-pr. Q.F. guns, and four submerged torpedo-tubes in two compartments—one forward and one aft. The magazines and shell rooms for the heavy guns being conveniently situated near the hoists, the longitudinal transport of heavy material will be reduced to a *minimum*. The vessel is adapted for ramming, the stem being arranged as a powerful spur, strongly supported. Great care has been bestowed upon the arrangement of the protective material. There is a main belt extending for a length of 250 ft. amidships, the total depth of this belt being 8 ft. 2 in., and it is intended that when the ship is floating at the normal water-line the lower edge of the armour shall be 5 ft. 6 in. below water, and the upper edge 2 ft. 8 in. above water. The *maximum* thickness of the belt is 9 in. Above the main belt the sides from lower to main deck are covered with armour of a thickness of 6 in. for a length of 250 ft. The protection of the vitals of the ship is rendered more secure by a heavy protective deck extending all fore and aft, and sloping away from the under side of the main armour belt. The forward conning-tower is composed of 14-in. armour, and the after tower of 3-in. armour. The whole of the armour plating is being manufactured of the highest quality procurable. The ship is propelled by two sets of three cylinder triple-expansion engines. Each of the sets is designed to develop 7,500 I.H.P., giving a combined I.H.P. of 15,000. Steam will be supplied by water-tube boilers of the Belleville economiser type, working at a pressure of 300 lb., which will be reduced at the engines to 250 lb. Each set of engines will be placed in a separate engine-room, divided by a longitudinal water-tight bulkhead which extends the whole length of the machinery space, and each engine-room is in all respects exactly similar and entirely independent of

Asahi. Hatsuse.

the other. The coal bunkers will have a total capacity of about 1400 tons.

Armoured cruisers launched. Yakumo. Idzumo.

Of armoured cruisers the Yakumo was launched at the Vulcan Yard, Stettin, on July 8th, 1899, and the Idzumo at Elswick, on September 19th. Two others of the same class are in hand—the Azuma at St. Nazaire, and the Iwate at Elswick—all being sisters, with certain minor differences, of the Asama and Tokiwa, which were fully described in the *Annual* last year. The following are particulars of the Idzumo:—Length between perpendiculars, 400 ft.; beam, 68 ft. 6 in.; depth, 41 ft.; draught, 24 ft. 3 in.; displacement, 9,750 tons. The armament consists of four 8-in. B.L. guns twin-mounted in barbettes; fourteen 6-in. Q.F. guns—ten in casemates (six on the main deck and four on the upper deck), the remaining four being on the upper deck protected by shields; seventeen 12-lb. Q.F. guns—eight on the shelter decks, two on the main deck forward, four on the bulwarks and three in the tops; four submerged torpedo-tubes—two forward and two aft. The vessel has a complete water-line belt of Harveyed nickel-steel armour, 7-in. thick amidships, reduced at the ends. Above this there is a citadel of 5-in. Harveyed nickel-steel armour enclosing the bases of the barbettes, and carried from the top of the water-line belt to the main deck. The barbettes are of Harveyed nickel steel 6 in. thick, the casemates are of nickel steel 6 in. thick, and the conning-tower is Harveyed nickel steel 14 in. thick. The machinery is of the twin-screw vertical triple-expansion type, to develop 14,500 I.H.P., and the speed guaranteed is $20\frac{3}{4}$ knots, the boilers being of the Belleville latest type. The cruiser has a bunker capacity for about 1,400 tons of coal. Accommodation is provided for an admiral, 52 officers, and 430 petty officers and men.

Trials Tokiwa.

The trials of the Tokiwa took place off the mouth of the Tyne, in May, 1899. Runs were made at 10, 15·6, 18·8, 21·2, and 22·73 knots, and during six hours the vessel ran at a mean speed of 20·85 knots with open stokeholds. At her last trial she ran for three hours with a pressure in her stokeholds scarcely exceeding $1\frac{1}{2}$ in. at a mean speed of 23·1 knots. The engines were supplied by Messrs. Humphrys, Tennant, and Co., and the cruiser has twelve single-ended cylindrical boilers.

Protected cruisers and gun-boats.

The two third-class cruisers and two torpedo-gunboats required to complete the programme will probably be built in Japan, where the Chihaya (875 tons), of the latter class, is in hand at Yokosuka.

Torpedo craft.

Six 31-knot destroyers of the Ikadsuchi class (described in the *Annual* last year) have now been launched by Messrs. Yarrow at Poplar, and have successfully undergone their trials. The following were the several speeds during a three-hours' run, with a load of 35 tons, under British Admiralty conditions:—

	Speed.	Air Pressure.
Ikadsuchi	31·32 knots.	2·5 in.
Inadsuma	31·037 „	2·2 „
Akebono	31·08 „	1·7 „
Sazanami	31·382 „	1·3 „
Oboro	31·362 „	1·3 „
Niji	31·15 „	0·9 „

Several of these destroyers have proceeded to Japan.

Six destroyers have also been built by Messrs. Thornycroft at Chiswick. The contract speed was 30 knots, with a load of thirty-five tons, which in every case was exceeded. The Shiranui steamed at 30·443 knots on the measured mile, and at 30·517 knots during a three-hours' trial, while the speeds of the Kagerou were 30·54 knots and 30·24 knots, and of the Usugumo 30·602 knots and 30·370 knots.

The following further particulars concerning the torpedo flotilla are taken from a paper contributed by Mr. Sassow, Director of Japanese naval construction, to the *Jiji-Shimpo* early last year, and doubtless much of the work has already been put in hand.

"Five first-class torpedo-boats and twenty-one second-class torpedo-boats have been already ordered by our Government, and there still remain seven more to be ordered from abroad. There are also eight torpedo-boats under construction at home, and twenty-two others will be built gradually at home. Among the boats (120-ton class) which were ordered abroad, one has been completed, and received in Japan, in sections; the second-class boats (80-ton class), in sections, have also been sent to Japan. Among the above-mentioned torpedo-boats, one first-class and two second-class are now under construction at the Mitsu Bishi Shipbuilding Co's. Works at Nagasaki; and two second-class boats at the Kawasaki Dockyard, Kobe. Two of these second-class torpedo-boats are under construction at Kuré Dockyard (the Second Naval Division), and four at Sassebo Dockyard (the Third Naval Division). The reason for entrusting the construction of torpedo-boats to the two private yards was that the naval dockyards were somewhat pressed with work, and, in addition, it was thought prudent, as a means of instructing the workmen of the private yards in connection with the design and construction of torpedo-boats, for future contingencies of the Service."

Mr. Sassow's survey of the possibilities of shipbuilding in Japan, embodied in the same article, is interesting, and deserves to be quoted.

Shipbuilding in Japan.

" *Yokosuka Dockyard.*—This dockyard was established during the Tokugawa Regency, by the Shogunate, in 1866. French officers, including naval constructors and engineers (M. Verney being the Chief Director), were engaged, together with a considerable number of leading workmen, for organising the work and for instructing the Japanese workmen. Several wooden ships of war have been built there. In 1875 the services of the greater part of the French employés were dispensed with, and the administration passed entirely into our hands. In 1884–85 some few ships of composite type were built, but this system of construction soon gave place to iron and latterly to steel. We are building entirely of steel at the present time. Our artisans, in all branches of shipbuilding and engineering, have now attained considerable skill.

"Hitherto the limit of size has been about 5,000 tons, which we could construct at Yokosuka; but it is intended to enlarge the dockyard, which will enable us to build cruisers of the largest class. The shipbuilding space at Yokosuka at the present time will not admit of our building a battleship, but in course of time it is expected that we may do so.*

"At present we labour under the difficulty of having to purchase all the material, even for cruisers, from abroad. This is bought by tender through merchants on the Admiralty list, and entails much trouble and delay. After the tender is accepted it takes fully seven months before we can calculate on getting delivery.

"In the event—which frequently occurs—of the material not passing the inspectors, fresh tenders are called for, and this involves a further loss of seven months before the material arrives. In the case of a vessel under construction, delay, caused by the rejection of material and waiting the arrival of a further supply, causes great annoyance. With such drawbacks to contend with, it would necessarily require six or seven years to accomplish what might be done in England in thirty months.

" *Battleship Construction.*—This would entail ordering the armour, from England, say, thousands of miles from Japan; everything would be made, to templates, holes for rivets, &c., and in the course of transport some damage might occur. These would have to be

* There is now at Yokosuka a first-class modern dock, where the Victorious was recently docked for cleaning purposes, and Rear-Admiral FitzGerald, who read a paper on the Japanese Navy before the Institute of Naval Architects (April, 1900), said "he never saw a similar operation more quickly, more quietly, nor more methodically performed in any English dockyard." Japan has dockyards also at Kuré and Sassebo, and a fourth, Maisuru, on the north-west coast of the main island, is also in course of construction.

returned for adjustment, causing great delay and expense: thus, we do not propose to build battleships in Japan for some time to come.

"*Dockyard Artisans.*—At Yokosuka there are 3,800 hands employed, fifty per cent. of whom are classed as good workmen, but no more than twenty per cent. of them are what you would class as first-class workmen. It is a most regretable fact that the Japanese workmen are more carried away in paying attention to outward things (show, or effect generally) than mastering their own business.

"*Steel Armour Plate Manufactory.*—Should such be established in Japan, it would hardly be able to manufacture plates within six years from starting. With the experience of six years even, they will probably find that it will be only after many years of further experience they are able to turn out thick plates of uniform quality."

MINOR NAVIES.

Argentine Republic.

The Garibaldi was lately at Genoa for a refit, and left on April 20th, 1900. Originally there were two under-water torpedo-tubes, and four above water. The first two were taken out and the holes stopped up before the vessel left Genoa, and now, as above-water tubes were considered to be more harmful than otherwise to the vessel herself when in action, the other four have been taken out also, and she has at present no torpedo-tubes at all.

Port Belgrano.

The semi-official *Revista Maritima Brazileira* recently stated that the naval and military authorities of the Argentine Republic had laid down a scheme of coast defences, and that the necessary works are to be carried out by the Ministry of Public Works, the responsibility for the defence of the littoral being confided to the navy. Attention is directed to the great activity displayed in the works of the new naval port of Belgrano at Bahia Blanca. A large mole is being built, and the channel of approach to the outer harbour has a depth of about 21 ft. at low tides and of 30 ft. with a mean rise, and the width is 164 ft. There will be accommodation for twenty vessels of the Garibaldi class, besides torpedo and other basins, a dry dock, engineering shops, electric works, ordnance and other storehouses, magazines, and every provision for repairing, supplying, and victualling ships, as well as for accommodating seamen and troops, with a naval school, an astronomical and meteorological observatory, a hospital, and other buildings. A railway line connects Port Belgrano with Punta sin Nombre, where fortifications are being erected to protect the mouth of the Bahia Blanca channel. For a distance of about fifteen miles below the entrance to the channel leading to the port a series of forts and batteries armed with Krupp 10-in. guns and mortars will give protection to the place. The naval estimates for 1900 include £6,100 for naval works and fortifications, and £176,496 for the naval arsenal. The shipbuilding estimate is £1,132,524, with provision in addition for laying down a transport.

Austria-Hungary.

Estimates 1900.

The Budget for 1900 provides, in the ordinary estimates, for completing the torpedo-cruiser Zenta and continuing work upon the Aspern and "C" of the same class, these three vessels being intended to replace the Greif, Helgoland and Fasana; also for laying down a "ram-cruiser," "E," of 7,000 tons, to replace the Radetzky. The extraordinary estimates include the last vote for six sea-going torpedo-boats, the fifth for the Kaiser Karl VI., the third for the coast-defence ship "I," the second for "II," and the first for "III"; these being sister-ships of about 8,300 tons, as well as votes for the gunnery and torpedo armament of the various vessels, for continuing work at the coaling and torpedo station at Teodo, enlarging the magazines and stores at Vallelunga and Fisella, and certain other work.

Trials. Zenta.

The torpedo-cruiser Zenta, constructed by the Stabilimento Tecnico Triestino, successfully passed her official trials in the Adriatic in April, 1899, attaining a speed of 20·9 knots mean with 7,800 I.H.P., being 8 per cent. above the guaranteed power. The vessel is 312 ft. long with 39 ft. 6 in. beam, and about 2,250 tons displacement. The triple-expansion twin screw engines are supplied with steam by eight Yarrow water-tube boilers of similar construction to those recently fitted in the Dutch cruisers Holland, Zeeland and Friesland.

Python.

The official trial took place, in the same month, of the torpedo-boat Python, 152 ft. 6 in. by 15 ft. 3 in., the last of four torpedo-boats built by Messrs. Yarrow and Co., of Poplar, for the Austrian navy. The speed realised was 24·34 knots, the revolutions being 350, and the steam pressure 170 lb. The load carried was 44 tons, representing the vessel fully equipped under service conditions.

Launch. Aspern.

The torpedo-cruiser Aspern, of the Zenta class (described in the *Annual*, 1898), but of 150 tons greater displacement, was launched at Pola on May 3, 1899.

A disastrous accident occurred on July 22nd, on board the torpedo-boat Adler, cruising off the Dalmatian coast, through the bursting of her boiler, by which one officer, three engineers, and a stoker lost their lives, and two others were injured. **Disaster. Adler.**

Brazil.

The coast-defence turret-ship Marshal Deodoro, built by the Forges et Chantiers de la Méditerranée at La Seyne, and described in the *Naval Annual* for 1897, has undergone successfully a 24 hours' official trial, attaining a mean speed of 14 knots. **Trials Deodoro**

Chili.

The following appeared in *The Times*, January 6th, 1900, from a correspondent:—"In view of the letters which appeared in *The Times* concerning the Chilian cruiser O'Higgins, it may interest the public to learn that the following communication has been received from a high officer in that ship, dated Tomé, November 20, 1899:—'It may probably be satisfactory to those who were more directly connected with the building of the O'Higgins to know that up to date all has gone well with her; she has been in constant service the whole time without a hitch. After a six weeks' stay in Valparaiso, replenishing stores, &c., we started out again for a 5,000-knot cruise ten days ago, and have got so far on the way.'" **Almirante O'Higgins.**

The training ship General Baquedano (2,330 tons), built at Elswick, has been completed, and, after attaining a speed of 13·75 knots at her trials, visited Brest, and left for Chili in November. **General Baquedano.**

The torpedo tables in Part II. have been re-arranged in accordance with more definite information. There are four destroyers (built at Birkenhead), and six boats of the Viper type, of which two steamed to Valparaiso, and four were sent out in sections, and put together at that port and at Talcahuano. **Torpedo craft.**

China.

In the *Annual* of 1899 a description was given of two torpedo cruisers of 871 tons, building at Foochow. One of these has been launched, and has received the name of Kien-Wei. The guns were supplied by Canet, and the machinery by the Forges et Chantiers de la Méditerranée. **Launch. Kien-Wei**

Trials.

The Hai Tien and Hai Chi, protected cruisers of 4,300 tons, have been completed at Elswick. The first-named steamed at a mean of 22·64 knots during her six hours' trial, and at 24·1 knots as the mean of four runs during the forced-draught trial. Herr Schichau has completed at Elbing the four destroyers Hai Lung, Hai Niu, Hai Ching, and Hai Hoha, of which the satisfactory trials were reported last year. (Hai Lung, 35·2 knots.)

Denmark.

Launch. Herluf Trolle.

On September 2nd the coast-defence armour-clad Herluf Trolle was launched from the royal dockyard, Copenhagen. Her displacement is 3,470 tons, length about 271 ft., beam 50 ft., and depth in water about 16 ft. 2 in.; 4,200 I.H.P.; maximum speed 15 knots. The armament will consist of two 9·4-in. Canet guns placed in the turret, four 5·9-in. Q.F., manufactured at Bofors, Sweden, and placed in the protected corners of the deckhouse, ten 2·2-in. Q.F., made in Denmark, and eight lighter guns. There are three submerged torpedo-tubes—one in the bows and two at the sides. The Herluf Trolle is built of steel, with double bottom and numerous watertight compartments; the perpendicular side armour extends from about 3 ft. above the water-line to the same distance below, and its thickness decreases towards the ends. The side armour ceases about 20 ft. from the bows, and the ends of it are connected by an armoured bulkhead, from which the armoured deck extends to the bows. The Estimates of 1900 include provision for completing the Herluf Trolle, and for proceeding with another ship of the same type.

Reconstruction. Odin.

The old central battery ship Odin is to be reconstructed. The mounting of the four 10-in. Armstrong M.L.R. guns will be modified, so as to give a larger arc of fire and greater rapidity of loading, and the casemate will be divided by a steel bulkhead, while additional protection will be provided in other parts of the ship, and 1·4-in. Maxims will be mounted on the bridge.

Netherlands.

Shipbuilding programme

The shipbuilding programme introduced by Minister Van der Wijck, with the Estimates of 1897, was presented in a new form with the Estimates of 1900 by Minister Roell. It is to provide for the defence of the Dutch East Indies, for general service, and the defence of the mother country. The auxiliary squadron for the East Indies is to consist of six modern ships, and for that purpose and the general defence of the Dutch possessions, including the West Indies, five armoured vessels and seven protected cruisers are considered necessary. For home defence, the Helder position calls for five large coast-defence armourclads, four gunboats, and six large torpedo-boats; Ymuiden and the Nieuwen Channel (defended chiefly by forts and booms), require six large torpedo-boats and a gunboat; Goeree, the Hollandsch Diep and the Volkerak, three small armourclads, six gunboats, and six large and six small torpedo-boats; the Zuiderzee and its approaches, three unarmoured shallow-draught monitors, five gunboats, three large and six small torpedo-boats (but if the Zuiderzee be drained, only two gunboats for transport work); and Nieuwe Merwede and Waal, two gunboats.

In order to complete this programme there are required three armoured ships of 3,400 tons (Kortnaer, Evertsen, Piet Hein), five others of the Koningin Regentes type (of which two are in hand), three coast-defence ships for inner waters (the Reinier Claeszen and two to be built). Seven protected cruisers (Koningin Wilhelmina and three each of the Friesland and Gelderland classes, all afloat), three new monitors for the Zuiderzee, twenty-one large and twelve small torpedo-boats, fourteen gunboats and five schooners, all the smaller vessels having yet to be built except two schooners; three schooners are for the fishery service. The building scheme extends until the year 1909, the total outlay contemplated being about £3,304,750, or at the rate of £350,000 per annum except in the last year.

New types.

The types of the new ships were described in the *Annual* of 1899. The following account of the machinery of the cruiser Noord Brabant is from the *Engineer* :—

"The engines are of the three-cylinder triple-expansion vertical type, and drive twin screws. They run inwards, *i.e.*, the right-handed propeller is on the port side of the vessel. The engine space is divided fore and aft by a water-tight bulkhead; the starting platforms are at the ship's sides, and are particularly roomy. The diameters of the cylinders are 33 in., 49 in. and 74 in.; the stroke 29 in. They were designed to develop 9,750 I.H.P., when making 145 revolutions per minute, and to propel the ship at a speed of 20 knots per hour. The cylinders are supported by six steel inverted Y columns of I section. The bed-plates, thrust block, cylinder-covers, pistons, and crosshead guide-blocks are likewise of cast steel, all being designed with a view to economise weight. The piston-rods, which are of nickel steel, are forged solid with the crosshead gudgeons. This enables both rod and gudgeons to be bored out hollow, it being so important to reduce the weight of this reciprocating mass, and thereby the vibration of the engine. The connecting-rods are also bored hollow from the crank pin and up to within a few inches of the fork, the result of this being that the total weight of the low-pressure piston has been reduced to 4,955 lb., with rod, guide-blocks, connecting-rod, and crank-pin brasses, made up as follows:—Piston, 1,335 lb.; connecting-rod complete, 2,420 lb.; piston-rod and guide blocks, 1,200 lb. Dividing this weight by the area of the low-pressure piston gives 1·15 lb. per square inch, which certainly is very low. All slide valve rods are of nickel steel, and they, as well as the piston-rods, work in metallic packings. The steam pipes are very noteworthy, being bored out of the solid."

Norway.

Launch. Norge.

The coast-defence armour-clad Norge was launched at Elswick on March 31st, 1900. She is the third warship built at the same establishment for the Norwegian navy, while a fourth, the Eidsvold, a sister of the Norge, is in hand. The vessels previously launched were the Tordenskjold and the Harald Haarfagre. The Norge is 290 ft. long, 50 ft. 6 in. beam, 16 ft. 6 in. draught and 3,847 tons displacement, with a guaranteed speed of 16½ knots. She is to be armed with two 8·2-in. Q.F. guns, six 5·9-in., eight 12-prs., six 3-prs., and two torpedo-tubes (submerged). The vessel has an armour belt of Harveyed steel six inches in thickness. The casemates, four in number, are of nickel steel armour, five inches thick. The barbettes are of nickel steel, six inches thick. The machinery is of the twin-screw vertical triple-expansion type, with Yarrow boilers, to develop 4,500 I.H.P., and give a speed of 16½ knots.

Portugal.

Trials. Dom Carlos I. São Rafael. São Gabriel.

In the *Annual* last year the second-class cruiser Dom Carlos I., built at Elswick, and the third-class cruisers São Rafael and São Gabriel, built at Havre, were described. All have since undergone their trials. Those of the Dom Carlos I. took place in April, 1899. Steam is provided by twelve Yarrow boilers. For six hours, with ½-in. air pressure, about 8,000 I.H.P. were developed, and a mean speed of 20·64 knots was obtained; and during the forced-draught trial, with a pressure not exceeding 2 in., 12,690 I.H.P. were realised, and a mean speed of 22·15 knots was reached in a considerable sea and half a gale of wind. With a smooth sea and no wind the vessel

"HARALD HAARFAGRE,"

NORWEGIAN COAST-DEFENCE SHIP.

could easily have steamed at 22½ knots. The gunnery trials passed off without any difficulty.

The third-class cruisers have triple-expansion engines of the type made at the Forges et Chantiers de la Méditerranée, with Normand-Sigaudy boilers giving 2,600 horse-power under natural draught. On June 7th, in the official four hours' trial, with 300 tons of coal on board, the mean speed with natural draught, without ventilators, was 15·75 knots, and with forced draught and 4,000 I.H.P. 17·53 knots as the mean of a two hours' run. The natural draught contract speed was 15 knots. The gunnery trials were also successful.

Launch. Rainha Amelia.

The third-class cruiser Rainha Amelia, constructed at Lisbon from the designs of M. Croneau, was launched on April 10th, 1899. The following are the principal particulars: displacement, 1,660 tons; length, 246 ft.; beam, 36 ft.; draught, 14 ft. 8 in.; protection, 1 in. steel deck; armament, four 5·9-in., two 3·9-in., and two 3-pr. Schneider-Canet Q.F., and four machine guns; two torpedo-tubes for 14-in. torpedoes; two triple-expansion vertical engines with Normand boilers, 3,000 I.H.P., net draught; 5,000 I.H.P., forced draught; 17·5 knots; range, 4,200 miles at ten knots.

Spain.

The Cuban Flotilla.

What remained of the Spanish naval force in Cuba, forming a small squadron under the command of Capt. Marenco, composed of the auxiliary cruisers Patriota and Rapido, the cruisers Magellanes, Marques de la Enseñada, the torpedo gunboats Vincentè Yanez Pinzon, Martin Alonzo Pinzon, Marques de Molins, and the gunboats Vasco Nuñez de Balboa and Nueva España, arrived at Cadiz in April, 1899. The gunboats Galicia, Filipinas, and Diego Velasquez were reported to have been left behind at Martinique, as being without fighting value.

Gunboats sold in the Philippines

The vessels sold by the Spanish Government to the United States in the Philippines were thirteen in number, twelve gunboats and one torpedo-boat. The gunboats were the Calamianes, Maraveles, Bulusan, Pampanga, Paragua, Samar, Albay, Manileno, Panay, Vasco, Urdañeta (afterwards destroyed in a river action by the rebels), and Guardoqui. The torpedo-boat was the Barcelo. Nine vessels were lost to the Americans in action—the Don Juan de Austria, Isla de Cuba, Isla de Luzon, Manila, El Cano, Callao, Leyte, Mindoro and Mindanao. Most of the vessels are gunboats of light draught.

Sweden.

New Coast Defence Ships.

Three coast-defence vessels, temporarily known as A, B, and C, of which a diagram is included in this volume,* are being built under the direction of Captain Lilliehöök, Chief Constructor of the Swedish Navy. They are in hand respectively at Stockholm (the Bergsund yard), at Gothenborg (the Motala and Lindholmen yard), and at Malmö (the Korkum yard). The three ships are generally of the same class as the Dristigheten, and the following are their dimensions: displacement, 3,670 tons; length at the water-line, 287 ft.; extreme beam, 49 ft. 3 in.; draught, 16 ft. 5 in. The protection for the armament is of Krupp and nickel steel, that on the 8·2-in. gun turrets varying in thickness from 5 in. to 7½ in. and that on the turrets of the 5·9-in. guns from 2½ to 6 in., while the citadel and conning-tower are protected by 7 in. of Krupp steel. All the guns have been manufactured at the Bofors Gallspånz works, Sweden, and are as follows: two 8·2-in., six 5·9-in. Q.F., ten 2·2-in. Q.F., and two 1·4-in. Q.F., and there will be two submerged torpedo-tubes forward for 17½-in. torpedoes. The I.H.P. will be 5,500 (½-in. air pressure), giving a speed estimated at 16½ knots. The engines are being built in the same places as the hulls of the ships, and the boilers will be of the water-tube type. The normal coal supply will be 375 tons, and the range at ten knots about 4,000 miles. The estimated cost of each vessel is about £270,000.

Refits.

The armourclads Svea, Gota, and Thule are to be reconstructed, the work involving an outlay of £275,200, of which half is to be expended in 1900 and the balance in 1901.

Turkey.

Refits.

Messrs. Ansaldo, of Sampierdarena, have now well in hand the reconstruction of the old central battery ironclad Messoudieh, built in the Thames in 1874. The following are the new particulars:—protection: belt, 12 in.; gun positions, 6 in. and 9 in.; deck, 1 in.; armament, two 9·2-in. B.L., 12 6-in. Q.F., 14 3-in. Q.F., 10 6-prs. Q.F., 2 3-prs. Q.F.; engines 11,000 I.H.P., water-tube boilers, speed 15 knots. The Assar-i-Tewfik went with the Messoudieh to Genoa, but particulars concerning her refit are not yet known.

The torpedo-boat Siamjevelot sank in the outside harbour at Beyrout, on April 21st, 1900, owing to a boiler explosion. Twenty-seven lives were lost.

John Leyland.

* Pt. II., Plate 87.

CHAPTER III.

Comparative Strength.

Ships in commission. England and France.

THE war in South Africa made necessary some slight redistribution of our naval forces. Owing to the fact that we were engaged in hostilities with countries which possessed no sea power, the duties thrown upon the Fleet—except in the remarkable work of the naval brigades—were not conspicuous. The Niobe, Thetis, Pelorus and Fearless were temporarily attached to the Cape of Good Hope Station; a patrol of the route followed by the transports was established, and the ships of the First Reserve assembled at Portland on March 1st for a month's gunnery and torpedo practice.

A list is given on the following page of the most important Squadrons in European waters. It calls for little comment. The Channel Fleet remains unchanged, except that the Niobe, which replaced the Blake, is absent, and that the Pactolus has relieved the Pelorus. In the Mediterranean, the Canopus, Ocean, and Renown have taken the places of the Hood, Anson, and Camperdown, the Squadron thus gaining by the addition of modern first-class ships for one of the first-class and two of the second. Further changes have been made in the French Squadrons. The new battleships Charlemagne and Gaulois have been commissioned in the Mediterranean, to be joined by their sister ship the St. Louis; and the Carnot and Masséna have left the Squadron there and proceeded to the Channel. The great reconstruction of the ships of the second line which is in progress has withdrawn many of them temporarily from active service, only the Formidable, Baudin, and Duperré being now in full commission, and the last of these is about to be put in hand, as well as the Marceau. When the work upon the various vessels has been completed the Squadrons will have received a considerable accession of strength.

Adopting the classification of the lists which are at the end of this chapter it appears that Great Britain has now in commission in

	GREAT BRITAIN.			FRANCE.			RUSSIA.
Class.	Mediterranean Fleet.	Channel Fleet.	Reserve Squadron of Coast and Port Guard Ships.	Mediterranean Fleet. Permanent Squadron.	Mediterranean Fleet. Reserve Division.	Northern Squadron.	Mediterranean.
Battleships . .	Canopus Cæsar Empress of India Illustrious Ocean Ramillies Renown Revenge Royal Oak Royal Sovereign Devastation (Gibraltar)	Hannibal Jupiter Majestic Magnificent Mars Prince George Repulse Resolution	Alexandra (C.G.) Benbow (C.G.) Collingwood (C.G.) Colossus (C G.) Howe (P.G.) Nile (P.G.) Rodney (C G.) Sans Pareil (P.G.) Thunderer (P G.) Trafalgar (P.G.) Conqueror } *Tenders* Hero }	Gaulois Charlemagne Bouvet Brennus Charles Martel Jauréguiberry	Amiral Tréhouart Bouvines Terrible Jemmapes Valmy Marceau*	Carnot Masséna Amiral Baudin Amiral Duperré Formidable Redoutable	Alexander II.
Coast-Guard Ships .	Rupert (Alexandria) Orion (Malta)	..	..	..	Achéron (Bizerta) Tempête (Bizerta)	Cocyte (Dunkirk)	Khrabry
Cruisers, 1st Class .	Andromeda Theseus	Diadem	Australia Galatea	Chanzy, Latouche-Tréville Pothuau	..	Dupuy-de-Lôme Bruix Guichen	..
Cruisers, 2nd Class .	Dido, Isis, Astræa Venus	Arrogant Furious	Severn	Du Chayla Cassard	..	..	..
Smaller Cruisers and Gunboats	4	Pactolus	Melampus	Galilée, Lavoisier Linois	..	D'Estrées Fleurus	1
Torpedo-Gunboats .	Halcyon, Hazard, Hebe, Dryad, Speedy Salamander *Torpedo Ram.* Polyphemus *Torpedo Depôt Ship.* Vulcan	..	..	Condor (Crete) Dunois *Torpedo Depôt Ship.* Foudre	Dague (Algeria) Flèche (Tunis) Mouette (Constantinople)	Cassini La Hire S^te^.Barbe (Dunkirk)	Ablek
Destroyers . .	8	..	..	11†	..	6‡	..

* At Toulon; not attached to Mediterranean Squadron. † Hallebarde and 10 sea-going boats, including 3 at Tunisian and Algerian ports. ‡ Durandal and 5 sea-going boats.

European waters eighteen first-class, seven second-class, and four third-class battleships, and France eight of the first-class, four of the second, and four of the third, not including the Jemmapes and Valmy. The battleship strength of the French Mediterranean Squadron is maintained, though it is at present weaker in cruisers owing to the departure of the D'Entrecasteaux, Guichen, and D'Assas, but the Northern Squadron has now three first-class and three second-class battleships, instead of being composed entirely of second- and third-class ships, as was the case after the redistribution referred to in the *Annual* last year.

Italy.

The Italians have at the present time an unusually large number of ships in commission in the Mediterranean. The Squadron consists of the first-class battleships Re Umberto, Sardegna, and Sicilia; the second-class ships Lepanto, Doria, Lauria, and Morosini, and the third-class battleship Dandolo. Only the Italia and Duilio remain in reserve out of the whole list of completed battleships. The Lauria joined the Squadron in January, 1900, the Morosini in March, and the Lepanto and Sardegna in April. The torpedo-gunboats Urania, Caprera, Goito and Calatafimi, with the Volta, and many torpedo-boats, are also attached to the Squadron.

Russia

It is understood that the Russians have in commission a "practice," or evolutionary squadron, with a training squadron, and a number of special vessels.

Germany

The German Squadron in home waters consists of the four first-class battleships of the Brandenburg class, three ships of the third-class (Baden, Bayern, and Oldenburg), and a few cruisers and dispatch vessels, and the Beowulf and some others of the coast-defence ships are also in commission.

The Far East

The table given on page 66 of ships in East Asian waters calls for little remark. The chief facts of note are the substitution in the British Squadron of the Goliath for the Victorious, in the French Division of the first-class cruiser D'Entrecasteaux for the old Vauban, and the addition to the Russian force of the new battleship Petropavlovsk. We have now several destroyers on the China Station, and it is said that a number of the new Russian destroyers are going out to those waters. With the exception of the Undaunted all our first-class cruisers are comparatively new to the station. The Terrible has lately arrived after the memorable service her captain, officers, and crew rendered in South Africa. The accession of strength to the United States Squadron is in consequence of the annexation of the Philippines, and many of the vessels are gunboats for river and local service in the islands. The Italians have on the China Station the third-class cruisers Liguria, Elba, and Calabria.

SHIPS IN COMMISSION.

EASTERN ASIA.

Class.	British	French.	Russian.	German.	United States.
Battleships	Goliath* Centurion Barfleur	..	Petropavlovsk Navarin Sissoi Veliky	Kaiser Deutschland	Oregon
1st-Cl. Cruisers	Orlando Undaunted(a) Aurora Endymion Terrible	D'Entrecasteaux	Rossia Rurik Ad. Nahimoff Dmitri Donskoi Vladimir Monomach	Kaiserin Augusta	Brooklyn
2nd-Cl. Cruisers	Bonaventure Hermione Pique	Descartes Pascal Jean Ba t	Ad. Korniloff	Irene Gefion Prinzess Wilhelm	Baltimore‡ Newark New Orleans
3rd-Cl. Cruisers	Alacrity Brisk	..	Zabiaka Razboynik	Arcona Cormoran	..
Sloops, etc. .	14†	2 5 (Cochin-China)	5	1	13‖
Torpedo-Gunboats	..	..	2		
Destroyers	5	..	..	..	..
Armoured Gunboats	..	Styx	Gremiastchy Otvajny		
Monitors .					Monadnock Monterey

(a) To be replaced by Argonaut.

* Delayed for machinery repairs.
† Including 4 river gunboats.
‡ Returning.
‖ This includes the Isla de Cuba, Isla de Luzon and Don Juan de Austria, captured from the Spaniards at Manila. The United States have besides several auxiliary cruisers and other vessels in commission in the Philippines.

EAST INDIES.

Class	British.	French.
2nd-Cl. Cruiser . . .	Highflyer	
3rd-Cl. Cruisers . . .	Marathon Cossack Pomone	Nielly D'Estaing
Sloops and Gunboats . .	3	3
Torpedo-Gunboats . .	2 (1 *in reserve*)	
Coast-Defence Ships . .	Magdala Abyssinia (*In reserve*)	

SHIPS IN COMMISSION.

ATLANTIC.

Class.	British. Cape.	British. America.	French.	United States.
Battleships . .	Monarch (*In reserve, Capetown*)			Indiana Massachusetts Texas
Coast-Defence Ships		Hotspur (*In reserve, Bermuda*)		
1st-Cl. Cruisers .	Niobe*	Crescent		New York
2nd-Cl. Cruisers .	Doris Forte Thetis*	Indefatigable Flora Hermes Tribune	Cécille D'Assas Suchet	Chicago
3rd-Cl. Cruisers .	Magicienne Philomel Pelorus* Fearless* Barracouta Barrosa Tartar Racoon	Pearl Proserpine Psyche	Troude	Detroit Montgomery
Sloops and 1st-Cl. Gunboats . .	5	6	1	4
Destroyer . . .		1		

* Temporarily detached from the Channel and Mediterranean Squadrons.

The additions made to the Squadron under the command of Sir Robert Harris have been alluded to, and are seen in the above Table. The Powerful, homeward bound from China, and the Terrible, proceeding thither to take her place, were also temporarily attached to the Squadron, but have since proceeded to their several destinations. Both of them played a great part in the early operations against the Boers. It was owing to the ingenuity and energy of Captain Percy Scott, of the Terrible, and to the skill of those associated with him, that the naval guns—the 12-pr., the 4·7-in., and finally the 6-in.—were provided with wheel mountings which enabled them to accompany the troops. In a chapter on "Comparative Strength," it may perhaps be permissible to say that Captain Hedworth Lambton of the Powerful proceeded to Ladysmith with a naval brigade, which took a glorious part in the defence with 4·7-in. and 12-pdr. guns. "Had it not been for these guns," said Sir George White, "the guns of the Boers would have been brought up to positions very much nearer to my defences of Ladysmith, and it would have enormously embar- Cape of Good Hope Station.

rassed my powers of resistance, and would have added enormously to the mortality of the garrison." Thus a preponderance of naval force told decisively against the Boers. A naval brigade with naval guns also took part in the operations of Sir Redvers Buller; another was with Lord Methuen, losing very heavily at Graspan, and a brigade with its guns has shared in the movements directed by Lord Roberts.

Pacific Station.

The following table of ships in commission in the Pacific calls for no explanation. A small increase of strength in those waters was caused by the Samoan difficulties.

SHIPS IN COMMISSION.

PACIFIC.

Class.	British.		United States.
	Australian Station.	Pacific Station.	
Battleship . .	..	..	Iowa
1st-Cl. Cruisers .	Royal Arthur	Warspite	..
2nd-Cl. Cruisers .	..	Arethusa Leander Phaeton	Philadelphia
3rd-Cl. Cruisers .	Katoomba (Sydney re- Mildura serves) Ringarooma Tauranga Wallaroo Mohawk Porpoise Pylades	..	Marblehead
Sloops and Gun-boats	3	2	2
Torpedo-Gunboats	1	..	..
Destroyer	..	1	..

The French have the second-class cruiser Protet and 2 sloops, and the Germans two small cruisers, in commission in the Pacific.

Comparative Tables.

In the chapter upon Comparative Strength in the *Annual* of 1899, Mr. Brassey explained the changes which had been introduced into the Comparative Tables. All first-class battleships which had appeared in "Section B" in previous years were removed to the second class and most of the second-class battleships were transferred to the third class, the less effective third-class vessels being struck out or placed in the coastguard list. The present writer has

"KEARSARGE,"
UNITED STATES BATTLESHIP.

deemed it right to preserve the comparative tables in the exact form which they had last year. A summary of them is printed on p. 80, and the tables themselves have been brought up to date by the inclusion of the ships put in hand since the *Annual* last appeared.

Difficulties of classification.

Mr. Brassey remarked in his chapter upon the difficulties of classification, and upon the fact that it is not easy to draw the line between some cruisers and some battleships. It certainly does not become any easier through the introduction of improved systems of manufacturing armour which make it possible to give to a ship sufficient protection with thinner plating, thus enabling cruisers in this matter to approximate even more closely to battleships. The new ships which have been designed for the Italian Navy, and of which a diagram will be found in Part II., have been described as battleships. As a matter of fact it is open to question whether the vessels would not better be described as cruisers. They are to have complete waterline protection of six inches of special steel, presumably equal to that of the Krupp process, and protecting the bases of all the turrets, which are six in number, each mounting two 8-in. guns. It may certainly be said that there are some cruisers fit to "lie in a line" with some battleships; but it may also be affirmed, as Mr. Brassey said last year, that unless the modern cruiser can fight the modern battleship she will not secure the command of the sea, and that the relative strength of navies must still be estimated by relative strength in battleships.

It will be seen from the table on p. 80 that in the list of first-class battleships we are in a satisfactory position, notwithstanding the fact that Mr. Goschen was compelled to announce, in the Memorandum accompanying the Navy Estimates, that delays in delivery of material, difficulties in securing adequate numbers of workmen, and other circumstances, had seriously affected progress in regard to contract-built ships, and that the work in the dockyards had also been affected. Although the position appears satisfactory, this cannot be held as any reason for delay, more especially as the Powers of the Continent are increasing their rate of production and providing for the laying down of larger numbers of ships.

Battleships completed.

Of first-class battleships we have 22 completed as compared with 9 for France, 6 for Russia, and 5 for Germany. In the list of second-class battleships, which, of course, are less important, we have only 11, as against 20 for France and Russia ; and again in the third-class we have only 12 compared with 16, though this deficiency is really due to the transfer of certain old vessels to the list of coastguard ships. The French are now completing their new battleships, and are busily engaged in re-arming and re-boilering some recent classes. Certain of these are now coming forward for commission.

Battle-ships in hand.

In course of construction, we have not less than 14 first-class battleships as against 2 for France, 7 for Russia, 9 for Germany, 7 for the United States, 4 for Italy, and 2 for Japan. In the previous chapter the conditions which have delayed the building of the United States battleships are alluded to, and the extensive ship-building programmes of France and Germany are described. Some uncertainty, as usual, attends Russian ship-building, but it is known that three new battleships have been put in hand. It must necessarily become a subject of consideration with the Admiralty as to how far our ship-building programmes may need to be expanded owing to the larger efforts that are being made by other Powers.

Com-parative strength.

Although it is not easy to say in exactly what state certain ships will be at the close of the year, it may be estimated that the strength in completed battleships will at that time probably be as follows:—

Battleships	England	France	Russia.	France and Russia.
First-class	24	10	6	16
Second-class	11	9	10	19
Third-class	12	15	1	16
Total . . .	47	34	17	51

It appears, therefore, that the Franco-Russian combination may have a slight advantage in regard to numbers, but that our large proportion of modern and powerful battleships probably makes the British Navy more than equal in strength to the navies of the two Powers.

John Leyland.

Comparative Tables of British, French, Russian, Italian, German, Japanese, and United States Ships.

TABLE I.—FIRST-CLASS BATTLESHIPS.

GREAT BRITAIN.

Launched	Name	Displacement.
		tons
1891	Empress of India	
1891	Hood	
1892	Ramillies	
1892	Repulse	14,150
1892	Resolution	
1892	Revenge	
1892	Royal Oak	
1891	Royal Sovereign	
1895	Renown	12,350
1894	Magnificent	
1895	Majestic	
1895	Prince George	
1895	Victorious	
1896	Cæsar	14,900
1896	Hannibal	
1896	Illustrious	
1895	Jupiter	
1896	Mars	
1897	Canopus	
1899	Glory	
1898	*Albion*	
1898	Goliath	12,950
1898	Ocean	
1899	*Vengeance*	
1898	*Formidable*	
1898	*Irresistible*	
1899	*Implacable*	
1899	*London*	15,000
1899	*Venerable*	
1899	*Bulwark*	
	Cornwallis	
	Duncan	
	Exmouth	14,000
	Russell	
	Albemarle	
	Montagu	
	36 ships.*	

FRANCE.

Launched.	Name.	Displacement.
		tons.
1891	Brennus	11,395
1894	Carnot	12,008
1893	Charles Martel	11,880
1893	Jauréguiberry	11,824
1895	Massena	11,924
1896	Bouvet	12,200
1895	Charlemagne	
1896	Gaulois	11,275
1896	St. Louis	
1898	*Iéna*	12,052
1899	*Suffren*	12,728
	11 ships.†	

RUSSIA.

Launched	Name	Displacement.
		tons
1892	Tri Sviatitelia	12,480
	Kniaz Potemkin Tavritchesky	
1894	Petropavlovsk	
1894	Poltava	10,960
1895	Sevastopol	
1898	Peresviet	
1898	Oslabya	12,674
	Pobieda	
	Retwisan	12,700
	Tsarevitch	12,900
	Borodino	
	Alexander III	13,000
	Orel	
	13 ships.	

ITALY.

Launched	Name.	Displacement.
		tons
1889	Re Umberto	13,825
1890	Sardegna	13,860
1891	Sicilia	13,375
1897	*E. Filiberto*	9,800
1897	*San Bon*	
	Rª Margherita	12,427
	Benedetto Brin	
	7 ships.‡	

GERMANY.

Launched.	Name.	Displacement.
		tons
1891	Brandenburg	
1891	Kurfürst Friedrich Wilhelm	10,100
1891	Weissenburg	
1892	Wörth	
1896	Kaiser Friedrich III	
1897	*Kaiser Wilhelm II*	
1899	*Kaiser Wilhelm der Grosse*	11,000
1900	*Kaiser Barbarossa*	
1899	*Kaiser Karl der Grosse*	
	C	
	D	
	E	11,[illegible]00
	F	
	G	
	14 ships.	

UNITED STATES.

Launched.	Name.	Displacement.
		tons.
1893	Indiana	
1893	Massachusetts	10,288
1893	Oregon	
1896	Iowa	11,410
1898	Kearsarge	11,525
1898	Kentucky	
1898	Alabama	
1898	Illinois	11,525
1898	*Wisconsin*	
	Maine	
	Missouri	12,500
	Ohio	
	Pensylvania	
	New Jersey	13,500
	Georgia	
	15 ships ‖	

JAPAN.

Launched.	Name.	Displacement.
		tons.
1896	Fuji	12,320
1896	Yashima	
1899	Asahi	15,000
1899	*Hatsuse*	
1898	Shiki-shima	14,850
	Mikasa	15,200
	6 ships.	

* 2 projected. † 2 projected. ‡ 2 projected. ‖ 2 projected.

TABLE II.—SECOND-CLASS BATTLESHIPS.

GREAT BRITAIN.			FRANCE.			RUSSIA.			ITALY.			GERMANY.			UNITED STATES.			JAPAN.		
Launched.	Name.	Displacement.	Launched.	Name.	Displacement.	Launched.	Name.	Displacement.	Launched.	Name.	Displacement.	Launched.	Name.	Displacement.	Launched.	Name.	Displacement.	Launched.	Name.	Displacement.
		tons.			tons.			tons.			tons.									
1888	Nile	11,940	1883	Baudin	11,911	1892	Georgi Pobiedo-nosetz . .	10,280	1880	Italia . .	14,387									
1887	Trafalgar .		1879	Duperré	11,209	1891	Navarin .. .	10,206	1883	Lepanto	14,400									
1886	Anson	10 600	1881	Courbet	10,808	1886	Catherine II. ...		1885	Andrea Doria ...										
1885	Benbow . .	10,600	1879	Dévastation ...	10,704	1887	Sinope	10,181	1884	Lauria	11,000									
1885	Camperdown .	10 600	1885	Formidable .	12,165	1886	Tchesmé		1885	Morosini . .										
1882	Collingwood .	9,500	1886	Hoche	10,997	1894	Sissoi Veliky .	8,880												
1885	Howe	10,300	1890	Magenta	10,851	1896	Rostislav													
1884	Rodney	10,300	1887	Marceau	10,850	1889	Nicolai I.	9,672												
1887	Sans Pareil .	10,470	1887	Neptune	10,983	1887	Alexander II. .	9,927												
1893	Barfleur . ..	10,500	1899	*Henri IV.*	8,948	1890	Dvenadzat Apostoloff .	8,076												
1892	Centurion																			
	11 ships.			10 ships.			10 ships.			5 ships.										

TABLE III.—THIRD-CLASS BATTLESHIPS.

GREAT BRITAIN.			FRANCE.			RUSSIA.			ITALY.			GERMANY.			UNITED STATES.			JAPAN.		
Launched.	Name	Displacement.	Launched.	Name	Displacement.	Launched.	Name.	Displacement.	Launched.	Name.	Displacement.	Launched.	Name.	Displacement.	Launched.	Name.	Displacement.	Launched.	Name.	Displacement.
		tons.			tons.			tons.			tons.			tons.			tons.			tons.
1875	Alexandra ...	9,490	1873	Friedland... ...	8,994	1872	Peter Veliky ...	9,891	1878	Dandolo	11,202	1880	Baden		1892	Texas	6,315	1882	Chin Yen ...	7,400
1882	Colossus	9,420	1876	Redoutable	9,437				1876	Duilio	10,138	1878	Bayern	7,441						
1882	Edinburgh ..		1885	Caiman	7,239							1877	Sachsen							
1871	Devastation .	9,330	1883	Indomptable ..	7,234							1878	Württemburg...							
1872	Thunderer ...		1885	Requin	7,822							1874	Deutschland ..	7,319						
1875	Dreadnought .	10,820	1881	Terrible	7,575							1874	Kaiser	7,531						
1868	Hercules ..	8,680	1875	Colbert*	8,924							1884	Oldenburg .	5,200						
1876	Inflexible	11,880	1873	Richelieu* ..	8,128															
1868	Monarch	8,320	1876	Trident*	8,857															
1870	Sultan	9,290	1880	Bayard	6,011															
1875	Superb .. .	9,170	1883	Duguesclin* ...	6,210															
1876	Temeraire .	8,640	1879	Vauban	6,208															
			1882	Turenne*... ..	6,349															
			1892	Bouvines ..	6,610															
			1893	Trehouart... ...	6,629															
	12 ships.			15 ships			1 ship.			2 ships.			7 ships.			1 ship.			1 ship.	

* Built of wood

Table IV.—Coastguard and Harbour Defence Ships.

GREAT BRITAIN.

Launched.	Name.	Displacement.
	C G	tons
1879	Agamemnon	8,660
1880	Ajax	
869	Audacious	
1869	Invincible	6,0 0
1870	Iron Duke	
1865	Bellerophon	7,550
1874	Neptune	9,3 0
1870	Swiftsure	6,910
1870	Triumph	6 64
1881	Conqueror	6,200
1885	Hero	
1872	Rupert	5,440
	H D	
1870	Abyssinia	2,900
1879	Orion	4,870
1870	Magdala	3,340
1871	Glatton	4,910
1871	Cyclops	
1871	Gorgon	3 560
1871	Hecate	
1871	Hydra	
1870	Hotspur	4,010
	21 ships	

FRANCE.

Launched.	Name	Displacement.
	C G	tons
1892	Jemmapes	6,592
1892	Valmy	
1877	Fulminant	5,965
1883	Furieux	5,019
1876	Tempête	4,869
1880	Tonnant	5,091
1875	Tonnerre	5,858
1878	Vengeur	4,709
	H D	
1885	Achéron	1,721
1887	Cocyte	1,714
1893	Phlégéton	1,798
1892	Styx	1,796
1885	Flamme	1,128
1884	Fusée	1,142
1888	Grenade	1,089
886	Mitraille	1,128
	16 ships.	

RUSSIA.

Launched.	Name	Displacement.
	C G	tons.
1894	Adm Senjavin	
1893	Adm. Oushakoff	4,126
1896	Gen. Adm. Apraxin	
	Adm. Boutakoff	
	H D	
1868	Adm Chicagoff	3,511
1867	Adm Greig	3,593
1868	Adm. Lazareff	3,556
1867	Adm. Spiridoff	3,500
1864	Charodeika	2,026
1875	Kniaz Pojarski	5,138
1873	Popoff	3,59
1892	Novgorod	2,706
1891	Gremyastchy	
1892	Grostjastchy	1,49
1895	Otvazny	
1868	Khrabry	
	16 ships.	

ITALY.

Launched.	Name.	Displacement.
	C G	tons.
1865	Affondatore	4,062
1863	Maria Pia	4,268
1863	San Martino	
	3 ships.	

GERMANY.

Launched.	Name.	Displacement.
	C G	tons
1890	Beowulf	
1891	Frithjof	
1893	Hagen	3,500
1892	Heimdall	
1892	Hillebrand	
1889	Siegfried	
1895	Aegir	3,600
1894	Odin	
	H D	
1878	Basilisk	
1876	Biene	
1878	Camaleon	
1879	Crocodil	
1881	Hummel	
1887	Mucke	1,109
1880	Natter	
1880	Salamander	
1877	Skorpion	
1876	Viper	
1876	Vespe	
	19 ships.	

UNITED STATES.

Launched.	Name.	Displacement.
	C G	tons.
1883	Amphitrite	
1876	Miantonomoh	3,990
1883	Monadnock	
1883	Terror	
1891	Monterey	4,084
1884	Puritan	6,060
	Arkansas	
	Connecticut	4,000
	Florida	
	Wyoming	
	10 ships.	

JAPAN.

Launched.	Name.	Displacement.
	C G	tons.
1890	Hei-Yen	2,000
	1 ship.	

Note.—**C G** = coastguard ships. **H D** = harbour defence ships.

TABLE V.—FIRST-CLASS CRUISERS.

GREAT BRITAIN.

Speed.	Name.	Displacement.
kts		tons.
16	Impérieuse	8,400
16	Warspite	
21½	Blake	9,000
21½	Blenheim	
19½	Crescent	7 700
20	Edgar ..	7,350
20	Endymion	7,350
19¾	Gibraltar ..	7,700
20	Grafton ..	7,350
20	Hawke ..	7,350
19¾	Royal Arthur	7,700
19¾	St George	7,700
20	Theseus .	7 350
22	Powerful	14,200
22	Terrible .	
20¼	Andromeda ..	11,000
20¼	Diadem ...	
20¼	Europa	
20¼	Niobe	
20¾	Amphitrite ..	11,000
20¾	Argonaut	
20¾	Ariadne ..	
20¾	*Spartiate* ..	
21	*Aboukir* .	12,000
21	*Cressy*	
21	*Hogue* .	
21	*Sutlej* ...	
21	*Euryalus*	
21	*Bacchante* .	
23	*Drake* .	14,100
23	*Good Hope* .	
23	*King Alfred* .	
23	*Leviathan* .	
23	*Monmouth* .	9 800
23	*Kent*	
23	*Bedford*	
23	*Essex* .. .	
	37 ships.*	

FRANCE.

Speed.	Name	Displacement.
kts.		tons.
20	Dupuy de Lôme	6 406
19	D'Entrecasteaux	8,114
23	Guichen ..	8,277
23	*Jeanne d'Arc* .	11,329
23	Chateau Renault	8,018
21	*Dupetit Thouars*	9 517
21	*Gueydon* .	
21	*Montcalm*. .	
21	*Condé*...	10,007
21	*Sully*	
21	*Gloire*	
21	*Desaix*	7,700
21	*Dupleix*	
21	*Kleber* .	
21	*Marseillaise*	10,014
21	*Amiral Aube*	
	16 ships.†	

RUSSIA.

Speed.	Name	Displacement.
kts		tons.
16½	Admiral Nahimoff	8,524
18	Pamyat Azova	6 675
18	Rurik .	10,933
20	Rossia .	12,130
20	*Aurora*	6,630
20	*Diana* .	
20	*Pallada*	
20	Gromoboi	12,236
21	*Bayan* . .	7,800
23	*Askold* .. .	6,500
23	*Bogatyr*	
23	*Waryag* .	
23	Boyarin . ..	
	13 ships ‡	

ITALY.

Speed.	Name	Displacement.
kts		tons.
20	Carlo Alberto .	6,396
20	Vettor Pisani . .	
20	*Giuseppe Garibaldi* ..	7,282
20	*Varese*	
20	*F Ferruccio* ...	
	5 ships ‖	

GERMANY.

Speed.	Name	Displacement
kts		tons
21	Kaiserin Augusta .	6,231
19	Fürst Bismarck	10,650
20½	*Prinz Heinrich*	8,868
20½	*B* . . .	
	4 ships.	

UNITED STATES

Speed.	Name	Displacement.
kts		tons.
21 9	Brooklyn	9,215
21	New York .	8,200
22·8	Columbia	7,375
23	Minneapolis . .	
22	*West Virginia*	12,000
22	*Nebraska*	
22	*California*	
	7 ships ‡	

JAPAN.

Speed.	Name	Displacement
kts		tons
21½	Asama	9,750
21½	*Tokiwa*	
21	*Azuma*	9 436
20	*Yakumo*	9,850
21	*Idzumo*	9,750
21	*Iwate* .. .	
	6 ships	

* 6 projected. † 4 projected. ‡ 2 projected. ‖ 1 projected. ‡ 6 projected.

Table VI.—Second-Class Cruisers.

GREAT BRITAIN.

Speed.	Name.	Displacement.
kts		tons.
16	Aurora	
16	Australia	
16	Galatea	
16	Immortalité ...	5,600
16	Narcissus... .	
16	Orlando	
16	Undaunted ...	
17	Amphion	
17	Arethusa	
17	Leander	4,300
17	Phaeton . .	
17	Forth	
17	Mersey	
17	Severn	4,050
17	Thames	
19	Astræa .. .	
19½	Bonaventure ..	
19½	Cambrian . ..	
19½	Charybdis ...	
19½	Flora...	4,360
19½	Forte...	
19½	Fox	
19½	Hermione... ...	
19¾	Æolus	3,600
20	Andromache ..	3,400
20	Apollo	3,400
19¾	Brilliant	3,600
19¾	Indefatigable ...	3,600

FRANCE.

Speed.	Name.	Displacement.
kts		tons.
18	Bruix	4,754
18	Chanzy	4,933
18	Charner	4,792
18	Latouche Tréville	4,756
19	Pothuau	5,360
19	Cécille	5,933
19	Tage	7,589
19½	Alger	4,382
19	Isly	4,477
19	Jean Bart... ...	4,109
19½	Bugeaud ..	3,740
19½	Chasseloup-Laubat... ..	3,758
19½	Friant	3,739
19½	Pascal	4,015
19	Descartes.. .	3,990
17	Sfax	4,728
20	Suchet	3,440
19	Cassard	3,952
19	D'Assas	4,000
19	Duchayla... ...	3,952
19	Catinat	4,065
19	Protet	4,055
23	*Jurien de la Gravière* ...	5,605

RUSSIA.

Speed.	Name.	Displacement.
kts.		tons.
16	Dimitri Donskoi	5,882
15	Vladimir Monomach	6,061
14	Minin	6,136
15	Gerzog Edinburgski	5,050
14	General Admiral	4,722
17½	Admiral Korniloff ...	5,817
20	Svietlana	3,828

ITALY.

Speed.	Name.	Displacement.
kts		tons.
19	Marco Polo ...	4,527
20	Vesuvio	3,373
18	Etna	3,474
17½	Fieramosca ..	3,542
19	Stromboli	3 420

GERMANY.

Speed.	Name.	Displacement.
kts.		tons.
20	Gefion .	4,207
20	Irene	4,400
19	Prinzess Wilhelm	
21	Freya .	
21	Hansa . .	
21	Hertha ..	5,650
21	Victoria Luise	
21	Vineta	

UNITED STATES.

Speed.	Name.	Displacement.
kts.		tons.
20	Albany	3,600
20	Baltimore . ..	4,413
18	Chicago	4,500
19	Newark	4,098
20	New Orleans ..	3,600
21½	Olympia	5,800
19½	Philadelphia ...	4,324
19	Cincinnati ...	
19	Raleigh	3,213
19½	San Francisco...	4,098
16½	*Denver*	
16½	*Des Moines* ...	
16½	*Chattanooga* ...	3,400
16½	*Galveston* ...	
16½	*Tacoma*	
16½	*Cleveland*... .	

JAPAN.

Speed.	Name.	Displacement.
kts.		tons.
17	Hashidate ..	
17	Itsukushima .	4,277
17	Matsushima ...	
18½	Naniwa . ..	3,700
18½	Takachiho ..	
22	Chitose ...	
22	Kasagi	4,760
23	Takasago... ...	4,160
23	Yoshino	4,160
23	*Unnamed*.. ...	4,160

19¾	Intrepid	3,600						
19¾	Iphigenia	3,600						
20	Latona	3,400						
20	Melampus ...	3,400						
20	Naiad	3,400						
19¾	Pique	3,600						
19¾	Rainbow	3,600						
19¾	Retribution .	3,600						
20	Sappho ...	3,400						
20	Scylla	3,400						
19¾	Sirius	3,600						
19¾	Spartan .	3,600						
20	Sybille	3,400						
20	Terpsichore	3,400						
20	Thetis .	3,400						
20	Tribune .	3,400						
19½	Diana .							
19½	Dido ...							
19½	Doris							
19½	Eclipse							
19½	Isis ..	5,600						
19½	Juno							
19½	Minerva . .							
19½	Talbot ...							
19½	Venus ..							
19	Arrogant ..							
19	Furious ..	5,750						
19	Gladiator ...							
19	Vindictive							
20	Hermes .. .							
20	Highflyer	5,600						
20	Hyacinth ..							
	60 ships.*		23 ships.	7 ships.	5 ships.	8 ships.	16 ships.	10 ships.

* 1 projected.

TABLE VII.—THIRD-CLASS CRUISERS.

GREAT BRITAIN.

Speed	Name.	Displacement
kts		tons
18	Iris	3,730
18	Mercury	
16¾	Fearless	1,580
16¾	Scout	
16½	Archer	1,770
16½	Brisk	
16½	Cossack	
16½	Mohawk	
16½	Porpoise	
16½	Racoon	
16½	Tartar	
17	Alacrity	1,700
17	Surprise	
17¾	Barham	1,830
17¾	Bellona	
16¾	Barracouta	1,580
16¾	Barrosa	
16¾	Blanche	
16¾	Blonde	
19	Magicienne	2,950
19	Marathon	
19	Melpomene	
19	Medea	2,800
19	Medusa	
19	Pallas	2,575
19	Pearl	
19	Philomel	
19	Phœbe	
19	Katoomba	
19	Mildura	
19	Ringarooma	
19	Tauranga	
19	Wallaroo	
20¼	Pelorus	2,135
20	Proserpine	
20	Pactolus	
20	Pegasus	
20	Perseus	
20	Pomone	
20	Prometheus	
20	Psyche	
20	Pyramus	
20	Pandora	2,200
20	Pioneer	
	44 ships	

FRANCE.

Speed.	Name.	Displacement.
kts.		tons.
18	Milan	1,738
19½	Coëtlogon	1,932
20½	Cosmao	1,954
20½	Forbin	1,820
20½	Lalande	1,926
20½	Surcouf	2,044
20½	Troude	2,025
20	Davout	2,291
20	Linois	2,345
20	Galilee	2,317
20	Lavoisier	
20½	D'Estrees	2,452
20½	*Infernet*	
	13 ships.	

RUSSIA

Speed	Name	Displacement
kts.		tons
15	Rynda	3,508
16½	Pamyat Merkuria	3,000
25	*Novik**	3,000
	3 ships.	

ITALY.

Speed	Name	Displacement.
kts.		tons.
21	Piemonte	2,600
19	Calabria	3,608
19½	Dogali	2,055
18	Elba	2,690
17½	Giovanni Bausan	2,922
19	Etruria	2,245
21	Lombardia	
18	Liguria	
18	Umbria	
20	Puglia	2,550
16	C. Colombo	2,675
	11 ships	

GERMANY.

Speed	Name.	Displacement.
kts		tons
16	Blitz	1,380
16	Pfeil	
16½	Buzzard	1,827
16	Condor	1,614
16	Cormoran	
16	Geier	
16	See-Adler	
15½	Falke	1,703
23	Greif	1,971
23	Hela	
19	*Gazelle*	2,650
21½	*Nymphe*	2,800
21½	*Niobe*	
20	*C*	
20	*D*	
20	*E*	
20	*F*	
	17 ships	

UNITED STATES.

Speed	Name	Displacement.
kts.		tons.
15½	Atalanta	3,000
15½	Boston	
18½	Detroit	2,089
18½	Marblehead	
18½	Montgomery	
	5 ships.	

JAPAN.

Speed.	Name	Displacement.
kts		tons
20	Akashi	2,700
19	Akitsushima	3,150
18	Idzumi	2,950
20	Yayeyama	1,600
20	*Miyako*	1,800
20	*Suma*	2,700
	6 ships	

* Several others of this class are said to be included in the programme.

TABLE VIII.—TORPEDO GUNBOATS.

GREAT BRITAIN.

Speed.	Name.	Displacement.
kts.		tons.
19	Grasshopper	
19	Sandfly	525
19	Sp der	
18¼	Rat lesnake	550
20	Assaye	
20	Boomerang	
20	Gleaner	
19	Gossamer	
20	Karrakatta	
20	Plassy	735
20	Salamander	
20	Seagull	
20	Sharpshooter	
20	Sheldrake	
20	Skipjack	
20	Spanker	
20	Speedwell	
19	Alarm	
19	Antelope	
19	Circe	
19	Hebe	
19	Jaseur	
20	Jason	810
19	Leda	
19	Niger	
19	Onyx	
19	Renard	
20¼	Speedy	
19	Dryad	
19	Halcyon	
19	Harrier	1070
19	Hazard	
19	Hussar	
18	Polyphemus	2640
	34 ships.	

FRANCE.

Speed.	Name.	Displacement.
kts.		tons.
17¾	Condor	1243
17	Epervier	1288
17¼	Faucon	1239
18	Fleurus	1310
17¼	Vautour	1235
18¼	Wattignies	1292
18¼	Bombe	420
18	Couleuvrine	435
18	Dague	408
18	Dragonne	410
18	Flèche	425
18	Lance	402
18	Sainte-Barbe	437
18	Salve	413
18¾	Leger	517
18½	Levrier	505
21½	Casabianca	960
21½	Cassini	958
21½	D'Iberville	967
23	Dunois	896
23	Lahire	
	21 ships.	

RUSSIA.

Speed.	Name.	Displacement.
kts		tons
20	Captain Sacken	742
20	Lieutenant Ilyn	714
22	Gaidamak	
22	Vzadnik	400
22	Griden	
21	Kazarsky	
22	Posadnik	462
22	Voevada	448
21	Abrek	535
	9 ships.	

ITALY.

Speed.	Name	Displacement.
kts		tons.
20	Aretusa	833
20	Calatafimi	833
21	Caprera	833
17	Confienza	768
20	Euridice	833
19	Goito	812
19½	Iride	833
19	Minerva	833
19	Montabello	814
19	Monzambano	840
19	Partenope	833
18	Tripoli	848
20	Urania	833
20	Saetta	400
20	Folgore	370
23	*Agordat*	1318
23	*Coatit*	
	17 ships.	

GERMANY.

Speed	Name.	Displacement.
kts		tons
19	Jagd	1230
19	Wacht	
22¼	Komet	931
21	Meteor	
	4 ships.	

UNITED STATES.

Speed.	Name.	Displacement.

JAPAN.

Speed.	Name.	Displacement.
kts		tons.
21	Tatsuta	875
21	*Chihaya*	875
	2 ships.	

EFFECTIVE FIGHTING SHIPS, BUILT AND BUILDING.

Class.	Great Britain. Built.	Great Britain. Building.	Great Britain. Total.	France. Built.	France. Building.	France. Total.	Russia. Built.	Russia. Building.	Russia. Total.	Italy. Built.	Italy. Building.	Italy. Total.	Germany. Built.	Germany. Building.	Germany. Total.	United States. Built.	United States. Building.	United States. Total.	Japan. Built.	Japan. Building.	Japan. Total.
Battleships—																					
1st-Class . .	22	14	36	9	2	11	6	7	13	3	4	7	5	9	14	8	7	15	4	2	6
2nd-Class . .	11	..	11	9	1	10	10	..	10	5	..	5	..	..	..	..	..	..	..	..	..
3rd-Class . .	12	..	12	15	..	15	1	..	1	2	..	2	7	..	7	1	..	1	1	..	1
Total Battleships	45	14	59	33	3	36	17	7	24	10	4	14	12	9	21	9	7	16	5	2	7
Coast-Guard Ships .	21	..	21	16	..	16	15	1	16	3	..	3	19	..	19	6	4	10	1	..	1
Cruisers—																					
1st-Class . .	22	15	37	4	12	16	5	8	13	2	3	5	2	2	4	4	3	7	1	5	6
2nd-Class . .	60	..	60	22	1	23	7	..	7	5	..	5	8	..	8	10	6	16	9	1	10
3rd-Class . .	44	..	44	12	1	13	2	1	3	11	..	11	10	7	17	5	..	5	4	2	6
Total Cruisers .	126	15	141	38	14	52	14	9	23	18	3	21	20	9	29	19	9	28	14	8	22
Torpedo-Gunboats .	34	..	34	21	..	21	9	..	9	15	2	17	4	..	4	..	..	..	1	1	2

CHAPTER IV.

The Tactics of Fast Craft.

Strategy cannot be taught in schools.

There are few suggestions for the bettering of the education of naval officers more plausible than that of starting a school of Naval Strategy and Tactics; but, like most other suggestions not adopted by the officers responsible for the management of naval matters, the advantages are more apparent than real. The principles of strategy should be studied from history, and history only. The actual details of the exact strategy that will be employed by us in our wars of the future—that outcome of years of thought of the finest brains of our naval and military experts—is naturally a confidential and secret subject. When war is declared the policy entrusted to our admirals will be reviewed by the world—to lecture on it now would be criminal. It is clearly undesirable for officers acquainted with the details of our strength, with our resources, as well as with our weaknesses, to publicly discuss the probable lines of strategy we shall adopt, instructing our neighbours with but small gain to ourselves; but, since in war time officers will undoubtedly find themselves called on to exercise considerable strategic judgment, a study of history and the old wars, both naval and military, aided by their knowledge of sea life and conditions, will stand them in good stead, and the accustomedness of their minds to assimilate the factors by constant and individual thought will be of more value than professional dicta. It must not be assumed because naval officers do not talk or write much on such professional subjects, that necessarily they neglect study and do not think; on the contrary, their life at sea, with fleets and in fleet conditions, and their intercourse with their seniors and other thoughtful minds, are to them a long career of learning by theory and practice.

The fleet the school for tactics.

If this is true of strategy, is it not infinitely more so of tactics? The only real school for tactics is the fleet at sea; a lecture-room will never help a tactician—a tactician is born, not made. The man who can practically anticipate the new formation of the enemy by the swing of the ships and their alteration in speed, and is able to meet the movement instantly by a preconceived disposition of his own vessels, is one to whom a short course of lectures will have taught but little, but whose brain and eye during his whole life have worked together when ships manœuvred. Anything of the nature of a cram in such subjects is useless. Tactical facility, apart from mere tactical knowledge, can only be firmly implanted in the mind by individual and life-long wrestling with the subject and its many branches. Other brains may help to classify the subject and propound the elements, but individual thought must supply the facility of recognition if ever quickness of grasp and appreciation is to become a second nature to the tactician. For these reasons I venture to think a good text-book would be of more value than a course of lectures. What, in plain language, naval officers want is to have their thoughts and reasoning powers guided, not warped, to avoid being merely the mirrors of academic ideas, to be encouraged to use their reason to detect and assimilate the teaching of their superior officers, and of their own practical experience, to ponder, think, and grow accustomed by thought to the varying conditions of naval armaments, to keep their minds ready to grasp each movement and manœuvre as recorded by the eye, and by practised thought to instantly appreciate its bearing on the existing conditions. An officer who engages in a battle merely equipped with the opinions of others and the manœuvring signal-book, without having his ideas formed and his captains in his confidence, without thorough and reciprocal feeling between himself and them, without knowledge of their capabilities, will fail, and fail most signally. Can any course of lectures teach this? or, to quote Captain Mahan, "The lesson . . . . . . is the danger of disgraceful failure to men who have neglected to keep themselves prepared, not only in the knowledge of their profession, but in the sentiment of what war requires."

What, however, I think are always welcomed by naval officers are papers or books dealing with their profession, especially with the principles rather than the details of strategy and tactics. These, whether in agreement or not with their experience and ideas, may present even commonplace subjects in some new aspect and increase their interest. One subject of great importance is the study of the use of fast craft, and I think a summary of the operations they may be called on to undertake, together with a few remarks on their use, may not be without interest.

Tactics as practised by ships in peace time must differ considerably from those that will be used in actual warfare. The evolutionary tactics of peace time have an immense value in practising officers in handling their ships at equal and varying speeds. They deal chiefly with undamaged ships and an unbroken line of battle, and with the relations of each ship of a fleet to the other, rather than with that more important branch of tactics, that of the entire fleet or of individual ships with the enemy.

The question of the tactics of a fleet or of single ships when opposed to others is one of the greatest importance, and has for all ages been a sealed book. The Dutch, English, and French schools evolved with experience different modes of attack, none of which were generally used, even by the fleets of their own country, and they were continually being modified by alteration in ship construction, till really the genius of an admiral was shown by the confidence with which he discarded the recognised tactical stagnation of the day, and evolved new methods often diametrically opposed to the old. The choice of the lee gauge in preference to the weather by the inferior fleet, and Nelson's bold tactics at Trafalgar against fleets whose value he well knew, are two well-known examples of absolute revolution against accepted tactical notions.

Differentiation of tactics.

In the future I venture to think that tactics will be divided into two broad and distinct parts—one the tactics of the line of battle, and the other, for want of a better name, I must call the tactics of fast craft. The germs of this division are barely discernible in the past, but the growth of speed as a factor in war has caused a broad distinction to grow up in the present and future. It is with the latter class of tactics that I will chiefly deal.

Fast craft.

By the term tactics of "fast craft" I do not mean any particular vessels from battleship to torpedo-boat, but merely the tactics that any ship or boat not in the line of battle may have to employ when her speed is superior to her opponent, although otherwise she may be inferior to her. First, second, or third class cruisers, destroyers, or even torpedo-boats, when together and operating against similar craft, must form line of battle and use tactics similar to those of a battleship action, but when opposed to superior ships, or when their high speed is required to be turned to account in scouting or for other purposes, then they enter upon a different field of action where armament is no longer solely opposed to armament, but where speed is the predominating factor and has at times to be pitted alone against armament to obtain desired results.

The factor of speed.

The essential feature, therefore, of fast craft fighting is the fact of having in opposition to them units, or combinations of units, superior in fighting strength, but speed is preponderatingly in favour of the smaller craft. It is this crucial point of speed that can equalise and at times make a puny fragile craft superior in duel to a battleship. It should be well noted that such an equality or superiority exists only at the time of the circumstance of position. No speed can make a torpedo-boat *equal* to a battleship, any more than a man who

G 2

presses the firing key of an observation mine can be supposed equal to the ship's company he may destroy. The potential worth of a battleship is a totally different thing to her actual worth in adverse and perhaps hopeless circumstances. Many hasty and hazy conclusions have been drawn by attempting to compare the relative value to a navy of vessels of different functions from data based on exceptional conditions, and it is the confusion of the accident for the practice of war that has led to much diversity of opinion as to the relative value of functionally different ships.

The question of the relative value of a battleship, cruiser, or torpedo-boat is an absurdity, unless the nature of the service which they have to perform is considered; and the extremely difficult question of the relative number of each required in a navy can only approximately be determined by a complex consideration of the various combinations which every imaginable desirable condition of strategy in warfare in any section of the globe may require. In war against one country battleships might be of little value, in another combination they might form the whole force desirable. In one war torpedo-boats might remain on the slips, under other conditions they could not be built too quickly; and the same for other classes of vessels. Different classes of vessels can therefore have no absolute relative value apart from the nature of the service that the strategy of the moment may demand.

Speed is the factor of warfare that determines the range of fire in an action. Strictly speaking of itself it cannot force an action. Strategy and political necessities are far more potent causes of action than speed can ever be, even with small craft against their larger enemies, since either may require the inferior force to attack the larger, even if annihilation ensues.

With avoiding actions *in toto* speed is the one useful factor; it is a greater security to a fast craft attacked by a slower than all the armament in the universe. Speed, however, is liable to derangement from internal causes, guns and armour practically only fail from external attack, and can therefore only give way under hostile gun fire; speed may, on the other hand, vanish from internal accidental causes; perhaps, without an enemy in sight, a fast craft may be deprived of her chief weapon of defence.

This consideration leads to two maxims—one that every care of the machinery and boilers should be studied and speed never unnecessarily forced, and the engine-room given every possible notice of large change in speed; and the other is that the warfare of a class whose defence may vanish as rapidly as the explosion of a magazine is one of great danger, and that whenever these vessels are used the

Admiral should be prepared for their loss. Not only are they pitted against vessels of superior armament—in fact, crushing superiority—but internal as well as external accident may cause their ruin. This point will be referred to again when considering fast craft used in offence; but we cannot too often school our ideas to the wholesale loss that we shall have to put up with with equanimity in future naval warfare.

The armament of fast craft is of two kinds, the gun, and the ram and torpedo, the ram being but a short range torpedo, or the torpedo considered to be a long distance ram, whichever we wish. The gun armament is chiefly of use against equal or inferior craft. In most of the smaller cruisers so much in the way of armament has been sacrificed to speed and coal capacity that their protection in the execution of their office as scouts must be left to larger and better protected classes. In fact they are what might be termed the abnormal development of a species all eye, stomach, and legs, and very few teeth. The torpedo armament of such a vessel is an important one, and for all fast craft one which cannot be insisted on too much, as in certain of her functions it makes her a deadly enemy to the largest ship afloat. **Armament**

There has been a considerable cry against above-water torpedo-tubes being placed in vessels. There is no danger in an above-water torpedo-tube—the danger lies, and that only under certain conditions, in the loaded and primed torpedo placed in it. There is no necessity because the tube happens to be placed in the ship that it should always have a torpedo in it, any more than that a man armed with a rifle and a revolver should hold the revolver in one hand and use his rifle with the other when the enemy is a thousand yards off.

In all fast craft torpedoes are an absolute necessity to the proper performance of all their functions. If no room exists for these tubes below water then they must be placed above water. In battleships above-water tubes are an anomaly and a thing of the past, in smaller vessels an immense possibility in the future. It is the torpedo that allows a small craft the chance under certain conditions of engaging a larger with advantage; the reason being that when once within close range, say five hundred yards, the torpedoes of the smaller vessel are able to destroy and sink the larger, so that once within this range both craft may temporarily be considered as of equal offensive strength, relatively equal in annihilating effect. **The necessity of torpedoes to fast craft.**

It is this fact that endows modern fast craft with powers unheard or undreamed of in naval history, and gives a chance of success, where in previous ages capture or annihilation was inevitable. As long as guns and armour (represented by the thick wooden sides of the vessels) were the only factors, then the larger ship, with equally well-trained crews, had absolute certainty of victory. Even frigates could under no conditions fight efficient line of battleships by day or night; and though small and slender as this modern condition of equality may seem, both from the difficulty of acquiring the desired position, and from its short duration, still it is one which must actively and potentially enter into the tactics of the future—actively, by ships being placed in the above conditions, and potentially, by the restraining and limiting force exercised on otherwise unhampered fleets.

Limiting and restraining factors.

The presence of numerous fast craft at a given protected spot is at night time a source of danger to any ship within striking distance, whose strategic value outweighs that of the craft. Usually destroyers or torpedo-boats, and perhaps catchers, would constitute the danger to a fleet; but it is quite possible that larger vessels might be used for a similar purpose if the particular object in view was considered sufficiently worth the expenditure of material, and if, when the chance offered, the existing conditions rendered their use advisable.

As surely as one battleship would be sacrificed in a single duel with another (in which, whatever might be the outcome in the bitter end, both must temporarily forfeit their services to their navies for some considerable time), as surely as such duels are to the advantage of one side or to the other, so surely it may be of advantage to risk ships, not mere boats, of a less required class, even to annihilation, to deal a blow and restore the balance in vessels of a more needed type. The sacrifice of ships of one class, to produce equality or superiority in a larger class by reduction of the enemy's number, is a new possibility in naval warfare, and had no counterpart in the olden days. Now it is not too much to say that the anchorages of fleets are restricted to defended harbours, and the presence of a blockading battle fleet in the immediate vicinity of a properly equipped harbour should be an impossibility. This restraint of position is due to fast craft. They prevent the use of many anchorages for shelter, coaling, or repair, they prevent stationary bases near an enemy's port, and they necessitate a higher speed and consequently waste of coal in certain waters.

We will now consider how fast craft may be called on to use their two weapons, armament and speed. There are three broad conditions.

Armament alone—when the action will practically be a duel against an approximately equal enemy and so really simulate a battleship action.

Armament and speed combined in attack; either purely in offence, or else in offence for necessity of defence, against a superior enemy.

Speeed only—which embraces the whole of the operations known as scouting and blockading.

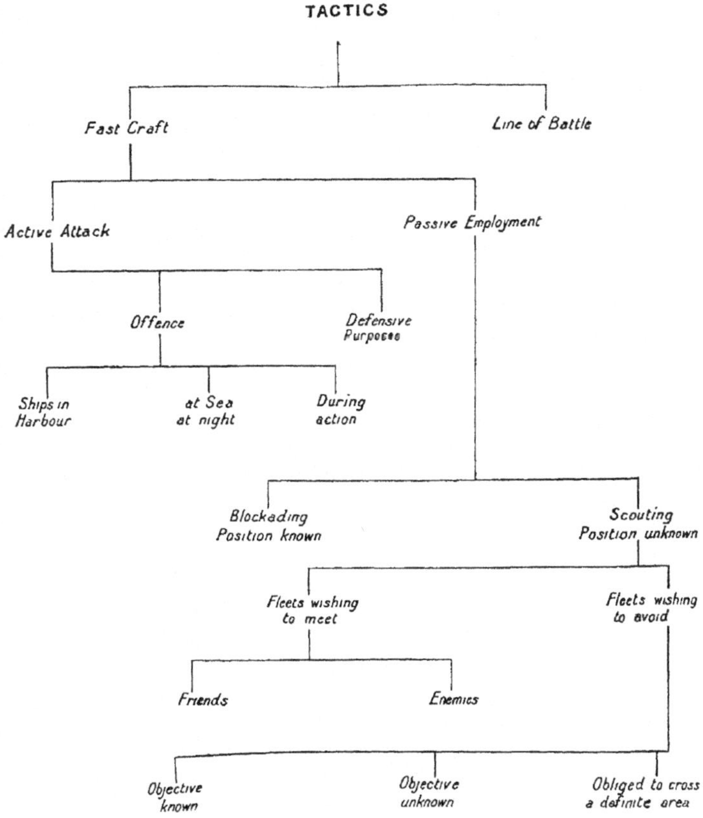

Fig. 1 gives in diagrammatic form the different functions of fast craft in war time, and is useful in localising the points under discussion.

Offensive action of fast craft.

The first branch to consider is the offensive action of fast craft in deliberate offensive attack.

I venture to say that there is no more exhilarating form of fight than the attack of large vessels by fast small ones, the cavalry charge of naval warfare. This is usually undertaken by torpedo-boats or destroyers at night, but it may equally be the duty of larger vessels under certain conditions. But whatever the craft employed, he who sends them forth to attack and destroy large worthy prey must do so with the full knowledge that they may never return. The craft must be hurled at the enemy with the same single-heartedness and unconcern that a shell is hurled on its deadly mission. The man who has any doubts, who in any way restricts their movements by desire of their return, or even mentions return before the object to be attained has been attempted to the legitimate full, had better never determine to order the attempt. In any other work they may be of use, in a half-hearted attack never. I do not for one instant mean that attacks on ships should be attempted because ships or boats are in their vicinity—war might wage for a year, and an admiral with many fast craft might not wish to use them offensively either from the nature or the position of the enemy, or from the fact that they were more useful to him in other spheres. But what I do say is that when the moment for their use has come, and it has been decided that an attack shall be made, then launch them forth as a stone thrown into the water, and should not one return to tell the tale of success, partial success, or failure, well, the shot has been expended, that is all. Sentiment can find no place in war. Vessels will have to be deliberately sacrificed and lost for objects worthy to be attained, and the man who orders a vessel on a course from which there is most probably no return must do so with no hasty half-saving orders, which most probably will ruin the whole plan. Peace manœuvres have done nothing towards teaching us how fast craft should be used for attack, since it never has been possible to simulate war conditions, owing to the fact that in peace manœuvres no real risks to the vessels themselves can be incurred. As regards torpedo-boats the Admiralty have always most generously shown an appreciation of the true spirit necessary for their handling in night attack by never damping the enterprise of young officers by taking serious notice of damage done to the boats under such conditions, and this has been of incalculable worth in fostering the right spirit in their handling.

Sacrifice is involved

But the fatal necessity of having to consider risk of loss to ships and boats robs such manœuvres of any value as standards of judgment as to the result of their use in war time.

Attack by fast craft on large vessels is of two distinct kinds:—

1. Attack on ships in harbour.
2. Attack on ships at sea.

Both of these are normally the function of the boat or destroyer when available, but under certain conditions they may be the duty of larger craft.

The disadvantages under which larger craft labour in this class of attack are, that their large size renders them good targets for the guns, and that their turning powers are poor compared with smaller craft. There are, however, occasions when the use of large craft may be desirable.

If the harbour is defended by a boom it may be necessary to carry the obstruction away, and then continue the attack—a potent threat to the weak point of breakwaters. If the harbour to be attacked is far away, as in the case of a suspected temporary coaling base, it may be of advantage to send one or more cruisers to reconnoitre the harbour at night and attack any ships found there. This might be done at distances where the use of smaller craft would be out of the question, and with considerable safety to the attacking craft, especially if the harbour were unfortified.

The actual attack of boats or destroyers on ships in a harbour has been so much discussed that no further remarks are necessary. With the boom destroyed they should prove the reason for their existence.

Attack on ships at sea

The next consideration, that of attack at sea, has two conditions of probability—a raid by the fast craft on the night preceding a fleet action, and a raid on the enemy in daytime during a fleet action.

Imagine the cruisers of a fleet sighting a superior hostile force during the afternoon, and an engagement imminent in which the preponderance of force lies with the other side; and suppose the balance would be brought more nearly even if they were reduced by four or five ships. What would be the tactics of the cruisers and other fast craft supposing the morrow's conflict of vital national importance? Is there any doubt but that action would if possible be delayed till the morrow (for of course no fleet except in desperate circumstances would force a night action, and discard the whole of the peace gunnery training of its crews, any more than a man would force a close action without first finding himself outmatched in gunnery at long ranges)? During the night would not the admiral hurl his fast craft at the enemy? Would they be kept for future scouting work which might never prove necessary? Would they be kept as silent spectators of the morrow's defeat? Would they not be let loose like hell hounds to tear into the enemy's fleet, torpedoing, ramming, destroying, and using every nerve and knot to destroy, and so pave the way for the morrow's victory? Suppose twenty first and second-class cruisers thus launched at the opposing fleet, suppose fifteen missing in the morning, but the fleets equalled, even if the balance were not changed, would not the loss be fully compensated for? Would not the loss be well worth the gain? I repeat, the method of the use of fast craft, cruiser or destroyer, would merely depend on whether they or the battleships were of most value on the morrow. It might from general strategical conditions be either, but so surely as the balance of the morrow's fight was uncertain, as surely as for national purposes of prestige, money supply, or strategical condition, a decisive victory by the line of battle was desirable, so surely would a portion or the whole of the fast craft be hurled at the enemy to do their worst. Who cannot imagine the scene, the confusion of the enemy firing alike on friend and foe, while the blast of the cruiser charge sweeps through the mass, sinking, being sunk, ramming, torpedoing, and being rent by shells, but under able heads and skilful handling rendering the morrow's victory assured?

Analogous to this class of tactics is that pursued by a cruiser when forced to attack offensively for defensive purposes. Unable to escape from her pursuing enemies, under cover of night, with full knowledge of her purpose, she can turn and rend her uncertain, and perhaps unprepared, foe. Warfare in the future will show some glorious actions, where pigmies in the last throes of distress will, if not themselves escaping destruction, destroy their enemies in their expiring gasp.

One other class of attack the fast craft may be called on to give is during a fleet action. The exact functions of fast craft not in the line of battle during a fleet action are uncertain, and largely depend on the tactics of the opposing side, but they will probably be of two distinct kinds—cruiser actions, where cruiser meets cruiser in individual contest, or in a miniature fleet action of several ships on each side; and secondly, the joint action of the smaller fast craft with the ships in the line of battle. This forms a branch of torpedo-boat work and not of fast craft in general, but for ships to have small vessels of fast speed and quick turning power under their shelter, ready to strike when the proper moment comes, gives them a valuable weapon which at times may be of enormous worth. Mind, there is no question of the ships looking after the boats or caring one farthing for them—they manœuvre, fire, and engage in absolute oblivion of their presence till the boats are wanted; then, if they have survived, they can strike.

To pass now to the second class of fast craft tactics where armament is not necessarily employed, but speed is made use of for vedette work in blockades or scouting work while cruising.

Here we must at once change our estimate of the value of a fast craft from mere speed to one in which a vessel possessing speed must also possess a large distance-steaming endurance, not necessarily at a high speed. The high speed is always available for use when required, but the second factor of endurance is co-equally important. Safety from superior types is still ensured by high speed; utility to the fleet, by both combined.

This class of tactics broadly has two divisions—first, when fast craft are employed to watch and report the actions of a fleet at a known position, or what is commonly called blockading; secondly, when employed to discover the locality and report the actions of a fleet whose position is unknown, or what is commonly called scouting.

Blockading. Blockading can no longer mean forcing a fleet to remain in a harbour, nor is it, nor has it ever been, the true strategy of the stronger naval force to make its enemy do so; it simply means observing an enemy at anchor and communicating any movement to the blockading fleet.

The accepted idea of a blockade—the one which was used in the old wars—consisted of a fleet of battleships, of sufficient numbers to engage the enemy if they came out, stationed off the port at a distance varying with the direction and force of the wind. An on-shore wind meant rest, an off-shore wind vigilance and activity. Inside of the battleships there was often an in-shore squadron of smaller vessels to observe the numbers and watch for signs of movement on the part of the blockaded vessels. Immediately we come to consider such a disposition of ships in the present day we are met by two fatal considerations—first, the necessary daily consumption of coal to the fleet remaining in the offing, and therefore their inferiority in this respect to the issuing fleet; and, secondly, the danger to the fleet from the attack of fast craft at night.

Difficulties of blockaders For a fleet to be in all respects ready to fight another it must be practically complete with coal. This condition cannot be maintained at sea, and only in harbour by constant coalings. The probability of the issuing fleet trying to escape must always be considered, in which case a high speed would have to be kept by the following ships; the escaping ships will probably be steaming to a base, the pursuing ships probably away from theirs, consequently coal is even of more importance to a blockading squadron than to the escaping ships.

It should be remembered that the radius of action of a fleet is an inverse function of the speed at which it steams. The faster the rate of steaming the less the total distance it can steam; as a matter of fact, by forcing the enemy to steam at full speed he can only traverse one-third of the distance that he could if allowed to proceed at eight knots; and by maintaining a close pursuit at a high speed an enemy may be forced to steam directly to a base to replenish. Again, close pursuit at high speed increases the chance of breakdowns, perhaps equally to both fleets, but the lame ducks of the pursued fleet must either fall into the jaws of the following vessels or an action must be forced, whereas the derelicts of the pursuing fleet (the numbers of such breakdowns being quite unknown to the enemy) can be left to effect repairs or be towed by cruisers, as may be most convenient. But unless the bunkers of blockading fleets are full, or nearly so, they will be forced to steam at a comparatively low rate of speed, and the advantages inherent to a pursuing fleet will be lost.

Power of fast craft.

Another great danger to the blockading fleet is the fast craft inside the harbour, which should keep the adjacent waters clear of the enemy's large vessels. The only protection to a large vessel against such attack are her nets, her invisibility, and her guns. Against modern high speed small vessels, used night after night for months, the first is the only one likely to give security, and even net defence does not protect the vessel completely, both her stern and bow being exposed to attack. Experience in manœuvres has distinctly shown the danger of fast boats to ships in the offing waters, and if we consider that instead of a few fine light summer nights, lasting for short manœuvres, the boats may ply their deadly game through long dark nights for perhaps many months, their danger will be a veritable scourge to a blockading fleet. These two conditions against blockade lead naturally to considering whether the risks of blockade are worth the undertaking, and if so how they can best be minimised.

Blockades in the olden days, while possible chiefly on account of the long sea-keeping endurance of the battleships, were needful on account of the wide range of action possessed by ships whose time limits of keeping the sea were practically only restricted by the renewal of provisions and water. It was therefore possible for a squadron leaving a port to proceed to any rendezvous, and strike at any objective, however far off. In these days coal intervenes, and though it is hardly true to say that a chased vessel must steer towards a coaling-station, there is a considerable amount of truth in the assertion. If the coaling-stations away from the main arsenals of the country are destroyed, an issuing fleet will have but a small radius of possible action, and the chief danger will be the effecting of junctions with other fleets to crush squadrons inferior in strength at moderate distances. If suitable anchorages are available, interior positions at anchor will more effectually prevent this than the more crude form of interior and containing positions at sea. Generally speaking, in European waters, we may be said to hold interior positions to any possible combinations of countries whose ports we could hope to blockade. For these reasons blockade will probably resolve itself into observation by fast craft, whose duty it will be to report the movements of the blockade squadron to the fleet at anchor, the actual tactics of observation employed being such as at night-time to place the larger fast craft outside the range of the small fast craft attack, and yet to insure knowledge of the movement of the other fleet.

Blockade on this principle with cruisers gives rise to two classes of tactics, one of observation, the other of keeping in touch after the fleet have left their harbour. The choice of the tactics of observation is determined by the fact of the risk run by the fast craft remaining off the harbour at night. This leads, if possible, to all observation being done in the daytime.

The actual details of the disposition of ships for the patrol in daytime by fast craft, of all waters in which vessels escaping from a harbour the previous night may be, are purely geometrical, and can be worked out by anyone with a chart and pair of compasses. Tactics of observation.

Fig. 2, however, shows the principle of such an organisation.

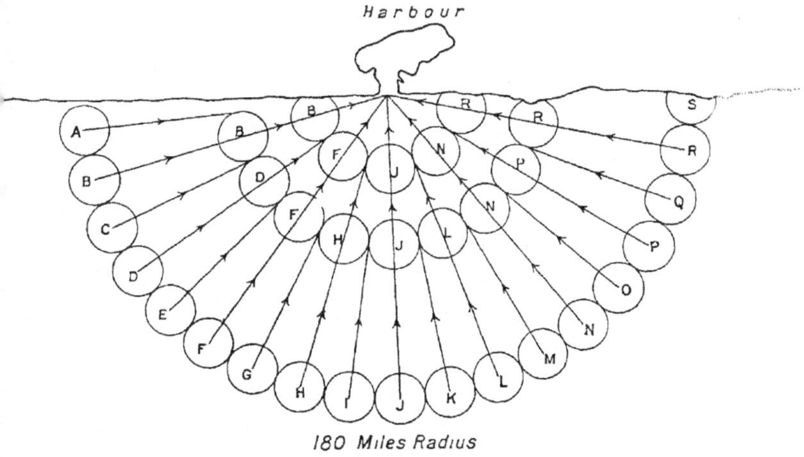

The employment of cruisers as scouts is a subject that, so far as I know, has never been systematically written about; the few articles that have been written apparently treat the subject from one point of view only—namely, that of a squadron wandering about endeavouring to find another, without regard to the probability of the frequency, or otherwise, of such conditions occurring in war time; and the points discussed have chiefly been of secondary importance—namely, geometrical distributions, regardless of the more important ones of communication and mutual support.

We may fairly assume that in future in rational warfare no fleet will wander away to sea without a definite object. Public opinion, with its invariable ignorance of strategy, will be no factor with an intelligent Admiralty. Again, the absolute necessity of coal prevents a squadron roving indefinitely, and limits the places it is bound to put into. So in future the objective of fleets will probably be known with greater certainty than in the older wars.

Objects of escaping ships.

A squadron must leave a port under one of the following conditions as regards its opponents or friends :—

I. Either wishing to meet another fleet ; or
II. Wishing to avoid another fleet.

Class I. divides itself into two conditions .—

A. Where the fleet is friendly.
B. Where the fleet is hostile.

If a fleet leaves a harbour intent on meeting a friendly one, it is difficult to imagine that a rendezvous should not have been previously arranged. If intelligence of the movements of a friend has been received sufficiently explicit to cause a fleet to proceed to sea to meet it, there is not much reason why the particular rendezvous should not also have been communicated. But even suppose this not to have been the case, general rendezvous must be arranged for war time, and these visited by cruisers from each fleet would be the most rational procedure for enabling the two fleets to meet. It is difficult to imagine two friendly fleets being so bereft of intelligence as to wander about the ocean looking for each other. Should an enemy superior in strength to both be found to be near a rendezvous, then alternative rendezvous visited by cruisers will still be the means for conveying intelligence and transmitting orders.

I. B. Where the fleets are enemies.

If two hostile fleets put to sea with the intention of finding each other they will most assuredly do so, and the functions of the cruisers will be reduced to merely giving intelligence of the near approach of the enemy, and if possible of their number. No special scouting arrangements will be required.

Going on to Class II., where the issuing fleet wishes to avoid another, this may be divided into four heads :—

A. The tactics of the cruisers of the superior fleet when her enemy leaves for a known objective.

B. When she leaves for an unknown objective.

C. When a particular area geographically suited for keen search has to be passed.

D. The tactics of the cruisers of the inferior fleet.

Conditions A., where the enemy's objective is known, may be simply disposed of, except in the case where reinforcements await them at their objective which would make them combined superior to the scouting fleet.

The best position for the scouting fleet to ensure falling in with the enemy is off the objective, since the area off a port can usually be more effectually watched than the sea between two places.

If, however, another fleet lies in the objective, the question of whether the two halves should be engaged in succession, or whether it is desirable that one only should be brought to action, is purely one of the strategy of the war operations, and will depend among other things on the relative numbers and strength and efficiency of the three fleets, and therefore the probable effective strength of the fleet after an engagement with either of the other two. Should it be desirable to engage both in succession, the offing of the port will still form a suitable position from which to direct the scouting operations; since information of the enemy's approach should be obtained sufficiently early to enable the advancing squadron to be engaged by itself.

Should, on the other hand, it be undesirable to engage the second fleet after a close action with the first, it is unlikely that the second fleet would remain in harbour till the first arrives, instead of meeting at a rendezvous. But should this be the case, the operations of the searching fleet must be necessarily so hampered by this consideration that the fact of knowing the objective will be of but small value to them.

If the objective of a fleet is unknown the whole problem becomes more open, but still in the majority of cases must be one of two or at most three places. The strategical exigencies of each side must be more or less known to the other, and at the most there can be very few alternatives of destination for a fleet leaving a port. If no narrow area suitable for observation is in the path of the fleet, it is possible it may have to be searched for. But it is a dangerous and doubtful expedient for a fleet to wander about expending coal looking for another. As a general rule cruisers and blockading vessels should do this, leaving the ships at a base; but since the eventuality *may* arise of a fleet cruising with her cruisers, using them to discover another, the question of their tactics should be discussed.

Tactics of cruisers with a fleet

The exact geometrical arrangement of the cruisers as regards the fleet is of secondary importance to their mutual arrangement with reference to each other with a view to communication and support; for obviously it is of little use to discover enemies' cruisers if the recognisance cannot be pushed home and sufficient information of the position of the battleships gained to bring them to an action.

The arrangement of front of the cruisers will probably consist of the cruisers in line abreast on each beam of the squadron. The reasons for this, though purely geometrical, are worth discussing more fully. Suppose the position of the enemy totally unknown; then any spot where the cruisers were placed, either in line ahead or in line abreast, would be equally likely or unlikely to put them in sight of the enemy's squadron totally regardless of the position of their own battle squadron. If arranged 100 miles ahead of it in line abreast they would stand the same chance of sighting the enemy as if 100 miles astern. Now since the position of the battle squadron cannot affect the fact of the cruisers sighting the enemy or not, its position should be that nearest to the enemy when the cruisers do sight it, or, in other words, in the centre of the line of cruisers.

The same reasoning applies if the position of the enemy is suspected, or if the line of approach of the enemy is known. The place for the pursuing squadron is in the centre of its cruisers. If the enemy is known to be ahead and the battle squadron is unable to arrive at a position in time, and it is desirable to send cruisers ahead to track and observe them, or, in other words, when there is no question of an action between the two battle squadrons, then the cruisers may be sent ahead, but otherwise, all that cruisers ahead of a searching squadron do is to give the looked-for squadron earlier knowledge of the approach of their enemy than is advisable or necessary.

Should the course of the fleet be suddenly altered it must be due to information received, and it may be urgent to effect the alteration as quickly as possible. The formation with the fleet in the centre of the cruisers is far more mobile than in the case of separation of the cruisers from the fleet, when in changing front each cruiser would have to move on the chord of the arc of a circle equal in radius to their distance ahead of the squadrons. In the more compact formation each cruiser not in sight of an enemy would move as quickly as possible, by mooring board, to its new position. In doing so it would have to close the fleet, which would facilitate transmission of orders, which would probably be required if enemies' vessels had been sighted.

Tactics of eluding cruisers.

Now we must consider the formation and tactics of the eluding fleets and cruisers. We have seen that the formation of cruisers on the beam of a battle squadron places that squadron nearest the enemy when it is sighted. For this reason it must be the worst formation for an evading fleet. Broadly speaking, there are two types of formation: one when the locality of the searching fleet is suspected with some degree of certainty, or if the *locus* of the line of advance is reasonably certain, then the cruisers might be detached along that *locus*, or to that locality, at a great distance ahead of the fleet to search for the enemy, and then communicate with the squadron by detaching a ship. The objection to this method is the very large distance the cruisers would have to be sent ahead to give sufficient warning of the enemy's approach.

Suppose the cruisers 200 miles ahead of the fleet when they sight the enemy. A fast vessel is despatched, steaming say 25 knots; it will take her six hours to reach the fleet, and meanwhile the fleets will have closed 150 miles and be only fifty miles apart, and the enemy's cruisers ought to be close on the messenger cruisers' heels; so, assuming prompt action on the part of the cruisers on each side, this method appears of little use.

The other and seemingly better way would be to range the cruisers in line ahead of the battleships, when with fifteen cruisers a distance of 220 miles could be covered, and assuming a signal down the line to take ten minutes to pass from ship to ship, two-and-a-half hours would transmit information and the fleet would only have closed sixty miles. The tactics of each cruiser after conveying its information to the next in the line should be to deceive the enemy's cruisers as to the line of bearing of their fleet, and do their best to prevent their forcing on to discover its whereabouts.

Diagram III. (see page 98) gives an example of two fleets meeting with cruisers arranged on the above principles.

If the evading fleet has to cross the supposed course of the other at right angles, or at any large angle, or has to pass through an area that has geographical advantage for minute search by the enemy's cruisers, then probably night-time and a close formation would be chosen for the attempt, for in this case the less area the fleet occupied the greater the measure of safety. The smallest area a fleet can occupy is when the ships are arranged in the form of a square. Probably two cables apart for the lines and ships is as close a formation as would be adopted at night without lights in moderate weather; twenty-five ships in such a formation would only occupy one square mile of water, and in such a formation would be more liable to escape detection than in a long drawn out line.

As regards the watching of a geographically suitable area through which an enemy must pass, the problem mainly consists in placing cruisers at equal intervals across it within sight of one another, remembering that the strength of the patrol is only the strength of its weakest link—namely, the distance visible at night, and to be efficient the patrol must be as effective at night-time as it is in the day

Tactics of rival cruisers in contact.

And now we have arrived at a portion of the subject of fast craft tactics which is of extreme importance—namely, the tactics of cruisers on sighting the enemy's fast craft. This naturally has two divisions: the tactics of the searching and the tactics of the evading vessels.

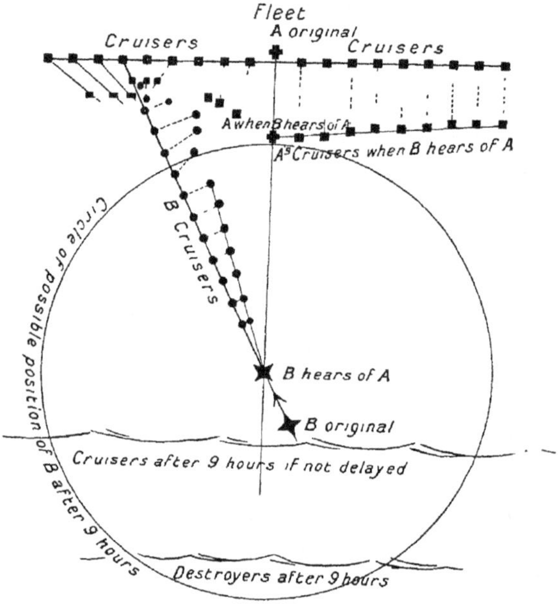

As regards the searching cruisers their duties are twofold: first, to press home and get in touch with the enemy's fleet; secondly, to maintain communication with their own fleet. This in the face of the opposition of the enemy's vessels appears almost an impossibility. Just at the time when every fast vessel is required by the searching fleet to locate the enemy, the hostile cruisers will seize the opportunity to engage and cripple them, and if at all comparable in numbers should most effectually do so.

The main aids to the searching cruiser are the knowledge of the enemy's objective, and the acumen of the captain. Early sighting the enemy's masts and consequent approximate knowledge of her course before she has time to alter it, and the relative bearing of the next cruiser sighted, are the main guides to the bearing of the enemy's fleet. Knowledge of the probable number of cruisers the enemy possesses may, according to their disposition, be a guide to its distance. But even suppose the bearing and distance of the enemy's fleet known, the double question arises: Retreat or evasion? which will the fleet try? Again conditions and surmises can give the only possible clue—nearness of the fleet, position as regards the flanks of the cruiser advance, distance from its base, reserve speed, nearness of its objective, time of day, length of night; all these factors bear on the motion of the enemy's fleet. These must be integrated, and the resultant course steered. These are the problems the captain of any cruiser may be called on to solve on the spur of the moment in war time; nothing but personal study and thought will help him in such a crisis; rules of thumb are useless, intuition will only come from constant thought on the subject.

The tactics of the evading fleet cruisers are simpler, since they can be preconcerted and formulated for whatever bearing the enemy's cruisers may be sighted on, and will vary whether the strategy of the evading fleet is retreat or evasion. Their two main weapons are deception and engagement—deceiving the advancing cruisers as to the bearing and alteration in course of their fleet, and engagement to delay them.

A few general remarks in conclusion. Smoke should be watched for on the horizon, as on the news of a fleet in sight the pursued fleet will probably stoke up, and may be seen by the following cruisers. Signals must be simple and rapid: two of the greatest importance might be made by signs—namely, cruiser or battleship in sight, followed by the bearing of the ships. Too much importance cannot be attached to the early sighting of a ship. Not only might a cruiser's masts escape observation altogether unless the signalmen are much exercised in long distance telescopy, but since the relative position of the masts gives an indication of the course of the cruisers, and, consequently, the probable course of the squadron, instant recognition is necessary, for the moment the evading cruiser sees her opponent she will alter course and act as a decoy to deceive the advancing enemy. Practice in long distance vision is a great necessity to our signalmen, and means more than many other accomplishments and might be made popular by competition.

A hundred years of peace have passed, with millions spent and generations devoted to the Navy, since our supremacy on the sea was disputed. In the future lies war—it may be in our time or it

H 2

may not; at all events, when it does come the main crucial question of command of the sea will probably be decided within the first month, and foreshadowed on the day of the first great sea fight. In that day the past store and present accumulation of matter and mind will be ruthlessly expended. It is in that momentous day and eventful month that the fast craft will be called on to strain every knot and nerve to pave the way of the battleship to victory—pave the way truly, for it is over the sunken hulls of many of the fast craft that the battleship will steer cheering to victory.

R. H. S. BACON.

CHAPTER V.

THE MANŒUVRES OF 1899.

AFTER a year's intermission manœuvres were organized for 1899 on a considerable scale, and with a programme which presented several features of novelty. The programme was as follows:— Programme

OBJECTS.

The principal object of the 1899 Manœuvres is to obtain information as to the most advantageous method of employing a considerable body of Cruisers in conjunction with a Fleet.

A subsidiary object is to throw some light on the relative advantages and disadvantages of speed and fighting strength.

Another subsidiary object is to obtain information relative to the working of Destroyers and Torpedo Boats.

GENERAL IDEA.

A British Convoy (C) of slow ships escorted by a fast Cruiser, on passage from Halifax to Milford Haven, is ordered to wait at a certain rendezvous the arrival of a protecting squadron.

N.B.—The slow ships cannot be taken in tow, must remain in company, and have no fighting value.

A hostile squadron (A) of fast ships, lying at Belfast, is sent to sea to intercept and capture the Convoy and bring it into Belfast.

After an interval, a superior British squadron (B) of slower ships is sent to protect the Convoy (C) which has been ordered to a pre-arranged rendezvous, cover it from the hostile squadron, and bring it into Milford

The whole of Ireland is hostile territory, and belongs to A.

The coast of England and Wales from the island of Islay to the Lizard, including the Scilly's and the Isle of Man, is British territory.

The hostile fleet (A) has Torpedo Boats at Waterford, Kingstown, and Belfast.

The British fleet (B) has Destroyers at Milford Haven, Holyhead, and Lamlash.

The fleets engaged were the Channel and Reserve Fleets, reinforced for the occasion by large contingents of cruisers. Their composition was as follows:— Fleets.

CHANNEL FLEET.	RESERVE FLEET.
FLEET "A."	FLEET "B."
MAJESTIC (flag).	*Alexandra (flag)*
HANNIBAL.	*Howe*
PRINCE GEORGE	*Colossus.*
MARS.	*Nile*
JUPITER	*Trafalgar.*
MAGNIFICENT (flag).	*Sans Pareil (flag).*
REPULSE.	*Rodney.*
RESOLUTION.	*Benbow.*
DIADEM.	*Collingwood.*
NIOBE.	*Thunderer.*
ANDROMEDA	*Europa.*
MERSEY.	*Argonaut.*
THAMES.	*St. George.*
PIQUE.	*Galatea.*

CHANNEL FLEET.	RESERVE FLEET.
FLEET "A"—*continued.*	FLEET "B"—*continued.*
TERPSICHORE.	*Forth.*
ARROGANT.	*Severn.*
MINERVA.	*Andromache*
SYBILLE.	*Sappho.*
CAMBRIAN.	*Vindictive.*
LATONA.	*Diana*
NAIAD.	*Sirius.*
ÆOLUS.	*Charybdis*
RETRIBUTION	*Melampus.*
ARETHUSA.	*Apollo*
FURIOUS.	*Spartan.*
PACTOLUS.	*Brilliant*
PELORUS.	*Mercury.*
	Gladiator.
	Juno.
	Rainbow

"A" Flotilla.		"B" Flotilla.	
SPANKER	. 81, 84, 71, 76, 77, 66, 79, 42	*Lena*	. *Crane, Chamois; Hunter, Flying Fish, Lightning, Star, Violet, Teazer, Fawn, Sylvia.*
GOSSAMER	. 63, 64, 65, 68, 72, 73, 74, 83.	*Niger*	. *Angler, Haughty, Cygnet, Contest, Janus, Mallard, Porcupine, Dasher.*
JASON	. 86, 45, 52, 53, 55, 57, 58, 49.	*Renard.*	*Bat, Ferret, Lynx, Panther, Seal, Shark, Thrasher, Wolf, Fairy, Gipsy.*

CONVOY.

Calliope, Curaçoa	Slow ships
A first-class Cruiser to be detached from B .	Escort.

The A fleet was commanded by Vice-Admiral Sir Harry H. Rawson, K.C.B., with Rear-Admiral Arthur D. Fanshawe as second in command; and the B fleet was commanded by Vice-Admiral Sir Compton Domvile, K.C.B., with Rear-Admiral Pelham Aldrich as second in command.

Rules

The rules and regulations drawn up for the conduct of operations are here appended:—

RULES AND REGULATIONS TO BE OBSERVED DURING THE MANŒUVRES OF 1899.

The ports of Belfast and Milford Haven are to be considered as strongly fortified, and proof against torpedo attack within the following limits:—

Belfast . Within a line drawn from Black Head to Orlock Point.
Milford . " " , St Ann's Head to Sheep Island.

After hostilities have commenced, ships outside these limits are liable to attack from any vessel of the enemy.

Destroyer and Torpedo Boat Stations are proof against attack.

All other ports in the territories of the respective Admirals are open and undefended, and can be used only by vessels of the side to which the ports belong at their own risk.

The Signal Stations marked on the special chart issued with the Instructions for the Distribution of Intelligence will alone supply information during the Manœuvres.

The Fleets, with their respective Flotillas, being at their respective Base Ports, will be warned by a telegram to "prepare for hostilities"; upon which the Flotillas, with their Depôt Ships, will be sent off to their stations. Each Fleet will be free to send out vessels of the Flotillas, but none of them are to lie off an enemy's port during this time of preparation in such a manner as would be calculated to precipitate hostilities during a period of strained relations.

Not later than 48 hours after the telegram to "prepare," the telegram to "commence hostilities" will be sent, but no attempt is to be made by Officers in Command on either side to put a vessel out of action until it is known to them for certain that hostilities have begun

All Battleships are to be considered of equal power: the superiority of one battle squadron over another is to depend alone upon which has the greater number of Battleships.

The following table will govern Battleships, Cruisers, and smaller vessels being put out of action:—

Number and Class of Vessel.	Can put out of Action.	At what Distance.	In what Time.
1 Battleship	Battleship	Within 2 miles*	One hour.
1 Battleship	Any Cruiser	" 3 miles	30 minutes.
1 1st Class Cruiser	1st Class Cruiser or Ship of lower class	" 1 mile	30 "
1 2nd Class Cruiser	2nd Class Cruiser or Ship of lower class	" 1 mile	30 "
1 3rd Class Cruiser	3rd Class Cruiser or Ship of lower class	" 1 mile	30 "
2 Cruisers of same class	One Cruiser of same class	" 1 mile	30 "
1 Torpedo-gun Boat	Torpedo-gun Boat	" 1 mile	30 "
1 Destroyer	Torpedo Boat	" ½ mile	5 "
2 Torpedo Boats	Destroyer	" ½ mile	5 "
Battleships, all Cruisers, Torpedo-gun Boats	Destroyer	" 1,000 yds.	3 "
	Torpedo Boat	" 1,000 yds.	3 "

* In the case of Squadrons this distance will be that between the nearest Ships.

The period of "action" is to be between the two guns which either ship may fire to mark it; the first is to be fired when the two ships are within the prescribed distance, and the second from the same ship at the expiration of the time allowed; no other guns than these are to be fired.

No ship can put two vessels out of action in the same time; each must have its separate time allowance.

No Cruiser of a lower class can count against a Cruiser of a higher class.

With squadrons of Cruisers composed of more than two of the same class, if the superiority is less than 2 to 1, one Cruiser on the more numerous side will be put out of action as well as all the Cruisers on the less numerous.

Ships put out of action under the rules can take no further part in the Manœuvres, but must return to their port—Milford or Belfast—flying the Blue Peter at the fore. They are to select a route as far as possible clear of the scene of operations, and are strictly enjoined not to communicate any information to the ships on either side which they may meet on the way.

No Battleships can be put out of action except by Battleships or if torpedoed.

As the 18-inch Torpedo cannot be fired at a ship in a peace exercise, a Destroyer is to fire a blue light at night or blow her whistle by day at the moment when the Torpedo would be discharged, the tube being trained and all adjustments made as if actually firing; the distance at the time of firing must be within 500 yards, and the number of torpedoes considered successful will be assessed by the umpires.

Any Vessel will be out of action if a Torpedo fired from a Torpedo Boat strikes her before the Torpedo Boat is herself out of action.

At the expiration of the period of hostilities, vessels which have returned to port will carry out such orders as they have received relative to Target Practice and either return to ports of assembly, or rejoin their Squadrons, as may be ordered by the respective Vice-Admirals.

Any points which may arise, not provided for by these rules, will be decided by the umpires according to what, in their opinion, would be probable in war.

Comments on the rules

The instructions to umpires were of the usual character and need not be set forth at length, though a provision that "any matter or claim not especially provided for in the rules is to be dealt with on its merits," deserves mention.

The rules governing battleships, cruisers and smaller vessels being put out of action appear to be well conceived for the purposes of manœuvres, and to represent with adequate approximation the probable conditions of actual warfare. An engagement between two opposing ships of the same class could only result in putting both out of action. No number of cruisers could put a battleship out of action, and no number of cruisers of a lower class could count against a single cruiser of a higher class. It may be conjectured that these rules were designed partly to counteract the tendency disclosed in former manœuvres to employ cruisers for the purpose of strengthening the fighting line, and not in the discharge of their more legitimate function of scouting; and partly as a rough and ready approximation to the real fighting value of different classes of ships. But, as owing to the course taken by the operations the rules in question scarcely took effect at all, they need not be discussed at length. In one respect, however, they were not without influence on Admiral Rawson's dispositions. His primary object was to find and capture the convoy which was to be "escorted by a fast cruiser," and in another place the escort was stated to be a first-class cruiser. This made it impossible for Admiral Rawson to seize the convoy by any number of second-class cruisers. Hence he was obliged so to dispose his cruisers that if the convoy was first sighted by a second-class cruiser the latter would be able speedily to summon either a first-class cruiser or a battleship to her assistance. As he had only three first-class cruisers in all, this condition was far from easy to satisfy.

Another rule which invites some attention is that governing torpedo attacks, especially by destroyers. Destroyers and torpedo boats only came within the destructive range of larger ships at 1,000 yards, and they were entitled to discharge their torpedoes—or in the case of destroyers carrying 18-inch torpedoes, not adapted for use in a peace exercise, to fire a blue light by night or to blow their whistle by day—within a range of 500 yards. The period required for the destruction of a destroyer or torpedo boat by a larger craft was three minutes. Now a destroyer steaming at twenty-four knots will cover 1·2 nautical miles in three minutes. Assuming her to be in pursuit of a battleship steaming away from her at fifteen knots, the latter will cover ·75 of a nautical mile in the same time. Therefore, in order to come within effective striking distance before she is put out of action, the destroyer must cover 1,000 yards *plus* ·75 of a mile—that is, ·494 *plus* ·75, or 1·244 nautical miles in less than three minutes. It thus appears that the battleship must steam less than fifteen knots or the pursuing destroyer more than twenty-four knots if the latter is to be enabled to fire her torpedo before she is put out of action. But if two destroyers attacked the same battleship from opposite directions, or if a dozen destroyers, coming from different directions, attacked a fleet of battleships in cruising formation, unprotected by an outlying cordon of cruisers and small craft, it is certain that, under the rules, the battleship or fleet must have succumbed to the attack without the possibility of effective reply. What might happen in actual war is a very different thing, but the difficulty of framing rules to do equal justice to both sides in a mimic conflict between destroyers and battleships, may serve to illustrate the tremendous menace of the former under conditions favourable to its operation. We shall see presently what dispositions were made by Admiral Rawson to neutralize this menace.

The conditions of the problem.

We may now consider what *data* Admiral Rawson had for locating the rendezvous at which the convoy would be found. The first and primary *datum* was that the convoy was "on passage from Halifax to Milford Haven." Now a merchant ship making this particular passage might be expected to make Cape Clear its landfall and to steer a great circle course to that point from Cape Race as a departure. This course cuts the meridians from 20° W. to 15° W. at different points lying approximately between lat. 51° N. and lat. 51° 30′ N. *Primâ facie*, therefore, it might be expected that the rendezvous would be found within an area defined by longitudes 15° and 20° W. and latitudes 51° and 52° N. On the other hand, it seems equally likely that a convoy leaving Halifax for a western port in England would deviate from the ordinary trade route in order to avoid interception by a hostile fleet known to be on the look-out for it. The great circle distance from Cape Race to Cape Wrath in the north and to Cape Finisterre in the south is approximately 100 miles greater than its distance to Cape Clear, and each of these extreme points is some 300 miles, more or less, further than the Fastnet from Milford Haven. Thus, by taking either of these extreme courses the convoy steaming at 9 knots would only lengthen its total passage by less than 48 hours, and this extreme excess would be gradually reduced as the course chosen approximated on either side to the ordinary trade route. There is thus nothing in the conditions so far considered to disallow the hypothesis that the rendezvous of the convoy might be placed as far north as Cape Wrath, or as far south as Cape Finisterre. If we pursue this hypothesis a little further we find that it gives some preference to the northern area over the southern. From Cape Wrath to Milford the convoy might be escorted almost entirely through landlocked waters affording great facilities to the operation of B's destroyers and offering great impediments to the free action of A's scouts. Assuming, therefore, that the convoy might be ordered to deviate widely from its natural course, it was more likely that it would be ordered to deviate to the northward than to the southward. It is true that when Lord Howe went out in 1784 to intercept a convoy making from America to Brest, he sought for it in the offing of Brest, and not either far to the southward or far to the northward. But this historical example to the contrary of the hypothesis above considered does not really help us much. Sailing ships enjoyed no such freedom in the choice of a course as steamships possess, and as a matter of fact Lord Howe never found the convoy he went out to intercept. He encountered Villaret Joyeuse, who managed to decoy him off its track.

So far, then, we have seen no reason for supposing that the convoy would certainly, or even most probably, be found within a region narrowly defined by its relation to the ordinary trade route. The next *datum* is the time to be occupied by the operations. This was roughly known to be from Saturday, July 29, to Saturday, August 5, but the exact hours at which hostilities would begin and end were not known to either side until shortly before they began. As a working hypothesis, however, we may assume that they would last from noon to noon on the days mentioned—a period of 168 hours. Next A was to be free to leave Belfast as soon as hostilities began, whereas B could only leave Milford "after an interval." The length of this interval, which was all important, was unknown to A, but inasmuch as A had to search over a large area for a convoy of which the exact rendezvous was known to B, it must necessarily be considerable. Its length, moreover, would be in large measure determined by the position of the rendezvous in relation to Milford and Belfast respectively. It must not be so great in any case that B would have no time to reach the rendezvous with his cruisers at a speed of at least 15 knots and to escort it back to Milford at a speed of 9 knots, while leaving a reasonable margin for contingencies. One contingency was that of bad weather, either stormy or foggy, which might materially reduce the speed of B's ships going and returning. Another was that B having reached the rendezvous might find the convoy already captured and carried off. He would then have to search for it, an operation which might occupy many hours, and having recovered it, he might have to take it back from a distance greater than that of the original rendezvous from Milford. It seems to follow from these considerations that a start of 20 hours out of 168 in all was probably the *maximum* that would be likely to be assigned to A. This is all that A would have to find the convoy, and having captured it, to withdraw it from the observation of B.

If, then, we assume, as a working hypothesis, that A's start would be approximately 20 hours, it will be seen that the rendezvous could not be at a much greater distance than 600 miles from Milford. The time assigned to B for the whole operation is now reduced to 148 hours. It would take his cruisers 40 hours to reach the rendezvous at a distance of 600 miles at 15 knots, and the convoy would return at 9 knots in a little less than 67 hours. This leaves a margin of about 40 hours for the contingencies mentioned above, and this is none too much, seeing that if A had happened to hit on the rendezvous without losing much of his start, he would still be from 12 to 20 hours ahead of B when the latter's cruisers first reached the rendezvous.

Thus a hypothetical start of 20 hours given to A gives us a sea distance of 600 miles from Milford as the exterior limit of the area within which the rendezvous would probably be found. This area is bounded by the arcs of three intersecting circles, one of which is drawn from Milford with a radius of 600 miles, another from Cape Clear with a radius of 600 miles *minus* the distance from Milford to Cape Clear, and a third from the exit of the North Channel with a radius of 600 miles *minus* the distance from Milford to that point. Every point in this exterior limit is at a sea distance of 600 miles from Milford, and the point at which the two latter arcs intersect, which may for convenience be called X, is at the same distance from Milford whether the course be measured round Cape Clear or through the North Channel. North of X the courses round Cape Clear are longer than those through the North Channel, and south of X these conditions are reversed. The critical importance of X in relation to the *data* involved thus becomes apparent.

Dimensions of the area of search.

The area thus defined extends from north of the Shetlands to south of Cape Finisterre, and if it be regarded as a semicircle its diameter is 1,200 miles. Its superficial extent is over 570,000 square miles, but if a third of this be deducted as representing land, we find that in order to search the whole of the remainder Admiral Rawson would have to cover no less than 380,000 square miles. This was obviously impossible in the time allowed. No amount of start that could possibly have been given him would have afforded him more than a bare chance of finding the rendezvous unless he could locate it beforehand within much narrower limits. But a little consideration will show that he might safely neglect the whole of the area south of lat. 51° N. If the rendezvous were placed anywhere south of this parallel it would be much nearer to Milford than to Belfast, and A's assumed start of 20 hours would therefore be practically reduced in that case by the time it would take his cruisers to steam from Belfast to Milford—that is, by at least 12 hours (180 miles at 15 knots) if they went past Milford, and by more if they went, as they probably would, through the North Channel. Assuming, then, that 20 hours or thereabouts was a fair, and not more than a fair, advantage to give to A, in consideration of the task imposed upon him, any position of the rendezvous which practically placed B 12 hours nearer to it than A would require the actual interval between their respective starts to be at least 32 hours if A was to retain the assumed advantage of 20 hours net. This, however, would so greatly reduce the margin for contingencies required by B as to be practically out of the question. A, therefore, might safely neglect the whole of the area south of lat. 51° N., and regard it as practically excluded by the conditions of the case.

Further consideration of the conditions.

By parity of reasoning, if the rendezvous were so placed that B's shortest course to it would lie past Belfast, the assumed advantage of 20 hours' start assigned to A would be practically secured to him if B was allowed to leave Milford within eight hours of the time when A was free to leave Belfast. If B's start was postponed beyond that time, it might be regarded as practically certain that the rendezvous would not be found north of the latitude of X, and quite certain that it would not be found north of the dotted line drawn on the annexed chart. Hence the advantage to A of ascertaining that B was still at Milford eight hours after he himself had left Belfast was so paramount that the obtaining of certain intelligence to this effect might well be regarded as an object to which all his dispositions could be, and perhaps ought to be, subordinated. In that case his best course would be, after having obtained an offing as far as possible beyond the reach of B's destroyers, to wait off Tory Island or thereabouts until he had learnt from his nearest signal station, either that B had left Milford within eight hours of his own start from Belfast, or that he was still at Milford at the end of the eight hours. In the former alternative he would search to the northward, in the latter to the southward, and while he was waiting for intelligence a considerable portion of the area south of the dotted line might be searched by his cruisers. If he ascertained that B had left Milford within the eight hours he would then scour the area north of the dotted line, keeping a sharp look-out for B's cruisers coming up astern. If he ascertained that B had not left Milford within the eight hours he might then assume that B's nearest course to the rendezvous would not lie past Belfast,

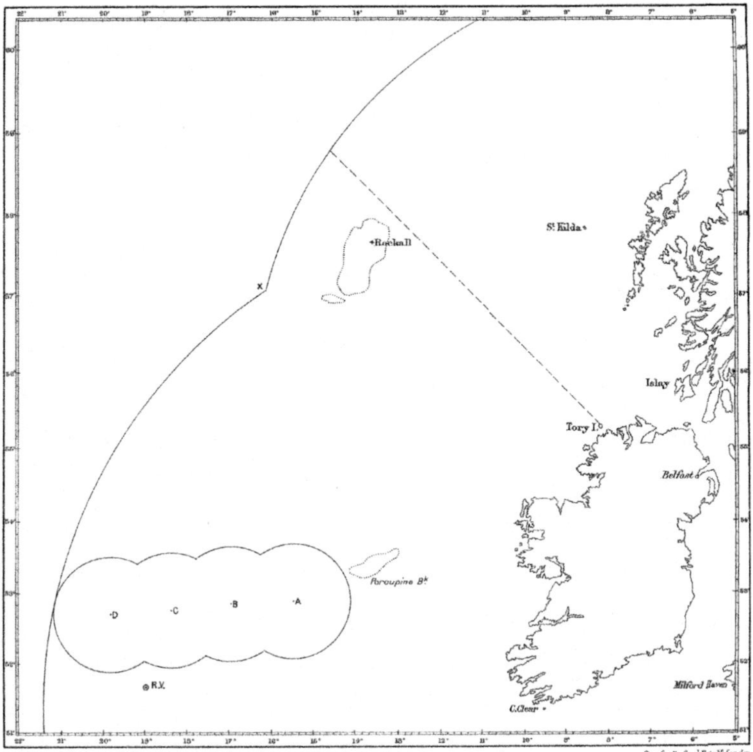

Stanford's Geogl Estabt London.

and by an extension of the same reasoning that the rendezvous would probably not lie very far to the north of a line drawn through the points marked A, B, C, D, on the chart, since any position to the north of that line would give A a shorter course to it than B, and therefore by that much increase A's hypothetical start of 20 hours. The points A, B, C and D on the chart are each equidistant from Belfast and Milford and are respectively at a sea distance of 400, 450, 500 and 550 miles from both. Circles are drawn round them with a radius of 50 miles. Within, or not far from, the area defined by these circles the rendezvous would most probably be found if the foregoing reasoning is correct. As a matter of fact, it was at the point marked R.V. on the chart in lat. 51° 40′ N., long. 19° W.

In the foregoing reasoning there is only one assumption not directly based on the *data* accessible to A from the outset, and that is the length of the interval between the starts of A and B. The length of this interval would, however, in any case he governed by two opposing conditions—one the necessity of giving A as long a period as possible within which he could search for the convoy without running the risk of molestation by B; the other the necessity of affording B a reasonable margin for contingencies over and above the time required to reach the rendezvous and escort the convoy back. The assumption of twenty hours seems to strike a fair mean between these limiting conditions, and, as a matter of fact, the actual interval was nineteen hours. A left Belfast at 10 a.m. on Saturday, July 29, the time assigned by the Admiralty for the beginning of hostilities, and B left Milford nineteen hours later, at 5 a.m. on Sunday, July 30.

Admiral Rawson's view of the situation.

The foregoing view of the situation did not present itself to Admiral Rawson; or, if it did, he rejected it. He held himself justified in neglecting the whole of the area of possible search which lay south of lat. 51° N., mainly for the reasons given above—namely, that if the rendezvous were placed in that region either his advantage of start must be reduced by the distance between Belfast and Milford, or the time allotted to Admiral Domvile must be less than was required for the due prosecution of the task allotted to him. But he did not apply similar considerations to the northern portion of the remaining area. On the contrary, he inferred from the regulations that on the whole the rendezvous was more likely to be placed well to the northward than anywhere in the neighbourhood of the position actually assigned to it. A subsidiary object of the manœuvres was defined to be "to obtain information relative to the working of destroyers and torpedo boats," and for this purpose a force of twenty-eight destroyers stationed at Milford Haven, Holyhead, and Lamlash was assigned to B, and a force of twenty-four torpedo boats stationed at Waterford, Kingstown, and Belfast was assigned to A. This disposition seemed to imply that the Irish Channel and its approaches were the field of operations especially appointed for the working of torpedo boats and destroyers; and from this assumption the further inference was drawn that the course of the operations, as contemplated by the Admiralty, would compel the B fleet to traverse the Irish Channel. The function of the torpedo boat is to attack a larger vessel with its torpedoes; it has little or no offensive capacity, except in the discharge of this function. Hence, unless larger vessels are likely to be found within its radius of action, or even unless it has definite intelligence that such larger vessels are actually within reach, it has little or no reason for exposing itself to the formidable menace of the destroyer, and every reason for remaining within the shelter of its station. The chances of a torpedo boat finding a single ship or even a fleet in the course of a roving search have been proved over and over again in manœuvres to be exceedingly small, and they are reduced almost to zero by the presence of a vigilant and active flotilla of destroyers within the area of search. Hence, in the absence of ships to be attacked the *rôle* of the torpedo boat is practically nullified; and in the absence of torpedo boats to be attacked the *rôle* of the destroyer is equally nullified. If, therefore, the B fleet was not likely to have to traverse the Irish Channel in the course of the operations, there seemed to be no strategic reason for placing destroyers and torpedo boats there, and little or no prospect of obtaining information of any value "relative to the working of destroyers and torpedo boats."

Criticism of this view.

So far the reasoning on which Admiral Rawson based his dispositions appears to be sound and cogent; but it does not necessarily sustain the conclusion which he drew from it—namely, that the shortest course to the rendezvous would be likely to take the B fleet through the Irish Channel on its exit from Milford. On the contrary, the more cautious inference would, perhaps, have been that matters would be arranged so as to give greater actuality and experimental value to the operations by deferring the conflict between destroyers and torpedo boats to a later stage of the proceedings, when both sides had been tried and hardened by the alarms and excursions which are so characteristic of torpedo operations, and so invaluable in the training they afford. This condition was as likely to be satisfied by a rendezvous placed in the neighbourhood of lat. 52° as by a rendezvous placed north of lat. 56°. It was at least an even chance that A would find the rendezvous and seize the convoy before B was in a position to interfere with him; and all that was necessary to satisfy the condition above defined would be that A, having found and seized the convoy, had brought it so near to the northern entrance to the Irish Channel that B, even if he recaptured it, would be compelled to take it through the Irish Channel on his way back to Milford. Now if we assume that the rendezvous was placed approximately equidistant from Belfast and Milford, and that A, having found it there, carried it off only three hours before B reached the rendezvous, A steaming with the convoy at nine knots would be at least 108 miles on his way back to Belfast before B could overtake him. But B, having reached the rendezvous and found the convoy gone, must needs spend some time in looking for it, since he would not be able to assume for certain that A had taken a direct course to Belfast. Every hour so spent would increase the distance from Milford and decrease the distance from Belfast at which B might expect to overtake the convoy, so that if A's start from the rendezvous was anything more than three clear hours in advance of B's, it was practically certain that B could only overtake and recapture the convoy in such a position as would compel him to take it back to Milford through the Irish Channel. Indeed, it may well be argued that if B found the rendezvous vacant his best course would be, not to search for the convoy, but to make at once for the entrance to the North Channel, and there await its passage. He would, it is true, be exposed in that position to the attack of A's torpedo boats, but, on the other hand, he would be able to employ his own destroyers in defence—and this, at any rate, would afford as favourable an opportunity of obtaining "information relative to the working of destroyers and torpedo boats" as his passage through the Irish Channel at the outset of the proceedings.

The foregoing reasoning is, of course, largely conjectural. But rational conjecture is all that an Admiral has to go upon when he has to discover an unknown rendezvous, and the hypothesis in question combines and harmonizes several converging lines of conjecture and satisfies all the known conditions in such a manner as to make it a good deal more probable that the rendezvous would be found at a point approximately equidistant from Belfast and Milford than at any other point in the area of possible search. Indeed, considering that the primary *datum* was that the convoy was "on passage from Halifax to Milford Haven," and that the great circle course for such a voyage passes through a series of points approximately equidistant from Belfast and Milford, it might almost have been assumed without too curious a consideration of possible alternatives that the rendezvous would be found in that neighbourhood. On the other hand, it is always hazardous in war to "make a picture," as Napoleon said of his

Generals. Admiral Rawson may have reflected that his predecessor in 1897 "made a picture" of an unknown rendezvous and found his perspective hopelessly at fault. But Admiral Stephenson's picture was based on a single analogy—that of the relative positions of Lough Swilly, Black Sod Bay, and Rockall to those of Plymouth, Brest, and Bantry Bay respectively. It was a questionable analogy at best, and it failed to harmonize all the conditions of the situation. The picture here drawn does not err in this respect. Its perspective might have proved all awry in the event, but even so it is no worse in this respect than the picture Admiral Rawson actually made for himself. For a picture he did make no less than his predecessor, and, it must be added, no more successfully. Reading between the lines of his instructions, he seemed to view the whole situation by the analogies of the English Channel. If the chart of the British Isles be laid on its side so that the East represents the North, Milford Haven may be supposed roughly to represent the Downs, Belfast Brest, and the south point of Islay the Land's End. On this analogy the fan-shaped area of sea radiating from the entrance to the North Channel and extending from the Butt of Lewis in the north down to lat. 51° N., and to a distance of some 300 or 400 miles from the coast of Ireland in that latitude, would represent the portion of the Atlantic on which all trade routes leading to the English Channel converge. Admiral Rawson held that the convoy might be located at almost any point in this area, but for the reasons given above he thought it most probable that it would be found in its northern portion. There were, moreover, certain points within the area which, being within the reach of ordinary soundings, seemed more likely to be chosen as a rendezvous than any locality where soundings would be much more difficult to obtain. Of course in clear weather, when observations are unimpeded, it is quite as easy for a ship to take up a position out of the reach of soundings as one which can be determined by soundings. But the eastern Atlantic is much beset by fogs, and, having regard to this fact, Admiral Rawson thought it likely that a position might be chosen for the rendezvous which could be found and held by means of soundings even if the weather rendered observations impracticable. St. Kilda, about fifty miles west of the Sound of Harris, in the Western Hebrides, is such a point in the northern portion of the area, Rockall Bank is another towards its middle, and Porcupine Bank, in its southern region, is a third. The position of all three is shown on the chart. The conjecture proved unfounded in all three cases, but it did not materially affect Admiral Rawson's dispositions except so far as it may have induced him to select the neighbourhood of Porcupine Bank as a convenient point for the concentration of his fleet during the prosecution of his search.

Admiral Rawson's scheme for the examination of the area to be searched was a modification and development of the method employed by Lord Walter Kerr for the solution of a similar but less complicated problem in the manœuvres of 1895 (see *Naval Annual*, 1896, Chap. XI., p. 185). What he had to do was first to find the convoy within a given but very extensive area, and next, having found it, to capture it. It was known that the convoy would be escorted by a "fast cruiser," and that the escorting cruiser would be a first-class cruiser. Now, before the convoy could be captured the escorting cruiser must be put out of action, and this could only be done, under the rules, either by a battleship or by one of the three first-class cruisers attached to the A fleet, no number of second-class cruisers being allowed by the rules to count against a first-class cruiser. Therefore it was necessary not only to make provision for an exhaustive examination of the area to be searched, but so to dispose the ships engaged in the search that if a second-class cruiser should happen to sight the convoy first, she should be able to summon either a battleship or a first-class cruiser for its capture without losing more time than, having regard to the probable start accorded to A, could be allowed for the operation without running the risk of interference from B. The first condition, that of finding the convoy, would be satisfied by any disposition of the ships at A's disposal that would enable them at the speed assigned to cover the area of search within the time allotted. The second, that of capturing the convoy, required such a disposition of the battle squadron and of the three first-class cruisers as would enable one or more of these ships, if summoned for the purpose, to reach the scene of action with as little delay as possible; and it also required that if the ships were dispersed each should know the exact position of all her consorts at any and every period of the operation. Subject to these conditions it was necessary to determine approximately the area that could be searched effectively in the time that Admiral Rawson, having regard to the probable length of his start and the time required for escorting the convoy back to Belfast, thought himself justified in allotting to the preliminary operation of search. This was found to be an irregular fan-shaped area radiating from a point well to the westward of the entrance to the North Channel, stretching to the northward beyond the outer Hebrides, to the southward beyond the Porcupine Bank, and passing in its exterior curve well beyond Rockall. Of course this area, if restricted to the northward, might have been extended much further to the southward;

The scheme of search

the fan, in fact, might have been turned on its radiating point until its northern or right-hand limb instead of passing near St. Kilda passed only a little to the north of Rockall, in which case its exterior curve would have extended much further to the southward and westward than it actually did. As the event showed, this would have given Admiral Rawson a much better chance of finding the convoy; but for the reasons already set forth he assumed from the outset that its more probable location would be somewhere to the north of the dotted line on the chart. Be this as it may, the problem now was to arrange for the thorough and, if possible, the exhaustive examination of the area here roughly defined. This would have been comparatively simple if the time at which he would start had been known. But it was not known beforehand, and was not to be disclosed by the Admiralty until a very few hours before the operations began. The principle involved is thus explained by a correspondent of the *Times*. "Let x be the number of miles at which two ships must be placed apart so as to obtain a *maximum* range of observation and yet to make certain that nothing can pass between them without being observed. The lineal range of observation will then be $2x$, because each ship will see as far on one side as on the other, and if the two ships steam on parallel courses for y miles they will search a rectangle whose sides are respectively $2x$ and y, having at each end two adjacent semicircles whose centres are the extremities of the two courses and whose radius is $\frac{1}{2}x$. For a given speed y is constant, but x is a variable dependent in all cases on the weather. If the courses are not parallel the area is rendered less regular, but the principle is unaffected. By the skilful juxtaposition of as many of such areas as there are ships to be employed it is theoretically possible to search exhaustively an area of sea approximately equal to their aggregate, the limits of this aggregate area being determined by the time allowed for the operation and the speed assigned to the ships employed, though if any of the courses cross each other the aggregate area will be to that extent reduced, with the corresponding advantage that the ground traversed by the superposed courses will be searched twice over at different times. This was in principle the method employed by Lord Walter Kerr in the manœuvres of 1895. Admiral Rawson developed and extended this method so as to adapt it to the known alternations of day and night without materially reducing the area to be searched. This is perhaps the practical limit of its flexibility. The alternations of fog and clear weather are incalculable beforehand, alike in occurrence and duration, and the only thing that can be done is to adjust the intervals between the ships to an average and not to an extreme range of vision. The whole area to be searched is the aggregate of the areas determined as above. . . . Each of these individual areas is represented by $2xy$, and if a line z be drawn across the courses from one boundary to the other the length of z will be independent of y, but will vary directly with x. It is possible so to adjust the courses for known alternations in the range of vision as to keep z constant in spite of local variations of x. But if x is to vary indefinitely according as fog or clear weather prevails, there must either be gaps in the field of observation, or its exterior limits must be curtailed. To organize a scheme of search which shall be equally effective in fog and clear weather is thus a geometrical impossibility. All that can be done is to adjust it to an average range of vision by night as well as by day and to leave the rest to chance."

Limitations of the method.

It is a characteristic of this method that the scheme of courses assigned to the several ships must be adjusted beforehand to the time of original dispersion. During the night the intervals between the ships must be reduced, and at the return of daylight they must again be enlarged; and in order to do this without materially reducing the length of z, as defined above, the courses must be arranged beforehand, so as to converge as night approaches and to diverge again as dawn returns. There is no insuperable difficulty in this, but it involves two conditions—a superposition of some of the courses in some regions of the area, and an exact adjustment of the whole scheme to the particular hour of the day or night assigned or assumed for the original dispersal. As this hour was not known beforehand, it was necessary to prepare several schemes, each adapted in all its details to different hours of dispersal. Of these schemes only one would be actually put in operation—that which was best adapted to the actual hour of dispersal, as determined at the last moment by the time fixed by the Admiralty for the operations to begin. The others would all be cancelled as soon as the actual time of dispersal was fixed. Seven such schemes were prepared, and each fully plotted down on a separate chart, and accurate tracings of each were supplied to each ship taking part in the operation. As soon as the time of dispersal was known, six of these were cancelled and the seventh alone became operative as the order of sailing of the fleet from the point at which the dispersal was to take place. The courses were arranged in accordance with the principles enunciated above. Adjacent courses were assigned to the battleships, and the courses of all the ships engaged were finally made to converge at a point distant by at least 100 miles from any point which the B fleet could have reached at the time of concentration, assuming it to have left Milford not less than twelve hours after A left Belfast and to have maintained a speed of twelve knots throughout. This point was situated not far from Porcupine Bank, and here, if all went well, the concentration would take place during the second night after the A fleet had left Belfast. By that time the greater part of an area of which the extreme northern apex was approximately in lat. 59° 40′ N., long. 8° 20′ W., while its extreme western apex was in lat. 52° N., long. 17° 30′ W., or thereabouts—its southern boundary being about lat. 52° N., and its western boundary a more or less regular curve running round from the northern to the western apex—would have been exhaustively searched, while nearly the whole force of the A fleet would be gradually concentrated at a convenient point with time in hand for the prosecution of a further search to the southward and westward, supposing the convoy not to have been discovered before the concentration was effected—as would actually have been the case, because, as a matter of fact, the position of its rendezvous lay outside the area of search.

Influence of the weather on the scheme.

It is an obvious criticism of this scheme of search that it was liable to dislocation should fog occur in any portion of the area under examination, and to complete frustration should fog prevail over the whole area. But this criticism applies to any scheme of search adapted to the examination of an extensive area within a limited time. The value of x would in ordinary weather be, perhaps, ten miles by day and five miles by night. In thick weather it could hardly be more than one mile, and might be little more than zero whether by day or night. There is no known method of search, perhaps there is no possible method so flexible as to be susceptible of adaptation from moment to moment to these extreme values of x. Fogs are prevalent enough in the eastern Atlantic, as Admiral Rawson found to his cost. But their prevalence is not so probable as the occurrence of clear weather during any given period of forty-eight hours, which was as much as Admiral Rawson required, or, indeed, could safely allow for the prosecution of his search for the convoy. He was, therefore, perfectly justified in basing his dispositions on the larger probability, though well knowing all the time that they might be completely frustrated—as, indeed, they actually were—by the prevalence of fog during the critical period. A more serious criticism might be that the area of search was injudiciously chosen. In sporting phrase, it may be said that Admiral Rawson drew the wrong cover, and that even in the best of weather he would have found it blank; but it must be added that he had so arranged matters that even in ordinary weather he would have had time to draw the right cover before the fox had quitted it. There was at the outset no certain criterion to show which was the right cover and which the wrong. It has been pointed out above that such a criterion might have been obtained, though at some sacrifice of time. On the other hand, it must be admitted that the time spent in obtaining it might have been quite as great as that required for completing the whole scheme of search in ordinary conditions of weather. Be this as it may, the fact remains that Admiral Rawson failed to find the convoy. How far his failure was due to faulty dispositions, and how far to untoward conditions of weather, which though not unforeseen could not be provided against, is a question by no means so easy to answer as might at first sight appear.

Proceedings of A fleet at commencement of hostilities.

The time fixed for the beginning of the operations was 10 a.m. on July 29. This was made known by the Admiralty in the course of the day before, and exactly at the time appointed Admiral Rawson's whole fleet crossed the line drawn from Black Head to Orlock Point, which marked the defended limits of his anchorage in Belfast Lough. Two divisions of his torpedo boats had been previously despatched in accordance with the regulations to Waterford and Kingstown, but the Kingstown division was recalled in time to accompany the fleet on its departure, together with the division stationed at Belfast. It was known to Admiral Rawson that Admiral Domvile had been instructed by the Admiralty to employ his destroyers primarily for the attack and destruction of his adversary's torpedo boats, and it was largely on this instruction that Admiral Rawson based his conjecture that the convoy would probably be found in such a position as would compel the B fleet to reach it by passing through the Irish Channel. But as one of B's destroyer stations was at Lamlash it seemed not impossible that the A fleet might be attacked by destroyers during its passage through the North Channel. For this reason the A fleet, with 16 torpedo boats in company, was disposed on its exit from Belfast in such a formation as would make it impossible for any destroyer to get within striking distance of a battleship without having previously been under fire for a longer time than would suffice to put it out of action. This formation is displayed in the annexed diagram, which is drawn to scale.† The leading division of cruisers was ordered to act independently in the event of its being attacked; but though it formed no essential part of the scheme of defence, it was so placed as to prevent any attack on the battle squadron by destroyers coming from right ahead. The most probable point from which an attack might be expected was indicated by the relative positions of Lamlash and the North Channel

† The writer desires to acknowledge his obligation to Admiral Rawson for permission to reproduce this diagram.

as the starboard beam; and here the defence was strongest, the outer guard of torpedo boats being disposed in pairs, because two torpedo boats could put a destroyer out of action in five minutes at a distance of a quarter of a mile, whereas a destroyer could only put both torpedo boats out of action at the same distance in ten minutes. The striking distance of a destroyer, with its torpedo, was 500 yards, and the range of fire against destroyers for all vessels other than torpedo boats was 1,000 yards. If circles were drawn on the diagram to represent these respective ranges of fire it would be found that no destroyer could get within striking distance of a battleship without having been under fire from the battleships themselves and from other vessels for more than the period of three minutes required to put it out of action, and in most cases for four and a half minutes. This disposition was not tested in practice, as no attack was made. But it served at any rate to familiarize the ships of the A fleet with a formation which must have been adopted in principle, if A, having succeeded in finding the convoy, had had to run the gauntlet of B's destroyers in the attempt to bring it back to Belfast.

Passage of the North Channel.

However, B's destroyers were not inactive, though they made no attack on A. One was observed as A left Belfast Lough, but she did not follow further than to enable her to ascertain whether A was making to the northward or to the southward, and then left as though to report proceedings at B's nearest signal station, afterwards returning at high speed and steaming past A in the direction of Lamlash. As the North Channel was approached six others were observed, but they made no attempt to attack; and before the North Channel was cleared the weather became so thick for an hour or more that the whole of the A fleet passed entirely out of the range of their observation. This was the last intelligence that Admiral Domvile received of the A fleet, until after his own return to Milford he learnt from some of his cruisers, left to scout off Cape Clear, that it had chased them on the afternoon of August 3. It appears that during the thick weather the observing destroyers entirely lost touch of the A fleet, and were not even able to ascertain whether it had kept its course towards the open or had doubled back into the Irish Channel. So completely was this the case that before he left Milford at 5 a.m. on the morning of July 30, Admiral Domvile had received circumstantial reports from the commanding officer of one of his own destroyers that the A fleet, or a considerable portion of it, had been observed during the night steering southward in the neighbourhood of Holyhead. It would seem that a flotilla of A's torpedo boats was mistaken by the officer in question for the main body of the A fleet, and reported as such to headquarters. There were also

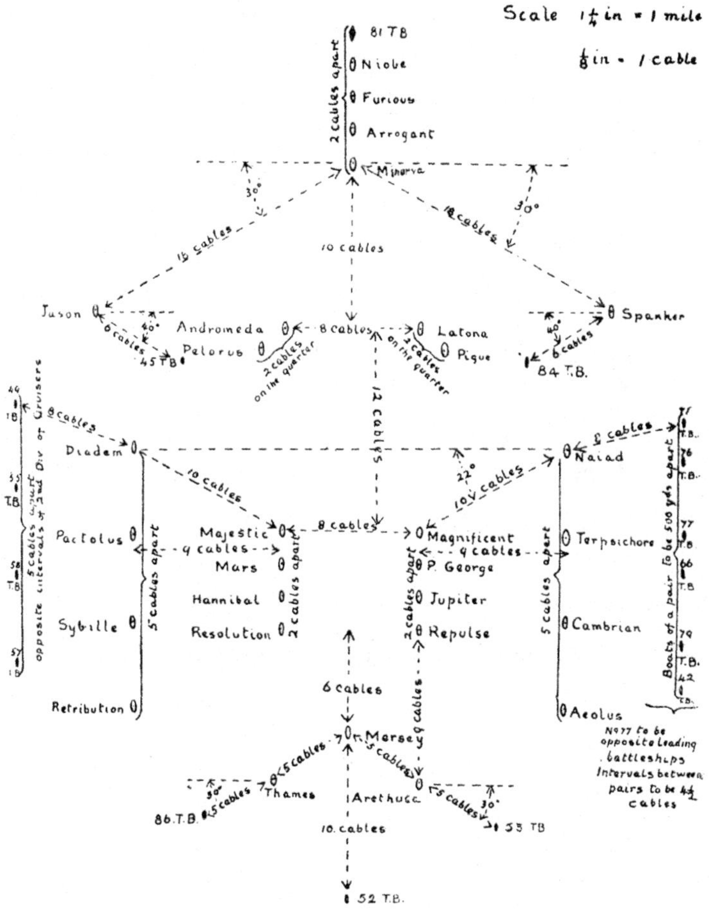

Scale 1¼ in = 1 mile
⅛ in = 1 cable
81 TB
Niobe
Furious
Arrogant
Minerva
2 cables apart
30°
30°
15 cables
10 cables
12 cables
Jason
Andromeda
8 cables
Latona
Spanker
40°
40°
6 cables
6 cables
Pelorus
45 TB
2 cables on the quarter
2 cables on the quarter
Pique
84 T.B.
12 cables
8 cables
8 cables
Diadem
Naiad
22°
10 cables
10½ cables
Majestic
8 cables
Magnificent
Pactolus
9 cables
9 cables
Terpsichore
Mars
P. George
Hannibal
Jupiter
Sybille
Resolution
Repulse
Cambrian
Retribution
Aeolus
2 cables apart
2 cables apart
5 cables apart
5 cables apart
5 cables apart opposite intervals of 2nd Div of Cruisers
Boats of a pair to be 500 yds apart
6 cables
9 cables
Mersey
5 cables
5 cables
Thames
Arethusa
30°
30°
5 cables
5 cables
86 T.B.
53 TB
10 cables
52 T.B.
No 77 to be opposite leading battleships. Intervals between pairs to be 4½ cables

other reports scarcely less circumstantial, about some very singular performances with rockets and other fireworks, supposed to have been carried on by the A fleet during the night of July 31 in the neighbourhood of the Isle of Man. Erroneous and misleading reports of this kind are no novelty in manœuvres; on the contrary, they are rather the rule than the exception. They may be paralleled by what occurred at a later date in the A fleet, when the ARROGANT reported that shortly after dawn on August 3 she had sighted a portion of the B fleet, and that though it was too dark to distinguish the number of ships present, her captain believed that there were three or four battleships and about six cruisers in company. As a matter of fact there were no battleships at all and not more than four cruisers. Occurrences such as these, which are only too frequent in manœuvres, suggest the reflection that inaccurate or imperfect observation at sea is often appreciably worse than no observation at all. The ARROGANT could not push her reconnaissance home because she was chased away by a superior force detached by the enemy for the purpose; and yet unless such a reconnaissance can be pushed home so as to leave no room for doubt as to the accuracy of the information obtained, an Admiral cannot but be gravely embarrassed by the receipt of intelligence which he can neither neglect nor trust. In order to render scouting operations really effective and informing it would seem that cruisers must work in company—possibly in groups of not less than three, as they were worked with signal success by Admiral Domvile in the manœuvres of 1897. A single cruiser seeking to observe an enemy's fleet can generally be chased away without being able to push its reconnaissance home. But if three cruisers are working in concert, one at least might expect to be able to effect its purpose without running any serious risk of capture. The subject of scouting, with some very significant illustrations of miscarriages and mistakes similar to those here noted, was fully and very instructively discussed in the official report on the manœuvres of 1891 (see *Naval Annual* for 1892, pp. 42—49). It is not altogether satisfactory to note that so little progress would seem to have since been made in overcoming the difficulties there shown to be characteristic of the operation.

The use of signal stations

If intelligence obtained from scouting cruisers is often precarious and indefinite, that obtained from direct observation by signal stations on shore is, as might be expected, nearly always inaccurate and misleading. In an experience of manœuvres extending from 1887 to 1899 the writer can recollect no single instance in which intelligence derived by an Admiral at sea from direct observations made by signal stations on shore was of any material service, and innumerable instances in which it was inaccurate, misleading and embarrassing. From the nature of the case a signal station being stationary cannot push its reconnaissance home. It can only see what comes fortuitously within its range of vision—and an enemy who knows his business will always take care that that is little enough—while its hastily mobilised staff is rarely trained either to observe accurately or to report what it sees with precision. The experience of other navies in this respect appears to be much the same as our own. "*Le service de nos sémaphores a été comme toujours exécrable*," was the comment of a French critic on the French manœuvres of 1897 quoted in the *Naval Annual* for 1898 (p. 140); and though Italian experience of the same year was a little less unfavourable, yet it appears that such success as the Italian signal stations achieved was more conspicuous in the transmission of intelligence than in its collection. This distinction is vital. The true function of a signal station in time of war will be to transmit authentic intelligence and instructions from headquarters to a fleet at sea. For this purpose a Marconi installation established out of range inland, and not visible from the sea, would be far more effective than any signal station placed on the coast. For the purpose of direct observation the value of any signal station is, as experience shows, quite insignificant; but for purposes of transmission its value can hardly be overrated. If Marconi installations were established in suitable places we might well be content to disestablish other signal stations, or to leave them only for the purpose of tempting hostile cruisers to waste ammunition in destroying them. The records of foreign manœuvres seem to show that certain Powers would not be proof against such a temptation. But it has more than once been pointed out in the *Naval Annual* that the calculus has yet to be invented which can express the objects of naval warfare in terms of the destruction of signal stations." From this point of view the notion which has lately been propounded on high authority, that it will be necessary in time of war to protect all signal stations by land defences of some sort against fugitive raids from the sea, seems to be quite preposterous.

The search begun

The fog which opportunely prevailed in the North Channel enabled Admiral Rawson to withdraw his main force from the observation of B's destroyers, and to make other dispositions conducive to the success of his search. Towards evening the division of cruisers which had led the fleet in the disposition described above was detached to carry out its share of the scheme of search, and subsequently the whole of the rest of the fleet—with the exception of three cruisers, the Thames, the Mersey and the Arethusa, detached to obtain intelligence at certain signal stations—was dispersed for the same purpose from a rendezvous situated in lat. 55° 52′ N., long. 8° 10′ W. The battle squadron was dispersed with the rest, but the ships were assigned adjacent courses so that in clear weather they would not have been beyond signalling distance from each other. The immediate destination of the squadron was a rendezvous situate in the neighbourhood of Porcupine Bank in lat. 53° 15′ N., long. 13° 43′ W., where, if all went well, the battleships would concentrate shortly after midnight on the morning of July 31, and be joined by the cruisers, to which the shorter courses had been assigned, the bulk of the remainder joining at another rendezvous in lat. 53° 58′ N., long. 13° 23′ W. in the course of the ensuing day. As it is impossible to detail the proceedings of a fleet thus dispersed, the proceedings of Admiral Rawson's flagship will henceforth be followed in the main. For a few hours after the dispersal the weather continued fairly favourable, but early in the morning of July 30 the flagship was enveloped in a thick fog, which lasted almost continuously throughout the day and following night, lifting only for a couple of hours towards evening. In the course of the morning the THAMES and MERSEY rejoined the flag from their respective signal stations, but brought no intelligence of the movements of the B fleet. Had it been otherwise, or had the ARETHUSA, which rejoined from Black Sod Bay on the morning of July 31, brought definite intelligence that the B fleet was still at Milford on the night of July 29, the whole course of the proceedings might have been altered. But the ARETHUSA only brought a rumour to that effect, and Admiral Rawson either decided not to rely on it, or must have failed to perceive its significance, since if it was certain that the B fleet was still at Milford at midnight on July 29, it was almost equally certain, for the reasons already given, that the rendezvous could not be so placed as to require Admiral Domvile to pass through the Irish Channel in order to reach it by the shortest course. Be this as it may, the vague information brought by the ARETHUSA had no effect on Admiral Rawson's dispositions. When the fog lifted towards evening on July 30, the MARS was discovered in her appointed station about eight miles on the starboard beam of the flagship, and she reported that the HANNIBAL, next beyond her, was also in her place. This seemed to show that in spite of the fog the ships had been able to keep to their appointed courses, and this inference was confirmed when the rendezvous was reached and all the battleships were found there together with all but one of the cruisers expected there. The missing cruiser was the TERPSICHORE, which had been given a somewhat irregular course and ordered to rejoin the flagship at an earlier rendezvous at 8 p.m. on July 30. She reached the rendezvous at the appointed time, but owing to the fog she was unable to find the flagship and did not rejoin until the morning of July 31, being followed shortly afterwards by the SYBILLE, which had completed her course with only a slight departure from the appointed track, but rejoined with engines temporarily disabled for a speed of more than eight knots. The ships as they rejoined each and all reported that they had experienced more or less fog, and had seen nothing whatever of the convoy. But it was impossible to infer from this that the convoy had not been located within their area of search, since just as the TERPSICHORE had missed the flagship, though punctual to her appointed rendezvous, so in the weather that prevailed any one of the ships might have passed close to the convoy without seeing it.

Its results on July 31.

Thus, on the morning of July 31 the situation was as follows:— Eight battleships and several cruisers, shortly to be followed by others, had in spite of the fog completed their appointed courses, and had concentrated at a preconcerted rendezvous at a time appointed beforehand. The course of the flagship had been far away from the land, but communication with the shore had been maintained at three separate points, though, unfortunately, the only intelligence obtained by this means was too vague to be trusted. This was a fine exercise in seamanship, a performance eminently to the credit of the Navy, but it had so far produced no results whatever conducive to the object in view—that is, to the discovery of the convoy. Forty-eight hours after he had left Belfast Admiral Rawson found himself in the neighbourhood of Porcupine Bank, still enveloped in fog and just as ignorant of the whereabouts of the convoy as he was when he started. Several of his cruisers were still engaged in the search for it, but in the weather that prevailed their prospect of finding it seemed to be no greater than that of the ships in company which had already completed their search. Two courses were now open to him—either to wait in the neighbourhood of Porcupine Bank until the ships to which the longer courses had been assigned had all rejoined, or to utilise the time which must elapse before they could rejoin in searching over again so much of the area which had been imperfectly searched in the fog as it was possible to search in the time still available. He chose the latter course. It was, as the event showed, a fatal choice, which finally extinguished any chance he might otherwise have had of finding the convoy before the B fleet could reach it. But with the *data* before him, it was not, perhaps, an injudicious choice—not such a choice as Villeneuve made when, having the opportunity of crushing Cornwallis off Brest and joining hands with Ganteaume, he turned away and took refuge at Cadiz. It was rather such a choice as Nelson made when he judged wrongly that Villeneuve had gone to Egypt, and decided to go thither in pursuit. The only *datum* that might have decided the matter in the opposite sense—certain information that the B fleet was still in Milford at midnight on July 29—was wanting. There was only a rumour to that effect brought by the ARETHUSA from Black Sod Bay; but rumours derived from signal stations are the sort of intelligence that no Admiral cares to trust. Moreover, Admiral Rawson had decided to search a given area, and circumstances beyond his control had largely frustrated his search. There was nothing as yet to show that his judgment was at fault in thinking that the convoy was more likely to be found within this area than outside. It thus seemed natural and logical to employ the time that remained in searching over again so much of the original area as still lay within his reach. To say that he was wrong is merely to be wise after the event. Nevertheless, it is impossible not to regret that when the alternatives were so evenly balanced, the intelligence brought by the ARETHUSA, vague and indecisive as it was, did not incline the search in the opposite direction.

Search to the northward renewed

At any rate, Admiral Rawson, having decided to search over again a portion of the area to the northward, made his dispositions accordingly during the forenoon of July 31. One division of cruisers was ordered to sweep along the exterior limit of the original area of search, from the latitude of Porcupine Bank to the latitude of Rockall, while the battleships took a parallel course nearer to the land in the same general direction, a rendezvous being fixed in the neighbourhood of the last mentioned locality. At the same time the MINERVA was again despatched to Black Sod Bay for intelligence, with orders to rejoin at the new rendezvous, and arrangements were made for keeping touch with the cruisers which had not yet completed their original search. The weather cleared for several hours on July 31, but before Rockall Bank was reached it had become as thick as ever. The renewed search for the convoy was fruitless, as of course it must have been, since the convoy was not located in any part of the area to be examined, and by noon on August 1 it became manifest that Admiral Rawson had completely failed to accomplish his object. The battleships were then at a rendezvous on Rockall Bank, completely enveloped in fog. More of the cruisers had rejoined either there or previously, and the majority had again been detached either to prosecute a fresh search, returning to a new rendezvous, or to replenish their bunkers, some in Black Sod Bay and some in Killery Bay, from colliers there stationed for the purpose—for three days' steaming, mostly at three-fifths power, had in many cases reduced their coal supply to a disquieting state of depletion. As the principal object of the manœuvres was "to obtain information as to the most advantageous method of employing a considerable body of cruisers in conjunction with a fleet," this fact is of quite capital significance. Of a considerable body of cruisers attached to a fleet some must be employed in maintaining communications with the shore or with the base. If the Admiral is to retain his freedom of action and the mobility of his fighting force, others must often be left at rendezvous which he finds it necessary to quit before detached cruisers ordered to concentrate there have arrived. If the latter run short of coal, as they are certain to do after a few days' steaming at high speed, they are no longer available until their bunkers have been replenished. It may thus easily happen that even a considerable body of cruisers may melt away to a very insignificant number before the Admiral, whose eyes and ears they are, has been many days at sea. As a correspondent of the *Times* remarked, writing on July 31, "Speed after all is a function of coal supply, and coal supply, even in the best of cases, is very apt to be exhausted when high speeds are maintained. Of the tactical value of speed—that is, of its importance in action—we are not likely to obtain any information during the present manœuvres; but it is clear from what we have already learnt that its strategic value is very apt to be overrated. Continuous steaming at three-fifths power very quickly exhausts the most capacious of bunkers, and perhaps the ship has not yet been designed which combines continuous rapidity of movement with a wide radius of action." At a later stage of the proceedings the difficulties here indicated were still more acutely felt. On August 3, when Admiral Rawson first sighted a portion of the B fleet, he was, owing to the causes above indicated, lamentably short of cruisers; and on August 5 the same correspondent wrote as follows:—"We have been at sea for a week, and during this period most of the ships have had to steam at high speed for the greater part of the time. What is the result? Many of the second-class cruisers have had to go to Killery Bay for coal, and while they were coaling they were practically out of action. On Thursday morning (August 3) the return made by the REPULSE of her coal consumption began to look disquieting, the NIOBE reported that she was running short of distilled water for her boilers, to which salt water is rapidly fatal, and the ARROGANT, that she had reached a point in her coal supply at which the coal which remained was so disposed in her bunkers that she could not supply it to her furnaces at a rate capable of maintaining a speed of more than fifteen knots." Thus, continuous high speed is much less compatible with an extended radius of action than is commonly supposed; and, perhaps, the chief lesson the manœuvres of 1899 have taught is, that the fundamental problem of cruiser tactics is to find the most advantageous mode of adapting all dispositions to the conditions imposed by these two incompatible factors.

Failure of the search.

Having reached Rockall Bank on August 1, Admiral Rawson there waited until the MINERVA returned from Black Sod Bay. She brought information, transmitted from Queenstown, that the B fleet left Milford at 5 a.m. on July 30, the battle squadron, with the bulk of the cruisers, steering for the south coast of Ireland, though it was reported, erroneously as subsequently appeared, that a detachment of cruisers seemed to be making for the Irish Channel. This satisfied Admiral Rawson that the convoy was so placed that the B fleet could reach it most rapidly by going round the south coast of Ireland, and, therefore, probably not so far northward as Rockall. He accordingly decided to return again to the southward, though with little expectation of finding the convoy, before B could reach it, unless, perchance, the cruisers not yet returned from their original search had already discovered and seized it. We have seen that if the convoy was placed anywhere within a distance of 600 miles from Milford, B's cruisers could reach it in forty hours at a speed of 15 knots; that is, starting at 5 a.m. on July 30 from Milford they would reach it at that speed about 9 p.m. on July 31, or earlier if its distance from Milford was less than 600 miles. It was thus certain on August 1 that unless the three cruisers not yet returned from their original search had sighted and seized the convoy before 9 p.m. on July 31, there remained no chance for A of capturing it. It may here be mentioned that the tardy receipt of the critical intelligence of B's departure from Milford, and of the course he had taken on leaving, was entirely due to the prevalence of fog on July 30. The MERSEY rejoined the flag early on the morning of that day. Before she could be given fresh instructions the fog became so thick that all signalling was stopped even while an attempt was being made to order her to go to Black Sod Bay for intelligence and return with all despatch to a rendezvous near Porcupine Bank. It is to be regretted, perhaps, that no more determined effort was made to give the MERSEY this important signal. She might have been ordered to close or even to send a boat. But to find a ship, even at a distance of half-a-mile, in fog so thick that the bows of a ship could hardly be seen from the bridge, is neither easy nor free from hazard. The whole incident illustrates the truth of Nelson's saying that five minutes may make all the difference between victory and defeat, and it may further be remarked that had the two ships been fitted with Marconi apparatus the whole course and issue of the manœuvres might have been different.

Leaving Rockall Bank on the afternoon of August 1, Admiral Rawson made again for a more southerly rendezvous, where he expected to find the cruisers, which, owing to his change of plan, had not yet rejoined the flag after completing their original search. These were the DIADEM, the RETRIBUTION, and the PACTOLUS, to which had been assigned the outermost and most southerly tracks of the whole scheme. If they had failed, like their consorts, the game was up. They had failed. The DIADEM was found at the rendezvous on the morning of August 2, but the PACTOLUS had already left for Killery Bay to coal. Of the RETRIBUTION nothing was known, except that on the afternoon of July 31 she had signalled to the PACTOLUS, her nearest consort in the scheme of search, that she had an engine disabled, but still hoped not to be behindhand in reaching the rendezvous for which she was making. This hope was disappointed, and with it vanished the last vestige of probability that the convoy might have been discovered. Even if the RETRIBUTION had seen it—which she did not—she would have been powerless to capture it without the assistance of the DIADEM, and the DIADEM had already returned to the rendezvous. The scheme of search was all but completed in spite of the fog, and this was really a very noteworthy achievement; but its results were *nil*, and owing to the fog the negative character of its results were ascertained too late for Admiral Rawson to organize a fresh scheme with any chance of success. On the morning of August 2 he recognized that the game was up, and made a general signal to that effect. Curiously enough the fog, which had played such havoc with his plans, lifted completely soon after he left Rockall Bank on the evening of August 1, and never troubled him again.

Admiral Rawson's further plans.

The sequel may be told in very few words. On August 2, Admiral Rawson issued the following signal:—"Unless B fleet has had a fog and missed the convoy the game is up, and they are back this evening at Milford. My proper move would therefore be to return quietly to Belfast, round the North of Ireland, so as not to give the destroyers a chance. But as I am certain the Admiralty wishes to gain some information as to torpedo boats and destroyers, I have ordered the whole of our torpedo boats to meet the squadron off Waterford, and I shall, with as many cruisers as I can collect, proceed up the Irish Channel to Belfast in the same formation as we left in. If B fleet is going back through the North Channel we shall meet it on the way. If they come out to meet me I shall let cruisers and torpedo boats at them and try and claim them." Such was the programme for the remainder of the operations. But its execution had not proceeded far, when, on the morning of August 3, the fleet being at that time at a rendezvous off Cape Clear, the ARROGANT reported that she had sighted a portion of the enemy's fleet at dawn in lat. 50° 58′ N. and long. 9 38′ W., and believed that it included three or four battleships The ensuing day and night were spent in a fruitless search for these phantom battleships. The search resulted only in the discovery of four of the enemy's cruisers, which speedily made off, and in the afternoon of August 4 the NIOBE, which had been sent to Berehaven for intelligence, returned with the information that the B fleet, with the convoy in company, had reached Milford, and that the Admiralty had, in consequence, declared hostilities to be at an end.

Proceedings of B fleet

As the two fleets never came into serious strategic contact, it has seemed expedient to keep the narrative of their respective proceedings entirely distinct. B's proceedings must, therefore, now be considered. It is obvious at once that the task imposed on Admiral Domvile was by far the easier of the two. He knew exactly where the convoy was, and if he could seize it before A appeared on the scene he had nothing to fear from the attack of the latter—unless, indeed, A could find an opportunity of "letting cruisers and torpedo boats at him," and could thereby so reduce the strength of his battle squadron as to establish his own superiority to it. This was not impossible under the rules; and we have Admiral Rawson's own authority, cited above, for saying that had an opportunity offered he would have attempted something of the kind. But no such opportunity presented itself, and as matters turned out Admiral Domvile never saw the A fleet, and knew nothing whatever about its proceedings after it had passed the North Channel until some time after his own return to Milford. He left Milford at 5 a.m. on July 30. He had four first-class cruisers in all, one of which—the *Galatea*—was detached to form an escort for the convoy. The remainder—the *Europa*, *Argonaut*, and *St. George*, accompanied by the *Vindictive*, *Gladiator*, *Diana*, and *Sappho*—were sent ahead at a speed of seventeen knots, on a direct course for the rendezvous of the convoy. The *Europa* was fitted with the Marconi apparatus, as were the *Juno* and the *Alexandra* (Admiral Domvile's flagship), so that by keeping the *Juno* in company, and placing her as a linking ship at extreme signalling distance by wireless telegraphy ahead of his fleet, Admiral Domvile might hope to be in communication with the *Europa* at a distance of at least sixty miles. The actual distance at which communication was first established appears to have been about eighty-five miles. Admiral Domvile, with his battle squadron and his remaining cruisers disposed as look-outs, proceeded at slower speed on a course which would take him clear of the south coast of Ireland, placing the *Juno* at a distance of twenty miles on his landward flank to keep him warned, by means of the Marconi apparatus, of any appearance of the enemy. His first rendezvous was in lat. 50° 35′ N., long. 9° 30′ W., and this was reached between 9 and 10 p.m. on July 30. Thence he steamed due west for 197 miles to a second rendezvous in lat. 50° 35′ N., long. 14° 40′ W., reaching it shortly after 3 p.m. on July 31. At this rendezvous, which was 176 miles from the convoy, course was again altered direct for the convoy, the *Juno* having been placed ahead so as to establish communication with the *Europa*. The advanced cruisers were expected to reach the convoy about noon on July 31, the distance being some 510 miles, and the speed seventeen knots. In the event of their reaching and finding it, the *Europa* was ordered to steam back at full speed along the line of advance of the fleet so as to get in touch with the latter as soon as possible. But they were somewhat delayed by fog, and the convoy, with its escort of cruisers, did not actually start on its homeward voyage until after 5 p.m. As the fleet steamed towards the convoy a message came out of space from the *Juno*, between 7 and 8 p.m. on July 31. It ran: "Communicated with *Europa* about sixty miles off convoy rendezvous. She was there with convoy, and has now returned to squadron. Convoy following at about nine knots. No enemy sighted by her or by us." Forty minutes later a second message was received, transmitted from the *Europa* by the *Juno*: "*Europa* alone. *Vindictive* follows two hours later. Afterwards remainder of division with convoy." Thus the convoy, with its escort and the main body of the B fleet, were approaching each other from opposite directions at a speed of some twenty knots. Before midnight they had effected a junction, and Admiral Domvile, now having the convoy in custody of a fleet superior to his adversary, had nothing further to do but to take it safely back to Milford. A glance at the chart will show that the fields of operation covered by the two opposing fleets lay wholly outside each other, and it will be seen from the narrative given above of the proceedings of the A fleet that at the time when B's cruisers were seizing the convoy the whole of the A fleet were steaming away to the northward in the direction of Rockall, with the exception of the three cruisers which had not at that time completed their original search on courses well to the northward of the rendezvous assigned to the convoy. Only one contingency could in the circumstances have given Admiral Rawson a slight chance of finding, and a still slighter chance of capturing the convoy. Had there been no fog on July 30 he might, on reaching his rendezvous near Porcupine Bank on the night of that day, have been able to satisfy himself from the reports of such of his cruisers as he found there that the convoy was not placed within the area originally selected for search. Then if without a moment's delay he had organized a fresh search to the southward and westward, he might perchance have lighted upon the convoy before B's cruisers had reached it—about 5 p.m. on July 31 —or, finding it later in their custody, he might have been able to drive them off and seize the convoy before they could recover touch with Admiral Domvile. Either chance was a slight one, the latter a very slight one—for it must be borne in mind that the *Europa* was at all times in touch with the B fleet at a distance of eighty or ninety miles—and it must be acknowledged that Admiral Rawson's failure to find the convoy, though rendered certain by the fog, must in any case have been rendered probable by his prepossession in favour of the extreme northern region of the area of possible search.

Wireless telegraphy.

"The Marconi system of wireless telegraphy was tried in the Naval Manœuvres of 1899, and proved very successful so long as only one ship was signalling. Signals were taken in at a distance of sixty miles." Such is the judgment of the Admiralty, as recorded in the First Lord's Statement explanatory of the Naval Estimates. The qualification, "so long as only one ship was signalling," is significant. Wireless telegraphy has been shown to be trustworthy for the transmission of signals between two ships sixty miles apart, but it is evident that the Admiralty are not yet satisfied that the system is applicable to the ordinary signalling of a squadron, where from the nature of the case many signals from different ships must often be made simultaneously. On the other hand, the limitation indicated applies to all signals available during a fog—that is, to all audible signals. Visible signals made simultaneously do not interfere with each other. Audible signals do. But as a whole fleet can be addressed in a fog by audible signals made from the flagship, the ships answering successively in the order of their fleet numbers, so the Marconi signals can be made available in like manner for many manœuvring purposes both in fog and clear weather. There is, moreover, another point to be considered. At present a Marconi signal made by any ship can apparently be taken in by any other ship, provided with the necessary apparatus, within a circumference of sixty miles. Thus, a cruiser giving important information to her own flag in one direction may quite unconsciously give the same information to an enemy in another direction. Ciphers may be used, of course, but unless a cipher is frequently changed it will very soon be deciphered by an enemy who knows his business, and to change ciphers daily on a system preconcerted beforehand is not very easy in practice, and is cumbrous and dilatory in any case. It may hereafter be found possible to direct the vibrations which transmit the message only in a given direction, as the beam of a searchlight is directed. But this at once limits the power of signalling to two ships each of which knows accurately the position of the other. It is true that the beam of vibration—if the expression may be permitted—may be slowly swept through an arc of the horizon, as the beam of a searchlight can be swept, in the hope of picking up the friendly ship with which it is desired to communicate. But if the position of the friendly ship is not known, the beam may have to be swept through a large arc of the horizon, or even through the whole circumference, with the chance of missing the friendly ship after all, and with no certainty of not betraying the presence of the transmitting ship to enemies on the look out for her. This has often happened in manœuvres through the incautious or injudicious use of a searchlight, and would be quite as likely to happen through the use of the Marconi system in the manner indicated. There is another method of wireless telegraphy known as the "Syntonic" system. In this system, unless the receiver is attuned to the transmitter beforehand, no effect is produced on it. This system does not appear as yet to have passed beyond the laboratory stage of experiment; but its principle, if susceptible of practical development, seems to promise far-reaching results.

Nevertheless, perhaps the most striking result of the manœuvres, as they worked out, was the proof they afforded of the great value and importance, within certain limits, of the method of wireless telegraphy. But curiously enough the strategic result of its employment on this occasion was *nil*. It might have been far otherwise if the two fleets had come into contact, but as they did not, all that happened was that Admiral Domvile got in touch with the convoy some six hours earlier than he would have done if no wireless telegraphy had been employed. But this was merely an accident of the form the operations took, and no one can doubt that the power of communicating in all weathers with a ship at a distance of sixty miles is a most advantageous extension of the methods previously available for signalling at sea. It still remains to be seen how far the apparatus is liable to be affected by electrical disturbances of the atmosphere, and how far it is sensitive to the concussion of heavy artillery.

Further dispositions and operations of the two fleets.

With the seizure of the convoy by Admiral Domvile's cruisers the manœuvres were practically at an end. There was nothing, and, from the nature of Admiral Rawson's dispositions as affected by the fog, there could have been nothing, to prevent the safe return of the convoy to Milford. Admiral Domvile had, however, made provision for a variety of contingencies not unlikely to arise. In the event of any accident, such as fog, or the menace of the enemy preventing the junction of the cruisers escorting the convoy with the main body of the B fleet, the escorting cruisers were instructed beforehand to take the convoy to a third rendezvous in lat. 49° 40′ N., long. 11° 3′ W. This rendezvous was at a distance of 149 miles from Admiral Domvile's second rendezvous, and 325 miles from the original position of the convoy, and so placed that a straight line joining it with the latter position would pass through the second rendezvous. Here the convoy with its escort, now consisting of four first-class cruisers, and therefore unassailable by any cruiser force that A could bring against it, was to await orders in a comparatively secure position which was little likely to be reached by A at all, and not at all likely to be reached by A's battle squadron without encountering the superior battle squadron of B. But nothing occurred to bring this third rendezvous into use, and the convoy was taken back to Milford, according to the original plan, without molestation or miscarriage of any kind. In company with the fleet that had rescued it it anchored in Dale Roads about 11 a.m. on August 3.

Contact established between them

It was not until early on this same morning that the A fleet first came in touch with a squadron of four of B's cruisers—the *Europa*, *Argonaut*, *Gladiator* and *Vindictive*—which Admiral Domvile had detached on the morning of August 1, and ordered to cruise off Cape Clear. This was the squadron which was erroneously supposed and reported by the ARROGANT to consist of some three or four battleships and about six cruisers. It succeeded in capturing two of A's cruisers, and it served to occupy Admiral Rawson's attention for about twenty-four hours, at the end of which time he learnt that the convoy was safe in Milford, and that hostilities had been ordered by the Admiralty to cease. There is little or nothing to be learnt from the sole contact established between the two fleets, except the lesson already insisted upon, that inefficient scouting is little better in any respect, and in some respects worse than no scouting at all. This is no reflection on the captain of the ARROGANT, who did all he could and reported the presence of the enemy, though he was unable to ascertain his numbers. Having no superiority of speed over the enemy's cruisers sighted, he could not push his reconnaissance home, and in the dim and uncertain light of an early dawn he fancied he saw more ships than were there. Perhaps the best course for an officer to take in the circumstances would be to report nothing but what he is quite certain he has seen. Not to have seen any battleships would not of course have proved that no battleships were about—indeed, as one of the ships observed was the *Europa*, she might have been in touch with her flagship at a distance of sixty miles. But to report his belief that he saw some battleships was almost to compel Admiral Rawson to go in search of them and to engage him in what proved to be nothing but a wild goose chase. Nelson was no better and no worse off than this when, having lost touch with Villeneuve, he went to look for him in the Levant, instead of following him through the Straits. In other words, the information obtained by scouting if inexact or in any degree conjectural may be just as embarrassing as the total absence of information due to the lack of appliances for scouting. It is not easy to resist the inference that the value and importance of scouting have possibly been somewhat exaggerated of late. Either we must greatly improve our methods, or we must acknowledge that the game is not often worth the candle. If two fleets want to find each other they will do so sooner or later, and generally sooner rather than later, without much assistance from their scouts. If one of them merely wants to evade the other, the best way to do it is not to put to sea at all. Admiral Rawson never had any reason for wanting to find the B fleet. All he wanted to find was the convoy, and could he have done so in circumstances which enabled him to seize it, his primary object would thenceforth have been not to find the B fleet, but, if it might be, to evade it. The information he received from the ARROGANT could only lead him to suppose that Admiral Domvile had for some unaccountable reason divided his battle squadron. Either this was a trap, or it was a strategic blunder not to be lightly attributed to a commander of Admiral Domvile's experience and proved capacity. In the former alternative there was little to be gained and much to be lost by following up a clue of information so inconclusive and conjectural as that obtained from the ARROGANT; the latter is hardly worth considering. Truly, the art of effective scouting and of searching reconnaissance at sea would seem to be still in its infancy. It may even be doubted whether its further growth in the direction suggested by the rather misleading experience of peace exercises is altogether to be desired.

General reflections.

But making abstraction of all incidental and accidental conditions, perhaps the real reason why Admiral Rawson failed, and must have failed, in his enterprise was that he was set to do with an inferior fleet what only a superior fleet could have hoped to accomplish. The sole advantage he had was the start accorded him. This was nineteen hours. But as the convoy was some sixty miles nearer to Milford than it was to Belfast, this start was reduced by three and a half hours, on the assumption that A's cruisers would steam as B's did, that is, at a speed of 17 knots. At least three hours would be required to get the convoy clear away from its rendezvous before B's cruisers could appear on the scene, so that in the most favourable circumstances the clear margin of time which A would have for the prosecution of his search would have been rather less than twelve hours. Even so his discovery of the convoy would probably have been only the beginning of his troubles, because the narrower the margin between the time of his discovering the convoy and the arrival of B's cruisers at its abandoned rendezvous the more certain was he to be overtaken by B's superior fleet before he could reach Belfast. On this point it is worth while to quote the comments of a correspondent of the *Times*:—"A fleet that knows itself to be inferior can never prosecute a vigorous offensive. The defensive is its proper *rôle,* and no student of naval history will blame a British Admiral for not knowing how to act on the defensive. Admiral Rawson is certainly not to be blamed for taking an inferior fleet to do what only a superior fleet could hope to accomplish. That was the task imposed on him by the Admiralty, and, according to all the teaching of naval war, his failure was predetermined. The enterprise could only have been strategically feasible if he had known exactly where the convoy was to be found, and taking advantage of his superior speed could have gone out and seized it and carried it into port before his superior adversary could interfere to prevent him. But that is essentially an evasive operation and not an offensive one, and the evasive or the defensive is the only attitude which the inferior fleet can properly assume. If by the conditions the evasive is denied to it and the offensive imposed on it, there can in the long run only be one result, and that is defeat. In clear weather it is not unlikely that the A fleet might have found the convoy, for to that extent the conditions of time and speed were probably favourable to Admiral Rawson; but with the capture of the convoy these conditions were reversed, and I confess I never could regard the prospect of A fleet getting its capture into Belfast as much better than a forlorn hope."

Conclusion.

It only remains to add that as soon as he had secured the convoy and made provision for its safe conduct to Milford, Admiral Domvile, rightly regarding the discomfited A fleet as still a "fleet in being," took instant measures for molesting it on its return to Belfast. On the morning of August 1 he despatched a squadron of four second-class cruisers—the *Melampus, Rainbow, Sirius,* and *Brilliant*—with orders to steam at full speed through the Irish Channel to Lamlash, and instruct the destroyers there stationed to make a sustained attack, first on A's torpedo boats, and subsequently on his returning fleet. The destroyers stationed at Holyhead were to be instructed by the advancing cruisers, through B's signal stations, to clear the Channel ahead of the latter and to concentrate betimes in the North Channel for a combined attack with the Lamlash destroyers on the A fleet. Admiral Domvile had reserved his destroyers from the outset for this purpose, and had for this reason ordered them not to attack the

A fleet on its exit from Belfast. In this he was doubtless well advised, though on the other hand it may be doubted whether, if Admiral Rawson, returning with the bulk of his cruisers in company, had been able to reproduce the organization in which he left Belfast, an attack by destroyers alone would have had much prospect of success. It was all in the game, however, and the dispositions made by both Admirals after the main object of the manœuvres had been accomplished showed that both were determined to play the game out to the end. But the operations having been brought to an end by the Admiralty as soon as the convoy reached Milford, neither side was able to play its last trick. It would not have been a decisive trick in any case, but it might have resulted in some very instructive experiences. The case would have been altogether different if A, having carried off the convoy, had found his return to Belfast blocked not only by B's torpedo boats, but by the superior fleet of the latter. In that case—the most probable result of A's seizure of the convoy—it seems certain not only that A must have surrendered his capture to his adversary, but that he could only have escaped overwhelming defeat by taking advantage of his speed to hasten back to Belfast. So impossible is it for an overmatched fleet to do anything but run away if only it can get the chance.

James R. Thursfield.

CHAPTER VI.

Marine Engineering.

As the *Naval Annual* goes to press the thoughts of a nation are in a measure diverted from the Navy to the military forces of the Crown. When war itself comes it overshadows preparation for war; and, as the fighting in South Africa is wholly on land, the army for a time almost fills the horizon. One brilliant feature of the operations has, however, been much before the public, and may be referred to here, since it has been dependent on engineering skill The outbreak of hostilities with the Boer Republics found our forces provided with artillery of a less powerful nature than that of the enemy. This was notably the case at Ladysmith, and the siege of that town gave an opportunity to the sea-service to show once more how high are the military qualities of the sailor, more especially his resourcefulness. The transportation of the Powerful's 4·7-in. guns to Ladysmith was a piece of engineering full of happy augury for a country that has to depend for existence on a Navy in which there is much engineering.

Third-class cruisers.

In the chronicle of a year during which nothing of great importance has occurred in the naval engineering department to claim first attention, it is as well, perhaps, to commence with the boiler, which is at the beginning of things mechanical in a war vessel. Here we see a step which is, it is to be hoped, a step towards improvement. Since the last issue of the *Naval Annual* trials have been made of some of the third-class cruisers which have been fitted with small tube or express type boilers. These vessels are nine in number, and have been supplied with four different types of boilers. The following particulars are taken from *Engineering* :—

	I.H.P. on economy trial.	Coal per I.H.P. per hour.	Type of boilers.
Pelorus	3559	2·20 lbs.	Normand.
Proserpine	3644	2·40 „	Thornycroft
Pactolus	3631	2·46 „	Blechynden
Pomone	3600	2·45 „	„
Perseus	3627	2 10 „	Thornycroft.
Prometheus	3557	2·01 „	„
Psyche	3637	2·31 „	„
Pegasus	3698	1·96 „	Reed.
Pyramus	3605	2·05 „	„

In a paper by Sir John Durston and Mr. H. J. Oram, read last spring before the Institution of Civil Engineers, some further particulars are given of these nine third-class cruisers, together with those of a number of other vessels in the Navy. This paper is an extremely valuable contribution to our knowledge of the construction and performance of the machinery in her Majesty's ships. From a table (Table IV.) attached it appears that the average fuel consumption of these nine cruisers with small tube boilers was 2·27 lbs. of coal per I.H.P. per hour. By another table we learn that the average economy of certain second-class cruisers with Belleville boilers having economisers was 2 lbs. of coal per I.H.P. per hour, whilst in the first-class cruisers with Belleville boilers and economisers the corresponding figure was 1·71 lbs. of coal. From these figures it may be inferred that the small tube boiler is as economical as the Belleville type, for the gain of the first-class vessels over those of the second-class—both with Belleville boilers—is about the same as the superiority of the second-class over the third-class. This view would be strengthened if we were to leave out of the list of the third-class vessels the two boats which were fitted with Blechynden boilers, and which showed an unusually high fuel consumption. Such a proceeding would be quite fair, as it is hardly just to pull down the average of good types of boiler by the inclusion of trial data of one not so good in this single element of design. It should be noted, however, in fairness to the Blechynden boiler, that the proportion of heating surface to horse-power with them was below the mean figure. Sir J. Durston's and Mr. Oram's Paper.

The average weight of boilers per I.H.P. for the nine third-class cruisers was 54·5 lbs. for full power trials. Taking the power developed on the thirty hours' economy trials, there were 107 lbs. of boiler weights per I.H.P. developed. Putting the figures the other way for convenience of reference, we find 40·99 I.H.P. per ton of boilers on full power; whilst there were, on economy trials, 20·95 I.H.P. to the ton of boilers. If we compare these figures with those of the second-class cruisers, we find 99 lbs. of boiler per I.H.P. on full power, and 141 lbs. per I.H.P. on the economy trial; or 22·48 I.H.P. and 15·95 I.H.P. per ton of boilers on the full power and economy trials respectively. It will be seen, therefore, that by the adoption of the small-tube type of boiler in place of the Belleville there is a saving of weight in full power runs of 44·5 lbs. on each unit of power developed, or, in other words, a ton of small tube or "express" boilers would give steam for 18·15 more I.H.P. than would a ton of Belleville boilers. Weight of boilers.

That is a great gain for a warship, and it remains to be seen what has to be paid for it. In another table attached to this most instructive paper we find that the boiler-rooms of the nine third-class cruisers of the Pelorus type each occupy 0·384 square foot of floor space for each unit of power developed at full speed trials, whilst in the Diadem the corresponding figure was 0·305 square foot. This is comparing a first-class cruiser of over 16,000 horse-power with a third-class cruiser of 7,000 horse-power, of course much to the disadvantage of the latter. Probably, however, when all allowances have been made, the Belleville boiler occupies less area than the small tube boilers as fitted to the third-class cruisers.

Durability of boilers.

Durability is, however, the chief point to be considered, for if we put that on one side I think there can be no second opinion as to the superiority of the small tube boilers. In Sir John Durston's and Mr. Oram's paper there is a passage on this question which I will quote in full:—

"The experience of the last few years enables the general conclusions to be arrived at as to the durability and probable life of the water-tube boilers of the large and small tube types adopted in the Royal Navy. It should be mentioned at first, with regard to the Belleville boiler, which is practically the only variety of large tube type at present in considerable use, that in the generator portion the pairs of tubes next the fire are $\frac{3}{8}$ in., and the pairs next above $\frac{5}{16}$ in. thick, while the upper and more numerous tubes are 0·192 in. thick. In the small tube type, however, the thickness in most cases is 0·104 in. but has been recently increased to 0·128 in. in the rows next the fire, and 0·116 in. in the others The ratio of thickness, *i e.*, $\frac{128}{116}$, is probably a rough approximation to the comparative durability of the tubes in the absence of a special cause of corrosion. In the Belleville boiler experience has shown that the life of the boiler-tubes will not be less than two commissions. Occasionally tubes are met with in which local pitting proceeds rapidly, and causes early failure, while, if by shortness of water or accumulation of salt or grease, overheating and burning occur, the failure of particular tubes will be rapid. As regards the small tube boiler, of which so many are fitted in torpedo-boat destroyers and the small cruisers, experience shows that the conditions are not so favourable The hard forcing to which they are subjected, the difficulty of thoroughly cleaning them, their comparative thinness, etc, renders their life under present conditions of preservation and service much shorter."

This was written at the commencement of last year, and no one can question the soundness of the reasoning, but how far the authors would be inclined to add something to their statements, in the light of more recent experience, is a matter open to speculation.

The records of the nine third-class cruisers, valuable as they are, possess one serious defect. It will be understood of course that when the Admiralty gave out orders for these vessels they were prompted by the laudable desire to form a test class for the small-tube type of water-tube boilers with a view to feeling their way towards its employment for more important vessels than torpedo craft. Unfortunately for the completeness of this good intention they left out one of the most important of the express boilers, presumably because Messrs. Yarrow & Co. had their yard already full of work for a considerable time ahead. Fortunately we are able to

Mr. F. T. Marshall's Paper.

supplement Sir John Durston's and Mr. Oram's paper by data from another paper read by Mr. F. T. Marshall at last year's Newcastle meeting of the Institution of Naval Architects. In this contribution were given details of the performance of twelve Yarrow boilers in the Portuguese cruiser Don Carlos I. This is a more powerful vessel than the third-class cruisers, as it has engines indicating 12,500 I.H.P. with forced draught, and 8,000 I.H.P. with the half-inch air pressure that is natural draught by Admiralty regulations.

Weight of Yarrow boiler.

The total weight of the twelve Yarrow boilers, with uptakes and funnels, was 217 tons, and they contained 30 tons of water. The total boiler room weights, including pumps, pipes, water in feed tanks, etc., was 329 tons. The wide difference between these two figures shows how easily one may be misled in an estimate if all features are not included in "boiler room weights." In this case, where nothing was left out, the weight in boiler rooms per I.H.P. with natural draught was 92·2 lbs., and with forced draught 59 lbs., whilst the I.H.P. per ton of boiler room weights was 24·3 at natural draught and 38 for forced draught. The figures here given are not strictly comparable with those of the third-class cruisers on account of the larger size of the engines of the Don Carlos, a fact which illustrates the objection to mixing up engine and boiler performances, as one has to do when accepting the indicator as an efficiency measuring instrument for the whole installation. It is to be regretted that Mr. Marshall was not able to present figures giving the coal consumption; but he states in a note to his paper that the fuel economy of the Yarrow boiler is fully equal to that of the Belleville type. On the whole it may be safely concluded that the inclusion of the Yarrow boiler amongst those of the nine cruisers would have maintained if not raised the average figures before quoted. This is the more satisfactory as the Yarrow boiler has straight tubes which lend themselves more easily to inspection and cleaning—a matter of considerable importance.

The Diadem's boilers

The case of the Diadem's boilers, already referred to, is not encouraging. This vessel was fitted with the economiser type of the Belleville boiler, and after being less than two years in commission she was taken to Chatham for an extensive retubing of her boilers, a large number of the economiser tubes being in a bad state. The deterioration appeared to come from the inside, and it took the form of a large number of pinholes. These tubes were solid drawn and of steel, and their defect comes within the classification in Sir John Durston's and Mr. Oram's paper—written before the discovery of the defects—of "unusual causes of early failure of tubes that may be anticipated." The deterioration was too extensive to be classed as a mere temporary or transitory circumstance, and the matter is one needing serious investigation.

That the source of the difficulty and its remedy will be discovered probably few engineers doubt. The advantages of the water-tube boiler are so manifest that defects which may be developed will not be allowed to cause its rejection now that its value has been proved by actual service. We may feel satisfied that, as we owe the introduction of the water-tube boilers into the Navy chiefly to the courage and foresight of Sir John Durston, every effort will be made to overcome such obstacles as they arise.

Older engineers will recognise the phase through which marine practice is now passing; indeed, it is remarkable how engineering history seems to repeat itself at every step in the march of improvement. After the jump of steam pressure from 30 lbs. to 60 lbs. was made, and when surface condensers were used, we had very much the same difficulties, and if the voice of the croaker, ever heard in the land, had been listened to then we should be still running our ships with boilers pressed to 30 lbs., even if allowed that "dangerous excess."

Mr. J. Dewrance's Paper. The nature of the damage to the Diadem's tubes indicates that the problem, as has been previously stated, is one for the chemist as well as for the engineer. In a paper recently read before the Institution of Civil Engineers Mr. John Dewrance has treated the subject of corrosion from a chemical as well as from an engineering standpoint. This paper, though brief, is one of considerable importance. At present its full text, and the interesting discussion that followed its reading, are not available, but the extracts that have been published in the engineering journals give some valuable information. One point the author insists upon is the beneficial effect of vigorous circulation in preventing internal corrosion. This has long been recognised as an ascertained fact, but Mr. Dewrance gives such good reason for it that a somewhat full reference to his instructive paper will here be made.

Causes of corrosion. In regard to oxidation, it was stated that iron and steel would not affect the decomposition of water at a temperature below red heat, and experiments were quoted showing that with air excluded this was a fact, whilst air dissolved in feed water was the only serious cause of corrosion, putting aside fatty acids, which should never be admitted to a boiler. It was found by experiment that air caused little more corrosion in sea water than in distilled water; and it was concluded that corrosion of highly heated surfaces in contact with sea water was due, not to the water, but to the salts contained in it. Sea water, evaporated until it crystallises, becomes acid, and hydrochloric acid is produced by heating magnesium chloride in a current of steam. The effect of rapid boiling is that water is dashed against the heated steel and evaporated to dryness in rapid succession. Each time this occurs with sea water the crystallising point is reached and a minute quantity of hydrochloric acid is produced on the surface of the steel. Furnace scale rolled into the plates in the course of manufacture becomes detached, and heat is transmitted more freely through the thinner part. Steam bubbles form in the cavity, giving rise to slight explosions as the water enters the cavity. The expelled water leaves behind a small quantity of chlorides, and these give off hydrochloric acid which cause corrosion.

Air in feed water.

This description applies to shell boilers, but the same may hold good with water-tube boilers. The need of a sufficiently active circulation is evident in whatever class of boiler may be under consideration. Mr. Dewrance said, "The evils due to air would cease if it could be excluded, but the evils due to froth were more persistent, the only palliative being to improve the circulation and limit the fire heat to enable the water to wet the heating surfaces as much as possible." In other words, if your circulation be low your furnace temperature must be low too.

In order to keep air out of the feed water Mr. Dewrance proposes placing the hot well between the condenser and the air pump, using the latter only for extracting air, and not pumping water also, as at present. The water from the condenser falls into the hot well, and from thence runs into the feed pump by gravity, as it should do in any case. It should be noticed, however, that the feed pump has not the atmospheric pressure to help it, and would be more likely to fail. There is another point to consider. Air in water plays a considerable part in evaporation. It is said that water from which all air has been extracted is not easily turned into steam. Moreover, it is very difficult to ensure that salt water shall never get into a boiler, through a leaky condenser or otherwise, so it is a wise precaution, at the least, to ensure such rapid circulation that bubbles of steam as formed are immediately swept onward, and the metal is never denuded of water long enough to become dry and form the objectionable hydrochloric acid.

Nickel steel.

Mr. Dewrance refers chiefly to steel in his remarks, and this is natural, because steel has superseded iron as a material for boiler plates. That is, the ordinary mild steel made in the open hearth or Siemens-Martin furnace, for Bessemer plates has not, I believe, been used in marine boiler work for many years in this country. There is, however, a description of steel which appears almost to defy the effects of corrosion, but it possesses one terrible defect; it is extremely costly. In the last issue of the *Naval Annual* something was said about the uses to which nickel steel might with great advantage be put if it were not so dear. Since then Mr. Yarrow has contributed a valuable addition to our knowledge of the subject in the form of a paper read at last year's summer meeting of the Institution of Naval Architects. He immersed, for many hours, pieces of tube of ordinary steel, and steel containing twenty to twenty-five per cent. of nickel, in a mixture one part hydrochloric acid and two parts water. The full details are given in the paper as published in the Transactions of the Institution for 1899, but it may be stated briefly that whilst the nickel steel corroded to the extent of losing 5 grammes in weight during 533 hours' immersion in the liquid, the mild steel lost 98 grammes under the same conditions. Another test gave 7 grammes and 100 grammes respectively as the loss. What were called "fire tests" were also made. The tubes were heated to bright red and straw colour and allowed to cool. The losses on one series of tests were 47 grammes for nickel steel and 145 grammes for mild steel. On another series of tests the figures were 52 and 143. As a result of the tests Mr. Yarrow laid it down as probable that nickel steel tubes in a boiler would last two and a half times as long as ordinary mild steel tubes. This estimate, however, appears to err largely on the safe side, a fact which may be attributed to the author's characteristic caution and a characteristic desire not to overstate his case. As a matter of fact, as was pointed out subsequently by Mr. Macfarlane Gray, the action was much more rapid when the deterioration had been long established, and as the conditions of test were much more severe than would be experienced in ordinary working, the ratio of durability might be more nearly estimated at six to one.

Cost of nickel steel.

The impossibility of keeping the "commercial element" out of a discussion of any subject of practical engineering is, of course, apparent. When the laboratory has said its last word the counting-house passes judgment. In regard to nickel steel for boilers, it would seem—for only experience can give assurance—that nothing remains to settle beyond the sordid detail whether it would pay. Last year I gave Mr. Riley's former estimate of £3 per ton for each 1 per cent. of nickel. That would bring the price of material to £75 per ton for the 25 per cent. steel, plus the price of the original plate, or, roundly, £80 per ton in ordinary times. Mr. Yarrow has stated that the price of a Belleville boiler would be increased "about 30 per cent. on the original price of the boilers so as to secure the increased durability due to nickel steel plates."

Mr. David H. Browne, in a paper read before the American Institute of Mining Engineers, stated that the addition of 3 per cent of nickel to steel would increase the price $20 per ton, which would represent half the price of the older statement of Mr. Riley, say roughly, £40 per ton for the steel as an estimate, allowing a margin for contingencies. Mr. Yarrow's experiments indicated that nickel steel could be worked in conjunction with ordinary steel in a boiler, and if the tubes that are most subject to deterioration from high temperature, *i.e.*, those near the fire, were alone made from 25 per cent. nickel steel, it seems very probable that it would pay to use the material to this extent, considering the saving in labour of retubing and the advantage of having the vessels containing such boilers more often in a fit state to go to sea. The further consideration of the commercial side of the question would, however, lead us too far afield. As Mr. Yarrow said at Newcastle, "Whether it is worth the increased price is a question for those who buy the boilers to decide."

Iron or steel for tubes.

Putting nickel steel on one side, however, it would be a pity if we had to abandon steel for iron as a material for water-tube boilers. Solid drawn steel tubes are mechanically such beautiful pieces of work that it would be sad if they proved chemically deficient. It is, however, well known that mild steel is more subject to corrosion than iron; and so much is this the case that many shipbuilders are putting in iron decks, and are using iron for floors or double bottoms where there is much heat from boilers, or in places not easily accessible for painting.

It is held by some that if Swedish iron or other charcoal-smelted pig were used for steel making, that corrosion or pitting of boiler tubes would not take place, or at any rate would not take place so readily. The foundation for this statement does not appear to rest on a very sure basis of observed data. What is much needed at present is a complete inquiry into the question of the best material for water-tube boilers, to be made by a commission of marine engineers, metallurgists, and chemists. If the Institution of Naval Architects, as representing marine engineering, would join with the Iron and Steel Institute and one of the chemical societies in instituting a research committee on these lines it would be doing excellent service to the country and acting strictly within the scope of the original scheme of the Institution.

Circulation in the Belleville boiler.

It is not only in the pinholing of economiser tubes that the Belleville boiler is proving defective. There has been corrosion in other parts that bears a curious likeness to the experience of years past when the old superheaters were in use. In the light of Mr. Dewrance's teaching the fact points to defective circulation in the Belleville boilers; but there are other indications in the same direction. It will be remembered that at the bottom of each element where the feed enters the steam-generating tubes, there is placed a species of non-return valve, which is intended to prevent the reversal of circulation, or perhaps, one should rather say, to prevent the sudden generation of steam in the lower tubes driving the entering feed water back into what are known as the feed collectors. Now it is evident that when the free passage of water is stopped by the closing of the check valve, circulation, in the bottom tubes at any rate, must be arrested. The fact was, of course, recognised by the Admiralty engineers as soon as the need for check valves was made known to them, but "though there might be arrestation," they were told, "yet it was of so momentary a nature that there was no time for damage to be done." This view, which was entirely reasonable, appeared to be borne out by experiment, but more extended experience appears to show that this system of steam generation does suffer from the absence of the most essential element of success in all steam boilers, vigorous and unimpeded circulation.

"Blistering."

Another prominent manner in which this is made apparent is in the "blistering," or the swelling out on the fire side of the feed collectors. It is stated that this effect is very apparent in the Diadem's boilers, and that those of other ships have suffered in the same way.

A lack of circulation would produce such a result. Remembering that evaporation does not take place smoothly and regularly in the pipes of water-tube boilers, especially large tube boilers, it will be easily understood that in the lower tubes of the elements of a Belleville boiler there will be, at times, sudden increases of evaporation. One of these periods of augmented steam generation may be small at first, but sufficient to close the check valve and stop circulation. Any water in the bottom tube then loses its forward motion and is therefore turned into steam with great rapidity. If the circulation were not stopped this would not matter, because the steam as formed would be swept upwards by the following current of water. When, however, the circulation is stopped or checked, the steam, when formed, has to lift the column of water that is above it in the element before it, the steam, can escape to the receiver. The lifting of the water in the generating tubes or elements naturally causes a back pressure which closes the check valve.

Check valves.

In the Belleville boiler the passage for water and steam, *i.e.*, the element or generating tubes, is long and tortuous, so that the natural circulation once checked is very difficult to start again. The check valve has therefore a tendency to remain closed an appreciable space of time until, in fact, a great part of the water in the element has been driven out and its place taken by steam. Then the far greater weight of water in the downcomer tubes overcomes the smaller weight of steam (with a little water) in the generating tubes or elements, the check valve opens and the circulation is resumed. In the meantime, however, the tubes have become filled with superheated steam, and thus the deleterious action described by Mr. Dewrance takes place, even if the boiler is not burnt and strained by unequal expansion due to excessive overheating. It is said that the greater amount of "make-up" water that is needed with Belleville boilers is caused by the variation in temperature due to the circumstances described. The tubes of the boiler are screwed into their junction-boxes, and it is supposed that leakage takes place here when the temperature becomes excessive. Naturally the water escaping would only appear as steam, which would pass up the chimney, its presence being unnoticed. An analysis of flue gases would throw light on this question. It is stated that the fresh water needed for making up waste on the Highflyer's trial was sixty-four tons as against nine tons for the Minerva. Loss of water.

That the circulation is often stopped is shown by the fact that the check valves are much damaged by wear, and they can also be heard in operation. The "blistering" of the feed distributor is also evidence in the same direction. It may be objected that Mr. Yarrow's experiments* show that when circulation is once started in a water tube boiler it is difficult to reverse its direction. That is true of boilers with small diameter tubes, which are also short and straight, as they were in Mr. Yarrow's experimental apparatus. A comparison of the Yarrow boiler, illustrated on p. 125 of the *Naval Annual* for 1896, with the Belleville boiler, illustrated on p. 217 of the issue of 1898, will at once show on how different a footing the two stand in regard to any check to circulation. In one case the water has to travel upward through only five or six feet of straight pipe, whilst in the other there may be 90 or 100-ft. run of pipe to traverse; but what is of more consequence, the direction of flow is reversed fourteen times by sudden bends. The danger to economiser tubes of the Belleville boiler was foreseen by the Admiralty authorities, and it was the practice at first to coat the interior with lime, and later on to galvanize them both inside and outside.

The adoption of the Belleville boilers for the fleet was almost a necessity, and subsequent events have amply warranted the discarding of the old return tube boiler, for there is not a navy in which the latter is now used for new designs.

At the time the change was made the Belleville boiler was the only steam generator of which there existed sea experience with big ships, and that experience was of a nature to show its superiority over the shell boiler for naval work. Since that time the Admiralty

* Reference to these experiments was made in the *Naval Annual* for 1896, p. 128, and a full account of them, with illustrations of the apparatus used, is to be found in *Engineering*, vol. LXI., p. 89.

engineering department has been gathering knowledge, both by means of trials made with vessels in the Royal Navy and from outside sources. The result has been that other types of water tube boiler have been shown to possess advantages, and confidence in them has thus been gained by extended use. Opinion has thus grown up that the Belleville boiler, in spite of its many ingenious devices to overcome initial defects, will not continue to hold the position it has occupied in the past. In saying this, however, it is but fair to add a tribute of admiration of the French engineers for the boldness and engineering skill they have shown in adapting the water tube system of steam generation to practical uses; in fact, they carried the idea to a successful issue after we had practically abandoned it. So far as the Belleville Company are concerned they have had their reward, but it would be hardly creditable to British engineers if an improvement could not be made on a water-tube boiler that has fourteen abrupt turns in its generating element and needs check valves to prevent a reversal of the circulation.

Sir J. Durston's and Mr. Oram's opinion.

At present we have probably in the Royal Navy, built and building, not very far short of a million horse-power in Belleville boilers. As was stated by Sir John Durston and Mr. Oram in their paper before the Institution of Civil Engineers, "Careful consideration and experience have led to the selection, at least for the present, of the Belleville type of large tube boiler for use in the larger class of warships. For such ships the question of durability of boilers is of primary importance."

Whether the experience gained since this was written by the two chief authorities in the engineering department of the Admiralty leads them to conclude that "the present" of then has now become the past, in regard to the exclusion of other types of water tube boiler, is a matter upon which there is no public intimation at the time of writing, but such incidents as the retubing of the Diadem's boilers must at any rate cause great anxiety to those who have the decision of these matters in their hands.

On the whole, it would seem that those who advocate small tube boilers have every reason to look forward hopefully to the future in regard to their adoption for bigger ships. The Yarrow boilers fitted in the Dutch cruisers, described in the 1896 issue of the *Naval Annual*, are, I hear, still running well, and indeed the government authorities of that country are so well satisfied with the result obtained that they are extending the use of small tube boilers for bigger ships.

Perhaps one of the chief things that has checked the more extended use of small tube boilers for large vessels has been their very success in torpedo craft. When 70 or 80 I.H.P. are obtained per ton of boiler—as in the destroyers—a false standard is raised for more serious work, and the tradition of such practice is apt to stick. However, designers are overcoming this weakness, as will be gathered from the figures before given.

Combustion.

It is also being recognised that it is desirable to get economical results rather by good combustion than by multiplying heating surface. We know how fatal the crowding in of tubes proved in the case of the cylindrical Navy boilers; and though there is no fear of analogous difficulties with water tube boilers, yet the neglect of adequate space for combustion can never lead to good boiler performance even if the heating surface be increased to any extent. It is, therefore, better to leave good clear room above the fire rather than cumber the flame-space with an extension of tubes. It is in this matter that the best designed of the small tube boilers have an advantage over the Belleville type; and though the latter has a mixer, in the shape of air jets injected at pressure above the fire, the device can only be described as one of those ingenious inventions designed to overcome initial defects, to which reference has already been made. The additional complication introduced by these devices is a point that has to be considered.

Large and small grates.

The ease with which a large grate area is secured in the design of water tube boilers is often urged as an advantage, and no doubt this is true, but the principle may easily be carried too far. With very big grates the fire has to be kept thin, and to maintain a thin fire and yet keep all the grate covered requires very skilled and careful stoking. It is hardly necessary to point out how soon steam generating capacity and fuel economy fall off when there are holes in the fire letting cold air through to the heating surface. The prejudice against forced draught is due to the evil results that followed the excessive use of this means of making a little boiler do the duty of a big one. But reasonable forced draught is an excellent thing, and leads to economy. Moderate grate area and moderate fan draught are conducive to economy. It is a mistake to suppose slow combustion in itself leads to a low percentage of fuel burnt for a given evaporation of water in the boiler. The eddying of the air and gases caused by forced draught makes the combustion more perfect, if only sufficient space be provided above the fire, so that the gases can be burnt before they come in contact with the heating surface, and are thus cooled below the temperature needed for their combustion. It is failure in this latter respect that causes the long column of flame one sometimes sees streaming away from the chimney tops of vessels on trial trips, or on other occasions when the fires are being urged. Mr. Yarrow gave a very good illustration of this during a boiler discussion at a meeting of one of the technical societies. He instanced an Argand burner which, with the chimney removed, gave a long, sluggish, dark flame, with a very plentiful deposit of carbon on any cold metallic substance introduced into the body of the flame. Directly the chimney was put on, thus causing an acceleration of the draught, the flame was shortened, the heat was greatly increased, and there was no smoke; a knife blade placed in the flame soon became red hot in place of being covered with residual carbon.

Boilers with sluggish or uncertain circulation cannot safely have the high furnace temperatures which tend to economy, both of weight in the boiler as well as in the coal burnt.

For this reason the large grate area that can be placed on a given floor area is an advantage to the Belleville boiler, with its comparatively slow and intermittent circulation. But in the small-tube boilers, with their vigorous circulation, the grates may be smaller, for the forced draught that can be safely and conveniently used with this type of generator enables more coal to be burnt on the smaller grate than on the bigger grate of the Belleville boiler, with the beneficial result of a hotter fire and more perfect combustion.

Bullfinch disaster.

One of the most memorable events in connection with Naval engineering of last year was of a very tragic nature. On July 21 the destroyer Bullfinch was on trial in the Solent, and when running between 29 and 30 knots, the high-pressure connecting-rod of the starboard engine parted at the fork end, and the cylinder itself was broken for two-thirds of its circumference. The result was terrible. Steam escaped through the cracked cylinder, and filled the engine-room, and eleven men were killed. The engineer to the vessel, Mr. W. A. Dathan, R.N., escaped by dropping on his hands and knees and crawling to a ladder, up which he ultimately climbed, and though he was so overcome that he would have fallen back had he not been caught, he was able to go forward and report the accident to the lieutenant in command. There is a lesson in this that is worth impressing. Steam is lighter than air, and dry air is a very bad conductor of heat. Escaping steam will therefore remain in the upper part of an engine-room, and that section may be at an exceedingly high temperature, whilst the lower part will be comparatively cool. It is quite probable that there would be greater safety in lying down on an engine-room floor and waiting until the steam cleared off, rather than penetrating to an upper zone by climbing a ladder. That of course does not apply if steam were escaping in a downward direction, so that the air and steam became well mixed. It took only two minutes in the Bullfinch to close the stop-valves and shut off steam, a fact due to the presence of mind of Mr. Tyacke, the contractor's engineer, who also behaved admirably, being lowered into the engine-room, which was full of steam, in order to render what assistance he could to the injured men.

Connecting-rod design.

The way in which the connecting-rod gave way and the cause that led up to it are the points of interest to the engineer. The rods were of steel made by Messrs. Cammell and Co., the design being that of the contractors, Earle's Shipbuilding Company, and approved by the Admiralty. Now there are two general ways of forming the little ends of marine connecting-rods: one is by having a long fork or jaws, and fitting the gudgeon-pin into it so that the pin moves with the fork; the other is to fit the jaws of the connecting-rod with brasses, the pin being in one with the piston-rod. The Bullfinch had the short-jawed connecting-rod with brasses. The rod was 3½ inches outside diameter and had a 2-inch hole bored through the centre, thus forming it into a tube. The rod gave way by a vertical split developing in the rod just below the fork. This proceeded a little distance and then the rod parted, with the disastrous results stated. The designers of the engines attributed the mishap to the material, whilst the steel makers were of opinion that the design of the rod was at fault. The Admiralty authorities were inclined to think that design and material should share the blame between them, a splitting of the difference which perhaps met the justice of the case. It was calculated that the metal would not be stressed to as much as 7000 lbs. to the square inch, whereas the minimum test was 30 tons and the maximum test 35 tons, which, it is presumed, the test pieces forged on the rods, in accordance with Admiralty requirements, withstood. It is, however, extremely difficult to estimate what the maximum stress on the rod might be. Of course it would be easy if the problem could be solved statically; that would simply need multiplying the number of square inches of section in the piston into the maximum steam-pressure in lbs. per square inch and adding a factor of safety. The number of revolutions at which the engine was running at the time was about 392 per minute, so that the direction of stress was changed about 13 times a second. It is hardly necessary to state that a quickly changing stress is far more trying to material than one of equal intensity that is constant. It was not, however, the tension and compression in line with the axis of the rod that caused the initial rupture, as the rod was split or cracked at first in the direction of its length, and only gave way transversely as a secondary result. The origin of the mishap has not yet been cleared up. In defence of the design, it may be said that the arrangement has been largely used for marine work, and so far as I am aware has never led to such results before. On the other hand, every care seems to have been taken in selecting and preparing the steel. The analysis was good, and a great part of the ingot from which it was made, about one half, was rejected. Moreover, the rod was not again heated after being annealed. There was a disposition in some quarters to blame the design because a greater cross section was not allowed, it being considered that a saving in weight of 25 lbs.—that due to boring the rod—was a very small gain not worth speaking of, in fact. No doubt this would be true were it a stationary part, say an engine pillar, under consideration, but a connecting-rod is a moving part, and when it moves so rapidly, as in the case of a destroyer's engine, the inertia strains become of enormous importance, so that the addition of metal may become under some circumstances, a positive source of weakness.

Limit of safety. The accident opens up a wide field for speculation and study which would take up too much space to deal with adequately here. The closeness to which practice was approaching the limit of safety in these remarkable vessels of excessive speed was a revelation to engineers, and probably many considered themselves fortunate in gaining the experience this accident afforded without its terrible consequences being brought more closely home to them. No doubt in future connecting-rod ends for this class of work will be made rather of a V shape than of a U form, and the gudgeon-pin will be fixed in the jaws of the rod and riveted in firmly so as to support the two arms of the fork. It has been suggested that a return should be made to iron rods in place of steel, but this appears quite a retrograde movement. It is also desirable to have two bolts for holding brasses in place of four, as it is easier in this way to make sure of each bearing its proper burden.

Auxiliaries. In the last issue of the *Naval Annual* some attention was drawn to the large amount of steam consumed by auxiliary machinery. It was suggested, or rather the suggestion was repeated, that a central electrical generating station on board ship to supply power to auxiliary machinery would be a remedy for the evil. In the last Annual Report of the Chief of the Bureau of Steam Engineering for the United States Navy the subject is treated at some length. As the conclusions arrived at are the result of a long period of experiment and observation by very competent authorities, they may be quoted with advantage somewhat fully. It may be said, to begin with, that Rear Admiral Melville and his colleagues have determined that "after carefully investigating the adaptability of electric motors to the driving of the numerous auxiliary engines on board ship . . . they conclude that the electric drive of the auxiliaries would not, under

Rear Admiral Melville's report.

existing conditions, be so satisfactory and economical on the whole as the steam drive."

Electric driving.

The advantages claimed for electric motors over small steam engines are stated in the report to be "greater ease of operation, avoidance of heat and much greater economy." Against these advantages, the report places the drawbacks of "much greater weight of the necessary electric outfit, the greater delicacy of the type of electric motors ordinarily used, the lack of ready adaptability to various conditions of service, a general denial of claims for economy as ordinarily presented, and the increase in the amount of space below the protective deck for the installation of the necessary dynamo rooms, this space being necessarily taken from coal bunkers."

It is further pointed out that "the use of an electric motor involves a total weight for the motive power at least three times that of the motor itself, because there is always the generator and its driving engine, besides the motors supplied by them; or, "in other words the electric drive of an auxiliary will weigh at least three times as much as a steam drive, assuming the motor to weigh no more than the engine it displaces, although usually it does weigh more." The Engineer-in-Chief of the United States Navy is further of opinion that greater economy can be reached by using compound engines for auxiliaries, by utilising the exhaust steam for heating the fresh water in the boilers, or by conducting the exhaust steam into the receivers of the main engines.

Exhaust steam.

In cases where the auxiliary engines are in the engine or boiler rooms, it would be easy to dispose of the exhaust steam in the way mentioned; but when the engines have to be placed some distance away, as they have in certain cases on war vessels, the difficulty of dealing with exhaust steam is great. Admiral Melville speaks of "the absurdity of an electric drive of auxiliary machinery on board ship situated closer to the main engines than the engines driving the dynamos." It may be absurd, or, under certain circumstances, it may not, for electric current is so much more conveniently transmitted than steam that it might easily pay better to take electric current 300 feet between the generating station and the electric motor rather than have a steam motor and carry the live steam 100 feet from the boilers to the motor, and the exhaust steam back through the same distance. It is this facility of conveying power—the light, flexible, and easily handled wire, compared with the rigid steam pipes—the absence of heat, and the facility with which electric motors can be started—without having to warm up engines—that constitutes their charm for outlying positions on board ship. To use them for continuously running auxiliaries in the machinery compartments is, of course, quite a different matter.

Capstan and steering engines.

The capstan and steering engines are not under the charge of the Bureau of Steam Engineering, over which Admiral Melville presides, and he points out that as these are placed some distance from the boilers, we should naturally expect to find them electrically driven. There are, however, special difficulties in the way of using electricity for steering gear motors, but even these have apparently been overcome, as a system of electric steering has been introduced in the German Navy. One of the most advantageous positions for electric power is in working ammunition hoists. I have known small compartments in war vessels all but unbearable from the heat given off by a steam engine working an ammunition hoist, whilst the long steam pipes leading to it have been reservoirs for condensed steam, the resultant water having to be got rid of, and the pipes heated up before the mechanism could be put in motion. In regard to eduction pipes of course it is always possible to let the exhaust steam escape into the atmosphere, and thus avoid long leads of big pipes to the condensers, but this is a thing that we cannot afford to do in the present day of highly pressed boilers, when salt water make-up is taboo. And here it may be worth considering how far the large quantity of make-up water needed may be largely due to the escape from numberless steam auxiliaries and their connections.

Space for auxiliaries.

Admiral Melville, however, not only puts before us his views generally, but, like the good engineer that he is, gives us concrete facts. "In the battleship Alabama," he says, "the space required for electric motors, where used, is approximately the same as that needed for steam engines to do the same work. The space required for the wiring, etc., is less than that necessary for steam piping, had that been used." The figures quoted are as follows: For generating sets there are needed 10,140 cubic feet for a total of 256 kilowatts. "If all the machinery were operated by electricity, and if the space required by the electric generators were increased in the ratio of the increase of necessary capacity in the generating room, the space that would be required in the generating rooms would be 50,700 cubic feet for a capacity of 1,280 kilowatts." It will be noticed that the Admiral simply states an hypothesis. "*If* all the machinery," and "*if* the space required," he says. But how far do these "ifs" apply to actual practice? In the Alabama the actual space needed for generating sets is about 39·6 cubic feet per kilowatt, and working this out by the light of simple arithmetic we find the answer to be, as above, 50,700 cubic feet for the 1,280 kilowatts needed for all the auxiliary machinery. But this supposes not only the power but the number of generating sets to increase with the increase in motors. It is as though we calculated the space needed for the machinery of an ocean liner by cubing the engine rooms, say, of a number of steam yachts.

Of course, this is not what Admiral Melville means, and no doubt he intends his figures as to weights to be subject to a discount. Still speaking of the Alabama, he says, that "consequent upon the extension of the use of electricity, would be an increase in the total weight of machinery equal to from 150 to 250 tons as a minimum." This increase in weight would be as much as the gain following the use of water tube boilers. The engineer-in-chief also tells us that "the increased space occupied by the larger generating rooms would accommodate 900 tons of coal, or 3,600 horse-power could be added to the power of the propelling engines, giving the ship in the first instance 45 per cent. greater coal endurance, or in the second instance 1·5 knots increased speed."

Mr. S. D. Greene on electric auxiliaries.

Another American engineer takes a somewhat opposite view. Mr. S. Dana Greene, in a paper read before the American Institute of Electrical Engineers, states that a modern first-class battleship needs about 2000 I.H.P. to drive its auxiliaries, and a first-class cruiser about 1200 I.H.P. for the same purpose. As all are not in use at the same time, the total power required is put at 1000 and 600 horse-power respectively. In regard to the latter consideration, Admiral Melville is of another opinion, and his word on such a question must certainly carry greater weight. He says:—"Our Naval machinery has to be designed so that in time of action everything can be ready for use; and, as a matter of fact, a very little study of the question will show it not only may, but almost certainly would happen that every auxiliary on the ship, except the capstan engine and some of the boat winches, would be used simultaneously." Admiral Melville's statement makes it necessary to apply corrections to Mr. Greene's figures, which, nevertheless, are of considerable interest. He refers to the trials made by Mr. White, a past assistant engineer of the United States Navy, by which it was found that the weight of steam used by the main engines of the Minneapolis was 33,620 lbs., and by the auxiliaries 10,416 lbs. per hour (see *Journal of American Society of Naval Engineers*, for February 1898). This was at the rate of 20·83 lbs. of steam per I.H.P. per hour for the main engines, and an average of 119 lbs. per I.H.P. per hour for the auxiliaries; the lowest figure being 55·06 lbs., and the highest 318·68 lbs. per hour.

Supposing electric transmission to be substituted, Mr. Greene assumes "an efficiency of 82 per cent. for engine and generators, and an average line and motor efficiency of 80 per cent.; giving a total efficiency of the system of 65·6 per cent. In other words, to develop 1000 horse-power at the motors would require 1500 I.H.P. at the engines, or about 900 kilowatts generator capacity." Seven sets of 150 kilowatts would be needed, one being in reserve. With the efficiency stated, allowing 30 lbs. of steam per effective horse-power at the motors (20 lbs. at the engine), and 25 per cent. margin for losses due to leakage, friction of gears, &c., leaves 37·5 lbs. of steam per horse-power per hour, as against 119 lbs. as shown by Mr. White's trials on the Minneapolis. Reducing the figures to a coal standard, Mr. Greene estimates the steam auxiliaries of the ship would account for 84 tons of coal per 24 hours; whilst if they had been electrically driven the consumption would have been 26·5 tons a day. The saving would therefore be, on these figures, 57·5 tons of coal a day.

Weight of auxiliaries. On the question of weight, Mr. Greene has also something to say. The present weight of steam auxiliaries of a first-class battleship is assumed at 100 tons. The weight of the seven 150 kilowatt sets would be 175 tons. The electric auxiliaries would weigh about the same as steam motors, and this would give a total of 275 tons for electric machinery as against 100 tons for steam. Allowing for a serving-in wire as against steam-pipes, the author assumes electric driving would necessitate a weight of from two-and-a-half to three times that of steam motors. Against that he claims a saving of from 10 to 20 per cent. in coal for a given steaming radius, which would amount to between 200 and 400 tons. The space that would be needed for this coal if combined with the existing dynamo room would be more than sufficient for the electric power station. The first cost of the electric station would certainly be greater than that needed for steam driven auxiliaries.

Efficiency of auxiliaries. It will be seen that Admiral Melville, as representing the naval engineer, and Mr. Greene representing the electrical engineer, arrive at different conclusions, although agreeing in some respects. The chief point on which they split is that of fuel economy. The admiral objects very emphatically to the practice of crediting the electric method with an economy claimed on data which give the highest figure for the efficiency of the generators and motors and also the most economical steam engine, and comparing these results with the uneconomical form of steam cylinders, which, for very good reasons, have until recently ordinarily been used with the steam-driven auxiliaries. If the American engineering bureau have removed the "very good reasons," and have made the motive part of steam-driven auxiliaries more economical—which I believe to be a fact—by so much must the balance claimed in favour of electricity be decreased. The efficiencies assumed by Mr. Greene are not extravagant, but even they may not always be reached on a ship's installation; that is a point on which further experimental data obtained under conditions of actual work would be of value. Admiral Melville points out that auxiliaries on board ship work at a wide range of power and speed, so that the motors are not likely to be always at their best. On the other hand, something might be done by raising the tension of the current for power purposes above that usual on board ship for lighting.

Importance of auxiliaries.

At present it would seem that the most one can say is that auxiliaries on warships have been too much neglected. They have increased in number and importance by successive steps, and it has hardly been realised until recently what a very large demand they make on the bunkers. Each auxiliary engine is not a big thing in itself and the designers (contractors) have generally been careful to produce something that will occupy little space and do its work without chance of breakdown, and they have thought little of the coal that would be burnt but never charged against them. This is notoriously the case with some steam pumps often seen afloat. From this we may argue that now attention has been called to the subject, and the magnitude of the question has been realised, a good deal will be done to bring about improved economy in the steam auxiliaries, so there will be less need for alteration. On the other hand, electricity for conveyance of power is a new branch of applied science, and one which must almost necessarily make great advances within the next few years. For purposes ashore, closely analogous to the driving of auxiliaries in ships, it is being more and more applied every day. Perhaps it is safest to say up to now that within the machinery bulkheads steam is the better agent, but that the undoubted advantages offered by electricity make it preferable for outlying positions where habitability and personal comfort have to be considered.

Control of electrical designs.

There is one thing, however, upon which both Admiral Melville and Mr. Greene are agreed, and from which I think few engineers will dissent. The chief engineer's report says:—

"The operation of electrical machinery is purely mechanical. That this may be done efficiently requires good mechanical ability at the generating engines. Electric difficulties and casualties are almost always questions of mechanical engineering. I know that it would conduce to the efficiency of the service, to the feasibility of a more extended use of electricity, and to an increase in the life of electrical apparatus, if the electric generating plant were placed in charge of this bureau. I therefore recommend that this change be authorized. I desire to call attention to the fact that it is almost the universal commercial practice to place electric generating plants in the charge of mechanical engineers."

The words of the United States Engineer-in-Chief are of equal force in their application to this country. In the early days of electric development there was nothing checked progress so much as the thoroughly bad mechanical engineering put into the designs by those who controlled the business. Though not engineers they were electricians, and therefore had some right to the position they occupied; but to put electrical engineering into the hands of those, neither electricians nor engineers, seems at least anomalous. As Mr. Greene says, "The problem is purely an engineering one."

Electric power in U.S. ships. The United States battleships Kearsarge and Kentucky represent the most recent complete practice in the application of electricity for power purposes. A very full descriptive paper of these installations was read by Mr. J. J. Woodward, U.S.N., before the American Society of Naval Architects and Marine Engineers at the New York meeting of last year. The details are too voluminous to give here, but English readers will find them fully set forth in a reprint of the paper which appeared in *Engineering*, commencing January 5th, 1900. Electricity is used in these ships for lighting, rotating the turrets, elevating, loading and handling ammunition for 13-in. and 8-in. guns, for operating hoists, deck winches, boat cranes, ventilating fans, and for steering.

The Parsons steam turbine. The trials of the past year have not brought forward any points of novelty. It was expected that the official trials of H.M.S. Viper, the destroyer built on the Tyne, and propelled by Parsons' steam turbine, would have been completed at a much earlier period, but up to the time of writing they have been put off on account of the weather being too stormy whenever a day was fixed for the purpose. In this respect Mr. Parsons has had continuous bad luck. A certain number of preliminary runs have been made, and wonderfully high speeds have been attained, but it is preferable to wait more definite information before discussing the properties of this most novel and interesting vessel. We are able to give, however, an illustration of the Viper, for which we are indebted to *Engineering*. This has been reproduced from a photograph taken when the Viper was steaming 35½ knots, although under what conditions is not stated. The vessel is 210 ft. long, 21 ft. wide, and 12 ft. 9 in. deep, her displacement being 350 tons. The horse-power at the speed stated is given at about 11,000 indicated. The revolutions are about 1200 per minute. There are four shafts, each having two propellers, or eight in all. *Engineering* gives the weights as follows:—

	tons	cwt.	qrs.
Boiler Room weights, with water	100	15	0
Engine Room weights, with auxiliary gear and water in condensers	52	6	1
Propellers, shafting, &c.	7	14	2
Total	160	15	3

From "Engineering."

THE TURBINE MACHINERY OF THE "VIPER."

Showing one of the two duplicate sets having high-pressure, low-pressure, and reversing turbines.

The boilers are of the Yarrow type.

The illustration on page 157, also taken from *Engineering*, give a view of the propelling engines and steam turbines.*

The Japanese destroyers. During the year the trials of a series of six destroyers built by Messrs. Yarrow and Co. for the Japanese Government have been completed. These vessels are interesting because they represent a squadron of the fastest vessels yet built.

Particulars of the official trials made, under Admiralty conditions will be found in the chapter on Foreign Navies, p. 53. The low air pressure, especially with the later boats, is a notable feature. On the Niji's three-hours' run the coal consumption was only 1·98 lb. per I.H.P. per hour.

Trials of Minerva and Highflyer. Another series of trials that have attracted some attention are those carried out with H.M. cruisers Minerva and Highflyer. The former is fitted with ordinary return tube boilers, and the latter with Belleville boilers. Only very scanty details have been made public at present. According to a statement made in the House of Commons the trials were carried out in a satisfactory manner. The coal consumption is given in terms of the power developed, always an unsatisfactory measure—although in this case the best that could be expected—and though the Highflyer made half a knot greater speed than the Minerva in the thirty hours' highest speed trial, developing 400 horse-power more than the sister ship, the Minerva burnt less coal per I.H.P., the figures being for engines only, 1·97 lb. per I.H.P. per hour for the Minerva, and 2·1 lb. for the Highflyer. On a sixty hours' run at 17 knots the Minerva's coal consumption was 1·95 lb. per I.H.P. per hour, and the Highflyer's 2·07 lb. per I.H.P. per hour. These figures were for engines only.

G. R. Dunell.

* The trials at length took place on May 4th, 1900, and the following is from the *Times*:—"Rejecting the first runs and taking the following six, it was found that the speed was just on 34¼ knots. The best pair of runs gave 34·67 knots. The boat had been in the water some time waiting for her trials and with a 'scribbed' bottom would undoubtedly have done better. The wind and rough water were also against high speed. The mean revolutions on the mile were about 1050, and the steam pressure ranged from 165 lb. to 175 lb. Unfortunately the relief valves were set rather light and a great quantity of steam escaped when the pressure was allowed to run up. The contract load was 40 tons, but 60 tons were carried. The displacement at trial draught was 370 tons."—Ed.

From "Engineering."

"VIPER,"

BRITISH TORPEDO-BOAT DESTROYER,

Steaming at 35½ knots.

CHAPTER VII.

The Recent Discussion on Naval Training.

In the autumn of 1899 it was announced that the so-called "masted" ships composing the Training Squadron were to be paid off, and that cruisers of ordinary type were to be put in commission to take their place. It was suggested that the immediate reason of this change was the necessity of increasing the number of effective fighting cruisers in commission. The crews of the Training Squadron amounted in the aggregate to about 1,400 officers and men of all branches of the Service—bluejackets, stokers, artificers, marines, domestics, etc. The total number of officers and men of all branches and boys under training voted by Parliament for the year 1899–1900 was 110,640. At first it would appear that a proportionately small variation, slightly exceeding 1 per cent., in the numbers which Parliament had sanctioned, viz., 109,240 instead of 110,640—a variation which would have followed as a matter of course had any member successfully moved to reduce the total by an amount equal to the crews of the Training Squadron—would have rendered it impossible to commission four additional cruisers, none being of the largest class. The emergency was undoubted. The country found itself involved in a difficult war, carried on at a great distance from the United Kingdom, which necessitated the sea-transport of a large body of troops, and, if not actual patrolling by, at any rate the keeping in immediate readiness of, additional cruisers. It looked as if the officers and men to man them could be found in only one way, viz., by taking those who were being trained in sea-going ships and, as the phrase goes, "in blue water." A reference, however, to the circumstances of the recent mobilisation of 1899 would have dispelled this apprehension. On that occasion enough officers and men were "mobilised," *i.e.*, withdrawn from half-pay and from harbour and shore service, to man 4 first-class and 26 second-class *boná fide* sea-going cruisers; and the Training Squadron was left untouched. It was, therefore, certain that the crews of the substituted cruisers could have been found, several times over if required, elsewhere than in the Training Squadron. The unavoidable inference was that the sea-training in the Squadron was

The abolition of the Masted Training Squadron.

held to be of less value than the training courses followed in harbour and shore establishments, and that the paying-off of the masted ships was not a temporary measure, but one which would not be reversed.

The discussion on the system of Training.

The step led to prolonged discussion, which, perhaps, is not yet finished. No one will dispute the great importance of the subject discussed, for it really is that of the whole training of naval officers and bluejackets. The *Annual* would not be complete were an account of the discussion omitted. It revealed not only a sharp difference of opinion as to the relative value of what was called "masts and sails training," and of harbour and shore training for seamen, but also a wide divergence of view as to the operation and effect of training in itself, of whatever kind it might be. It was made clear that those who took part in the discussion, whether naval officers or not, might be ranged in two schools, of which one thought that "masts and sails training" was useless because it trained seamen in the handling of appliances which they would not use in battle; and the other held that it was, and indeed had incontestably proved itself to be, useful in developing qualities essential to naval efficiency. As the discussion proceeded it was observed that these two schools differed on a fundamental point. The advocates of the "masts and sails," or older system, looked to the object of training, and, in the words of one of them, at "the result produced on the *personnel* far more than the incidental knowledge acquired thereby." The antagonists of the older system were in favour of another which would be devoted to simply making the trained familiar with the use of the modern appliances of war; in other words, to improving their incidental knowledge. It is absolutely necessary to keep clearly in sight this difference between the two schools.

The object of Training.

In considering the matter we have to begin by asking the question—"What is the object of training?" To this there can be only one answer—"To produce officers and men who will prove efficient in war." Another question at once presents itself, viz., "Of what does this efficiency consist?" There is not likely to be any difference of opinion as to the way in which this should be answered. There will be nearly, probably quite, universal agreement in saying that it consists of courage, readiness of resource, skill in using weapons on actual service, fortitude in difficulties, obedience. The list might be extended, and still agreement as to its composition might be looked for. In a leading article in the *Times* the qualities constituting naval efficiency were thus enumerated—"Self-reliance and resource, quickness of eye and steadiness of nerve, calmness and self-possession in emergency, steadfastness in danger, helpfulness in difficulties, and a quick sense of comradeship." No one came forward to dispute the accuracy of this summary of essential naval qualities; nor has anyone denied that they were developed and fostered by the older system of training, as we may conveniently designate the "masts and sails" system.

Objection raised to the training under masts and sails.

The objections raised to the continuance of that system were two-fold. It was asserted that the Training Squadron of masted ships had ceased to be of practical use even for the purpose for which it had been specially instituted. It was remarked that, "in any difficulty, under the present system," the course taken was "simply to furl sails and use steam"; and that "an awkward situation is thus easily saved without any strain upon the nerves." Without doubt this was a strong condemnation. What, however, those who urged it and agreed with it altogether failed to see was that what it condemned was, not the system, but the way in which the system had come to be worked. Now if there is anything that is certain, it is that systems of training and education, though of undisputed excellence, are liable to be worked occasionally in an ineffective manner. It is a commonplace of the history of education that methods, good in themselves, have been found very often to be so put into practice that much of their merit disappears. It is well known that this has been specially the case with "gunnery" and similar systems of training, as is demonstrated by the frequent appearance of new drill books intended to rectify applications of the method, but leaving the general principles unaltered. The argument brought against the Training Squadron was, therefore, not properly an argument for its abolition, but for its improvement. If accepted as good ground for abolition, no system of training would be safe.

The objection to the older system took also another form. This went beyond the particular case of the Squadron. It was maintained that, as battles will not be fought under sail, as, indeed, fighting ships do not carry sails, it is a waste of time to train men in the use of that which they will not have to handle in war. It was noticed at the time that this argument was not impartially applied. It would be impossible to lay too much stress upon this striking fact, for fact it is. The training question had been made a subject for argument. Each side showed perfect readiness to allow it to be settled by an appeal to reason. Even those whose rank and experience would have entitled them to express their opinions absolutely did not do so, but supported their particular contention with arguments just like anybody else. An argument, if used at all, must be followed to its logical conclusion. The one directed against training with sails because of their future non-employment in battle, if valid at all, is drilled is that composed of the various Colonial Corps, most of which were hastily raised. It is certain that adaptability to the conditions of war has not been found to be directly proportionate to perfection in drill; and our partially drilled Colonial troops have not proved to be inefficient.

Drill not to be neglected.

It does not follow that drill should be neglected. What is necessary is to confine it to its proper place in the process of producing belligerent efficiency. Drill is a necessary evil. If human beings were perfect creatures drill would be needless, as its operations would then be executed without preparation. A great object of drill is to bring the capacities of many individuals to the same level and to produce a mechanical uniformity. Drill is declared to be perfect when a body of men of varied physical and mental capacity have been brought to execute an operation with a uniformity which is complete. Drill encourages instantaneous compliance with recognised words of command; a frequently adduced illustration of which is the case of the man who dropped the dinner he was carrying from the cook-house and smartly brought his hands to his sides when a playful acquaintance shouted in a tone of command the word, "Attention!" In certain circumstances this readiness to comply almost mechanically with orders given in an established and formal shape is advantageous; but it does not come into play unless the formal order is given. Thus it checks and may altogether destroy initiative. As is the case with all necessary evils, the aim should be to have, not the greatest amount of it possible, but the least amount that can be put up with. Drill is the repetition of the same processes over and over again. Up to a certain point it is indispensable in order to make men properly familiar with the handling of the weapons used in battle. As soon as that point has been reached no further addition to the desired familiarity is possible, and all that is necessary, or in fact can be effected, is to prevent the familiarity from diminishing. We learned what comes of neglecting drill in the earlier months of the Second American War, when crews which had been made to polish, instead of being made to work their guns, were defeated, and, it may be said, deservedly. Practice should not be confounded with drill. No one will dispute the value of target-practice and the certainty of inefficiency if it be omitted. Philip Bowes, Vere Broke, and Exmouth showed us long ago how important attention to it is. Firing practice, however, cannot, like drill, be a thing of daily or almost hourly recurrence.

But it cannot replace other training.

If the older system of training is to be abandoned, what is there to put in its place? No one has suggested that it should be replaced by anything but drill, and it is not easy to see what else can replace it.

Additional time devoted to firing practice is not likely to be considerable. At any rate the addition will not be so considerable as to affect in any way a system of training on the old lines. It is forgotten that, owing to the large number of guns carried, more hours were spent in target-practice by the crew of a line-of-battle ship or a frigate than by the crew of a modern armour-clad or cruiser. The navies of to-day may be challenged to produce an instance of a squadron devoting more time to practice than did that commanded by Lord Exmouth in 1816. The challenge may be repeated as regards many ships in the Mediterranean and the Channel at a much later date. It is certain that masts and sails do not interfere with practice with weapons. If there is to be more practice it will not be because the older system of training has been abolished.

Drill cannot produce the necessary qualities.

The abolition will surely lead to more drill; that is to say, to more repetitions of the same operation, which operation has been formulated on paper in the study, and been fenced with rules against the introduction of variety. Is it seriously contended that such a system of training will produce or foster the qualities which are essential to the efficient man-of-war's man, on the quarter-deck or before the mast? The opponents of the older system should look forward to the time when, the last person acquainted with spars and sails having disappeared, the *personnel* of the Navy will consist of those who have been trained in drill alone. What evidence can they bring forward in the smallest degree tending to establish the belief that such a *personnel* will show real adaptability for naval warfare? It is not too much to say that they are bound to bring forward not evidence alone but proof of this. The change they advocate is great and revolutionary. Where is the wisdom in urging a change which has not been proved certain either to effect improvement or even to keep things as good as they are?

People who have been long engaged in drilling, or who have been associated with drill establishments, are naturally inclined to attach undue importance to drill. There is a tendency, to be observed in every country, to consider as the end what is in reality but the means; and this tendency is especially marked in those by whom the work of drill is elaborated and conducted. They are apt to believe not only that drill is everything, but also that everything must give way to it. As a class they are represented by the Russian Grand Duke who said he hated war: it spoilt the soldiers' drill. They cannot, therefore, be accepted as judicious advisers on the subject of training in general. They should, of course, be heard, but as witnesses, not as judges. As was remarked in a leading article in the *Times*, such persons are "very often the worst possible judges" of an educational system. "They are apt to think that it consists in imparting knowledge, not in developing faculty; in teaching a man to do directly this or that, not in so training all his faculties, mental, moral, and physical, as to make it twice as easy for him to learn to do this, that, or anything else within the range of his native capacity." Now, in their statement of the case against sails there is not a trace of any intention on the part of the opponents of the older system to substitute for it one with any object but that of simply imparting knowledge. As to developing faculty, they never even mention it. The country can see what it has to choose between—a system which, by the admission of its antagonists, did foster the qualities essential to naval efficiency, and another which makes no pretence of doing more than merely impart knowledge of a special and limited kind. We have sufficiently recent experience of serious war to be able to understand where the adoption of the latter would be but too likely to land us.

Dangers of a change.

It was contended during the discussion in question that the abolition of the older training would probably, or certainly, reduce the already much shrunken time spent afloat and "in blue water" by our young seamen. That this contention was not unfounded was demonstrated by the letter of a correspondent of the *Times*, who advocated the abolition of the training-brigs. If this correspondent's views are approved we must expect demands for the abolition of the masts in the stationary training-ships. These having been abolished, consistency—as no doubt will be triumphantly pointed out—will compel the abolition of the ships themselves, and the transfer of the boys under training to buildings on shore. Except that they will be greatly more expensive, and as purely educational establishments less efficient, it is not easy to make out how such places, when established, will differ from Board schools.

It has several times been asked, and very much with an air of putting a question to which there could be no answer, where the instructors are to be got if the older system is continued. A correspondent, who knows both the Navy and the Mercantile Marine, replied that if they could not be found anywhere else, they could be found in the merchant service. In these days, in which we can perceive that one result of a long peace is to engender in both armies and navies a considerable amount of professional self-sufficiency, the reply of this correspondent may be to many unpleasantly startling. It will afford some relief to their feelings if they will just cast a retrospective glance over the training arrangements of the Navy for the last thirty years or so. They will find that whenever any particular course of instruction has been considered desirable, it has been the unvarying custom of the British Navy to go outside its own ranks for instructors. When an improved sword drill was thought desirable, it was to the Royal Artillery that the Navy went for an instructor. Practice with rifled small arms was introduced into the Navy by officers and sergeants of the Royal Marines. When it was decided that the instruction should be given by naval officers and bluejackets, the first instructors belonging to the Service were sent to the Army School at Hythe, which accounts for the early naval name of musketry instruction, viz., "Hythe Course." The first instructors in torpedo were sent to acquire the necessary knowledge at Woolwich and Chatham under the Royal Engineers. The first gymnastic instructors in the Navy were not bluejackets. The first bluejacket instructors qualified themselves for their work at Aldershot and the Military Gymnastic School at Portsmouth. However, it is probably quite unnecessary to look for the instructors required anywhere but in the Navy itself. When a thing is in fashion there is no difficulty in finding instructors to impart a knowledge of it. Masts and sails training happens just now to be out of fashion. Old fashions, however, often reappear: and he is wise who makes allowance for the reappearance. There were advocates for the "secondary armament" of ships even in the days when it was the fashion to declare that the Devastation was the proper type of fighting ship of the future. Who does not now wish that these advocates had been listened to with attention? If we are to abolish the old training system, it is surely not too much to ask that the one substituted for it should give reasonable promise of developing faculty as well as simply imparting knowledge; and we should bear in mind that the abolition demanded is in no way justified by the impossibility of finding instructors.

CHAPTER VIII.

NAVAL BRIGADES.

The Navy and inland war. IT is impossible for this country to engage in any war which is not in some measure naval. Even where our seamen and marines take no direct share, it is still the case that the navy is at the back of the army. We are in a position to send punitive expeditions into the hills of the north-west frontier of India because we have been a sea power. The fleet has its indirect and, on the surface, invisible share in all our struggles, while the wars in which it did not take a direct and visible part have been with us the exception and not the rule. Our enemy may not only have no ships, but no sea coast, and yet the navy acts in the field against him. It was present with its guns in the Indian Mutiny, and in the operations against the Maories. In Upper Burma and in the Sudan there were great rivers, and the case is not equally strong. To-day we are again fighting with a foe who has neither ships nor sea coast, and is not accessible by river, and yet the seamen and marines are side by side with the soldiers, in some cases at great distances from their proper element. There is a naval brigade with Lord Roberts while these words are being written, which may be further inland before they are printed, illustrating as fully as Captain Peel's men did in the Indian Mutiny how completely our Empire, to its most remote frontier, is permeated by the fleet.

This share of the navy in the war with the Boers has been so taken as to give rise to some controversy. I have no intention to write a word tending to revive a dispute, which happily died down early. Yet the questions, Under what circumstances men who have been carefully prepared to fight in ships, and on the sea, ought to be landed to do work ashore as artillery and infantry; how expeditions landed for such purposes should be composed; and by whom commanded? are of great and permanent interest. Moreover, we must consider them, since our Empire is of such a nature that our fleet must needs be from time to time called upon to do the work of an army. Generations have passed since the British Empire ceased to be accurately definable as insular. Since the earlier part of the last century it has been a power with large land frontiers. To-day no

State has so many borders in such widely scattered parts of the world as the British Empire. When Campbell wrote that our "march" (using the word in its ancient sense of border) "is o'er the mountain waves," his words were only poetically true. Yet they could pass, for our march along "the steep" was not then formidably attacked. To-day they would not be even poetically true. The score of land frontiers which our growth over sea has given us to defend are liable to be assailed by enemies who range all the way from the organised armies of European rivals to the casual inroads of savages. Now, it has never been our policy to maintain such forces in every quarter of the Empire as can be relied upon to protect the point assailed from all sudden assaults of its inland neighbours. Whether we could do so is a problem which need not be discussed here. We must begin by recognising the fact that it has been in the past, and is to-day, the case that many British possessions are open to foreign attack or subject to internal rebellion, which can only be met at first by calling on a part of our forces available everywhere and at short notice. This is, and from the nature of things must be, the man-of-war, which is always within twenty-four hours of everything. To take an image from another art, it is the function of the Royal Navy to supply first aid in case of accident. It is the naval brigade we look to for immediate protection when some King Coffee Calcalli is raging at the back of a West African coast settlement, and when Natal has suddenly to be guarded against the rapid advance of an enemy little if at all less dangerous than a European army, once more it is the naval brigade we look to for the first reinforcement. When there is a riot, threatening if neglected to grow into worse, in a West Indian island, the restraining force has to be called for from the ships. The naval brigade is in constant request, and it acts in a wide variety of circumstances.

The call upon the Navy in emergencies.

Though sailors have often been called upon to take part with soldiers in fighting ashore, it is not so easy to find historical guidance in answering the questions put above as might appear probable. The formation of war fleets in the modern sense of the word is comparatively recent. In the ancient world, and in the Middle Ages, men fought not a little on the sea, and crossed it much for the purpose of fighting. Yet the sea officer and the seaman did not form the dominating element. The rowers of the Athenian triremes must have been sailormen of a kind, and now and then we hear of such a man as Phormio who defeated the Corinthians off Naupactus, and who must have been an admiral in the full sense of the word. No doubt, too, the Carthaginians were seamen, and some of their commanders handled their fleets in what our admirals of the eighteenth century would have called "an officer-like manner." But we know little about these old sea fighters, or of the trireme which was their line of battleship. Only this is very certain, that whatever its construction may have been, it was something so complicated and calling for such extreme skill on the part of its crew that it fell into disuse long before the Roman Empire came to an end. It was as great a mystery to the men of the sixth century as it is to us. In later times the warship became a mere transport of one or two banks of oars (the two-banked boat which was employed by the Malays till recently being a very simple affair) rowed by slaves, and carrying soldiers. All through the Middle Ages similar conditions prevailed. There was fighting at sea, but it was pretty much akin to war on land. The mariners' share in the work was chiefly to lay his ship alongside the enemy, and give the knights, spearmen, or archers a chance to handle their weapons. In Edward the Third's battle "with the Spaniards on the sea" the actual fighting was done by the king himself, his sons, nobles, and knights, with their squires and other followers. They were the elements of his army at Creçy or Poitiers. Even the earlier Norsemen did not aim at capturing an enemy's ships mainly, though they would do that when occasion served. Their object was to land somewhere, haul their ships up on the beach, stockade or embank them, and then march inland, leaving them under charge of a guard. When they were beaten their camp afforded them refuge, and they could launch their ships again and escape. In such conditions there was no opening for what we understand as a naval brigade—that is, a force of men whose business it is to fight at sea, but who are employed on land for a temporary purpose.

Great principles unvarying.

Yet the great principles of war have always been the same. The proper use of a fleet may be illustrated by the campaigns of Roger de Lauria, Luria, or Del Oria (for his name is given in all these forms), the Sicilian Norman who fought for the kings of Aragon in the fourteenth century, as fully as by those of De Ruyter, Hood, or Suffren. So it happens that the mediæval wars afford one very striking example how not to use a naval brigade. This was the disaster which came upon Johann Wittenborg, the admiral of the Hansa League, at Helsingborg in 1362. The League was at war with Waldemar of Denmark, and Wittenborg, who was Burgomaster of Lübeck, had been sent out to sweep the coasts of the Danes, and to expel them from Scania—that is, a part of southern Scandinavia lying between Sweden and Norway, which was then a Danish possession. It now belongs to Sweden. Wittenborg met no Danish fleet at sea. So after ravaging in the usual mediæval way,

Wittenborg at Helsingborg, 1362.

and plundering Copenhagen, he went on to Scania, expecting to be joined by the Norwegians and Swedes. They did not keep touch. The Hansa admiral came too readily to the conclusion that King Waldemar was not to be feared on the water, and landed his men to attack the town of Helsingborg. But though he did not know it, the Danes had still "a fleet in being," and their King was the man to employ it to some purpose. When the siege had lasted for sixteen days, and the Hansa crews were busy with their catapults, Waldemar swooped down upon their unguarded ships and made an example of them. He carried off twelve of the largest, and all Wittenborg's stores. The unhappy Burgomaster had to make his way back to Lübeck with a mere remnant of the fleet and army he had led forth. The Hansa was intolerant of errors of judgment, and Wittenborg's mistake cost him his head. It was hard measure, far harder than was meted out to Byng, for the Lübecker had really erred from error of judgment, and not from want of will to try his hardest. Yet grim as the treatment was, it was excellently adapted to instil a due respect for "the fleet in being" into the minds of the admirals of the Hansa, and to teach them not to violate elementary principles by turning a sea force into a land one till it was quite certain that they were not subject to attack from the water. The permanence of the essential conditions of war, under all superficial changes due to the development of weapons or growth of organisations, is illustrated by the fact that more than four centuries after these now all but forgotten events on the coast of Scania, Nelson was mildly rebuked by the Admiralty for committing something not unlike the mistake of Wittenborg. In his zeal for the cause of the King and Queen of Naples he weakened his ships to supply a naval brigade for the occupation of Capua. No misfortune followed because the French had no force, or, rather, no spirit to use what they had. Yet the Admiralty was right on the general principle. If, when Bruix did come round from Brest, he had had more confidence in himself and his crews, if he had struck boldly at our scattered fleet with his own united force, if he had found the squadron at Naples weakened by the absence of the naval brigade at Capua (and these conditions might have been met together), then all Nelson's unrivalled power of fighting battles, all his energy and skill, would hardly have saved him from disaster.

Dangers of depleting fleets of men.

It may be laid down as a rule to which there can be no exception that a naval brigade should never be of such strength that, in order to form it, there is a necessity to weaken the crews of the ships from which it is drawn so far as to leave them unable to meet an enemy's vessels. This is quite another thing from saying that the ships are

Limitations to the employment of seamen ashore.

never to be hazarded. The noblest man-of-war is but a means to an end—which end is the service of the State. It may at a given moment be the lesser consideration. Then it must be sacrificed altogether: sunk in a fair way to block an enemy's passage, or blown up because there is a peremptory need for the crew elsewhere and there is danger that in their absence the foe may endeavour to capture it. But so long as there is still the wish and the chance to use it on its proper element it must not be unmanned. The sailor is for the sea first and foremost, and only for the land on occasion, and subject to the prevailing obligation to stay by the ships. There is an excellent statement of the orthodox doctrine in the "Journal of the Capture and Recovery of Nevis," printed in Ekin's Naval Battles. It was made by Captain Everett of the Solebay, or by one of his officers. The Solebay had been forced on shore at Nevis by the French during what are known as Hood's operations in the Basseterre of St. Kitt's. Captain Everett blew her up after landing his men. "Some unthinking people" then suggested that he and his crew could remain to take possession of the island, and were answered that "it was of more consequence to His Majesty's service to join the English fleet with his men than to take such a *paltry island*." Captain Everett, or his officer, stated the case very well.

The danger incurred by disembarking the crews when there is an enemy on sea within striking distance is so palpable that there are but few examples of commanders who have repeated the fatal mistake of Wittenborg. Indeed I have met with no equally glaring instance of the error and its punishment, unless it be the still earlier destruction of the ships of the King of the French, Philip Augustus, at Damme, by William Longsword, in the reign of King John. Here, too, the circumstances were rather different. Yet it is not uncommon to meet cases when some of the most famous of the Elizabethan adventurers did undoubtedly put themselves in positions in which they might very well have incurred the fate of the Burgomaster of Lübeck. Here again we must not run the comparison with a modern fleet too close. Though our ships, whether belonging to the Queen or to the towns, were used in the Armada year for merely naval purposes, the enterprises of the Elizabethans, and more especially the voyages to the Isles and the West Indies, bore a much closer resemblance to the raiding expeditions of the Middle Ages and the Norsemen than to the operations of fully-developed fleets. They went out to land, and were full of soldiers, the sailors being there to handle the ships. The seaman element was increasing, but it had not yet become thoroughly predominant. How far this was the case is shown by the well-known fact that far into the seventeenth century it was not thought necessary that admirals and captains should be bred to the sea. Where the superiority is with the soldier, to whom the ship is but a transport, and whose aim is some feat to be achieved on the land, it is natural to find a certain readiness to risk the vessels when once they have carried him to the shore, or, at least, some want of understanding of the full scope of the peril which comes from neglecting them. We need not make too much of the lesson contained in the story of such a man as Oxenham. He was a desperado who knowingly risked the gallows, and who, we may be sure, accepted the penalty of failure as a matter of course. Yet it remains the fact that the actual cause of his ruin was the discovery by the Spaniards of the vessel he had hidden on the eastern coast of Darien. With his retreat cut off, it was a matter of days when he would fall into the hands of the enemy, or perish in the forest. Drake might well have incurred the same end, if the Spaniards had found his vessels, while he was in the mountains with the Cimarrones. Indeed, after the famous capture of the *recua* with the treasure, his followers were thrown into no small terror by the fear that their ships were gone.

Drake and the Spaniards, 1585 and 1594.

Drake's reliance on the naval weakness of King Philip, and the mismanagement of his officers, was uniformly justified by the result. Therefore we cannot fairly say that he was rash in his expedition of 1585, or even in the later and unfortunate voyage of 1594. He had calculated that the Spaniards would not interrupt him by attacks from the sea; they did not, and so he was not overbold. Yet both those voyages are full of examples of what ought not to be done with a fleet. In 1585 Drake sailed with twenty-five ships and 2300 men, of whom the majority were military adventurers. He lost some 300 or so before he reached the West Indies. When the town of San Domingo was attacked, a force of 1,200 men, equal to about the whole of the then population of the place, was landed. This would leave not more, perhaps even less, than 800 men with the ships. Now if King Philip, who was early informed of the sailing of the expedition, had had his navy in an efficient state, the English would have been in imminent peril of renewing the experience of the Lübeckers in 1362. We may be sure that the smaller of the twenty-five ships, helpless little victuallers, and craft of about the size of a herring lugger, would have run at once. The larger would have made a fight, but they would infallibly have shared the fate of the Frenchmen at the Terceiras a little earlier. Then Christopher Carleil and the 1,200 men on shore would have been cut off. So again in 1594, if the Spanish ships which Baskerville, who succeeded to the command on the death of Drake, met on his way home, had come down on our vessels while their soldiers were landed for the march to Panama, and had been well handled, the voyage would have ended not in repulse, but in disaster. To be sure, if the Spanish navy had been efficient, no such expeditions would ever have been undertaken. The feats of the Elizabethans in the West Indies from Drake to Cumberland, those of their baser successors the Buccaneers, and the achievements of the Frenchmen Pointis, at Carthagena, and Duguay Trouin, at Rio de Janeiro, all tell the same tale. It is that in dealing with the Spaniards and the Portuguese, who "are wanting in everything at the critical moment," as the Duke of Wellington used to say, very much may be ventured with safety.

Naval Brigades in the Civil War.

The seventeenth century saw first an arrest, and then a very rapid development of the navy as a fighting force wholly separate from the army. The peace of King James's reign stopped the process of formation which had gone well towards completion under the great Queen. It even threw the navy back, but the recovery was very rapid in the Civil War, and in the war with the Dutch, from whom we learnt much by example, and even more by opposition. The vital share of the navy in the Civil War gave an opening for the use of naval brigades in the proper sense—that is to say, of detachments from the crews serving in the fighting on shore, but with an ever present sense of the vital importance of attending to the safety of the ships. Though the King was weak at sea, he had generally some vessels at his command, and the "Tories" who sailed on his behalf from Irish ports and the Channel Islands were not despicable foes. The journals kept by Sir William Penn during his service for the Parliament on the coast of Ireland, and published in Mr. Grenville Penn's "Memorials," show the nature of the training which the navy then received. It was just such work as has been done by not a few officers now living. There are constant notices of measures taken in combination with the military officers, of guns lent, of men landed, for this purpose or for that, commonly with some such proviso as "provided always and by all means (no excuse to the contrary) they repaired on board every night; as also, at all times of the day they be ready to do the like, if necessity should require." A great change had happened in the half century since the taking of Cadiz by Howard, Essex and Raleigh. Then the admiral and the sea captains were still gallant gentlemen, whose business it was to fight, but whose connection with the sea was temporary and accidental. By 1646 the sea officer is a quite different and a professional man. The official formation of a corps of sea officers began with Charles II., and was even then a slow process; but the type was formed, and the Crown only recognised what already existed. At Cadiz admirals and captains are seen hurrying eagerly ashore to be abreast of one another in the storming of the town, to the neglect of the sailor-like business of destroying the Spanish ships which were trying to find refuge up the bay. In 1646 the naval officer never loses thought of his vessels, and whatever he does in combination with the soldier is done subject to the obligation to think first and always of them.

Under the Common-wealth.

It is, therefore, natural that we should find an undeniable naval brigade used in the wars of the Protector. One was formed in the jarring, and on the whole most ill-handled, fleet sent out to attack the Spaniards in the West Indies in 1655. This fleet carried a considerable body of soldiers, and it recruited more men in the West Indies. But neither the troops it carried from home, nor those raised at Barbadoes and the Leeward Islands, were of trustworthy quality. The soldiers were largely disbanded men of the King's regiments who had never been well disciplined, and had lost much of what military quality they ever had. At Barbadoes the new levies belonged to the floating element of adventurers who abounded in the West Indies, and who afterwards formed the bulk of the Brothers of the Coast and the Buccaneers. Of such men it may safely be asserted that they "came in on the plundering account"—to use a phrase of the time. Neither element was to be trusted. A regiment of sailors was formed and put under the command of Goodson, who, from the little that is known of him, appears to have been both an energetic man and a convinced Puritan. During the disastrous attack on San Domingo, when most of the soldiers and the West Indians behaved very ill, the sailors of Goodson's regiment set an excellent example. But for them and a few of the soldiers who did their duty, the fugitives would have been cut to pieces by the handful of horsemen from whom they were flying in panic. It is to be noted that these sailors had been specially picked and drilled after the fleet reached Barbadoes. Neither then nor for long afterwards was it thought necessary to give all men-of-war's men training as "small arm men." In Elizabethan times the Earl of Cumberland, though an experienced and seemingly a careful commander, had actually waited till he was in the West Indies before dividing his men into companies and practising them for his attack on Porto Rico. A large proportion of his following would consist of adventurers who had already formed part of the corps which were being constantly raised, and again disbanded, in those times. We may fairly suppose that the sailors chosen to form Goodson's regiment were precisely those who had had some practice already. Therefore we need not suppose that there was a total lack of experience either in Cumberland's followers, or in Goodson's regiment in 1655. Indeed, the success of both bears witness to their capacity. Still, we see that our ancestors of the sixteenth and seventeenth centuries were not yet convinced of the need for timely preparation.

The soldier as seaman replaced by the seaman as soldier.

The reign of Charles II. was a time of much bad execution, but it was also a period of not a little good design. It saw the beginning of the regular corps of officers, and of the attempt to make the sea-service as honourable as the land, which Coventry said was the object of the Duke of York. To this time also belongs the first effort to provide the navy with a permanent corps of "small arm men" in the shape of the admiral's regiment. There was, indeed, no novelty in the employment of soldiers in ships. On the contrary, the novelty lay in the growing superiority of the seamen and the sea officers. But in former times the soldier had been merely soldier. He went to sea as an incident in his service, but not to form part of a regular naval force. Now, however, we hear of the soldier as assigned to the service of the fleet. For a long time his position remained somewhat ambiguous. The admiral's regiment, which was disbanded by the Revolution of 1688, the regiments of Torrington and Pembroke, of King William's reign, the regiments of Queen Anne, and the other corps of unstable existence and ill-defined position, which were the predecessors of the present regiment of marines, raised in the beginning of the Seven Years' War, were designed not only to provide the fleet with small arm men, but to recruit the body of seamen. This was so much the case that they could be rated A.B. after two years' service at sea. The conflicts to which this use of the marines led belong to the history of the corps; but its foundation, or rather successive foundations and destructions, till the present regiment was reached, may be noticed here as part of the process by which the navy was rendered ambidexterous. It was supplied with a regular body of small arm men, and with the means of drilling the sailors.

The Dutch Wars.

The wars of Charles II. do not afford many examples of the use of naval brigades. As against the Dutch our work was at sea. Even the burning of the Dutch East India Company's storehouse and yards at Terschelling was all but purely naval work. Yet the retaking of St. Helena in the second war by Munden and his ships was an instance of the use of a naval brigade. The sailors had to land and turn the position of the enemy who held the port by climbing up the cliff, which is still called Hold Fast Tom in memory of the feat. Here work was done ashore by sailors landed for the purpose, and done both in proper circumstances and the right way. There was pressing need to recover the post, which as being the one resting-place we had of our own on the route to the East Indies was of inestimable value in those days of long voyages and ever threatening scurvy.

Therefore Munden was justified by the importance of the stake in risking something while the chance that he would be interrupted by the Dutch from the sea was not so very great. It might have deterred the stamp of timid man who has formed the opinion that war can be conducted without incurring any risk, but Munden was not of that class.

William III. and Anne.

It would appear at first sight that the long wars of King William and Queen Anne ought to supply us with numerous examples of the use of naval brigades. They were marked by repeated combined expeditions to the coast of France, and by almost yearly voyages against the French possessions in the West Indies. Yet they are on the whole rather barren. The attacks on the French coast were either futile, or, as in the case of the repulse at Brest, rescued from futility by disaster. Moreover, the forces landed were composed of soldiers sent for the purpose. As for the expeditions sent to the West Indies, they present, until we reach the last years of Queen Anne, the most dreary spectacle in all the history of the Royal Navy in so far as it is known to me. Here also it was the practice to send troops, commonly special regiments formed by drafts from those at home, to be landed. When the sailors went ashore it was rather in emulation of plunder than in honourable rivalry of service with the soldiers. The squadrons were frequently joined by Jamaica privateers, "who came in on the plundering account." This was the class of persons which produced Captain Kidd of piratical memory, and in the atmosphere of the West Indies, still redolent as it was of the Brothers of the Coast and the Buccaneers, seems to have set the fashion to soldier and sailor alike. Nobody who has looked into the history of the operations will, I think, be inclined to dispute the accuracy of Burchett's opinion that they brought the country nothing but loss and discredit. There is little to be learnt from a scene in which all the sweepings of the corruption of Charles II.'s reign were put down to work far from home, in surroundings which were at any rate semi-piratical, and with an encouraging sense that their great distance from control at home promised immunity for all excesses.

A Brigade at the taking of Gibraltar.

In Europe and under the eye of the great officers of the time, from Russell to Leake, things were better. A naval brigade, formed however of marines, had an active share in the taking of Gibraltar. The mole was occupied by the sailors after the bombardment. But the Spaniards, wanting in everything at the critical moment as their manner is, had left such a weak garrison in the place that the difficulty of the capture was not in proportion to its importance. Still the whole story of the taking of Gibraltar does illustrate very well the proper use of a fleet for operations against the shore. A place known to be ill garrisoned was bombarded, taken, garrisoned and stored by the fleet. When the Count of Toulouse came down to retake it, part of the stores and men were brought back into the ships to help to fight the battle of Malaga. When the French retired after the action, which they none the less claimed, and continue to claim, as a victory, the men and stores borrowed were restored as far as death and wounds and the expenditure of powder would allow. The operations on shore were kept subordinate to the effective use and safety of the fleet; and as it was not defeated they proved successful. Here, again, it may be pointed out that timidity and an overpowering sense of the perils which might supervene if the enemy appeared at sea would have prevented the Allies from acting against Gibraltar at all while there was any prospect that the Count of Toulouse might come. But Rooke and his Dutch colleagues acted as the Duke of Wellington did at Ciudad Rodrigo and at Badajos, when he knew that armies superior to his own were marching to the relief of those places. They struck in the interval allowed them, and the men they had to spare for the garrison did not so weaken them but that they were able to fight a drawn battle, which left the town in their possession. This, then, is an example of the legitimate use of a naval brigade—a measurable risk incurred for a sufficient object. The landing of more than half the fighting crews of Drake's ships at San Domingo was only accidentally legitimate. It was safe because there was no enemy, and he might have left his ships at anchor without a man in them and not have incurred appreciable danger. Yet he can hardly have had the right to consider that this was certain, though he was justified in holding that it was highly probable. Another instance of the correct use of a naval brigade was the landing of 2400 seamen and marines to work the batteries during the siege of Barcelona in 1705 by the English ships, and of 600 by the Dutch. This draft reduced the ships to their "middle complement." The help was given to the Prince of Hesse and to Peterborough on the distinct understanding that if the French fleet approached the men were to come on board again. In this case a risk was run since possible conditions of wind and light might have laid the Allied fleet open to attack in circumstances which would have made it hard to recover the men in time. But considering their own strength, what the Allies had seen of the methods of the French, and what they knew of the effect the financial distress of King Lewis had had on the condition of his navy, it was a reasonable risk they were entitled to run.

A legitimate example.

By the end of the reign of Queen Anne the navy was constituted in all essentials. Many improvements were made in detail, and it went through a moral and intellectual change of a very thorough kind between the war of the Austrian Succession and the American War of 1779–1783. But in outward things the alterations were less between the death of the Queen and the end of the first third or so of this century than have taken place in the last forty years. There will, therefore, be no need to diverge again into the constitution of the navy in dealing with the employment of naval brigades. A ship's company consisted of the "prime seamen," who were never secured in large numbers except by the direct or indirect action of the press, and of whom a captain rarely had as many as he desired; of marines, or soldiers serving as marines; and of the miscellaneous element of "waisters," landsmen, ordinary seamen and boys, who again were diluted by drafts, or volunteers, from the jails. With these elements, in which the bad or doubtful often predominated over the good, our officers had not only to fight at sea but to co-operate continually with the troops on shore. Their own exertions, a ferocious discipline, the example of the good men, and a natural pugnacity not always wanting to the bad, enabled them to achieve an extraordinarily high level of general success. The navy was never at a lower level of spirit and intelligence than when Vernon sailed to Carthagena with Wentworth. Yet it is striking to see how superior the sailors, officers or men, were to the soldiers in spirit, in self-reliance, and in adaptability. It would be quite beside the question to go here into the dispute between Vernon and Wentworth. Yet the open quarrel to which the sailors and soldiers came in that ill-managed business cannot be altogether passed over when we are dealing with the use of naval brigades. If they were to be employed with full effect it was necessary that they should work harmoniously with the soldiers. Unfortunately, there was a standing feud between the two dating back at least to the middle of the seventeenth century. The soldiers and sailors of the expedition of 1655 came to blows at Jamaica. In King William's reign military officers commonly declined to go on expeditions to the West Indies because they were unwilling to subject themselves to the "arrogance" of the sea commanders. The sailors retaliated by resenting the pretensions of the military men. We have to keep this mutual hostility well in mind in order to understand how two English gentlemen in the position of Vernon and Wentworth could come to the terms they did. The fact is that the sailor expected the soldier to be an ass—that is, an animal to which it was necessary to appeal with a cudgel. The soldier expected the sailor to be a brute. We will not inquire how far these respective estimates were justified in the particular case. It is enough to

Unfortunate co-operation.

Vernon and Wentworth at Carthagena.

point out that while they were held, harmonious co-operation was hardly to be expected, and while that was true, it followed that the amount of good work done by naval brigades was necessarily limited.

Pondicherry, 748.

Another instance of unfortunate co-operation occurred at the end of the same war. This was the abortive siege of Pondicherry in 1748. Boscawen, who commanded, was a man of more ability and also of incomparably more temper and judgment than Vernon. Moreover, he was in a much better position. Pondicherry was not naturally so strong a place as Carthagena, and he was commander-in-chief of the force of 5220 men landed. Of this number 880 were marines and 1100 were sailors "who had been taught the use of small arms." There was here the unity of command which Vernon believed would have secured success at Carthagena. Yet the siege was a complete failure. The reason for the ill success is not far to seek. Although Boscawen was in the position and had the responsibility of a general, he had not the full power, and he did not show that he had any of the knowledge. That he had "conspicuous valour and anxiety for the public service," as Beatson has it, is unquestionably true. But he was told to rely on his engineers for professional advice, and they were incompetent. Indeed, it was not till later times that we began to form a regular corps of engineers. An officer of more pliability of mind than Boscawen would have adapted himself to the circumstances, and would have found means to make good the deficiencies of the engineers. Nelson did when the same task was set him at Bastia. But Boscawen, though unquestionably an excellent seaman and a stout fighter, was nothing else, and as much must be said of the very great majority of the admirals of his generation. The navy was weak on the intellectual-military side. Even so energetic a man as Boscawen's contemporary Pocock was content to endeavour to carry out the Fighting Instructions, which were so drawn up as to make a decisive battle between equal fleets as good as impossible against an adversary who did not choose to meet us half-way. Therefore, it was only natural that when an officer who had been trained as a seaman to fight according to certain precise rules, was suddenly called upon to interpret the great underlying principles of war into unfamiliar terms of soldiering, he should have been at a loss. He did not even know that there were principles to apply, but only instructions to obey. Nelson, who had the happiness to rise in a generation in which Rodney's battle of Dominica and Howe's victory of the 1st of June had swept the cobwebs out of the eyes of naval officers, was able to interpret, but Boscawen was not, and probably would not have been, even if he had been unhampered by directions to take the advice of his engineers. So on this occasion the naval brigade had once more to take part in a costly failure. The sailors alone contributed 265 men to the total loss of 1065.

The Spanish and Austrian Succession Wars, which had dragged languidly along from 1739 to 1748, present, on the whole, a melancholy scene. The best they did was to rouse the nation, and the navy with it, to a sense of the pressing need for reform, not so much in machinery as in spirit. The results of the great movement which pulled the navy up from the Slough of Despond of Walpole's last years were seen in the Seven Years' War. Yet this began badly enough with the tragedy of Byng. We have not to deal with that story here, except to point out that one detail of it has something to tell us, at least indirectly, on the subject of naval brigades. When Byng left home he carried with him Lord Robert Bertie's regiment of Fusiliers to reinforce the garrison of Mahon. Marines were sent ashore to make room for these soldiers. It follows that if the admiral had landed Lord Robert and the regiment, he would at once have found his crews dangerously reduced. It would be difficult to quote a better example of how "not to do it" when a fleet has to co-operate with troops. The garrison could not have been reinforced without crippling the ships, though it was known that the French had a squadron at sea; of this Byng had every right to complain, though he had no right to make the laches of the Admiralty an excuse for putting his hand to the plough as if he feared it would burn his fingers, instead of setting himself to make good the follies of his superiors by his own exertions, and in the interest of his country. But the miserable failure to relieve Mahon was an inheritance from earlier and worse times. Already better times were beginning in the East Indies.

Right co-operation.

Clive and Watson.

The expedition which punished Surajah Dowlah for the capture of the company's factory at Calcutta and for the horrors of the Black Hole may, not fantastically, be said to mark the opening of a new era in the relations of soldiers and sailors. Clive and Watson, who commanded the squadron, had their differences of opinion, and on one occasion at least the soldier had some reason to think that the admiral was presuming on his rank as king's officer, but neither allowed his personal feelings to interfere with the Service. Although Clive was throughout the more conspicuous of the two leaders, the success of the enterprise was largely due to Watson, who co-operated loyally. If his place had been held by such a man as Griffin, who had commanded with signal incapacity in the East Indies during the previous war, the result might well have been very different.

Griffin was a bad specimen of the mere "Tarpaulin" and brutal "Wapineer Tar," though a gentleman by birth. He would certainly have been arrogant and quarrelsome, on the look-out for every shred of excuse for hectoring a soldier officer. The military men of an earlier generation would have been equally on the watch for something to resent. Clive and Watson made the least of all possible causes of difference between them and the utmost of every reason for promoting the public service. The major part of the work which fell to the squadron was thoroughly naval. The ships attacked forts and carried the soldiers up the river. But naval brigade work was done also. Historians of the campaign record how one "Strachan, a common sailor belonging to the Kent," being fired with native valour and also "grog" made of arrack (rice spirit), precipitated the storm of the fort of Budge-Budge by rambling into the breach during the night, when he saw himself double. It was not the most heroic way of promoting a feat of war; but at least Strachan's excesses did not prevent him from going in the right direction. The prominence given to this absurd story comes of the universal love of gossip. There was a small naval brigade of fifty men, commanded by Lieutenant Hayter and seven midshipmen, in the little army which finally ruined the Nabob of Bengal at Plassy, while the whole work of protecting the communications by land as well as by water fell to the squadron. The campaign was in fact an admirable example how to combine ships and men, and men from the ships with their guns, in order to attain a common object.

As long as the French continued to offer resistance to our advance on the coast of Coromandel the navy had its share of the fighting on shore as well as the whole task of protecting our own communications and cutting those of the French at sea. Towards the close of the war two combined expeditions took place which were as harmonious and therefore as successful as the campaign of Clive and Watson, though neither left equally permanent results.

Cornish and Draper at Manila.

Naval guns landed.

The first was the capture of Manila by Admiral Cornish and Colonel Draper. The second was the capture of Havana by Pocock and the Earl of Albemarle in 1762. A battalion of sailors was landed at Manila to assist the marines who were already attached to the troops, and the army drew its battering train from the ships' guns. Eight 24-pdrs. and two 18-pdrs. were landed by the admiral, "as, to save time, we brought only the land-carriages and platforms from Madras." It was no new thing when the army had to look to the navy to supply it with guns of position at Ladysmith.

Keppel's Brigade and guns at Havana.

The capture of Havana proved a more arduous business than the taking of Manila; but it was equally harmonious and successful. The "happy and perfect unanimity which subsisted between the land and the sea services" is recorded by the chief engineer, whose journal is printed in Beatson, with a faint air of surprise which suggests doubts whether he did not expect a different scene. Only twenty-two years had passed since the notorious quarrel of Vernon and Wentworth, but many things had happened in the interval, and notably the great ministry of the elder Pitt had instilled a new spirit into all officers. Besides doing much in its own way and on its own element the navy worked hard to serve the army. Keppel, acting under the direction of Pocock, "landed a considerable body of seamen, who were extremely useful in landing the cannon and ordnance stores of all kinds, manning the batteries, making fascines, and in supplying the army with water, there being no water or wells on the Cavannos. The admiral landed cannon of different calibres from the ships, two mortars from the Thunder bomb on the east side, two from the Grenada bomb on the west side, with old cables for erecting defences, old canvas for making sandbags, and ammunition; afforded every other assistance that was in his power; and set himself the example of the cordiality and harmony which subsisted between the two corps." This desire to make everybody understand how well the navy and the army had pulled together was obviously shared by Sir George Pocock. In his letter to the Secretary of the Admiralty announcing the fall of the town he says: "It will be as needless, as almost impossible, for me to express or describe that perfect harmony that has uninterruptedly subsisted between the fleet and army from our first setting out. Indeed, it is doing injustice to both to mention them as two corps, since each has endeavoured, with the most constant and cheerful emulation, to render it but one; uniting in the same principles of honour and glory for their King and country's service." The navy had been engaged in the work on which Collingwood described himself as having been engaged at Bunker Hill, namely, in "supplying the army with what was necessary for them." The sailor was, in fact, the handy man who was to be relied on to fight, build, dig, fetch, carry, and, generally speaking, make good the want of a proper corps of sappers and miners, commissariat or transport. Granted the absence of an enemy on the water, which was the case at Manila and Havana, this way of making a completely-equipped army by combining the two corps into one is perfectly legitimate. It is also to be noted that the Spanish sailors were doing much the same work on the other side. The guns in the batteries were mostly fought by them, and the defence of the Moro was directed by a naval officer, Don Luis de Velasco, who was killed in the storm.

The great French War.

Collingwood's account of his share in the battle of Bunker Hill is in fact a compendious description of much of the service rendered by the navy in the war of 1779–1783, till the intervention of France and Spain made it a naval war of the first magnitude, and even afterwards. The San Juan expedition, an essentially foolish business which all but deprived this country of Nelson, might be quoted as an example how not to employ a naval brigade or any other force. But both in this and the greater revolutionary war which followed, the instances of the services rendered by the navy to the army become so numerous that it is impossible to give a list of them. The naval campaigns of the first period and Wellington's six years of fighting in the Peninsula present so many and such glorious objects to our national pride that we forget much of the events of those years. One has to go out of the way a little to be able to realise to what an extent the British army of that period was afloat from Sweden to Egypt, with intervals of disembarkation here, there, and everywhere, often followed by swift re-embarkation. In all this the navy had a share which brought it in the way of doing much service on land. One passage is indeed familiar, because it belongs to the actions of the only admiral whose personality and life have ever taken hold of the imagination and affections of his countrymen—namely, the sieges of Bastia and Calvi, and, in general, the operations ashore in Corsica during 1794.

Lord Hood in the Mediterranean.

It is a story which has the further merit of illustrating very well how and in what circumstances a fleet can serve the purpose of an army. The occupation of Toulon had not done so much injury to the French Republican Navy as we had intended that it should, but it did enough to make our enemy totally helpless in the Mediterranean for the time being. Lord Hood could therefore calculate on being left free for a space from all interruption by French squadrons. It happened that the island of Corsica presented him with an almost ideal opportunity to deliver a blow at the enemy. The Corsicans were in revolt, and had driven the French from all the island except a few fortified coast towns. It may be added that these garrisons and their officers must not be supposed to have been of the same quality as the troops of the veteran armies of the Republic and the emperor were a few years later. They were full of the anarchy of the Revolution. We, on the other hand, had a body of soldiers at our disposal, not experienced as yet, but well-drilled, obedient, and quite ready to fight. This little army had been too small to hold Toulon, but it was amply sufficient to "discuss" the remains of the French forces in Corsica. In combination with the fleet it would have had no more difficulty in performing the task than would have been enough to give zest to the meal. Lord Hood was perfectly justified in expecting the co-operation of the military authorities in a resolute attempt to make an end of the remaining French garrisons. Everybody knows that he did not get what he might fairly have counted on. Sir Gilbert Elliot, who had been sent from home to act as commissioner at Toulon, and was now in Corsica, speaks bitterly in one of his letters of the "high lounge" of the soldiers with whom he had to deal, and compares it with the brisk zeal of the naval men. He was unquestionably thinking of the refusal of Dundas to lend his help in the siege of Bastia. Apologies may be laboriously made for the general, but they will never be anything more than the plausible excuses which are always available by or on behalf of the unenterprising man who is reluctant to act when difficulties have to be encountered. Even if the general had been right in considering the task of taking the town impossible, his troops would have been better employed in making the attempt than in doing nothing. But, as the result shows, it was not impossible. The navy has, however, some ground to be grateful to the memory of the general who gave it its most complete triumph over the army. When we say "the navy," it is necessary, however, to bear in mind that the majority of the men employed in the siege of Bastia were not sailors, but soldiers assigned to serve with the fleet as marines. Here, by the way, it may be noted that the help rendered ashore by naval brigades has its counterpart in the service which soldiers have given to the fleet at sea. They have often taken the place of the marine—that is, of the soldier attached to the fleet—and in conceivable circumstances might do so again. The presence of a draft of artillery or infantry in the modern battleship would be no more surprising or unprecedented than is the presence of a naval brigade at Ladysmith. When, during the Commonwealth and the Protectorate, it was found difficult to man the ships, the want of sailors was supplied by detachments of soldiers. Both Mahé de la Bourdonnais and Suffren in the course of their campaigns on the coast of Coromandel made up their crews with the help of soldiers, who were not always even of European race; and these two were by far the most successful of the French naval officers who have fought against us. Of course this was not because they employed soldiers, white or black, but because of their personal qualities. Yet their example shows that good use could be made of soldiers in ships, and that at a time when the trade of sailorman was far more alien from the shore life than it is in the steamer of to-day.

Nelson at Bastia. The force at Lord Hood's disposal for the siege of Bastia on the land was in the main military in training, and not a little of the work was done by military officers. The navy contributed the enterprise, the will, and the vigorous direction exercised by Lord Hood at sea and by Nelson on land. Then it contributed a proportion of sailors who were employed in "supplying the army with what they needed," and in those other ways already described by the engineer in the siege of Havana in the passage quoted above. In that period much of what we did in the world was effected by the process defined by Dugald Dalgetty as "laying the head of the sow to the tail of the grice," or, in less picturesque language, by making shift with what came to hand. At this the sailors, whose whole existence tended to make them handy, were likely to be particularly excellent. The view that you are not to be expected to achieve the result unless you are provided with everything needed for the purpose and a margin to guard against accidents was held, consciously or unconsciously, by Dundas. So, if the decision had rested with him, Bastia would not have been taken. The navy was taught to do with less, and so it tried and succeeded, the soldiers helping cheerfully, as they might be trusted to do when the example was set them, and they were no longer left to the direction of mere pedantry. Of all the services the navy rendered the army none was more signal than the stimulus given by the cutting rebuke implied in the taking of Bastia. I could not cite evidence to show that the lesson was appreciated, and yet it must have worked. At any rate, I do not think it would be possible to quote anything quite so bad as the languor of the military officers in those early days in Corsica from the following years of the long war. The triumph of the sailors was not very gently carried, and the relations between them and the soldiers were of the kind called strained for some time afterwards. Indeed, the naval men never, even towards the end of the Peninsular war, when the army had taken the first place and was abundantly energetic, quite ceased to look upon the military officer as essentially wooden.

Later instances. Space would fail for any attempt to describe the work done on shore by naval brigades round the coast of the Peninsula during the war. Its enterprises were innumerable. Phillimore's 'Life of Admiral Parker' gives a picture of it as it was seen on the north coast of Spain. But the most vivid account is in the seventeenth chapter of Dundonald's autobiography of a seaman, where he tells the tale of his defence of the Trinidad fort at Rosas. Much of the navy's share in those operations did not call for the employment of naval brigades. Our frigates and other light vessels were constantly engaged in giving refuge to the guerrillero bands of the Spaniards on the coast of the Bay of Biscay, and in landing them where they could resume their raids on the French. Again, the measures by which we made the coast road of Catalonia useless to the French were not at first of a kind to call for the landing of seamen and marines. We began by sweeping the road with that *feu d'enfer* which Marbot describes as having blocked the Corniche on the Italian Riviera years before. It was a most effectual method for a time. When, for example, the French were first repulsed from Gerona, and Souham had to retreat to Mataro, the fire of the English frigates made it impossible for him to march by the coast road. He was compelled to sacrifice all his artillery and take to the hills. When later on the French raised batteries, and made it less safe for our light vessels to approach the coast, then the navy fell back on the use of landing parties of seamen and marines to turn these places and destroy them. But feats of this character were performed all over the world wherever there were Frenchmen, or allies of France, to be attacked. The surprise of Fort Marrack in Java by Lieutenant (afterwards Lord) Lyons in 1811 was a conspicuous example. In these days there is less gratification than there was once in recalling the memory of the share the navy took in the occupation of Washington and the burning of its public buildings in 1814. Yet Rear-Admiral Cockburn's co-operation with General Ross of Bladensburg was as hearty, as indispensable, and as successful as had been the help given by Watson to Clive, by Cornish to Draper, and by Pocock to the Earl of Albemarle. But how often have we fought when the navy took no share? Except in the interior of India and on the North-West Frontier, the occasions have been rare in which it was not represented. Even in India Captain Peel's men and guns had their active and important part in the suppression of the mutiny. The memory of the naval brigade in the Crimea is still fresh in the minds of all. China has afforded a less famous field, but there also the naval brigades have been busy.

The frequent necessity of Naval help.

If we are justified in estimating the future from the past, and in the long run we have no other means of forecasting what is to come, it must be taken for granted that some portion of the navy will be with most armies which we have occasion to employ. A power which has the sea for its base everywhere, not only must rely on its navy for the power to conduct all wars beyond its own borders but has a peculiarly powerful motive to call upon its fleet for purely military work. Nations which have much less reason to lean on their navy do the same on occasions. The French, for instance, made use of a naval brigade for the defence of Paris. Napoleon had long before made regiments out of the crews of his ships and employed them in Germany. I do not say he called on his sailors, because the men who filled the long list of ships he kept idle in port did not as a rule deserve the name, at any rate in the later days of his reign. His example also is not much to the present purpose, for when he called away the crews of his ships it was because he had renounced all hope of making a legitimate use of his fleet, and was at his wit's end to fill up his armies. The situation can hardly arise with us, if only for the very sufficient reason that long before our navy has fallen to the level of impotence reached by Napoleon's we shall be ruined. Yet even his example shows that a fleet is always a potential part of the land forces of every State. It is only necessary that the crews should be able to use weapons which can be made available ashore as well as afloat, and this they always have been and always must be. When the French summoned part of their sailors to defend Paris they were doing exactly what we did when Captain Peel's brigade was sent into the interior of India. They did not use their bluejackets as soldiers because their fleets were shut into port by a German blockade. At sea they had an overwhelming superiority all through the war. What they did was to draw on the superfluous strength of their navy to make good the deficiency in their army. This is precisely the course we followed when Admiral Cornish's soldiers were landed at Manila, or when a division of marines was sent to aid Wellington to hold the lines of Torres Vedras, and to garrison Lisbon. The circumstances were more fortunate for us than for the French in 1870, but in other respects they were very similar.

The need must recur.

The call upon the navy for service ashore will hardly be less pressing in the future than it has been in former years. In exact proportion as our territorial obligations multiply will be the necessity for a mobile armed force which will be available everywhere and rapidly. And these obligations do increase. Neither so long as the conditions remain the same is there any reason why we should be reluctant to spare part of the ships' companies from time to time, and even for considerable periods, for the shore. The navy exists to serve the general interests of the Empire anywhere and everywhere, and its men may legitimately be used inland at a crisis so long as their absence does not imperil the ships. The general principle is clearly laid down in the letter of the Admiralty to Nelson to which reference has been made above. "Although in operations on the sea coast it may frequently be highly expedient to land a part of the seamen of the squadron to co-operate with and to assist the army, when the situation will admit of their being immediately re-embarked if the squadron should be called away to act elsewhere, or if information of the approach of an enemy's fleet should be received, yet their Lordships by no means approve of the seamen being landed to form a part of an army to be employed at a distance from the coast, where, if they should have the misfortune to be defeated, they might be prevented from returning to the ships, and the squadron be thereby rendered so defective as to be no longer capable of performing the services required of it; and I have their Lordships' commands to signify their directions to your Lordship not to employ the seamen in like manner in future." But when the Admiralty wrote thus in 1799 to Nelson, my Lords had in their minds a situation in which the appearance of a French fleet in the neighbourhood of a British squadron was highly probable. Nor, I think, is it possible to doubt that if Bruix had been a bolder man he might have profited by the scattered state of our fleet in the Mediterranean, and by the preoccupation of Nelson with the service of the Neapolitan sovereigns, to give our chiefs, though they included so great a man as Nelson, so great an officer as St. Vincent, and so stout an admiral as Lord Keith, a tolerably severe lesson. With that possibility before them My Lords were quite right in directing admirals to go by the rules which had been adopted by Penn on the coast of Ireland, and by the allied admirals at Barcelona—namely, never to allow the sailors to go further from the ships than they could be instantly summoned back. The case is altered where there is no fear of what Admiral Colomb defined as the "naval threat." Where you are secured against interruption from the sea there is not the same motive for caution. Then the question whether you will or will not send a large part of your crews up "to Capua" must be answered negatively or affirmatively according to the importance of the holding of Capua.

A danger.

There is, however, another aspect of the question which must be taken into account. We hear a good deal, and with only too much truth, of the tendency of all things naval to the shore in these days. If we are prepared to multiply the use of naval brigades there may seem to be a danger that naval officers will turn their thoughts more than ever to the land. Since the epoch of great naval wars came to an end they have had a strong temptation to seek every opportunity for being landed since their own element continues to offer them so few chances of distinction. And this will continue to be the case until great naval wars begin again, of which there is little present prospect. The peril is not a very grave one. A naval officer will be none the worse for the moral and intellectual training he undergoes, even in such purely soldierly work as has come in his way at Ladysmith or the Modder. Nor is it likely that he will be long enough away from the sea to grow rusty in his profession. Moreover, he is exercising a part of it when he fights on land. The sailor of to-day is a trained "small arm man," and has had the education of an infantry soldier. He was nowise out of place in the squares at Tamai and Tamanieb. There is a possibility that when work of this nature has to be done the naval officers will desire to have all their share in doing it they possibly can, since it must needs offer to many of them the best chance they are likely to meet of getting promotion in time to feel safe of clearing the dreaded age limit which bars the road to the highest ranks. There has been, and will be again, some controversy, and also there is some angry feeling close to this matter. Therefore one has to touch it very lightly. Still there are certain general considerations which are sufficiently obvious. Our fleet includes an element which is essentially military in the shape of the marines. Other States prefer as a rule to make no visible distinction in dress between the men who are added to their crews to serve as soldiers, and the bluejackets. Neither do they allow of any division of command. They have but one corps of combatant officers. This has not been our system, and although proposals may be heard occasionally for putting the various classes of officers in our navy into a species of Medea's cauldron, and seeing if we cannot combine them into something new and better, no such experiment is likely to be made. Our organisation, though deplorably wanting in that paper uniformity and mechanical "logic" which have always distinguished the French, can at any rate say this for itself, that it has stood the strain of more than two centuries (counting from the reign of Charles II., when the modern navy began to take shape), and has during that long period achieved a higher level of general success than any other fighting force in the world. It has always been admirable except in times of corrupt administration and a low moral level in the nation, and there are inward evils in the character of the men themselves against which no organisation is a safeguard, and for which no change in organisation is a remedy.

The Marines.

With this soldier element in the fleet there ought to be no reason why the seaman element should be unduly drawn upon for service on shore. The navy's contribution to the army on land has always consisted largely of marines, and ought to continue to be so composed. If it is made a rule that when both marine and bluejacket are landed the officers of each line shall be in strict proportion to the men, there will be no fear that sea officers proper will be drawn away in excessive numbers. Again, logic of a better kind than the French would seem to require that when the men of the fleet, marines and bluejackets, are soldiering, then the officer who is trained first and foremost to be a soldier should not be in the subordinate position which he inevitably has to take on shipboard.

DAVID HANNAY.

PART II.

BRITISH AND FOREIGN

ARMOURED AND UNARMOURED SHIPS.

PART II.

ALPHABETICAL LIST OF BRITISH AND FOREIGN ARMOURED AND UNARMOURED SHIPS.

The lists of ships were subjected to important modifications (ed. 1896). The order of the columns was rearranged so as to correspond in the British and Foreign Lists. A column was introduced for complements in place of that for coal endurance, and the place in the foreign lists where a ship has been built was added. The calibre of all foreign guns is now given in inches.

The maximum draught at normal displacement has been given wherever it was possible to ascertain it.

As every nation is constantly rearranging the armament of individual ships, it is only possible to publish the latest accessible information.

The vessels which in the British Official *Navy Lists* are called First-Class Gunboats, and in the French Lists are known as Aviso Torpilleurs, are called in these lists Torpedo Gunboats. Torpedo-boats of all classes below Torpedo Gunboats are placed in a separate list.

Storeships, Harbour Service Ships, and Training Ships are not included in these lists.

The ships of those Powers whose Navies are of small importance will be found at the end of Part II.

The sketches of many ships have been reduced to half scale, so as to enable more sketches to be given without increasing the size of the book.

The following abbreviations are used throughout the Alphabetical List, occurring mainly in the first column, showing the class of ship, and in the armour column:—

- a.c. Armoured cruiser.
- a.g.b. Armoured gunboat.
- b. Barbette ship.
- br. Broadside ship.
- c.b. Central-battery ship.
- c.d.s. Coast-defence ship.
- c. Composite-built hull.
- comp. (in armour column). Compound or steel-faced armour.
- c.t. Conning-tower.
- corv. Corvette.
- cr. Cruiser.
- d.v. Despatch vessel.
- g.b. Gunboat.
- g.v. Gun-vessel.
- H.S. Harveyed or similar hard-faced steel.
- K.S. Krupp steel.
- I. Iron hull.
- shd. Sheathed.
- S. Steel hull.
- 2 S. Twin screw.
- t. Turret-ship.
- *t.* Trial-speed and I.H.P. at trials (in speed and I.H.P. columns).
- to.cr. Torpedo-cruiser.
- to.g.b. Torpedo-gunboat.
- to.r. Torpedo-ram.
- w. Wooden hull.

ARMAMENT ABBREVIATIONS.—As breech-loading rifled guns are now the most numerous in all fleets, it must be understood that all guns are of that description, unless it be otherwise indicated.

- l. Light guns under 15 cwt., including boats' guns.
- M.L.R. Muzzle-loading rifled guns.
- Q.F. Quick or rapid-firing guns; unless otherwise indicated all guns following that first marked as Q.F. in the armament column are also quick-firers.
- f. tu. or b. tu. Fixed or bow tube for discharging Fish Torpedoes.
- sub. Submerged tube for do.
- B.L. To 6-in. guns indicates that separate cartridges are used, but it must be observed that though this service classification is retained for the latest pattern 6-in. (Vickers) gun, which has no metal cartridge, that gun attains the full Q.F. rate of fire.

BOILERS.—It has been thought desirable to indicate particulars of the water-tube boilers adopted in the principal fleets. The following abbreviations have, therefore, been given in the column devoted to indicated horse-power. Where no reference occurs the boilers are of the cylindrical type:—

- W.T. Water-tube boilers, where the type is not known or not yet decided.
- B. Belleville.
- Bl. Blechynden.
- B. & W. Babcock and Wilcox.
- D'A. D'Allest.
- E. Earle.
- Du T. Du Temple.
- Nic. Niclausse.
- Nor. Normand.
- N.S. Normand-Sigaudy.
- R. Reed.
- T. Thornycroft.
- Y. Yarrow.

GREAT BRITAIN.—Armoured Ships.

Class.	NAME.	Material of Hull.	Displacement.	Length.	Beam.	Draught.	Propellers.	Indicated Horse-Power.	Where Built.	Maker of Engines.	Date of Completion.	Cost ‡	Armour. Side.	Armour. Bulk-head.	Armour. Gun Position	Armour. Backing. Deck Plating.	Armament. Guns.**	Armament. Torpedo Tubes.	Speed.	Coals that can be carried in Bunkers.*	Complement.
			tons.	ft. in.	ft. in.	ft. in.	no.					£	in.	in.	in.	in.			knots.	tons.	
a.c.	Aboukir	S. shd.	12,000	440 0	69 6	26 3	2	21,000 B	Glasgow	Fairfield	Bldg.	722,985	6 K.S.	5 K.S.	6 K.S.	3–2	2 9·2-in. 12 6-in. B.L., 17 smaller Q.F.	2	21·0	800 / 1600	615
a.c.	Achilles	I.	9820	380 0	58 3½	27 3	1	4000	Chatham	Penn	1864	444,646	4½	4½	4½	18–10	14 9-in. M.L.R., 2 6-in., 8 3-pr. Q.F., 16 M., 2 l.	..	12·7	750	707
t. 2nd c.	Agamemnon	I.	8660	280 0	66 0	24 0	2	4500	Chatham	Penn	1883	504 065	18–15	16	16	18–9 3	4 12·5-in M.L.R., 2 6-in. Q.F., 6 6-pr., 8 3-pr., 5 M., 2 l.	2	12·1	960	410
t. 2nd c.	Ajax	I.	8660	280 0	66 0	24 0	2	4500	Pembroke	Penn	1883	518,357									
a.c.	Agincourt	I.	10,600	400 0	59 5	27 9	1	4000	Birkenh'd	Maudslay	1868	465,477	5½	4½	5½	10	17 9-in., M.L.R., 2 20-pr., 10 3-pr. Q.F., 7 M., 5 l.	2	12.0	750	687
b. 1st cl.	Albemarle	S.	14,000	405 0	75 6	26 6	2	18,000	Chatham	Thames Iron Works	Bldg.	†	7	14	11–6	4–6 2–1	4 12-in., 12 6-in. B.L., 12 12-pr., 6 3-pr., 2 M., 2 l.	4	19·0	900 / 2000	750
b. 1st cl.	Albion	S.	12,950	390 0	74 0	26 0	2	13,500 B	Blackwall	Maudslay	Bldg.	802,910	6 H. S.	6 H. S.	12–5 H. S.	.. 3–2	4 12-in., 12 6-in. Q.F., 18 smaller Q.F., 2 l.	5 (4 sub.)	18·25	800 / 1850	700
c.b. 2nd c.	Alexandra	I.	9490	325 0	63 8	26 6	2	7000	Chatham	Humphrys	1877	514,324	12	8–5	8	12–10 1½–1	8 10-in. M.L.R., 4 9·2-in., 6 4·7-in. Q.F., 4 6-pr. Q.F., 6 3-pr., 18 M., 3 l.	4	14·3	680	600
b. 1st cl.	Anson	S.	10,600	330 0	68 6	27 3	2	11,500	Pembroke	Humphrys	1889	x 724,765	18 comp.	16 comp.	14–12 comp.	10–15 3–2½	4 13·5-in., 6 6-in. Q.F., 12 6-pr., 10 3-pr., 7 M., 2 l.	5	16·9	900 / 1200	515
c.b. 3rd c.	Audacious	I.	6010	280 0	54 0	23 8	2	3300	Glasgow	Ravenhill	1869	246,482	8	5	6	10	10 9-in. M.L.R., 8 4-in., 4 6-pr. Q.F., 6 3-pr., 6 M., 3 l.	4	11·6	500	492
a.c	Aurora	S.	5600	300 0	56 0	24 6	2	8500	Pembroke	Thomson	1889	x 284,550	10 comp.	16 comp.	4½	6 3–2	2 9·2-in., 10 6-in. Q.F., 6 6-pr., 10 3 pr., 6 M., 3 l.	2	18·0	900	484
a.c.	Australia	S.	5600	300 0	56 0	24 6	2	8500	Glasgow	Napier	1888	259,390						4			
a.c.	Bacchante	S. shd.	12,000	440 0	69 6	26 8	2	21,000 B	Glasgow	Clydebank	Bldg.	†	6 K.S.	5 K.S.	6 K.S.	.. 3–2	2 9·2-in. 12 6-in. B.L., 17 small Q.F., 9 l.	2	21·0	800 / 1600	615

b. 1st cl.	**Barfleur**	S. shd.	10,500	360	0	70	0	25	6	2	13,163	Chatham	Greenock Foundry	1894	599,089	12 *comp.*	12 *comp.*	9	.. 2½–2	4 10-in., 10 4·7-in. Q.F., 8 6-pr., 12 3-pr., 7 M., 2 l.	7 (2 sub.)	18·5	1240	284
a.c.	**Bedford**	S.	9800	440	0	66	0	24	6	2	22,000	Glasgow	Fairfield Shipbg. Co.	Bldg.	..	4	..	..	..	14 6-in. B.L., 13 small Q.F.	..	23·0	800	..
c.d.s. *c.b.*	**Belleisle**	I.	4870	245	0	52	0	21	0	2	2600	Poplar	Maudslay	1878	240,000 (purchas'd)	12	9	9	16–9 3–1	4 12-in. M.L.R., 6 6-pr. Q.F., 6 M., 2 l.	2	11·9	510	284
c.b. 3rd c.	**Bellerophon**	I.	7550	300	0	56	1	26	7	1	4000	Chatham	Penn	1866	322,701	6	5	6	10	10 8-in., 4 6-in., 6 4-in., 4 6-pr. Q.F., 12 M., 4 l.	2	12·4	650	572
b. 1st cl.	**Benbow**	S.	10,600	330	0	68	6	27	3	2	11,500	Blackwall	Maudslay	1888	760,820	18 *comp.*	16 *comp.*	14–2 *comp.*	12–15 3–2½	2 16·25-in., 10 6-in. Q.F., 12 6-pr., 10 3-pr., 7 M., 2 l.	5	16·75	1200	509
b. 1st cl	**Bulwark**	S.	15,000	400	0	75	0	26	9	2	15,000 B	Devonp't	Hawthorne	Bldg.	1,018,949	9	12	12–5	3–2	4 12-in. 12 6-in. B.L., 24 small Q.F., 8 M.	2	18·0	900 / 2000	750
b. 1st cl.	**Cæsar**	S.	14,900	390	0	75	0	27	6	2	12,000	Portsm'th	Maudslay	1897	865,533	9 H. S.	14–9 H. S.	14–6 H. S.	.. 4–2½	4 12-in., 12 6-in. Q.F., 18 12-pr., 12 3 pr., 8 M., 2 l.	5 (4 sub.)	17·5	1850	757
b. 1st cl.	**Camperdown**	S.	10,600	330	0	68	6	27	3	2	11,500	Portsm'th	Maudslay	1889	¤ 769,456	18 *comp.*	16 *comp.*	12 *comp.*	10–15 3–2½	4 13·5-in., 6 6-in. Q.F., 12 6-pr., 10 3-pr., 7 M., 2 l.	4	16·9	900 / 1200	515
b. 1st cl.	**Canopus**	S.	12,950	390	0	74	0	26	0	2	13,500 B	Portsm'th	Greenock Foundry	1900	869,598	6 H. S.	12 H. S.	12–5 H. S.	3–2	4 12-in., 12 6-in. Q.F., 18 smaller Q.F., 2 l.	5. (4 sub.)	18·25	800 / 1850	700
b. 1st cl.	**Centurion**	S. shd.	10,500	360	0	70	0	25	6	2	13,214	Portsm'th	Greenock Foundry	1893	608,098	12 *comp.*	12 *comp.*	9 *comp.*	.. 2½–2	4 10-in., 10 4·7-in. Q.F., 8 6-pr., 12 3-pr., 7 M., 2 l.	7 (2 sub.)	18·25	1240	622
b. 1st cl.	**Collingwood**	S.	9500	325	0	68	0	26	10	2	9500	Pembroke	Humphrys	1886	624,000	18 *comp.*	16 *comp.*	12 *comp.*	17–10 2½	4 12-in., 6 6-in., Q.F., 12 6-pr. 8 3-pr., 6 M., 2 l.	4	6·50	1200	480
t. 2nd c.	**Colossus**	S.	9420	325	0	68	0	26	3	2	5500	Portsm'th	Maudslay	1886	646,786	18–14 *comp.*	16–13 *comp.*	16 *comp.*	22–10 3–2½	4 12-in., 5 6-in. Q.F., 4 6-pr., 10 3-pr., 6 M., 4 l.	2	4·2	970	388
t. 3rd c.	**Conqueror**	S.	6200	270	0	58	0	24	0	2	6000	Chatham	Humphrys	1882	418,433	12–8½	11½	12	13 2½–1¾	2 12-in., 4 6-in. Q.F., 6 6-pr., 12 M., 2 l.	6	15·3	650	330
b. 1st cl.	**Cornwallis**	S.	14,000	405	0	75	6	26	6	2	18,000 B	Blackwall	Thames S. Co.	Bldg.	†	7 K.S.	..	11–6 K.S.	..	4 12-in., 12 6-in. B.L., 12 12-pr. Q.F., 6 3-pr.	4	19	900 / 2000	750
a.c.	**Cressy**	S. shd.	12,000	440	0	69	6	26	3	2	21,000 B	Glasgow	Fairfield	Bldg.	723,012	6 K.S.	5 K.S.	6 K.S.	3–2	2 9 2-in. 12 6-in. B.L., 17 smaller Q.F.	2	21·0	800 / 1600	615
c.d.s. *t.*	**Cyclops**	I.	3560	225	0	45	0	16	4	2	1200	Blackwall	Elder	1871	154,026	8–6	9–8	9	11 1½	4 10-in. M.L.R., 4 3-pr. Q.F., 5 M., 1 l.	..	9·9	250	196

* When two set of figures are given the upper shows the coal capacity at load draught.
¤ Includes Hydraulic Machinery, Gun Mountings, &c.
† Details of cost incomplete.
** See note relating to the new Vickers 6-in. gun on the previous page.
‡ Unless mentioned this does not include armament.

GREAT BRITAIN.—Armoured Ships—*continued.*

Class.	NAME.	Material of Hull.	Displacement.	Length.	Beam.	Draught.	Propellers.	Indicated Horse-Power.	Where Built.	Maker of Engines.	Date of Completion.	Cost.	Armour. Side.	Armour. Bulk-head.	Armour. Gun Position	Armour. Backing. Deck Plating.	Armament. Guns.	Armament. Torpedo Tubes.	Speed.	Coals that can be carried in Bunkers.	Complement.
			tons.	ft. in.	ft. in.	ft. in.	no.					£	in.	in.	in.	in.			knots.	tons.	
t. 2nd c.	Devastation	I.	9330	285 0	62 3	27 6	2	7000	Portsm'th	Maudslay	1873	353,848	12–10	12–10	14	18–16 3–2	4 10-in., 6 6-pr. Q.F., 8 3-pr., 5 M., 2 l.	2 (2 sub.)	14·0	1800	410
a.c.	Drake	S.	14,100	500 0	71 0	26 0	2	30,000 B	Pembroke	Humphrys	Bldg.	*	6 K.S.	12 K.S.	12–5 K.S.	3–2	2 9·2-in., 16 6-in. B.L., 14 12-pr. Q.F., 3 3-p.	2	23·0	1250 / 2500	..
t. 2nd c.	Dreadnought	I.	10,820	320 0	63 10	26 9	2	6500	Pembroke	Humphrys	1875	592,573	14	13	14	18–15 3–2	4 12·5-in. M.L.R., 6 6-pr. Q.F., 12 3-pr., 7 M., 2 l.	2	13·7	1200	453
b. 1st cl.	Duncan	S.	14,000	405 0	75 6	26 6	2	18,000 B	Blackwall	Thames S. Co.	Bldg.	*	7 K.S.	..	11–6 K.S.	..	4 12-in., 12 6-in. B.L., 12 12-pr. Q.F., 6 3-pr.	4	19·0	900 / 2000	750
t. 2nd c.	Edinburgh	S.	9420	325 0	68 0	26 3	2	5500	Pembroke	Humphrys	1886	642,333	18–14 *comp.*	16–13 *comp.*	16 *comp.*	22–10 3–2½	4 12-in., 5 6-in. Q.F., 4 6-pr., 10 3-pr., 6 M., 2 l.	2	14·2	970	388
b. 1st cl.	Empress of India	S.	14,150	380 0	75 0	27 6	2	13,000	Pembroke	Humphrys	1893	838,087	18–5 *comp.*	16 *comp.*	17–6 *comp.*	.. 3	4 13·5-in., 10 6-in. Q.F., 16 6-pr., 12 3-pr., 8 M., 2 l.	7 (2 sub.)	17·5	1800	740
a.c.	Essex	S.	9,800	440 0	66 0	24 6	2	22,000 B	Pembroke	John Brown	Bldg.	*	4	..	..	..	14 6-in. B.L., 13 small Q.F. guns	..	23·0	800	..
a.c.	Euryalus	S shd.	12,000	440 0	69 6	26 3	2	21,000 B	Barrow	Vickers	Bldg.	*	6 K.S.	5 K.S.	6 K.S.	4 1½–1	2 9·2-in., 12 6-in. B.L., 14 12-pr. Q.F., 3 3-pr., 8 M.	2	21·0	800 / 1600	615
b. 1st cl.	Exmouth	S.	14,000	405 0	75 6	26 6	2	18,000 B	Laird	Laird	Bldg.	*	7 K.S.	..	11–6 K.S.	..	4 12-in., 12 6-in. B.L., 12 12-pr. Q.F., 6 3-pr.	4	19.0	900 / 2000	750
b. 1st cl.	Formidable	S.	15,000	400 0	75 0	26 9	2	15,000 B	Portsm'th	Earle	Bldg.	1,012,780	9	12	12–5	.. 3–2	4 12-in., 12 6-in. B.L., 24 small Q.F., 8 M.	..	18·0	900 / 2000	750
a.c.	Galatea	S.	5600	300 0	56 0	24 6	2	8500	Glasgow	Napier	1889	258,390	10 *comp.*	16 *comp.*	4½	6 3–2	2 9·2-in., 10 6-in., Q.F., 6 6-pr. 10 3-pr., 6 M., 3 l.	4	18·1	900	484
c.d.s.	Glatton	I.	4910	245 0	54 0	19 5	2	2000	Chatham	Laird	1872	219,529	12–10	12	14	20 3–1½	2 12-in. M.L.R., 3 6-pr. Q.F., 4 M., 1 l.		11·0	540	192
t. *b.* 1st cl.	Glory	S.	12,950	390 0	74 0	26 0	2	13,500 B	Laird	Laird	Bldg.	844,057									
b. 1st cl.	Goliath	S.	12,950	390 0	74 0	26 0	2	13,500 B	Chatham	Penn	1900	860,788	6 H. S.	12 H. S.	12–5 H. S.	3–2	4 12-in., 12 6-in. Q.F., 18 smaller Q.F., 2 l.	5 (4 sub.)	18·25	800 / 1850	700

a.c.	**Good Hope, late Africa**	S.	14,100	500	0	71	0	26	0	2	30,000 B	Glasgow	Fairfield	Bldg.	*	6 K.S.	12 K.S.	12–5 K.S.	.. 3–2	2 9·2-in., 16 6-in. B.L., 17 small Q.F.	2	23·0	1250 2500	..
c.d.s.	**Gorgon**	I.	3560	225	0	45	0	16	4	2	1200	Jarrow	Ravenhill	1872	138,567	8–6	9–8	10	11 1½	4 10-in. M.L.R., 4 3-pr. Q.F., 5 M., 2 l.	..	9·9	270	192
t. *b.* 1st cl.	**Hannibal**	S.	14,900	390	0	75	0	27	6	2	12,000	Pembroke	Harland	1897	867,403	9 H. S.	14–9 H. S.	14–6 H. S.	.. 4–2½	4 12-in., 12 6-in. Q.F., 18 12-pr., 12 3-pr., 8 M., 2 l.	5 (4 sub.)	17·5	1850	757 ..
c.d.s.	**Hecate**	.	3560	225	0	45	0	16	4	2	1200	Poplar	Ravenhill	1872	140,593	8–6	9–8	10	11 1½	4 10-in. M.L.R., 4 3-pr. Q.F., 4 M., 1 l.	..	9·9	270	205
t. *a.b.* 3rd c.	**Hercules**	I.	8680	325	0	59	0½	26	6	1	8500	Chatham	Penn	1868	361,134	9–6	6–5	9	12–10	8 10-in. M.L.R., 2 9-in. do., 2 7-in. do., 2 6-in. Q.F., 6 4·7-in., 9 6-pr., 13 3-pr., 7 M., 2 l.	4	14·6	610	638
t. 3rd c.	**Hero**	S.	6200	270	0	58	0	24	0	2	6000	Chatham	Rennie	1888	397,271	12 *comp.*	11½ *comp.*	12 *comp.*	13½ 2⅓–1¾	2 12-in., 4 6-in. Q.F., 7 6-pr., 12 M., 2 l.	6	15·2	620	330
a.c.	**Hogue**	S. shd.	12,000	440	0	69	6	26	3	2	21,000 B	Barrow	Vickers	Bldg.	724,492	6 K.S.	5 K.S.	6 K.S.	4 1½–1	2 9·2-in. 12 6-in. B.L., 17 small Q.F.	2	21·0	800 1600	745
t. 1st cl.	**Hood**	S.	14,150	380	0	75	0	27	6	2	13,000	Chatham	Humphrys	1893	830,536	18 *comp.*	17 *comp.*	18–6 *comp.*	.. 3	4 13·5-in., 10 6-in. Q.F., 10 6-pr., 12 3-pr., 8 M., 2 l.	7 (2 sub.)	17·5	900 1800	730
c.d.s. *t.*	**Hotspur**	I.	4010	235	0	50	0	21	10	2	2500	Glasgow	Napier	1871	171,528	11 & 8	8	10–8½	15–10 2–1	2 12-in. M.L.R., 2 6-in. Q.F., 6 12-pr., 4 3-pr. Q.F., 8 M., 2 l.	..	11·25	300	232
b. 1st cl	**Howe**	S.	10,300	325	0	68	0	27	3	2	11,500	Pembroke	Humphrys	1889	667,022	18 *comp.*	16 *comp.*	11½ *comp.*	15–12 3–2½	4 13·5-in., 6 6-in. Q.F., 12 6-pr., 10 3-pr., 7 M., 2 l.	1	16·8	1200	515
c.d.s.	**Hydra**	I.	3560	225	0	45	0	16	4	2	1200	Glasgow	Elder	1872	141,372	8–6	9–8	9–10	11–9 1½	4 10-in. M.L.R., 4 3-pr. Q.F., 5 M., 2 l.	..	9·9	250	196
t. *b.* 1st cl.	**Illustrious**	S.	14,900	390	0	75	0	27	6	2	12,000	Chatham	Penn	1898	885,945	9 H. S.	14–9 H. S.	14–6 H. S.	.. 4–2½	4 12-in., 12 6-in. Q.F., 18 12-pr., 12 3-pr., 8 M., 2 l.	5 (4 sub.)	17·5	1850	757
a.c.	**Immortalité**	S.	5600	300	0	56	0	22	6	2	8500	Chatham	Earle	1889	278,500	10 *comp.*	16 *comp.*	4½	6 3–2	2 9·2-in., 10 6-in. Q.F., 6 6-pr., 10 3-pr., 6 M., 3 l.	2	18·0	900	484
a.c.	**Impérieuse**	S. shd.	8400	315	0	62	0	27	4	2	10,000	Portsm'th	Maudslay	1886	x 536,814	10 *comp.*	9 *comp.*	4½ *comp.*	10 4–2	4 9·2-in., 10 6-in. Q.F., 8 6-pr., 10 3-pr., 6 M., 2 l.	6	16·7	900 1130	544
t. 2nd c	**Inflexible**	I.	11,880	320	0	75	0	26	4	2	6500	Portsm'th	Elder	1881	795,268	24–16	22–14	17 *comp.*	17–25 3	4 16-in. M.L.R., 8 4-in., 4 6-pr. Q.F., 2 3-pr., 15 M., 2 l.	4 (2 sub.)	12·8	1300	485
c.b. 3rd c.	**Invincible**	I.	6010	280	0	54	0	23	2	2	3500	Glasgow	Napier	1870	239,441	8	5	6	10	10 9-in. M.L.R., 6 4-in., 17 light Q.F. & machine guns.	4	12·5	500	492

* Details of cost incomplete.

x Includes Hydraulic Machinery, Gun Mountings, &c.

GREAT BRITAIN.—Armoured Ships—*continued.*

Class	NAME.	Material of Hull.	Displacement.	Length.	Beam.	Draught.	Propellers.	Indicated Horse-Power.	Where Built.	Maker of Engines.	Date of Completion.	Cost.	Armour. Side.	Armour. Bulk-head.	Armour. Gun Position.	Armour. Backing. Deck Plating.	Armament. Guns.	Armament. Torpedo Tubes.	Speed.	Coals that can be carried in Bunkers.	Complement.
			tons.	ft. in.	ft. in.	ft. in.	no.					£	in.	in.	in.	in.			knots.	tons.	
c.b. 3rd c.	Iron Duke	I.	6010	280 0	54 0	23 3	2	3500	Pembroke	Ravenhill	1871	196,479	8	5	6	10	10 9-in. M.L.R., 4 5-in., 16 light Q.F. & Machine Guns.	4	12·5	500	492
b. 1st cl.	Implacable Irresistible London	S.	15,000	400 0	75 0	26 9	2	15,000 B	D'port Chatham Portsm'h	Laird Maudslay Earle	Bldg.	1,002,909 986,731 1,023,671	9 K.S.	12 K.S.	12–5 K.S.	.. 3–2	4 12-in. 12 6-in. B.L., 24 small Q.F., 8 M.	2	18·0	900 2000	755
a.c.	Kent	S.	9800	440 0	66 0	24 6	2	22,000	Ports'mth	Hawthorn	Bldg.	*	4	..	..	..	14 6-in. Q.F., 13 smaller guns.	..	23·0	800	..
a.c.	King Alfred Leviathan	S.	14,100	500 0	71 0	26 0	2	30,000 B	Barrow Clydeb'k	Vickers Clydeb'k	Bldg.	*	6–2 K.S.	5 K.S.	6-5 K.S.	.. 4–6	2 9·2-in., 16 6-in. B.L., 14 12-pr., 3 3-pr., 9 l.	2	23·0	1250 2500	900
b 1st cl.	Jupiter	S.	14,900	390 0	75 0	27 6	2	12,000	Clydeb'nk	Thomson	1897	901,216									
b 1st cl.	Magnificent	S.	14,900	390 0	75 0	27 6	2	12,000	Chatham	Penn	1895	912,291									
b 1st cl.	Majestic	S.	14,900	390 0	75 0	27 6	2	12,000	Portsm'th	Barrow	1895	910,632	9 H. S.	14–9 H. S.	14–6 H. S.	.. 4–2½	4 12-in., 12 6-in. Q.F., 18 12-pr., 12 3-pr., 8	5 (4 sub.)	17·5	1850	757
b 1st cl.	Mars	S.	14,900	390 0	75 0	27 6	2	12,000	Birkenh'd	Laird	1897	902,631									
a.c.	Minotaur	I.	10,690	400 0	59 4½	27 3	1	4000	Blackwall	Penn	1867	456,830	5½	4½	5½	10	17 9-in. M.L.R., 4 4·7-in. Q.F., 8 3-pr. 8 M., 3 l.	2	12·0	750	701
t. 3rd c.	Monarch	I.	8930	330 0	57 6	26 7	1	8216	Chatham	Maudslay	1869	354,575	7–6	5–4½	8	12	4 12-in. M.L.R., 2 9-in. do., 1 7-in. do., 4 12-pr. Q.F., 10 3-pr., 6 M., 2 l.	2	15	630	598
a.c.	Monmouth	S.	9800	440 0	66 0	24 6	2	22,000	Glasgow	London & Glasgow Shipbg. Co.	Bldg.	..	4	..	..	..	14 6-in. B.L., 13 small Q.F.	..	23·0	800	..
b. 1st cl.	Montagu	S.	14,000	405 0	75 6	26 6	2	18,000	Devonport	Laird	..	*	7	14	11·6	4–6 2–1	4 12-in., 12 6-in. B.L., 12 12-pr., 6 3-pr., 2 M., 2 l.	4	19·0	900 2000	750

a.c.	**Narcissus**	S.	5600	300	0	56	0	24	6	2	8500	Hull	Earle	1889	257,390	10 *comp.*	16 *comp.*	4½	6 3–2	2	9·2-in., 10 6-in. Q.F., 6 6-pr., 10 3-pr., 6 M., 3 l.	4	18·1	750 900	484
a.c.	**Nelson**	I. shd.	7630	280	0	60	0	26	6	2	5500	Glasgow	Elder	1880	390,855	9–6	9–6	6	13–10	4	10-in. M.L.R., 8 9-in.do., 4 4·7-in. Q.F., 6 6-pr., 14 3-pr., 7 M., 3 l.	2	13·6	1150	580
a.c.	**Northampton**	I. shd.	7630	280	0	60	0	25	9	2	4500	Glasgow	Penn	1878	395,804	9–6	9–6	6	13–10 3–2	4	10-in. M.L.R., 8 9-in. do., 4 4·7-in. Q.F., 8 3-pr., 10 M., 3 l.	2	12·6	1150	580
t. 2nd c.	**Neptune**	I. shd.	9310	300	0	63	0	26	1	1	6000	Poplar	Penn	1878	600,000 (purchas'd)	12–9	8–6	11–13	11–13 3–2	4	12·5-in. M.L.R., 2 9-in. do., 6 6-pr. Q.F., 8 3-pr., 11 M., 2 l.	2	13·4	670	412
t. 1st cl.	**Nile**	S.	11,940	345	0	73	0	27	6	2	12,000	Pembroke	Maudslay	1890	819,717	20–16 *comp.*	18–14 *comp.*	18 *comp.*	6 3	4	13·5-in., 6 6-in. Q.F., 8 6-pr., 12 3-pr., 7 M., 3 l.	6 (2 sub.)	16·7	1200	558
a.c.	**Northumberland**	I.	10,780	400	4	59	5	27	1	1	4381	Millwall	Penn	1868	471,352	5½	4½	5½	10	7	9-in. M.L.R., 20 8-in.do., 1 6-in., 1 5-in., 6 4·7-in. Q.F., 10 3-pr., 6 M., 5 l.	..	13·3	756	701
b. 1st cl	**Ocean**	S.	12,950	390	0	74	0	25	5	2	13,500 B	Devonport	Hawthorn	1900	881,248	6 H. S.	12 H. S.	12–5 H. S.	3–2	4	12-in., 12 6-in. Q.F., 18 smaller Q.F., 2 l.	5 (4 sub.)	18·25	800 / 1850	700
c.d.s	**Orion**	I.	4870	245	0	52	0	21	4	2	2600	Poplar	Maudslay	1882	292,229	12–7	9–5	8	16–9 3–1	4	12-in. M.L.R., 6 6-pr. Q.F., 6 M., 2 l.	4	11·9	520	284
a.c.	**Orlando**	S.	5600	300	0	56	0	24	8	2	8500	Jarrow	Palmer	1888	266,812	10 *comp.*	16 *comp.*	4½ *comp.*	6 3–2	2	9·2-in., 10 6-in. Q.F., 6 6-pr., 10 3-pr., 7 M., 3 l.	2	18·1	900	484
c.d.s.	**Penelope**	I.	4470	260	0	50	0	17	6	2	2700	Pembroke	Maudslay	1868	186,848	6–5	4½	5	10–11	8	9-ton M.L.R., 4 3-pr. Q.F. 11 M., 4 l.	..	11·0	470	265
c b. *b.* 1st cl.	**Prince George**	S.	14,900	390	0	75	0	27	6	2	12,000	Portsm'th	Humphrys	1896	885,037	H. S.	14–9 H. S.	14–6 H. S.	.. 4–2½	4	12-in., 12 6-in. Q.F., 18 12-pr., 12 3-pr., 2 l.	5 (4 sub.)	17·5	1850	757
b. 1st cl.	**Renown**	S. shd.	12,350	380	0	72	0	23	[illegible]	2	12,000	Pembroke	Maudslay	1896	696,425	8–6 H. S.	10–6 H. S.	10 H. S.	.. 3–2	4	10-in. 29-ton, 10 6-in. Q.F., 14 12-pr., 15 smaller Q.F. and Machine Guns.	5 (2 sub.)	18·0	1800	674
b. 1st cl.	**Ramillies**	S.	14,150	380	0	75	0	27	6	2	13,000	Glasgow	Thomson	1893	874,255										
b 1st cl.	**Repulse**	S.	14,150	380	0	75	0	27	6	2	13,000	Pembroke	Humphrys	1894	841,274	18–5 *comp.*	16 *comp.*	17 *comp.*	.. 3	4	13·5in., 10 6-in. Q.F., 16 6-pr., 12 3-pr., 8 M., 2 l.	7 (2 sub.)	17·5	1800	730
b. 1st cl.	**Resolution**	S.	14,150	380	0	75	0	27	6	2	13,000	Jarrow	Palmer	1893	852,755										

* Details of cost incomplete.

x Includes Hydraulic Machinery, Gun Mountings, &c.

GREAT BRITAIN.—Armoured Ships *continued.*

Class.	NAME.	Material of Hull.	Displacement.	Length.	Beam.	Draught.	Propellers.	Indicated Horse-Power.	Where Built.	Maker of Engines.	Date of Completion.	Cost.	Armour. Side.	Armour. Bulk-head.	Armour. Gun Position	Armour. Backing, Deck Plating.	Armament. Guns.	Armament. Torpedo Tubes.	Speed.	Coals that can be carried in Bunkers.	Complement.
			tons.	ft. in.	ft. in.	ft. in	no.					£	in.	in.	in.	in.			knots.	tons.	
b. 1st cl.	**Revenge**	S.	14,150	380 0	75 0	27 6	2	13,000	Jarrow	Palmer	1895	852,755									
b. 1st cl.	**Royal Oak**	S.	14,150	380 0	75 0	27 6	2	13,000	Birkenh'd	Laird	1894	877,378	18–5 *comp.*	16 *comp.*	17 *comp.*	.. 3	4 13·5-in., 10 6-in. Q.F., 16 6-pr., 12 3-pr., 8 M., 2 l.	7 (2 sub.)	17·5	900 1800	730
b. 1st cl.	**Royal Sovereign**	S.	14,150	380 0	75 0	27 6	2	13,312	Portsm'th	Humphrys	1892	824,583									
b. 1st cl.	**Rodney**	S.	10,300	325 0	68 0	27 3	2	11,500	Chatham	Humphrys	1888	669,278	18 *comp.*	1[illegible] *comp.*	11 *comp.*	15–12 3–2½	4 13·5 in., 6 6-in. Q.F., 12 6-pr., 10 3-pr., 6 M., 2 l.	4	16·75	1200	515
c.d.s. *t.*	**Rupert**	I.	5440	250 0	53 0	23 7	2	6000	Chatham	Portsm'th	1874	232,677	11–9	12	14–12	14–10 3–2	2 9·2-in., 2 6-in. Q.F., 4 6-pr., 6 3-pr., 2 M., 2 l.	4	14·0	480	293
b. 1st cl.	**Russell**	S.	14,000	405 0	75 6	26 6	2	18,000 B	Jarrow	Palmer	Bl'g.	‡	7 K.S.	..	11–6 K.S.	4–6 2–1	4 12-in., 12 6-in. B.L., 12 12-pr. Q.F., 6 3-pr.	4	19	900 2000	750
t. 1st cl.	**Sans Pareil**	I.	10,470	340 0	70 0	27 3	2	14,000	Blackwall	Humphrys	1889	*x* 719,442	16–18 *comp.*	16 *comp.*	18 *comp.*	6 3	2 16·25in., 1 10-in., 12 6-in. Q.F., 12 6-pr., 12 3-pr., 8 M., 2 l.	6 (2 sub.)	17·2	1200	583
c.d.s. *t.*	**Scorpion**	I.	2750	224 6	42 4½	16 11	1	1000	Birkenh'd	Laird	1865	110,573	4½	..	5	10–8	4 9-in. M.L.R., 6 M., 1 l.	..	8·5	320	151
c.b. 3rd c.	**Sultan**	I.	9290	325 0	59 0½	27 6	1	8000	Chatham	Thomson	1871	357,415	9–6	6–4½	9–8	12–10	8 10-in. M.L.R., 4 9-in.do., 4 4·7-in. Q.F., 9 6-pr., 13 3-pr., 7M., 2 l.	4	14·0	810	661
c.b. 2nd c.	**Superb**	I.	9170	332 3	59 0	26 5	1	8500	Blackwall	Maudslay	1880	443,000 (purchas'd)	12–10	10–5	10	7–12 1½	12 10-in. M.L.R., 10 6-in. Q.F., 22 light Q.F. and Machine Guns.	4	15·0	970	654
a.c.	**Sutlej**	S. shd.	12,000	440 0	69 6	26 3	2	21,000 B	Clydeb'nk	Clydebank Company	Bldg.	733,625	6 K.S.	5 K.S.	6 K.S.	.. 3–2	2 9·2-in. 12 6-in. B.L., 17 small Q.F.	2	21·0	800 1600	615

c.b. 3rd c.	**Swiftsure** .	I. shd.	6910	280	0	55	0	26	0	1	3500	Jarrow	Maudslay .	1872	257,081	8–6	6–4	6	10	10 9-in. M.L.R., 8 4-in., 4 3-pr. Q.F., 12 M., 3 l.	4	12·6	540	497
c.b. & *b* 2nd c	**Temeraire** .	I. shd.	8540	285	0	62	0	27	2	2	6500	Chatham	Humphrys	1877	454,969	11–8	8–5	10–8	12–10 1½–1	4 11-in. M.L.R., 4 10-in. do., 6 4-in. Q.F., 4 6-pr., 10 3-pr., 8 M., 4 l.	2	13·8	620	592
t. 2nd c.	**Thunderer** .	I.	9330	285	0	62	3	27	0	2	7000	Pembroke	Maudslay	1877	358,542	12–10	12–10	14–12	18–16 3–2	4 10-in., 6 6-pr. Q.F., 8 3-pr., 4 M., 2 l.	2	14·0	1600	592
t. 1st cl.	**Trafalgar** .	S.	11,940	345	0	73	0	27	6	2	12,000	Portsm'th	Humphrys	1890	862,794 *z*	20–16 *comp.*	18–14 *comp.*	18 *comp.*	6 3	4 13·5 in., 6 6-in. Q.F., 8 6-pr., 12 3-pr., 6 M., 3 l.	6 (2 sub.)	16·7	1200	572
c.b. 3rd c.	**Triumph** .	I. shd.	6640	280	0	55	0	26	2	1	3500	Jarrow	Maudslay	1873	258,322	8–6	6–4	6	10	10 9-in. M.L.R., 4 5-in., 8 6-pr. Q.F., 8 3-pr., 5 M., 3 l.	4	12·6	550	497
a.c.	**Undaunted** .	S.	5600	300	0	56	0	22	6	2	8500	Jarrow .	Palmer .	1889	256,055	10 *comp.*	16 *comp.*	4½ *comp.*	6 3–2	2 9·2-in., 10 6-in. Q.F., 6 6-pr., 10 3-pr., 7 M., 3 l.	4	18·1	750 900	484
b. 1st cl.	**Victorious** .	S.	14,900	390	0	75	0	27	6	2	12,000	Chatham	Hawthorn	1897	868,313	9 H. S.	14–9 H. S.	14–6 H. S.	.. 4–2½	4 12-in., 12 6-in. Q.F., 18 12-pr., 12 3-pr., 8 M., 2 l.	5 (4 sub.)	17·5	1850	757
b. 1st cl.	**Venerable** .	S.	15,000	400	0	75	0	26	9	2	15,000 B	Chatham	Maudslay	Bldg.	1,010,863	7 K S.	..	11–6 K.S.	4–6 2–1	4 12-in., 12 6-in. B.L., 24 small Q.F., 8 M.	2	18·0	900/2000	755
b. 1st cl.	**Vengeance** .	S.	12,950	390	0	74	0	26	0	2	13,500 B	Barrow .	Vickers .	Bldg	814,619	6 H.N.S.	12 H.N.S.	12–6 H.S.	4 2–1	4 12-in., 12 6-in. Q.F. & 18 small Q.F.	4	18·25	800/1850	750
a.c.	**Warspite** .	S.	8400	315	0	62	0	27	4	2	10,000	Chatham	Penn .	1888	529,332	10 *comp.*	9 *comp.*	8 *comp.*	10 3–2	4 9·2-in., 10 6-in. Q.F., 4 6-pr. Q.F., 9 3-pr., 6 M., 2 l.	6	16·7	1130	535
c.d.s.	**Wivern** .	I.	2750	224	6	42	4	17	0	1	1000	Birkenh'd	Laird .	1865	116,514	4½	..		8–10	4 9-in. M.L.R., 8 M., 1 l.	..	8·5	300	151
b.	2 new ships . (Programme 1900–1901.)	S.	Design not settled.									Chath'm D'port												
a.c.	2 new ships (Programme 1900–1901.)	S.										P'sm'th P'broke												
cr. 2nd cl.	2 new ships . (Programme 1900–1901.)	S.										Contract												

The following, which appear in the Official Navy List, are lent to India and Australia:—

c.d.s. *t.*	**Abyssinia*** (*Indian Marine.*)	I.	2900	225	0	42	0	14	6	2	900	Poplar	Dudgeon	1870	116,549	7–6	8–7	10–8	11–9 1½–1	4 8-in. 14-ton, 7 M., 2 l.	..	9·0	92	193
c.d.s. *t.*	**Cerberus†** (*Victorian Marine.*)	I.	3480	225	0	45	0	15	3	2	1660	Jarrow	Maudslay	1870	117,556	8–6	9–8	10–9	11–9 1½–1	4 10-in. 18-ton M.L.R., 4 M.	..	9·75	120	196
c.d.s. *t*	**Magdala*** . (*Indian Marine.*)	I.	3340	225	0	45	0	15	3	2	1400	Blackwall	Ravenhill	1870	132,400	8–6	9–8	10–9	11–9 1½–1	4 8-in. 14-ton, 7 M., 2 l.	..	10·0	120	194

z Includes Hydraulic Machinery, Gun Mountings, &c. * At Bombay. † At Melbourne.

GREAT BRITAIN.—Cruising Ships, &c.

Class.	NAME.	Material of Hull.	Displacement.	Length.	Beam.	Draught.	Propellers.	Indicated Horse-Power.	Where Built.	Maker of Engines.	Date of Launch.	Cost.	Armour. Gun Position.	Armour. Deck.	Armament. Guns.	Armament. Torpedo Tubes.	Speed.	Normal Coal Supply.	Complement.
			tons.	ft. in.	ft. in.	ft. in.	no.					£	in.	in.			knots.	tons.	
2nd cl. Cr.	Active	I. shd.	3080	270 0	42 0	21 4	1	2400	Blackwall	Humphrys	1869	126,156	..	..	10 6-in., 2 64-pr. M.L.R., 9 M., 2 l.	..	15·10	410	339
" "	Æolus	S. shd.	3600	300 0	43 0	17 6	2	9000	Devonp'rt	Hawthorn	1892	208,450	2	2–1	2 6-in. Q.F., 6 4·7-in. 8 6-pr. 1 3-pr., 4 M., 1 l.	4	19·75	400	273
Dsp. Ves.	Alacrity	S.	1700	250 0	32 6	14 0	2	3000	Jarrow	Palmer	1885	77,969	..	..	10 6-pr. Q.F., 2 M.	..	17·00	400	114
T. G. B.	Alarm	S.	810	230 0	27 0	8 9	2	3884	Sheerness	Penn	1892	59,346	2	..	2 4·7-in. Q.F., 4 3-pr. do.	3	19·25	100	91
2nd cl. G. B.	Albacore	C.	560	135 0	26 0	10 6	1	500	Birkenh'd	Laird	1883	28,556	..	..	2 5-in., 2 4-in., 2 M.	..	11·0	85	59
Sloop	Alert	S. shd.	960	180 0	32 6	11 6	1	1400	Sheerness	Sheerness	1894	60,309	..	..	6 4-in. 25-pr. Q.F., 4 3-pr., 2 M.	..	13·25	130	101
"	Algerine	S.	1050	185 0	32 6	11 3	2	1400	Devonp'rt	Devonport	1895	63,904	·22	..	6 4-in. 25-pr. Q.F., 4 3-pr., 3 M.	..	13·0	160	106
2nd cl. Cr.	Amphion	S.	4300	300 0	46 0	20 6	2	5000	Pembroke	Maudslay	1883	160,500	..	1½	10 6-in. Q.F., and 14 smaller, and M.	4	16·6	550/1000	309
" "	Andromache	S.	3400	300 0	43 0	16 6	2	9000	Chatham	Earle	1890	186,280	2	2–1	2 6-in. Q.F., 6 4·7-in., 8 6-pr., 1 3-pr., 4 M., 1 l.	4	20·0	400	273
"	Apollo	S.	3400	300 0	43 0	16 6	2	9000	Chatham	Earle	1891	186,361							
1st cl. Cr.	Andromeda	S. shd.	11,000	435 0	69 0	25 3	2	16,500 B	Pembroke	Hawthorn	1897	575,146	3	3–6	16 6-in., 17 small Q.F., 2 12-pr. boat.	3 (2 sub.)	20·5	1000	600
T. G. B.	Antelope	S.	810	230 0	27 0	8 9	2	3621	Devonp'rt	Yarrow	1893	61,397	2	..	2 4·7-in. Q.F., 4 3-pr.	3	19·25	100	91
3rd cl. Cr.	Archer	S.	1770	225 0	36 0	14 6	2	3500	Glasgow	Thomson	1885	87,583	..	..	6 6-in. Q.F., 8 3-pr., 2 M., 1 l.	3	16 5	475	172

1st cl. Cr. .	Argonaut .	S. shd.	11,000	435	0	69	0	25	3	2	18,000 B	Fairfield .	Fairfield .	1898	535,557	3–6 H. S.	4	16 6-in. Q.F., 14 12-pr., 11 smaller Q.F., and Machine Guns.	3 (2 sub.)	20·75	1000	677
,, ,, .	Ariadne .	S. shd.	11,000	435	0	69	0	25	3	2	18,000 B	Clydeb'nk	JohnBrown	1898	543,705							
,, ,, .	Amphitrite .	S. shd.	11,000	435	0	69	0	25	3	2	18,000 B	Barrow .	Vickers .	1898	546,227							
2nd cl. Cr. .	Arethusa .	S.	4300	300	0	46	0	20	6		5000	Glasgow .	Napier .	1882	145,198		1½	10 6-in. Q.F., 8 3-pr., 6 M., 2 l.	4	16·6	1000*	309
,, ,, .	Arrogant .	S.	5800	320	0	57	6	21	0	2	10,000 B	Devonp'rt	Earle .	1896	278,878	3 N.S.	1–2 N.S.	4 6-in., Q.F., 6 4·7-in., 9 12-pr.,3 3-pr., 1 12-pr. boat, 5 M.	2	19·1	500	480
,, ,, .	Astræa .	S. shd.	4360	320	0	49	6	19	0	2	9112	Devonp'rt	Devonport	1893	244,831	2	2–1	2 6-in. Q.F.,8 4·7-in.,8 6-pr. 1 3-pr.,4 M., 1 l.	4	19·75	400	312
3rd cl. Cr. .	Barham .	S.	1830	280	0	35	0	13	3	2	4700 T	Portsm'th	Hawthorn .	1889	113,302	2	2–1	6 4·7-in. Q.F., 4 3-pr., 2 M.	2	18·6	140	169
,, ,, .	Bellona .	S.	1830	280	0	35	0	13	3	2	4700	Newcastle	Hawthorn .	1890	94,195	2	2–1	6 4·7-in. Q.F., 4 3-pr., 2 M.	2	17·8	140	169
,, ,, .	Barracouta	S.	1580	220	0	35	0	14	0	2	3000	Sheerness	Palmer .	1889	96,315	2	2–1	6 4·7-in. Q.F., 4 3-pr., 2 M.	2	16·5	160	159
,, ,, .	Barrosa .	S.	1580	220	0	35	0	14	0	2	3000	Portsm'th	Palmer .	1889	79,238	2	2–1	6 4·7-in. Q.F., 4 3-pr., 3 M.	2	16·5	160	159
,, ,, .	Blanche .	S. shd.	1580	220	0	35	0	14	0	2	3000	Pembroke	Earle .	1889	91,112	2	2–1	6 4·7-in. Q.F., 4 3-pr., 2 M.	2	16·5	160	159
,, ,, .	Blonde .	S. shd.	1580	220	0	35	0	14	0	2	3000	Pembroke	Earle .	1889	90,059	2	2–1	6 4·7-in. Q.F., 4 3-pr., 2 M.	2	16·5	160	159
Sloop .	Basilisk .	S. shd.	1170	195	0	28	0	12	6	2	2000	Sheerness	Rennie .	1889	58,013	..	..	8 5-in., 8 M.	..	14·7	160	138
,, .	Beagle .	S. shd.	1170	195	0	28	0	12	6	2	2000	Portsm'th	Rennie .	1889	56,474							
1st cl. Cr. .	Blake .	S.	9000	375	0	65	0	25	9	2	20,000	Chatham .	Maudslay .	1889	440,471	6	6–3	2 9·2-in., 10 6-in. Q.F., 16 3-pr Q.F., 7 M., 2 l.	4 (2 sub.)	21·5	1500	570
,, ,, .	Blenheim .	S.	9000	375	0	65	0	25	9	2	21,411	Blackwall	Humphrys	1890	425,591							

Includes Gun Mountings, &c.

* Bunker capacity.

GREAT BRITAIN.—Cruising Ships, &c.—*continued.*

Class.	NAME.	Material of Hull.	Displacement.	Length.	Beam.	Draught.	Propellers.	Indicated Horse-Power.	Where Built.	Maker of Engines.	Date of Launch.	Cost.	Armour. Gun Position.	Armour. Deck.	Armament. Guns.	Armament. Torpedo Tubes.	Speed.	Normal Coal Supply.	Complement.
			tons.	ft. in	ft. in.	ft. in.	no.					£	in.	in.		knots.	tons.		
2nd cl. Cr.	**Bonaventure**	S. shd	4360	320 0	49 6	19 0	2	9000	Devonp'rt	Hawthorn	1892	247,128	2	2–1	2 6-in. Q.F., 8 4·7-in., 8 6-pr., 1 3-pr. 4 M., 1 l.	4	19·5	400	312
T. G. B.	**Boomerang** (Australia)	S.	735	230 0	27 0	8 3	2	3500	Elswick	Bellis	1889	49,962	2	..	2 4·7-in. Q.F., 4 3-pr., 1 M.	3	20·0	100	91
2nd cl. Cr.	**Brilliant**	S. shd.	3600	300 0	43 8	17 6	2	9164	Sheerness	Hawthorn	1891	204,228	2	2–1	2 6-in. Q.F., 6 4·7-in., 8 6-pr., 1 3-pr., 4 M., 1 l.	4	19·7	400	273
3rd cl. Cr.	**Brisk**	S.	1770	225 0	36 0	14 3½	2	3500	Glasgow	Thomson	1886	87,583	..	..	6 6-in. Q.F., 8 3-pr., 2 M., 1 l.	3	16·5	325	172
1st cl. G. B.	**Bramble**	S.	700	180 0	33 0	8 0	2	1300	Liverpool	Fawcett & Co.	1898	50,461	..	..	2 4-in., 4 12-pr. Q.F.	..	13·5	50	85
„ „	**Britomart**	S.	700	180 0	33 0	8 0	2	1300	Liverpool	Fawcett & Co.	1898	50,401	..	..	2 4-in., 4 12-pr. Q.F.	..	13·5	50	85
Sloop	**Buzzard**	C.	1140	195 0	28 0	11 6	2	2000	Sheerness	Barrow	1887	58,700	..	..	8 5-in., 8 M.	2	14·50	160	138
3rd cl. Cr.	**Calliope**	S. shd.	2770	235 0	44 6	19 11	1	4020	Portsm'th	Rennie	1884	120,000	..	1½	4 6-in., 12 5-in. 38 cwt., 9 M., 2 l.	2	14·6	550	293
„ „	**Calypso**	S. shd.	2770	235 0	44 6	19 11	1	4000	Chatham	Rennie	1883	119,500							
2nd cl. Cr.	**Cambrian**	S. shd.	4360	320 0	49 6	19 0	2	9000	Pembroke	Hawthorn	1893	236,919	2	2–1	2 6-in. Q.F., 8 4·7-in., 8 6-pr., 1 3-pr., 4 M., 1 l.	4	19·5	400	312
3rd cl. Cr.	**Champion**	S. shd.	2380	225 0	44 6	19 3	1	2000	Glasgow	Elder	1878	113,983	..	1½	4 6-in., 8 5-in., 4 3-pr., Q.F., 6 M., 2 l.	2	12·75	470	265

3rd cl. Cr.	Cleopatra	S. shd.	2380	225	0	44	6	19	3	1	2000	Glasgow	Humphrys	1878	113,924	..	1½	4 6-in., 8 5-in., 4 3-pr. Q.F., 6 M., 2 l.	2	13·0	470	265
" "	Comus	S. shd.	2380	225	0	44	6	19	3	1	2000	Glasgow	Elder	1878	113,974	..	1½	10 6-in., 7 M., 2 l.	2	12·75	470	265
" "	Cordelia	S. shd.	2380	225	0	44	6	19	3	1	2000	Portsm'th	Rennie	1881	104,500	..	1½	10 6-in., 9 M., 2 l.	2	12·75	470	265
Sloop	Condor	S. shd.	980	180	0	33	3	11	6	2	1400	Sheerness	Thames	1898	65,185	..	..	6 4-in. Q.F., 4 3-pr.	..	13·25	130	130
2nd cl. Cr.	Charybdis	S. shd.	4360	320	0	49	6	19	0	2	9000	Sheerness	Earle	1893	237,344	2	2–1	2 6-in. Q.F., 8 4·7-in, 8 6-pr., 1 3-pr., 4 M., 1 l.	4	19·5	400	312
T. G. B.	Circe	S.	810	230	0	27	0	8	9	2	3500	Sheerness	Penn	1892	61,979	2	..	2 4·7-in. Q.F., 4 3-pdr.	3	19·25	100	91
2nd cl. G. B.	Cockchafer	C.	465	125	0	23	6	9	6	1	360	Pembroke	Maudslay	1881	77,000	..	..	2 64-pr. M.L.R., 2 20-pr., 2 M.	..	9·8	40	61
3rd cl. Cr.	Cossack	S	1770	225	0	36	0	14	3½	2	3500	Glasgow	Thomson	1886	*x* 87,583	..	..	6 6-in. Q.F., 8 3-pr., 2 M., 1 l.	3	16·5	325	172
1st cl. Cr.	Crescent	S. shd.	7700	360	0	60	0	23	9	2	12,000	Portsm'th	Penn	1892	383,068	6	5–1	1 9·2-in., 12 6-in. Q.F., 12 6-pr., 5 3-pr., 7 M., 2 l.	4 (2 sub.)	19·7	850	560
3rd cl. Cr.	Curaçoa	S. shd.	2380	225	0	44	6	19	3	1	2000	Glasgow	Humphrys	1878	112,931	..	1½	4 6-in., 8 5-in., 1 3-pr. Q.F., 9 M., 2 l.	2	13·0	470	265
G. V.	Curlew	S.	950	195	0	28	0	10	6	2	1200	Devonp'rt	Penn	1885	*x* 49,963	..	..	1 6-in. 3 5-in., 7 light Q.F. and M.	1	14·5	250	103
Sloop	Daphne	C.	1140	195	0	28	0	11	6	2	2000	Sheerness	Greenock F'udry Co.	1888	*x* 57,600	..	..	8 5-in., 8 M.	..	14·0	160	138
1st cl. Cr.	Diadem	S. shd.	11,000	435	0	69	0	26	0	2	16,500 B	Govan	Fairfield	1896	550,127	4½–2	4–2½	16 6-in. Q.F., 14 12-pr., 11 light Q.F. and M.	3 (2 sub.)	20·5	1000	600
2nd cl. Cr.	Diana	S. shd.	5600	350	0	54	0	21	0	2	9600 B	Govan	Fairfield	1895	249,332							
" "	Dido	S. shd.	5600	350	0	54	0	21	0	2	9600 B	Glasgow	London and Glasgow Co.	1896	252,278	3	2½	5 6-in. Q.F., 6 4·7-in., 1 12 pr., 11 smaller Q.F. and M.	3 (2 sub.)	19·5	550	470
" "	Doris	S. shd.	5600	350	0	54	0	21	0	2	9600 B	Barrow	Barrow	1896	254,029							
T. G. B.	Dryad	S.	1070	250	0	30	6	9	0	2	3500	Chatham	Maudslay	1893	73,491	2	..	2 4·7-in. Q.F., 4 6-pr.	3	19·0	100	120

x Includes Gun Mountings, &c.

GREAT BRITAIN.—Cruising Ships, &c.—*continued.*

Class.	NAME.	Material of Hull.	Displacement.	Length.	Beam.	Draught.	Propellers.	Indicated Horse-Power.	Where Built.	Maker of Engines.	Date of Launch.	Cost.	Armour. Gun Position.	Armour. Deck.	Armament. Guns.	Armament. Torpedo Tubes.	Speed.	Normal Coal Supply.	Complement.
			tons.	ft. in.	ft. in.	ft. in.	no.					£	in.	in.			knots.	tons.	
1st cl. G. B.	Dwarf	S.	700	180 0	33 0	8 0	2	1300	Glasgow	London and Glasgow Co.	1898	51,139	..	..	2 4-in. Q.F., 4 12-pr.	..	13.5	50	85
2nd cl. Cr.	Eclipse	S. shd.	5600	350 0	53 0	20 3	2	9600 B	Portsm'th	Portsm'th	1894	279,345	3	1½–3	5 6-in. Q.F., 6 4·7-in., and 15 smaller, 4 M., 1 12-pr. boat.	3 (2 sub.)	19·5	550	477
1st cl. Cr.	Edgar	S.	7350	360 0	60 0	23 9	2	12,000	Devonp'rt	Elder	1890	401,083	6	5–1	2 9·2-in., 10 6-in. Q.F., 12 6-pr., 5 3-pr., 7 M., 2 l.	4 (2 sub.)	20·5	850	544
" "	Endymion	S.	7350	360 0	60 0	23 9	2	12,000	Hull	Earle	1891	350,459							
Sloop	Egeria	C.	940	160 0	31 4	14 3	1	700	Pembroke	Humphrys	1873	42,882	..	..	4 20-pr., 2 M., 1 l.	..	11·3	100	122
"	Espiègle	S. shd.	1070	185 0	33 0	11 3	2	1400	Sheerness	Wallsend Slipway Co.	Bldg.	‡	..	..	6 4-in. Q.F., 4 3-pr.	..	13·25	160	160
1st cl. Cr.	Europa	S. shd.	11,000	435 0	69 0	26 0	2	16,500 B	Clydeb'nk	Thomson	1897	561,126	4½–2	4–2½	16 6-in. Q.F., 14 12-pr., 11 light Q.F. and Machine Guns.	3 (2 sub.)	20·5	1000	600
Sloop	Fantôme	S. shd.	1070	185 0	33 0	11 3	2	1400	Sheerness	Devonport	Bldg.	‡	..	..	6 4-in. Q.F., 4 3-pr.	...	13·25	160	160
3rd cl. Cr.	Fearless	S.	1580	220 0	34 3	14 6	2	3200	Barrow	Barrow	1886	£ 87,452	..	..	4 4·7-in. Q.F., 8 3-pr., 2 M., 1 l.	3 (1 sub.)	16·7	450	147
2nd cl. G. B.	Firebrand	C.	455	125 0	23 6	10 0	1	360	Glasgow	Thomson	1877	22,800	..	..	2 5-in., 2 4-in., 2 M.	..	10·17	40	61
2nd cl. Cr.	Flora	S. shd.	4360	320 0	49 6	19 0	2	9000	Pembroke	Barrow	1893	241,819							
" "	Forte	S. shd.	4360	320 0	49 6	19 0	2	9000	Chatham	Chatham	1893	240,816	2	2–1	2 6-in. Q.F., 8 4·7-in., 8 6-pr., 1 3-pr., 4 M., 1 l.	4	19·5	400	312
" "	Fox	S. shd.	4360	320 0	49 6	19 0	2	9000	Portsm'th	Portsm'th	1893	244,078							
" "	Forth	S.	4050	300 0	46 0	20 0	2	5700	Pembroke	Hawthorn	1886	201,952	2	3–2	2 8-in., 10 6-in. Q.F., 3 6-pr., 8 3-pr., 6 M., 2 l.	2	16·8	900	326
" "	Furious	S. shd.	5800	320 0	57 6	21 0	2	10,000 B	Devonp'rt	Earle	1896	280,772	3	1–2	4 6-in. Q.F., 6 4·7 in., 9 12-pr., 3 3-pr., 1 12-pr. boat, 5 M.	2	19·0	500	480
" "	Gladiator	S. shd.	5750	320 0	57 6	21 0	2	10,000 B	Portsm'th	Maudslay	1896	287,642							

1st cl. Cr.	Gibraltar	S. shd.	7700	360	0	60	0	23	9	2	12,000	Glasgow	Napier	1892	347,634	6	5–1	2 9·2-in., 10 6-in. Q.F., 12 6-pr., 5 3-pr., 7 M., 2 l.	4 (2 sub.)	19·7	850	544
T. G. B.	Gleaner	S.	735	230	0	27	0	8	3	2	3600	Sheerness	Sheerness	1890	63,798	2	..	2 4·7-in. Q.F., 4 3-pr.	3	19·0	100	91
,,	Gossamer	S.	735	230	0	27	0	8	3	2	3600	Sheerness	Sheerness	1890	54,490							
1st cl. G. B.	Goldfinch	C.	805	165	0	31	0	11	7½	1	1200	Sheerness	Sheerness	1889	40,889	..	..	[illegible] 4-in., 2 [illegible]-pr. Q.F., 2 M.	..	13·0	105	76
1st cl. Cr.	Grafton	S.	7350	360	0	60	0	23	9	2	12,000	Blackwall	Humphrys	1892	351,851	6	5–1	2 9·2-in., 10 6-in. Q.F., 12 6-pr., 5 3-pr., 7 M., 2 l.	4 (2 sub.)	20·0	850	560
T. G. B.	Grasshopper	S.	525	200	0	23	0	8	9	2	2700	Sheerness	Maudslay	1887	x 34,065	..	..	1 4-in., 6 3-pr. Q.F.	4	17·0	80	67
,,	Halcyon	S.	1070	250	0	30	6	9	0	2	3500	Devonp'rt	Hawthorn	1894	75,091	2		2 4·7-in. Q.F., 4 6-pr.	3	19·0	100	120
,,	Harrier	S.	1070	250	0	30	6	9	0	2	3500	Devonp'rt	Hawthorn	1894	73,428							
1st cl. Cr.	Hawke	S.	7350	360	0	60	0	23	9	2	12,000	Chatham	Elder	1891	365,491	6	5–1	2 9·2-in., 10 6-in. Q.F., 12 6-pr., 5 3-pr., 7 M., 2 l.	4 (2 sub.)	20·0	850	544
T. G. B.	Hazard	S.	1070	250	0	30	6	9	0	2	3500	Pembroke	Elder	1894	74,076	2		2 4·7-in. Q.F., 4 6-pr.	3	19·0	100	120
,,	Hebe	S.	810	230	0	27	0	8	9	2	3566	Sheerness	Sheerness	1892	73,433	2		2 4·7-in. Q.F., 4 3-pr.	3	19·25	100	91
T. D. S.	Hecla	I.	6400	391	7	38	9	24	3	1	2400	Belfast	Harland & Wolff	1878	126,190	..		4 7·5-in., 14 M.	4	13·0	2,200	277
2nd cl. Cr.	Hermes	S. shd.	5600	350	0	54	0	20	6	2	10,000 B	Glasgow	Fairfield	1898	278,349							
,, ,,	Highflyer	S. shd.	5600	350	0	54	0	20	6	2	10,000 B	,,	,,	1898	278,186	3	1½ 3	11 6-in. Q.F., 15 smaller Q.F.	..	20·0	600	477
,, ,,	Hyacinth	S. shd.	5600	350	0	54	0	20	6	2	10,000 B	,,	London and Glasgow Co.	1898	282,761							
,, ,,	Hermione	S. shd.	4360	320	0	49	6	19	0	2	9000	Devonp'rt	Thomson	1893	223,267	2	2–1	2 6-in. Q.F., 8 4·7-in., 8 6-pr., 1 3-pr. 4 M., 1 l.	4	19·5	400	312

‡ Details not yet complete.

x Includes Gun Mountings, &c.

GREAT BRITAIN.—Cruising Ships, &c.—*continued.*

Class.	NAME.	Material of Hull.	Displacement.	Length.		Beam.		Draught.		Propellers.	Indicated Horse-Power.	Where Built.	Maker of Engines.	Date of Launch.	Cost.	Armour.		Armament.		Speed.	Normal Coal Supply.	Complement.
																Gun Position.	Deck.	Guns.	Torpedo Tubes.			
			tons.	ft.	in.	ft.	in.	ft.	in.	no					£	in.	in.			knots.	tons.	
T. G. B. .	Hussar .	S.	1070	250	0	30	6	9	0	2	3500	Devonp'rt	Hawthorn .	1894	72,886	2	..	2 4·7-in. Q.F., 4 6-pr.	3	19·0	100	120
Sloop . .	Icarus .	C.	970	167	0	32	0	13	6	1	1200	Devonp'rt	Barrow .	1885	x 52,104	..	..	8 5-in., 4 3-pr. Q.F., 4 M., 1 l.	..	12·2	150	126
2nd cl. Cr.	Indefatigable	S. shd.	3600	300	0	43	8	17	6	2	9000	Glasgow .	London and Glasgow Co.	1891	181,024							
,, ,,	Intrepid .	S. shd.	3600	300	0	43	8	17	6	2	9000	Glasgow .	London and Glasgow Co.	1891	181,157	2	2–1	2 6-in. Q.F., 6 4·7-in., 8 6-pr., 1 3-pr., 4 M., 1 l.	4	19·75	400	273
,, ,,	Iphigenia .	S shd.	3600	300	0	43	8	17	6	2	9000	Glasgow .	London and Glasgow Co.	1891	181,879							
,, ,,	Iris . .	S.	3730	300	0	46	0	22	0	2	6000	Pembroke	Maudslay .	1877	213,186	..	..	13 5-in., 4 3-pr. Q.F., 8 M., 1 l.	3	18·0	780	450
,, ,,	Isis . .	S. shd	5600	350	0	54	0	21	0	2	9600 B	Glasgow .	London and Glasgow Co.	1896	252,067	3	2½	5 6-in. Q.F., 6 4·7-in., 9 12-pr., 11 lighter Q.F. and M.	4 (2 sub.)	20·0	550	470
,, ,,	Juno . .	S. shd.	5600	350	0	54	0	21	0	2	9600 B	Barrow .	Barrow .	1895	254,097							
T. G. B. .	Jaseur .	S.	810	230	0	27	0	8	9	2	3711	Barrow .	Barrow .	1892	48,238	2	..	2 4·7-in. Q.F., 4 3-pr.	3	19·25	100	91
,, .	Jason .	S.	810	230	0	27	0	8	9	2	3540	Barrow .	Barrow .	1892	49,253 x							
,, .	Karrakatta (Australia)	S.	735	230	0	27	0	8	3	2	3500	Elswick .	Bellis .	1890	47,649	2	..	2 4·7-in. Q.F., 4 3-pr.	3	20·0	100	91

3rd cl. Cr.	Katoomba (Australia)	.	2575	265	0	41	0	15	6	2	7500	Elswick	Hawthorn	1889	116,719	2	2-1	8 4·7-in. Q.F., 8 3-pr., 4 M., 1 l.	4	19·0	300	217
G. V.	Landrail	S.	950	195	0	28	0	10	6	2	1200	Devonp'rt	Penn	1886	49,963	..	..	1 6-in., 3 5-in., 4 3-pr. Q.F., 3 M.	..	14·5	250	46
1st cl. G. B.	Lapwing	C.	805	165	0	31	0	11	7½	2	1200	Devonp'rt	Devonport	1889	39,952	..	..	6 4-in. 25-cwt., 2 3-pr. Q.F., 2 M.	..	13·0	105	76
2nd cl. Cr.	Latona	S.	3400	300	0	43	0	16	6	2	9000	Barrow	Barrow	1890	171,068	2	2–1	2 6-in. Q.F., 6 4·7-in., 8 6-pr., 1 3-pr., 4 M., 1 l.	4	20·0	400	273
" "	Leander	S.	4300	300	0	46	0	20	6	2	5000	Glasgow	Napier	1882	148,453	..	1½	10 6-in. Q.F., 14 lighter Q.F., and M.	4	16·6	550 / 1000	309
T. G. B.	Leda	S.	810	230	0	27	0	8	9	2	3597	Sheerness	Penn	1892	62,145	2	..	2 4·7-in. Q.F., 4 3-pr.	3	19·25	100	91
G. V.	Linnet	C.	756	165	0	29	0	10	11	2	870	Blackwall	Rennie	1880	35,663	..	..	2 90-cwt. M.L.R., 4 6-pr. Q.F., 2 M.	..	11·80	180	92
1st cl. G. B.	Lizard	C.	715	165	0	29	0	11	10	1	1000	Belfast	Harland	1886	52,770	..	..	6 4-in., 4 M.	..	13·0	105	76
3rd cl. Crs.	Magicienne, Marathon	S. shd.	2950	265	0	42	0	17	6	2	9000	Glasgow	Hawthorn	1888	136,000	..	1½	6 6-in. Q.F., 9 6-pr., 1 3-pr., 3 M., 1 l.	4	19·0	400	218
1st cl. G. B.	Magpie	C.	805	165	0	31	0	11	7½	1	1200	Pembroke	Earle	1889	*x* 38,700	..	..	6 4-in., 4 M.	..	13·0	105	76
3rd cl. Crs.	Medea, Medusa	S.	2800	265	0	41	0	16	6	2	9000	Chatham	Humphrys	1888	*x* 141,700	..	1½	6 6-in. Q.F., 9 6-pr., 1 3-pr., 3 M., 1 l.	4	19·0	400	218
2nd cl. Cr.	Melampus	S.	3400	300	0	43	0	16	6	2	9000	Barrow	Barrow	1890	171,635	2	2–1	2 6-in. Q.F., 6 4·7 in., 8 6-pr., 1 3-pr., 4 M., 1 l.	4	20·0	400	273
Sloop	Melita	C.	970	167	0	32	0	13	6	1	1200	Malta	Malta Dock Yard	1888	60,179	..	..	8 5-in., 8 M., 1 l.	..	12·50	150	125
3rd cl. Cr.	Melpomene	S. shd.	2950	265	0	41	0	17	6	2	9000	Portsm'th	Palmer Co.	1888	*x* 142,000	..	1½	6 6-in. Q.F., 9 6-pr., 1 3-pr., 3 M., 1 l.	4	19·0	400	218
2nd cl. Cr.	Mercury	S.	3730	300	0	46	0	20	2	2	6000	Pembroke	Maudslay	1878	213,252	..	..	13 5-in., 12 light Q.F., and M.	4	16·8	780	291

x Includes Gun Mountings, &c.

GREAT BRITAIN.—Cruising Ships, &c.—*continued.*

Class.	NAME.	Material of Hull.	Displacement.	Length.		Beam.		Draught.		Propellers.	Indicated Horse-Power.	Where Built.	Maker of Engines.	Date of Launch.	Cost.	Armour.		Armament.		Speed.	Normal Coal Supply.	Complement.
																Gun Position.	Deck.	Guns.	Torpedo Tubes.			
			tons.	ft.	in.	ft.	in.	ft.	in.	no.					£	in.	in.			knots.	tons.	
2nd cl. Cr.	**Mersey**	S.	4050	300	0	46	0	19	6	2	6000	Chatham	Humphrys	1885	154,000	4	3–2	2 15-ton, 10 6-in. Q.F., 3 6-pr., 8 3-pr., 5 M., 2 l.	4	17·3	900	327
3rd cl. Cr.	**Mildura** (Australia)	S.	2575	265	0	41	0	15	6	2	7500	Elswick	Hawthorn	1889	116,062	2	2–1	8 4·7-in. Q.F., 8 3-pr., 4 M., 1 l.	4	19·0	300	217
2nd cl. Cr.	**Minerva**	S. shd.	5600	350	0	53	0	20	6	2	9600	Chatham	Chatham	1895	244,046	3	1½–3	5 6-in. Q.F., 6 4·7-in., 9 12-pr., 11 smaller Q.F., and M.	3 (2 sub.)	20·3	550	437
3rd cl. Cr.	**Mohawk**	S.	1770	225	0	36	0	14	6	2	3500	Glasgow	Thomson	1886	£ 87,583	..	..	6 6-in. Q.F., 8 3-pr., 2 M., 1 l.	3	16·5	475	172
Sloop	**Mutine**	S.	980	180	0	33	0	11	6	2	1400 B	Laird	Laird	Bldg.	63,204	..	..	6 4-in. Q.F., 4 3-pr.	..	13·25	130	130
2nd cl. Cr.	**Naiad**	S.	3400	300	0	43	0	16	6	2	9000	Barrow	Barrow	1890	171,445	2	2–1	2 6-in. Q.F., 6 4·7-in., 8 6-pr., 1 3-pr., 4 M., 1 l.	4	20·0	400	273
1st cl. Cr.	**Niobe**	S. shd.	11,000	435	0	69	0	26	0	2	16,500 B	Barrow	Vickers	1897	552,692	4½–2	4–2½	16 6-in. Q.F., 14 12-pr., 11 light Q.F., and M.	3 (2 sub.)	20·5	1000	600
Sloop	**Nymphe**	C.	1140	195	0	28	0	12	6	1	2000	Portsm'th	Greenock F'ndry Co.	1888	£ 57,600	..	..	8 5-in., 8 M.	..	14·0	160	138
T. G. B.	**Niger**	S.	810	230	0	27	0	8	9	2	3784	Barrow	Barrow	1892	48,177	2	..	2 4·7-in. Q.F., 4 3-pr.	3	19·25	100	91
"	**Onyx**	S.	810	230	0	27	0	8	9	2	3548	Birkenh'd	Laird	1892	53,961							
3rd cl. Cr.	**Pallas**	S.	2575	265	0	41	0	15	6	2	7610	Portsm'th	Hawthorn	1890	148,828	2	2–1	8 4·7-in. Q.F., 8 3-pr., 4 M., 1 l.	4	19·25	300	217
" "	**Pearl**	S.	2575	265	0	41	0	15	6	2	7500	Pembroke	Earle	1890	151,693							
1st cl. G. B.	**Partridge**	C.	755	165	0	30	0	11	4	1	1200	Devonp'rt	Devonport	1888	37,800	..	..	6 4-in., 4 M.	..	13·25	105	76
" "	**Peacock**	C.	755	165	0	30	0	11	4	1	1200	Pembroke	Barrow Co.	1888	37,600							
Sloop	**Pelican**	C.	1130	170	0	36	0	15	3	1	800	Devonp'rt	Humphrys	1872	56,221	..	..	2 6-in., 6 5-in., 4 M., 1 l.	..	10·6	130	145

3rd cl. Cr.	**Pactolus**	S.	2135	300		36	6	17	0	2	7000 B	Elswick	Penn	1897	139,736							
" "	**Pandora**	S.	2200	305		36	9	13	6	2	7000 R	Portsm'th	Portsm'th	1900	165,046							
" "	**Pegasus** *a*	S.	2135	300		36	6	17	0	2	7000 B	Jarrow	Palmer	1897	135,071							
" "	**Pelorus**	S.	2135	300		36	6	17	0	2	7000 B	Sheerness	Thomson	1896	164,840							
" "	**Perseus**	S.	2135	300		36	6	13	6	2	7000 B	Hull	Earle	1898	127,992							
" "	**Pioneer**	S.	2200	305		36	9	13	6	2	7000 B	Chatham	Fairfield	1899	149,080	·22	2	8 4-in., 8 3-pr. Q.F., 2 l.	2	20·0	250	224
" "	**Pomone**	S.	2135	300		36	6	13	6	2	7000 B	Sheerness	Penn	1897	149,568							
" "	**Prometheus**	S.	2135	300		36	6	13	6	2	7000 B	Hull	Earle	1898	127,975							
" "	**Psyche**	S.	2200	305		36	9	17	6	2	7000 B	Devonp'rt	Devonport	1898	155,563							
" "	**Proserpine**	S.	2135	300		36	6	17	0	2	7000 T	Sheerness	Devonport	1896	159,136							
" "	**Pyramus** *b*	S.	2135	300		36	6	13	6	2	7000 R	Jarrow	Palmer	1898	135,096							
Sloop (Surveying)	**Penguin**	C.	1130	170	0	36	0	16	1	1	700	Glasgow	Hawthorn	1876	52,111	..	..	2 64-pr. M.L.R., 2 M., 1 l.	..	11·0	150	145
2nd cl. Cr.	**Phaeton**	S.	4300	300	0	46	0	20	6	2	5000	Glasgow	Napier	1883	145,198 *x*	..	1½	10 6-in. Q.F., 14 light Q.F., and M.	4	16·6	550 / 1000	309
1st cl. G. B.	**Pheasant**	C.	755	165	0	29	0	11	4	1	1200	Devonp'rt	Devonport	1888	37,800	..	..	6 4-in., 4 M.	..	13·25	105	76
3rd cl. Cr.	**Philomel**	S.	2575	265	0	41	0	15	6	2	7500	Devonp'rt	Earle	1890	156,102	2	2–1	8 4·7-in. Q.F., 8 3-pr., 4 M., 1 l.	2	19·0	300	217
" "	**Phœbe**	S.	2575	265	0	41	0	15	6	2	7500	Devonp'rt	Devonport	1890	161,154							
Sloop	**Phœnix**	S. shd.	1050	185	0	32	6	11	3	2	1400	Devonp'rt	Devonport	1895	63,930 *x*	·22	..	6 4-in. Q.F., 4 3-pr., 3 M.	..	13·0	160	106
1st cl. G. B.	**Pigeon**	C.	755	165	0	30	0	11	4	1	1200	Pembroke	Barrow	1888	37,800							
" "	**Pigmy**	C.	755	165	0	30	0	11	4	1	1200	Sheerness	Barrow	1888	37,700	..	..	6 4-in., 4 M.	..	13·25	105	76
" "	**Plover**	C.	755	165	0	30	0	11	4	1	1200	Pembroke	Barrow	1888	37,700							
2nd cl. Cr.	**Pique**	S. shd.	3600	300	0	43	8	17	6	2	9000	Jarrow	Palmer	1890	184,108	2	2–1	2 6-in. Q.F., 6 4·7-in., 8 6-pr., 1 3-pr., 4 M., 1 l.	4	19·75	400	273
T. Ram	**Polyphemus**	S.	2640	240	0	40	0	20	0	2	5500	Chatham	Humphrys	1881	174,450	..	3–2	6 6-pr. Q.F., 2 M.	5	18·0	300	..

a Trial speed 21·2 kts., I.H.P. 7134. *b* Trial speed 20·7 kts., I.H.P. 7303. *x* Includes Gun Mountings, &c.

GREAT BRITAIN.—Cruising Ships, &c.—*continued.*

Class.	NAME.	Material of Hull.	Displacement.	Length.		Beam.		Draught.		Propellers.	Indicated Horse-Power.	Where Built.	Maker of Engines.	Date of Launch.	Cost.	Armour.		Armament.		Speed.	Normal Coal Supply.	Complement.
																Gun Position.	Deck.	Guns.	Torpedo Tubes.			
			tons.	ft.	in.	ft.	in.	ft.	in.	no.					£	in.	in.			knots.	tons.	
3rd cl. Cr.	Porpoise	S.	1770	225	0	36	0	14	6	2	3500	Glasgow	Thomson	1886	87,583	..	..	6 6-in. Q.F., 8 3-pr., 2 M., 1 l.	3	16·5	475	172
1st cl. Cr.	Powerful	S. shd.	14,200	500	0	71	0	29	0	2	25,000 B	Barrow	Barrow	1895	674,879	6	3–6	2 9·2-in., 12 6-in. Q.F., 18 12-pr., 12 3-pr., 9 M., 2 12-pr. boat	4	22·1	1500*	840
3rd cl. Cr.	Pylades	C.	1420	200	0	38	0	15	9	1	1400	Sheerness	Laird	1884	62,000	..	1¼	14 5-in., 8 M., 1 l.	..	12·6	400	170
3rd cl. Cr.	Racoon	S.	1770	275	0	36	0	13	6	2	4500	Devonp'rt	Harland	1887	x 91,606	..	..	6 6-in. Q.F., 8 3-pr., 2 M., 1 l.	3	17·5	475	176
2nd cl. Cr.	Rainbow	S. shd.	3600	300	0	43	8	17	6	2	9681	Jarrow	Palmer	1891	184,086	2	2–1	2 6-in. Q.F., 6 4·7-in., 8 6-pr., 1 3-pr., 4 M., 1 l.	4	19·7	400	273
,, ,,	Raleigh	I. shd.	5200	298	0	49	0	24	7	1	4200	Chatham	Humphrys	1873	193,386	..	..	4 6-in., 8 5-in., 12 M., 4 l.	2	13·9	550	571
2nd cl. G. Ves. (Surveying Service.)	Rambler	C.	835	157	0	29	6	13	7	1	650	Glasgow	Elder	1880	37,038	..	..	2 20-pr., 1 M., 1 l.	..	10·66	40	160
3rd cl. Cr.	Rapid	C.	1420	200	0	38	0	15	9	1	1400	Devonp'rt	Maudslay	1883	68,226	..	1¼	2 6-in., 10 5-in., 4 M., 1 l.	..	12·6	400	171
1st cl. G. B.	Rattler	C.	715	165	0	29	0	11	0	1	1200	Elswick	Hawthorn	1886	x 38,734	..	..	6 4-in., 4 M.	..	13·6	105	76
T. G. B.	Rattlesnake	S.	550	200	0	23	0	8	0	2	2700	Birkenh'd	Laird	1886	x 35,425	..	..	1 4-in., 6 3-pr. Q.F.	4	18·5	100	67

2nd cl. G. B.	Raven	C.	465	125	0	23	6	10	0	1	360	Poplar	Rennie	1882	21,050	..	..	2 64-pr. M.L.R., 2 20-pr., 2 M.	..	9·5	40	62
1st cl. G. B	Redbreast	C.	805	165	0	31	0	11	7½	1	1200	Pembroke	Earle	1888	38,700	..	..	6 4-in., 4 M.	..	13·0	105	76
” ”	Redpole																					
2nd cl. G. B.	Redwing	C.	461	125	0	23	6	10	0	1	360	Pembroke	Maudslay	1880	22,200	..	..	2 20-cwt., 2 M.	..	9·68	40	..
T. G. B.	Renard	S.	810	230	0	27	0	8	9	2	3500	Birkenh'd	Laird	1892	53,848	2	..	2 4·7 in. Q.F., 4 3-pr.	3	19·25	100	91
2nd cl. Cr.	Retribution	S. shd.	3600	300	0	43	8	17	6	2	9000	Jarrow	Palmer	1891	183,975	2	2–1	2 6-in. Q.F., 6 4·7-in., 8 6-pr., 1 3-pr., 4 M., 1 l.	4	19·75	400	275
Sloop	Rinaldo	S.	980	180	0	33	0	11	6	2	1400 B	Laird	Laird	Bldg.	63,180	..	..	6 4-in. Q.F., 4 3-pr.	..	13.25	130	130
3rd cl. Cr.	Ringarooma (Australia)	S.	2575	265	0	41	0	15	6	2	7500	Glasgow	Thomson	1890	128,076	2	2–1	8 4·7-in. Q.F., 8 3-pr. Q.F., 4 M., 1 l.	4	19·0	300	216
1st cl. G. B.	Ringdove	C.	805	165	0	31	5	11	7½	1	1200	Devonp'rt	Devonport	1889	39,753	..	..	6 4-in., 2 3-pr. Q.F., 2 M.	..	13·0	105	76
1st cl. Cr.	Royal Arthur	S. shd.	7700	360	0	60	0	27	9	2	12,000	Portsm'th	Maudslay	1891	402,414	6	5–1	1 9·2-in., 12 6-in. Q.F., 12 6-pr., 5 3-pr., 7 M., 2 l.	4 (2 sub.)	19·7	850	567
Sloop	Rosario	S. shd.	980	180	0	33	0	11	6	2	1400 B	Sheerness	Governm't	1898	77,962	..	..	6 4-in., 4 3-pr. Q.F.	..	13·25	130	130
3rd cl. Cr.	Royalist	C.	1420	200	0	38	0	15	9	1	1400	Devonp'rt	Maudslay	1883	68,173	..	1¼	2 6-in., 10 5-in., 4 M., 1 l.	..	12·6	400	171
1st cl. Cr.	St. George	S. shd.	7700	360	0	60	8	23	9	2	12,000	Hull	Maudslay	1892	377,204	6	5–1	2 9·2-in., 10 6-in. Q.F., 12 6-pr., 5 3-pr., 7 M., 2 l.	4 (2 sub.)	19·7	850	559
T. G. B.	Salamander	S.	735	230	0	27	0	8	3	2	3500	Chatham	Maudslay	1889	57,911	2	..	2 4·7-in. Q.F., 4 3-pr. Q.F.	3	20·0	100	91
” ”	Sandfly	S.	525	200	0	23	0	8	9	2	2700	Devonp'rt	Maudslay	1887	x 36,167	·22	..	1 4-in., 6 3-pr. Q.F.	4	19·0	80	67

c Includes Gun Mountings, &c

* Bunker capacity, 3000.

GREAT BRITAIN.—Cruising Ships, &c.—*continued.*

Class.	NAME.	Material of Hull.	Displacement.	Length.	Beam.	Draught.	Propellers.	Indicated Horse-Power.	Where Built.	Maker of Engines.	Date of Launch.	Cost.	Armour. Gun Position.	Armour. Deck.	Armament. Guns.	Armament. Torpedo Tubes.	Speed.	Normal Coal Supply.	Complement.
			tons.	ft. in.	ft. in.	ft. in.	no.					£	in.	in.			knots.	tons.	
2nd cl. Cr.	Sappho	S.	3400	300 0	43 0	16 6	2	9861	Poplar	Penn	1891	171,853	2	2–1	2 6-in. Q.F., 6 4·7-in., 8 6-pr., 1 3-pr., 4 M., 1 l.	4	20·47	400	273
3rd cl. Cr.	Satellite	C.	1420	200 0	38 0	15 9	1	1400	Sheerness	Humphrys	1881	62,900	..	1¼	2 6-in., 6 5-in., 4 M., 1 l.	..	12·6	400	167
,, ,,	Scout	S.	1580	220 0	34 0	14 6	2	3200	Glasgow	Thomson	1885	£ 87,516	..	..	4 4·7-in. Q.F., 8 3-pr., 2 M., 1 l.	3 (1 sub.)	16·7	450	147
2nd cl. Cr.	Scylla	S.	3400	300 0	43 0	16 6	2	9280	Poplar	Penn	1892	171,593	2	2–1	2 6-in. Q.F., 6 4·7-in., 8 6-pr., 1 3-pr., 4 M., 1 l.	4	20·62	400	273
T. G. B.	Seagull	S.	735	230 0	27 0	8 3	2	3500	Chatham	Maudslay	1889	56,922							
,, ,,	Sharpshooter	S.	735	230 0	27 0	8 3	2	3500 B	Devonp'rt	Bellis	1888	£ 50,029	2	..	2 4·7-in., 4 3-pr. Q.F.	3	20·0	100	91
Sloop	Shearwater	S. shd.	980	180 0	33 0	11 6	2	1400	Sheerness	Thames Co.	Bldg.	65,400	..	..	6 4-in., 4 3-pr. Q.F.	..	13·25	130	130
T. G. B.	Sheldrake	S.	735	230 0	27 0	8 3	2	3500 B & W	Chatham	Maudslay	1889	57,800	2	..	2 4·7-in., 4 3-pr. Q.F.	3	20·0	100	91
,, ,,	Skipjack	S.	735	230 0	27 0	8 3	2	3500	Chatham	Laird	1889	59,531							
,, ,,	Spanker*	S.	735	230 0	27 0	8 3	2	3920 Du T.	Devonp'rt	Bellis	1889	50,000	2	..	2 4·7-in., 4 3-pr. Q.F.	3	20·0	100	91
,, ,,	Speedwell	S.	735	230 0	27 0	8 3	2	3500	Devonp'rt	Laird	1889	52,000							
1st cl. Cr.	Spartiate	S. shd.	11,000	435 0	69 0	26 0	2	18,000 B	Pembroke	Maudslay	1898	569,841	4½–2	4–2½	16 6-in., 17 small Q.F., 2 12-pr. boat.	3 (2 sub.)	20·75	1000	600
2nd cl. Cr.	Severn	S.	4050	300 0	46 0	19 6	2	6000	Chatham	Humphrys	1885	£212,621	4	3–2	2 8-in., 10 6-in. Q.F., 3 6-pr., 2 3-pr., 10 M., 2 l.	..	17·3	900	327

2nd cl. Cr.	Sirius	S. shd.	3600	300	0	43	8	17	6	2	9000	Elswick	Maudslay	1890	186,649	2	2–1	2 6-in. Q.F., 6 4·7-in., 8 6-pr., 1 3-pr., 4 M., 1 l.	4	19·75	400	273
„ „	Spartan	S. shd.	3600	300	0	43	8	17	6	2	9000	Elswick	Maudslay	1891	186,351							
1st cl. G. B.	Sparrow	C.	805	165	0	31	0	11	7½	1	1200	Greenock	Greenock	1889	39,000	..	..	6 4-in., 2 3-pr. Q.F., 2 M.	..	13·0	105	76
T. G. B.	Speedy	S.	810	230	0	27	0	8	9	2	4703	Chiswick	Thornycrft	1893	58,927	2	..	2 4·7-in. Q.F., 4 3-pr.	3	20·21	100	91
„ „	Spider	S.	525	200	0	23	0	8	9	2	2700	Devonp'rt	Maudslay	1887	x36,300	·22	..	1 4-in., 6 3-pr. Q.F.	4	19·0	80	67
2nd cl. G. B.	Starling	C.	465	125	0	23	6	10	0	1	360	Poplar	Rennie	1882	21,100	..	..	2 64-pr. M.L.R., 2 20-pr., 2 M.	..	9·5	40	61
„ „	Stork	C.	465	125	0	23	6	10	0	1	360	Poplar	Rennie	1882	21,150	..	..	1 M., 2 l.	..	9·5	40	..
D. V.	Surprise	S.	1650	250	0	32	6	14	0	2	3000	Jarrow	Palmer	1885	x78,764	..	..	4 5-in., 4 6-pr. Q.F., 2 M.	..	17·0	400	114
Sloop	Swallow	C.	1130	195	0	28	0	11	6	2	1500	Sheerness	Rennie	1885	59,797	..	..	8 5-in., 8 M.	..	13·5	280	135
2nd cl. G. V.	Swift	C.	756	165	0	29	0	10	11	2	870	Blackwall	Rennie	1879	34,670	..	..	2 90-cwt. M.L.R., 4 6-pr. Q.F., 2 M.	..	11·81	180	92
2nd cl. Cr.	Sybille	S.	3400	300	0	42	0	16	6	2	9496	Stephenson.	Hawthorn.	1890	174,670	2	2–1	2 6-in. Q.F., 6 4·7-in., 8 6-pr., 1 3-pr., 4 M., 1 l.	4	20·0	400	273
„ „	Talbot	S. shd.	5600	350	0	53	6	21	0	2	9600	Devonp'rt	Devonport	1895	273,856	3	1½–3	5 6-in. Q.F., 6 4·7-in., 9 12-pr., 1 3-pr., 4 M., 1 l.	3 (2 sub.)	20·0	550	433
3rd cl. Cr.	Tartar	S.	1770	225	0	36	0	14	3½	2	3500	Glasgow	Thomson	1886	x87,583	..	..	6 6-in. Q.F., 8 3-pr., 2 M., 1 l.	3	16·5	325	177
„ „	Tauranga (Australia)	S.	2575	265	0	41	0	15	6	2	7500	Glasgow	Thomson	1889	128,101	2	2–1	8 4·7-in. Q.F., 8 3-pr. Q.F., 4 M., 1 l.	4	19·0	300	212
2nd cl. Cr.	Terpsichore	S.	3400	300	0	43	0	16	6	2	9000	Glasgow	Thomson	1890	173,341	2	2–1	2 6-in. Q.F., 6 4·7-in., 8 6-pr., 1 3-pr., 9 M., 1 l.	4	20·0	400	275
1st cl. Cr.	Terrible	S. shd.	14,200	500	0	71	0	27	0	2	25,000 B	Glasgow	Thomson	1895	681,419	6	3–6	2 9·2-in., 12 6-in. Q.F., 18 12-pr., 12 3-pr., 9 M., 2 12-pr. boat.	4	22·4	3000	810
2nd cl. Cr.	Thames	S.	4050	300	0	46	0	19	6	2	5700	Pembroke	Penn	1885	205,452	4	3–2	2 8-in., 10 6-in. Q.F., 3 6-pr., 8 3-pr., 6 M., 2 l.	2	16·8	900	326
1st cl. G. B.	Thistle	S.	700	180	0	33	0	8	0	2	1300	Glasgow	London and Glasgow Co.	Bt. g.	51,105	..	..	2 4-in. Q.F., 4 12-pr.	..	13·5	50	85
1st cl. Cr.	Theseus	S.	7350	360	0	60	0	23	9	2	12,000	Blackwall	Maudslay	1892	347,577	6	5–1	2 9·2-in., 10 6-in. Q.F., 12 6-pr., 5 3-pr., 7 M., 2 l.	4 (2 sub.)	20·0	850	544
2nd cl. Cr.	Thetis	S.	3400	300	0	43	0	16	6	2	9000	Glasgow	Thomson	1890	173,146	2	2–1	2 6-in. Q.F., 6 4·7-in., 8 6-pr., 1 3-pr., 4 M., 1 l.	4	20·0	400	273
„ „	Tribune	S.	3400	300	0	43	0	16	6	2	9000	Glasgow	Thomson	1891	173,006							

x Includes Gun Mountings, &c.

GREAT BRITAIN.—Cruising Ships, &c.—*continued.*

Class.	NAME.	Material of Hull.	Displacement.	Length.	Beam.	Draught.	Propellers.	Indicated Horse-Power.	Where Built.	Maker of Engines.	Date of Launch.	Cost.	Armour. Gun Position.	Armour. Deck.	Armament. Guns.	Armament. Torpedo Tubes.	Speed.	Normal Coal Supply.	Complement.
			tons.	ft. in.	ft. in.	ft. in.	no.					£	in.	in.			knots.	tons.	
1st cl. G. B.	Thrush	C.	805	165 0	31 0	11 7½	1	1200	Greenock	Greenock F'ndry Co.	1889	39,000	..	..	6 4-in., 2 3-pr. Q.F., 2 M.	..	13·0	105	76
Sloop	Torch	S. shd.	960	180 0	32 6	11 6	1	1400	Sheerness	Sheerness	1894	60,564	..	..	6 4-in. Q.F., 4 3-pr., 2 M.	..	13·25	130	..
2nd cl. Cr.	Venus*	S. shd.	5600	350 0	54 0	21 2	2	9600	Glasgow	Elder	1895	249,938	3	2½	5 6-in. Q.F., 6 4·7-in., 9 12-pr., 7 3-pr., 4 M., 1 12-pr. boat	3 (2 sub.)	19·5	550	470
Sloop	Vestal	S. shd.	980	180 0	33 0	11 6	2	1400 B	Sheerness	Governm't	Bldg.	94,301	..	..	6 4-in., 4 3-pr. Q.F.	..	13·25	130	130
2nd cl. Cr.	Vindictive	S.	5800	320 0	51 0	20 6	2	10,000	Chatham	Chatham	1896	290,458	3	1–2 N.S.	4 6-in. Q.F., 6 4·7-in., 9 12-pr., 3 3-pr., 1 12-pr. boat, 5 M.	2	19·5	500	450
2nd cl. Cr.	Volage	I. shd.	3080	270 0	42 0	22 0	1	2400	Blackwall	Ravenhill	1874	132,817	..	..	10 6-in., 2 64-pr. M.L.R., 10 M., 2 l.	..	12·8	420	339
T. D. S.	Vulcan	S.	6620	350 0	58 0	23 0	2	12,032	Portsm'th	Humphrys	1889	370,447	2	5–2½	8 4·7-in. Q.F., 12 3-pr., 16 M., 1 l.	6 (2 sub.)	20·0	1000	433
3rd cl. Cr.	Wallaroo (Australia)	S.	2575	265 0	41 0	15 6	2	7500	Elswick	Hawthorn	1889	115,995	2	2–1	8 4·7-in. Q.F., 8 3-pr., 4 M., 1 l.	4	19·0	300	218
1st cl. G. B.	Widgeon	C.	805	165 0	30 0	11 7½	1	1200	Pembroke	Rennie	1889	39,315	..	..	6 4-in., 2 3-pr. Q.F., 2 M.	..	13·0	105	76
2nd cl. cr.	1 New Ship		(Design not yet settled)					..	Chatham										
Sloop	2 New Ships		(Design not yet settled)					..	Sheerness										

* Venus: trial, 20·1 knots.

River Gunboats.—Herald, Mosquito (1890), 82 tons; Jackdaw, Heron, Robin, Nightingale, Sandpiper, Snipe (1897), 85 tons; Woodcock, Woodlark (1897), 122 tons, 2 6-prs., 4 Maxims; 4 recent boats in the Niger Protectorate. Recent Egyptian boats: Molik, Sultan, Sheik, 140 tons, 4 12-prs., 4 Maxims.

Royal Naval Reserved Merchant Cruisers.

—	Name.	Owners.	Length.	Breadth.	Draught of Water for the Admiralty List.	Gross Tonnage.	Indicated Horse-Power.	Ocean Speed.
			Feet.	Feet.	Feet.	Tons.		Knots.
Ships in receipt of an Annual subvention and permitted to fly the blue ensign.	Campania	Cunard Company	610	65	26	12,950	30,000	21
	Lucania	" "	610	65	26	12,950	30,000	21
	Himalaya	Peninsular and Oriental Co.	465½	52	22½	6,898	10,000	17
	Australia	" " "	465½	52	22½	6,901	10,000	17
	Victoria	" " "	466	52	22½	6,091	7,000	16
	Arcadia	" " "	466	52	22½	6,188	7,000	16
	Majestic 2 s.	White Star Company	565	58	24½	9,965	16,000	20
	Teutonic 2 s.	" "	565	58	24½	9,984	16,000	20
	Empress of India 2 s.	Canadian Pacific Railway Co.	440	51	24½	5,905	10,000	16
	Empress of China 2 s.	" " "	440	51	24½	5,905	10,000	16
	Empress of Japan 2 s.	" " "	440	51	24½	5,905	10,000	16
Ships held at the disposition of the Admiralty without subsidy.	Etruria	Cunard Company	501½	57	26	8,120	14,500	19½
	Umbria	" "	501½	57	26	8,128	14,500	19½
	Servia	" "	515	52	26	7,392	10,000	16½
	Aurania	" "	470	57	27	7,269	9,500	17
	Britannic	White Star Company	455	45	25	5,004	5,200	16
	Germanic	" "	455	45	25	5,008	5,200	16
	Cymric	" "	..	..	..	12,551	6,700	..
	Britannia	Peninsular and Oriental Co.	466	52	22½	6,525	7,000	16
	Oceana	" " "	466	52	22½	6,188	6,000	16
	Peninsular	" " "	410½	48	..	5,287	4,972	15
	Oriental	" " "	410½	48	..	5,284	4,972	15
	Valetta	" " "	420½	45	22½	4,904	5,000	15
	Massilia	" " "	420½	45	22½	4,902	5,000	15
	Rome	" " "	430	44½	22½	5,545	5,500	15
	Carthage	" " "	430	44½	22½	4,879	5,000	15
	Ballarat	" " "	420	43	22½	4,778	4,500	14
	Parramatta	" " "	420	43	22½	4,756	4,500	14

There are also numerous ships on the Admiralty List complying with Admiralty conditions as to subdivision which have no national tie. They are suitable for receiving an armament, but there is no arrangement with Owners, except the promise of preference for occasional State employment.

GREAT BRITAIN, COLONIES, &c.—Cruising Ships, Gunboats, &c.

To what Government belonging.	Class of Ship.	Name.	Material of Construction.	Propellers.	Where Built.	When Launched.	Length.	Breadth.	Draught of Water.	Displacement.	Indicated Horse-Power.	Speed.	Coal Stowage.	Armament.
							ft. in.	ft. in.	ft. in.				tons.	
INDIA	T. G.B. .	Assaye .	Steel	2	Elswick	1891	230 0	27 0	8 3	735	3,500	19·0	100	2 4·7-in. Q.F., 4 3-pr. do., 1 f. tu. & 3 l. car.
INDIA	D. V. . .	Lawrence .	Steel	Pad.	B'kenh'd	1886	212 2	32 2	18 3	1,154	1,277	13·5	270	Four 4-in. B.L.R., 4 6-pr. Q.F., 4 M.
INDIA	T. G. B. .	Plassy .	Steel	2	Elswick	1890	230 0	27 0	8 3	735	3,500	19·0	100	2 4·7-in. Q.F., 4 3-pr. do., 1 f. tu. & 3 l. car.
QUE'NS-LAND.	Gun-vessel	Gayundah	Steel	2	Glasgow	1884	115 0	25 0	10 0	450	400	10·0	..	One 8-in. 11½-ton; one 6-in. 4-ton; one 3-pr. Q.F.; 2 M.
QUE'NS-LAND.	Gun-vessel	Paluma .	Steel	2	Glasgow	1884	115 0	25 0	10 0	450	340	10·0	..	One 8-in. 11½-ton; one 6-in. 4-ton; one 3-pr. Q.F.; 2 M.
SOUTH AUS-TRALIA	Cruiser .	Protector .	Steel	2	..	1884	188 0	3 0	12 0	920	1,640	14·0	..	One 8-in. 11½-ton; five 6-in. 4-ton; five Gatlings.

The five second-class Cruisers, and the two Torpedo-Gunboats of the Australian Auxiliary Squadron, are included in the list of Ships of the Royal Navy, as well as the armour-clads Abyssinia, Cerberus, and Magdala.

ARGENTINE REPUBLIC.—Armoured Ships.

Class.	NAME.	Material of Hull.	Displacement.	Length.	Beam.	Draught.	Propellers.	Indicated Horse-power.	Where Built.	Date of Launch.	Cost.	Armour. Belt.	Armour. Battery, or Turret.	Armour. Deck Plating.	Armament. Guns.	Armament. Torpedo Tubes.	Speed	Normal Coal Supply.	Complement.
			metric tons.	ft. in.	ft. in.	ft. in.					£	inches	inches.	inches.			knots.	tons.	
c.b.	Almirante Brown	S.	4267	240 0	50 0	20 6	2	4500	Poplar	1880	190,000	9 (cp.)	8 (cp.)	1½	10 5·9-in. Q.F. (Canet), 4 4·7 in., 8 2·4 in., 2 M.	2	13·75	650	350
c.d.s.t.	Andes	I.	1556	186 0	44 0	9 6	2	750	Birkenhead	1875	85,600	6	9	1	2 11-in., 2 4·7-in., 4 M.	..	9·5	120	120
c.d.s.t.	Plata	I.	1558	186 0	44 0	9 6	2	750	Birkenhead	1874	85,600								
a.c.	Garibaldi	S.	6840	328 0	59 8	24 0	2	13,384	Sestri Ponente	1895	681,240	6 H.S.	6 H.S.	1½	2 10-in., 10 6-in. Q.F., 6 4·7 in., 10 2·2 in., 10 1·4 in., 2 M.*	..	19·9 (*l*)	1000†	500
a.c	General Belgrano	S.	7182	328 0	59 8	24 0	2	13,000	Leghorn	1897	—	6 H.S.	6 H.S.	1½	2 10-in., 14 6-in. Q.F., 2 3-in., 10 2·2-in., 8 1·4-in., 2 L., 2 M.*	4 sub.	20·1	1000	500
c.d.s.b.	Independencia	S.	2336	230 0	44 4	13 0	2	3000	Birkenhead	1891	176,600	8 (cp.)	8 (cp.)	2	2 9·4-in., 4 4·7-in. Q.F., 4 3-pr., 4 M.	2	14·4	340	225
c.d.s.b.	Libertad	S.	2336	230 0	44 4	13 0	2	3000	Birkenhead	1890	176,600								
a.c.	Pueyrredon	S.	6882	328 0	59 8	24 0	2	13,000 B	Sestri Ponente	1898	—	6 H.S.	6 H.S.	1½	2 10-in., 10 6-in. Q.F., 6 4·7 in., 10 2·2-in. 10 1·4-in., 2 M.	4 sub.	20·1	1000	500
a.c.	San Martin	S.	6882	328 0	59 8	24 0	2	13,000	Leghorn	1896	664,600	6 H.S.	6 H.S.	1½	4 8-in., 10 6-in. Q.F., 6 4·7, 12 2·2, & 10 1·4 in., 2 L., 2 M.*	4 sub.	19·6 (*l*)	1100	500

* Armament of Garibaldi, San Martin, General Belgrano and Pueyrredon, and Q.F. guns of Libertad and Independencia are Armstrong.

† Bunker capacity.

ARGENTINE REPUBLIC.—Cruising Ships, &c.

Class.	NAME.	Material of Hull.	Displacement.	Length.	Beam.	Draught.	Propellers.	Indicated Horse-power.	Where Built.	Date of Launch.	Cost.	Armour. Gun Position.	Armour. Deck.	Armament. Guns.	Armament. Torpedo Tubes.	Speed.	Normal Coal Supply.	Complement.
			metric tons.	ft. in.	ft. in.	ft. in.	no.				£	inches.	inches.			knots.	tons.	
g.v.	Argentina	S.	820	192 0	27 0	13 0	1	850	Trieste	1883	25,500	..	..	4 6-in., 6 7-c.m. Krupp, 4 M.	..	12·0	220	120
cr.	Buenos Aires	S. shd.	4780	396 0	47 2	19 0	2	17,000	Elswick	1895	383,000	4½	1–5	2 8-in. Q.F. (Armstrong), 4 6-in. Q.F., 6 4·7-in. Q.F., 16 3-pr., 6 1-pr.	5	23·2*	1000†	429
to.g.b.	Espora	S.	520	210 0	25 0	8 0	2	3500	Birkenhead	1890	..	..	..	3 3-in. Q.F., 4 3-pr., 2 M.	5	20·0	100	124
cr.	Nueve de Julio	S.	3570	354 0	44 0	19 6	2	14,350	Elswick	1892	293,000	4½	4½	4 6-in. Q.F. (Armstrong), 8 4·7-in., 12 3-pr., 12 1-pr.	5	22·74	770†	300
cr.	Patagonia	S. & W.	1442	220 0	32 10	12 9	2	2400	Trieste	1885	100,000	..	1½	1 10-in., 3 6-in., 6 L., 10 M.	..	13·0	350	210
to.g.b.	Patria	S.	1070	250 0	31 0	10 0	2	4500	Birkenhead	1893	87,000	..	..	2 4·7-in. Q.F., 4 8-pr., 2 3-pr., 2 M.	5	20·75	288	159
g.v.	Paraná	I.	550	142 8	25 0	11 9	1	475	Birkenhead	1874	..	..	..	2 6-in., 2 4·7-in.	..	11·0	..	..
cr.	25 de Mayo	S.	3200	325 0	43 0	16 0	2	13,800	Elswick	1890	260,000	4½	4½	2 8·2-in. (Armstrong), 8 4·7-in. Q.F., 12 3-pr., 12 1-pr.	6	22·43	600†	185
g.v.	Uruguay	I.	550	142 8	25 0	11 9	1	475	Birkenhead	1874	..	..	..	2 6-in., 2 4·7-in.	..	11·0	..	..

Messrs. Laird have completed a training-ship (cruiser), Presidente Sarmiento, 2750 tons, 2000 I.H.P. (Niclausse boilers), and 13 knots speed, with nineteen guns and three torpedo-tubes; launched 1897. There are several other small gunboats; also the torpedo-ram Maipú (1063 tons, 1750 I.H.P.), built in England in 1880. The Florio Company has sold to the Argentine Government the steamships Arno, Regina Margherita, and Sempione to be converted into cruisers; and the Spanish firm of Pinillos, Saluy & Co.; the Barcelona (4020 tons register), and Cadiz (4218 tons), which have been re-named Pampa and Gaucho.

* Natural draught. † Bunker capacity.

AUSTRIA-HUNGARY.—Armoured Ships.

Class.	NAME.	Material of Hull.	Displacement.	Length.		Beam.		Draught.		Propellers.	Indicated Horse-power.	Where Built.	Date of Launch.	Cost.	Armour. Belt.	Armour. Gun Position.	Armour. Deck Plating.	Armament. Guns.	Armament. Torpedo Tubes.	Speed.	Normal Coal Supply.	Complement.
			metric tons.	ft.	in.	ft.	in.	ft.	in.					£	inches.	inches.	inches.			knots.	tons.	
c.d.s.	Budapest	S.	5550	305	0	55	9	21	0	2	9185 B	Trieste	1896	349,600	10·6 H.S.	10·6 H.S.	2½ H.S.	4 9·4-in., 6 5·9-in. Q.F., 14 47-m.m. Q.F., 2 M.	4	17·8 (*t*)	500	..
c.b.	Custoza	I.	7060	302	3	58	0	24	6	1	4440	Trieste	1872	414,400	9	7	1½	8 10·2-in. (Krupp), 11 Q.F., 8 l.	2	14·0	584	567
a.c.	Kaiser Karl VI.	S.	6250	367	6	56	0	20	4		12,800 B	Trieste	1898	368,124	10·6 H.S.	9·8 H.S.	1½ H.S.	2 9·4-in., 8 5·9-in. Q.F., 18 1·8, 2 M.	4	20·0	800	450
c.b.	Don Juan de Austria	I.	3550	240	3	50	0	20	0	1	2700	Trieste	1875	..	8	6	1	8 8·2-in. (Krupp), 11 Q.F. & M., 6 l.	4	13·0	380	440
c.d.s.	Erzherzog Albrecht	I.	5940	285	2	56	3	22	0	1	3600	Trieste	1872	357,600	9	7	1½	8 9·4 in. (Krupp), 11 Q.F., 8 l.	2	13·0	453	535
a.c.	Kaiserin Maria Theresia	S.	5270	351	0	52	6	21	4	2	9755	Trieste	1893	304,187	4	4	2	2 9·4-in., 8 5·9-in. Q.F., 18 1·8 Q.F., 2 2·7-in., 2 M.	4	19·0	740	450
c.b.	Kaiser Max	I.	3566	240	3	50	0	20	0	1	2700	Trieste	1875	21[illegible],600	8	6	1	8 8·2-in. (Krupp), 11 Q.F. & M., 6 l.	4	13·0	380	440
Riv. Mon.	Körös	S.	448	177	0	29	6	4	0	2	1250	Buda Pesth	1892	.	2	3	¾	2 4·7-in. Q.F., 2 l., 1 M.		10·0	..	..
b.	Kronprinz Rudolph	S.	6940	295	0	62	4	25	8	2	7500	Pola	1887	330,000	12	10	2½	3 12-in. (Krupp), 6 4·7-in. Q.F., 11 smaller & M., 2 l.	4	16·0	600	492
b.	Kronprinzessin Stephanie	S.	5150	278	10	55	9	21	6	2	8800	Trieste	1887	300,000	9	8	1	2 12-in. (Krupp), 6 5·9-in., 11 Q.F. & M., 2 l.	4	17·0	400	510
Riv. Mon. "	Leitha Maros }	I. & S.	310	166	0	27	6	3	7	2	320	Buda Pesth	1871	20,000	1·7	2	1	1 4·7-in. Q.F., 2 M.	..	8·0	20	54
c.d.s.	Monarch	S.	5550	305	0	55	9	21	0	2	8900	Pola	1895	339,062	10·6 H.S.	10·6 H.S.	2½ H.S.	4 9·4-in., 6 5·9-in. Q.F., 14 47-m.m. Q.F., 2 M.	4	17·4 (*t*)	500	..
c.b.	Prinz Eugen	I.	3566	240	6	50	0	20	0	1	2700	Pola	1877	..	8	6	1	8 8·2-in. (Krupp), 11 Q.F. & M., 6 l.	4	13·0	380	440
Riv. Mon.	Szamos	S.	448	177	0	29	6	4	0	2	1250	Buda Pesth	1892	..	2	3	¾	2 4·7-in. Q.F., 2 Q.F., 1 M.	..	10·0		
c.b.	Tegetthoff	I. & S.	7390	286	11	71	1	24	10	2	8800	Trieste	1878 1893	..	14	14	3	6 9·4-in. (Krupp), 5 5·9-in. Q.F., 15 smaller do., 2 M.	4	16·3	670	578
c.d.s.	Wien	S.	5550	305	0	55	9	21	0	2	8480	Trieste	1895	337,850	10·6 H.S.	10·6 H.S.	2½ H.S.	4 9·4-in., 6 5·9 Q.F., 14 47-m.m. Q.F., 2 M.	4	17·6 (*t*)	500	..
c.d.s.b.	Unnamed I., II.	S.	8300	354	3	65	8	23	4	2	11,000 B.	..	Bldg.	480,000	8·6 H.S.	8·2 H.S.	2½ H.S.	3 9·4-in., 12 5·9-in. Q.F., 24 smaller.	..	18·0	..	..
c.d.s.b.	Unnamed III.	S.	8300	354	3	65	8	23	4	2	11,000 B.	..	Pro.	480,000	8·6 H.S.	8·2 H.S.	2½ H.S.	3 9·4-in., 12 5·9-in. Q.F., 24 smaller.	..	18·0	..	..

AUSTRIA-HUNGARY.—Cruising Ships, &c.

Class.	NAME.	Material of Hull.	Displacement.	Length.	Beam.	Draught.	Propellers.	Indicated Horse-power.	Where Built.	Date of Launch.	Cost.	Armour. Gun Position.	Armour. Deck.	Armament. Guns.	Armament. Torpedo Tubes.	Speed	Normal Coal Supply.	Complement.
			net tns.	ft. in.	ft. in.	ft. in.					£	in.	in.			knots.	tons.	
corv.	Aurora	C.	1370	190 6	32 10	16 1	1	1000	Trieste	1873	..	..	..	2 4·7-in. (Wahrendorf), 5 l., 2 M. or Q.F.	..	11·0	160	200
to. cr.	Aspern ("B.")	S.	2400	301 10	39 6	14 2	2	7000 W.T.	Pola	1899 Bldg.	155,000	..	..	8 4·7-in. Q.F., 12 1·8 in.	1	20·0	..	..
to. cr.	"C." (Ersatz Fasana)	S.	2300															
to. g. b.	Blitz	S.	360	193 6	22 4	8 0	..	3500	Elbing	1888	..	..	..	9 Q.F.	..	21·0	250	61
cr. 3rd cl.	Donau	C.	2344	230 0	42 8	19 8	1	1800	Pola	1893	..	..	..	10 4·7-in. (Uchatius), 4 M., 1 l.	..	12·0	320	..
cr. 2nd cl.	"E." (Ersatz Radetzky)	S.	7000	..	..	..	..	..	..	Pro.	..	..	..	.. ..	..	..	..	..
cr. 3rd cl.	Erzherzog Friedrich	W.	1590	173 10	39 5	16 9	1	800	Venice	1874	..	..	..	10 5·9-in. (Wahrendorf), 1 l.	..	9·0	200	261
cr.	Frundsberg	C.	1370	190 6	32 10	16 1	1	1000	Trieste	1873	..	..	..	2 5·9-in. (Wahrendorf), 5 l., 2 M. or Q.F.	..	11·0	160	209
cr. 2nd cl.	Kaiserin Elizabeth	S.	4064	321 6	47 6	18 7	2	9000	Pola	1890	..	3·5	2¼	2 9·4-in. (Krupp), 6 5·9-in. do., 11 Q.F., 2 l.	5	19·0	660	450
cr. 2nd cl.	Kaiser Franz Joseph I.	S.	4030	321 6	47 6	18 7	2	9000	Trieste	1889	..	3·5	2¼	2 9·4-in. (Krupp), 6 5·9-in. do., 11 Q.F., 2 l.	5	19·0	660	450
to. g. b.	Komet	S.	360	193 6	22 4	8 0	..	3500	Elbing	1888	..	..	..	9 Q.F.	..	21·0	..	61
cr. 2nd cl.	Laudon	C.	3430	253 0	46 0	20 8	1	2600	Trieste	1873	..	..	..	15 5·9-in. (Krupp), 7 Q.F. & M., 2 l.	..	13·0	450	497
cr. 3rd cl.	Leopard	S.	1582	224 0	34 0	14 0	2	6000	Elswick	1886	200,000	..	..	2 4·7-in. Q.F., 10 Q.F. & M.	4	18·3	250	148
to. g. b.	Lussin	S.	1011	200 4	26 3	12 2	2	1830 Dürr.	Trieste	1883	..	..	1½	2 5·9-in. (Krupp), 7 M., 1 l.	..	14·0	200	142

to. g. b.	**Magnet**	S.	510	219	10	26	10	8	0	2	6000 W.T.	Elbing	1896	51,052	..	..	6 1·8-in. Q.F.	3	26·0*t*	105	..
to. g. b.	**Meteor**	S.	350	187	0	22	4	8	0	2	3500	Elbing	1887	..	..	..	9 Q.F.	1	23.1	120	61
cr. 3rd cl.	**Panther**	S.	1582	224	0	34	0	14	0	2	6000	Elswick	1885	..	..	..	2 4·7-in. Q.F., 10 Q.F. & M.	4	18·5	250	148
to. deps.	**Pelican**	S.	2470	279	0	39	4	15	6	..	4600	Elbing	1891	..	..	..	2 5·9-in. (Krupp), 8 Q.F.	4	18·0	..	..
to. g. b.	**Planet**	S.	500	210	0	28	0	8	3	2	3500	Jarrow	1889	..	..	..	10 Q.F.	1	19·6	..	61
cr. 3rd cl.	**Saida**	C.	2500	233	4	42	8	19	1	1	1800	Pola	1878	..	..	..	11 5·9-in. (Uchatius), 1 l.	..	12·0	320	299
to. g. b.	**Satellit**	S.	540	220	6	26	9	9	2	2	4000	Elbing	1893	..	..	1½	9 Q.F.	..	21·87	..	61
to. g. b.	**Sebenico**	S.	900	187	0	26	3	12	2	2	1390	Pola	1882	..	..	..	7 Q.F., 5 l.	..	14·0	200	142
cr. 3rd cl.	**Spalato**	S.	850	179	6	26	3	12	2	2	1200	Trieste	1879	..	..	..	7 Q.F., 5 l.	..	14·0	150	142
to. cr.	**Tiger**	S.	1684	233	0	32	10	15	5	2	5260	Trieste	1887	..	..	..	4 4·7-in., 10 Q.F.	..	18·0	300	190
to. g. b.	**Trabant**	S.	530	210	0	23	0	8	3	2	3500	Trieste	1890	..	..	..	10 Q.F.	1	20·0	..	61
to. v.	**Zara**	S.	850	179	6	26	3	12	2	2	1200	Pola	1879	..	..	..	7 Q.F., 5 l.	..	14·0	150	142
to. cr.	**Zenta**	S.	2250	312	0	39	6	14	2	2	7800 Y.	Trieste	1897	143,780	..	..	8 4·7 Q.F., 12 1·8-in., 2 M.	1	20·9*t*	..	..
corv.	**Zrinyi**	C.	1370	190	0	32	10	16	1	1	1000	Trieste	1871	..	..	..	2 5·9-in. (Wahrendorf), 5 l., 2 M. or Q.F.	..	11.0	160	209

Four screw gunboats, between 540 and 870 tons displacement and 250 and 950 indicated horse-power.

BRAZIL.—Armoured Ships.

Class.	NAME.	Material of Hull.	Displacement.	Length.		Beam.		Draught.		Propellers.	Indicated Horse-power.	Where Built.	Date of Launch.	Cost.	Armour. Belt.	Armour. Gun Position.	Armour. Backing. Deck Plating.	Armament. Guns.	Armament. Torpedo Tubes.	Speed.	Normal Coal Supply.	Complement.
			tons.	ft.	in.	ft.	in.	ft.	in.	no.				£	inches.	inches.	inches.			knots.	tons.	
t. River	Alagoãs	W.	340	120	0	28	0	4	10	2	180	Brazil	1886	..	4½	4½	14½	1 7-in. M.L.R. (Whitworth), 2 M.	..	7·0	..	43
c.d.s., t.	Bahia	I.	1000	178	0	35	0	8	6	2	1640	Birkenhead	1865	..	4½	5½	10⅝	2 7-in. M.L.R. (Whitworth), 2 M.	..	6·0	..	125
t. River	Maranhao	S.	470	137	0	34	7	6	5	2	700	Rio de Janeiro	Bldg.	..	5 H.S.	..	..	2 4·7-in. Q.F., 1 2·5-in., 5 M.	..	12·0	..	..
c.d.s., t. c.d.s., t.	Marshal Deodoro Marshal Floriano	S.	3162	267	6	48	0	13	2	2	3400	La Seyne	1898 1899	..	13¾ H.S.	7¾-8¾ H.S.	1½	2 9·4-in., 2 5·9-in. howitzers, 4 4·7-in. Q.F., 2 M., 4 6-pr. and 2 1-pr.	2 (sub.)	15·0	236	200
t. River	Pará	S.	470	137	0	34	7	6	5	2	700	Rio de Janeiro	Bldg.	..	5 H.S.	..	..	2 4·7-in. Q.F., 1 2·5-in., 5 M.	..	12·0	..	..
t. River	Piauhy	W.	340	120	0	28	0	4	10	2	180	Brazil	1887	..	4½	4½	14½	1 7-in. M.L.R. (Whitworth)	..	7·0	..	43
t.	Riachuelo	S. shd.	5700	305	0	52	0	19	0	2	7300	Poplar	1883 R. 1895	365,000*	11 comp.	11 & 10 comp.	10 2″	4 9·2-in. (Whitworth, altered by Armstrong), 6 4·7-in., Q.F., 2 3-pr., 15 M.	5	16·71	800	450
t. River	Rio Grande	W.	340	120	0	28	0	4	10	2	180	Brazil	1888	..	4½	4½	14½	1 7-in. M.L.R. (Whitworth)	..	7·0	..	43
t.	24 de Maio (ex Aquidaban)	S shd.	4950	280	0	52	0	18	0	2	6200	Poplar	1885 R. 1897	345,000*	11	11½ & 10 cp.	10 2″	4 9·4-in. (Canet), 4 5·5-in., 2 Q.F., 13 M.	5	15·0	600	350

* Exclusive of guns and ammunition. The ship is undergoing reconstruction at Elswick.

Floating batteries, Brazil (1518 tons) and Lima-Barros (1444 tons).

BRAZIL.—Cruising Ships, &c.

Class.	NAME.	Material of Hull.	Displacement.	Length.	Beam.	Draught.	Propellers.	Indicated Horse-power.	Where Built.	Date of Launch.	Cost.	Armour. Gun Position.	Armour. Deck.	Armament. Guns.†	Armament. Torpedo Tubes.	Speed.	Normal Coal Supply.	Complement.
			tons.	ft. in.	ft. in.	ft. in.	no.					inches.	inches.			knots.	tons.	
cr.	Almirante Tamandare.	S.	4735	294 6	46 0	18 4	2	7500	Brazil	1890	..	..	1½	10 6-in. Q.F., 2 4·7-in., 8 M.	8	17·0	750	450
"	Andrada (ex America)	shd. S.	2600	252 8	34 0	18 0	1	3600	Bergen	1890	..	..	..	2 4·7-in. 2 14-pr. Q.F., 6 6-pr., 6 1-pr.	5	17·0	..	..
"	Barroso	shd. S.	3600	330 0	43 9	16 10	2	7500	Elswick	1896	..	4½	3	6 6-in. Q.F., 4 4·7-in., 10 6-pr., 4 1-pr., 4 M.	3	20·0	700	..
"	Benjamin Constant	shd. S.	2750	236 0	46 0	18 0	1	2800	La Seyne	1892	..	..	2	4 6-in. Q.F., 8 4·7-in., 8 M., 4 L.	4	14·0	260	287
to.cr.	Caramuru	shd. S.	1030	249 6	30 9	10 2	2	6000	Kiel	1896	..	..	½	2 3·0-in. Q.F., 6 2·2-in., 2 1·4-in.	3	22·5	..	..
to.g.b.	Gustavo Sampaio	S.W.	500	197 0	21 0	7 9	2	2500	Elswick	1893	..	..	..	2 20-pr. Q.F., 4 7-pr. Q.F.	3	18·0	150	95
cr.	Parnahyba (Torpedo training.)	C.	838	170 6	26 8	11 2	1	900	Havre	1878	..	..	..	5 4·7-in., 4 M.	..	10·0	..	..
"	Paysandu (ex Guanabára)	W.	1900	200 0	41 2	16 4	1	3000	Brazil	1877	..	..	..	9 70-pr. M.L.R. (Whitworth), 6 M., 2 L.	..	13·0	..	250
cr.	Primeiro de Março	C.	726	167 3	26 3	10 6	1	750	Brazil	1881	..	..	..	7 4·5-in. M.L.R. (Whitworth), 4 M.	..	9·0	..	..
"	Quinze de Novembro (ex Republica)	S.	1300	210 0	35 0	13 0	2	3300	Elswick	1892	..	..	2·1	6 4·7-in. Q.F., 4 6-pr., 6 M.	4	17·0	170	100
to.cr.	Tamoyo	S.	1030	269 0	28 10	9 10	2	6500	Kiel	1898	..	..	..	2 3·9-in. Q.F., 6 2·2-in., 2 1·4-in., 2 M.	3	23·0	293	110
"	Timbira	S.	1030	249 6	30 9	10 2	2	7000	Kiel	1896	..	..	½	2 3·9-in. Q.F., 6 2·2-in., 2 1·4-in., 2 M.	3	22·5	250	110
g.v.	Tiradentes	S.	800	165 0	30 0	11 0	2	1200	Elswick	1892	..	..	..	4 4·7-in. Q.F., 3 6-pr., 4 M.	2	14·5	110	107
cr.	Tonelero (ex Trajano)	shd. W.	1414	200 0	30 0	15 6	2	2400	Brazil	1873 / 1895	..	..	..	7 4·7-in. Q.F., 4 M.	..	13·0	..	..
"	Trinidade (ex Liberdade)	L.	250	161 8	21 8	10 10	2	280	Brazil	1884	..	..	..	2 L., 1 M.	..	10·0	..	..
to.cr.	Tupy	S.	1030	249 6	30 9	10 2	2	7000	Kiel	1896	..	..	½	2 3·9-in. Q.F., 6 2·2-in., 2 1·4-in., 2 M.	3	22·5	250	110

† All the Q.F. guns above 6-pr. in Brazilian Navy are Armstrong.

Ten screw gunboats, 200 tons to 400 tons, and eight paddle gunboats, 120 tons to 100 tons.

CHILI.—Armoured Ships.

Class.	NAME.	Material of Hull.	Displacement.	Length.	Beam.	Draught.	Propellers.	Indicated Horse-power.	Where Built.	Date of Launch.	Cost.	Armour. Belt.	Armour. Gun Position.	Armour. Deck Plating.	Armament. Guns.	Armament. Torpedo Tubes.	Speed.	Normal Coal Supply.	Complement.
			tons.	ft. in.	ft. in.	ft. in.	no.				£	inches.	inches.	inches.			knots.	tons.	
c.b.	Almirante Cochrane	W.I. shd.	3500	210 0	45 9	19 8	2	2920	Hull .	1874	..	9	8	8"	6 8-in. (Armstrong), 4 6-pr. Q.F., 4 3-pr., 7 M.	3	13·0	350	242
a.c.	Almirante O'Higgins	S. shd.	8500	411 9	62 6	22 0	2	16,000 B	Elswick .	1897	..	7	6	2	4 8-in. Q.F., 10 6-in., 4 4·7-in., 10 12-pr., 10 6-pr., 4 M.	8 (2sub.)	21·5 (t.)	1260	..
b.	Capitan Prat . .	S. shd.	6900	328 0	60 8	21 10	2	12,000	La Seyne .	1890	391,000	12	10½ 4	3"	6 9·4-in. (Canet), 8 4·7-in. Q.F. (Canet), 6 2·2-in., 4 1·8-in., 10 1·4-in., 5 M.	4	18·3	775	485
a.c.	Esmeralda . .	S. shd.	7020	436 0	53 2	22 3	2	16,000	Elswick .	1896	..	6 H.S.		2	2 8-in. Q.F., 16 6-in., 8 12-pr., 2 3-pr., 4 M.	3 (2sub.)	22·8 (t.)	1850	..

The Huascar, 1800 tons, launched at Birkenhead in 1865, is now a floating battery.

Cruising Ships, &c.

Class	NAME	Material of Hull.	Displacement.	Length.	Beam.	Maximum Draught.	Propellers.	Indicated Horse-power.	Where Built.	Date of Launch.	Cost.	Armour. Gun Position.	Armour. Deck.	Armament. Guns.	Armament. Torpedo Tubes.	Speed.	Normal Coal Supply.	Complement.
			tons.	ft. in.	ft. in.	ft. in.	no.					inches.	inches.			knots.	tons.	
to.g.b. / to.g.b.	Almirante Condell / Almirante Lynch	S	750	240 0	27 6	10 6	2	(4500) B	Birkenhead .	1890	..	..	..	3 14-pr. Q.F., 4 3-pr., 2 M. . . .	5	21·0	210	..
,,	Almirante Simpson	S.	812	240 0	27 6	10 6	2	4700 Nor.	Birkenhead .	1896	..	4½	..	2 4·7-in. Q.F., 4 3-pr. . . .	3	21·0‡	200	..
cr.	Blanco Encalada .	S. shd.	4400	370 0	46 6	18 6	2	14,500	Elswick . .	1893	..	..	4–1¾	2 8-in., 10 6-in. Q.F., 12 3-pr., 10 1-pr.*	5	22·78	†900	..
,,	General Baquedano (Training)	S.	2330	240 0	45 9	18‡ 0	1	1500 B	Elswick . .	1898	..	..	..	4 4·7-in. Q.F., 2 12-pr., 2 6-pr., 2 M., 1 L.	1	13·7	200	302
gv.	Magellanes . .	C.	800	190 0	28 0	14 9	2	1230	London .	1874	..	..	..	2 6-in., 1 7-in. M.L.R., 6 M., 2 L. .	..	11·0	300 / 300	100
cr.	Ministro Zenteno .	S. shd.	3600	330 3	43 9	16 10	2	6500	Elswick .	1896	..	..	..	8 6-in. Q.F., 10 6-pr., 4 1-pr.* . . .	3	20·0‡	800	..
gv.	Pilcomayo . .	W.	790	171 0	27 4	..	1	180 (nom.)	Birkenhead .	1874	..	..	..	2 70-pr. B.L.R. (Armstrong), 2 40-pr., 3 M.	..	9·0	125	125
cr. / ,,	Presidente Errázuriz / Presidente Pinto .	S. shd.	2080	268 0	35 9	19 6	2	5400	La Seyne .	1890	..	..	3½	4 6-in. Q.F. (Canet), 2 5-in., 4 2·2-in., 6 M.	3	19·0	200	171

Two Gunboats of 145 tons displacement and one of 180 tons. * Armstrong. † Bunker capacity. ‡ Mean draught.

CHINA.—Cruising Ships, &c.

Class.	NAME.	Material of Hull.	Displacement.	Length.	Beam.	Draught.	Propellers.	Indicated Horse-power.	Where Built.	Date of Launch.	Cost.	Armour. Gun Position.	Armour. Deck.	Armament. Guns.	Armament. Torpedo Tubes.	Speed.	Normal Coal Supply.	Complement.
			tons.	ft. in.	ft. in.	ft. in.	no.					inches.	inches			knots.	tons.	
cr.	**Foo-Ching**	S.	2500	253 0	36 2	18 0	2	2400	..	1893	..	4½	4–2	3 5-in. Krupp, 4 M., 2 l.	..	16·0	..	..
to.g.b.	**Fei-Ying**	S.	850	257 2	28 6	12 6	2	4500 Y	Stettin.	1895	..	2	..	2 4-in. Q.F., 6 3·4-in., 4 smaller.	3	21·8	75	90
cr.	Hai-Chi.	S.	4300	396 0	46 8	18 6	2	17000	Elswick.	1898	..	6	5	2 8-in. Q.F., 10 4·7-in., 12 3-pd., 4 1·4-in., 6 M.	5	24·0	300	374
,,	**Hai-Shen**									1898								
,,	**Hai-Shew**	S.	2050	314 8	41 0	16 0	2	8000	Vulcan, Stettin.	1897	..	2	3	3 6-in. Krupp Q.F., 8 4-in., 6 1·4-in. Hotchkiss, 6 M.	3 (1sub.)	20·7 (*l.*)	220 (500)	244
,,	**Hai-Yung**									1897								
,,	**Hai-Tien**	S.	4300	396 0	46 8	18 6	2	17000	Walker.	1897	..	6	5	2 8-in. Q.F., 10 4·7-in., 12 3-pd.	5	24·1 (*t.*)	300	374
,,	**Hi-Ying**	S.	2200	253 0	36 2	18 1	2	2400	..	1895	..	..	..	2 8-in. Armstrong, 8 4·7-in. Q.F., 4 M.	1	21·0	..	..
,,	**Huang-Tái**	C.	2110	260 0	36 0	20 0	1	1600	..	1886	..	..	..	3 7-in. Krupp, 7 40-pr., 6 M.	2	15·0	360	300
,,	**Kai-Chih**	C.	2110	260 0	36 0	20 0	1	1600	..	1882	..	..	..	2 8·2-in., 6 5·9-in., 6 M., 5 l.	..	14·5	360	300
,,	**Kien-Wei**	C.	2110	260 0	36 0	20 0	1	1600	..	1899	..	..	..	2 8·2-in., 6 5·9-in., 6 M., 5 l.	..	14·5	360	300
,,	**King-Ching**	C.	2100	250 0	36 0	20 0	1	2400	..	1886	..	..	..	3 7-in. Krupp, 7 40-pr., 6 M.	2	14·5	360	300
to.g.b.	**Kwang-Ting**	C.	1000	235 0	27 6	11 4	2	3400	..	1890	..	..	1	3 4·7-in. Q.F., 4 M., 2 l.	4	16·0	..	120
cr.	**Nan-Schuin**	S.	2200	253 0	36 2	18 1	2	2400	Kiel.	1884	..	..	..	2 8-in. Armstrong, 8 4·7-in. Q.F., 9 M.	1	14·5	600	250
,,	Nan-Thin	S.	2200	253 0	36 2	18 1	1	2400	Kiel.	1883	..	..	.	2 8-in. Armstrong, 8 4·7-in. Q.F., 9 M.	1	15·0	600	250
,,	Pao-Min	S.	1480	213 0	36 0	14 0	1	2400	..	1884	..	..	..	2 6-in. Armstrong, 6 5-in., 2 l.	..	9	300	200
g.b.	Tien-Sing	W.	200	105 0	20 4	7 0	2	340	..	1875	..	3	..	1 7-in. (Krupp)	..	10·0	..	..
cr.	Unnamed	C.	2110	260 0	36 0	20 0	1	1600	..	Bldg.	..	..	..	2 8·2-in., 6 5·9-in., 6 M., 5 l.	..	14·5	360	300

The displacement of German-built ships in metric tons.

Torpedo-gunboat Pei-Ting (349 tons), four gunboats of 411 tons, two of 300 tons, four of 215 tons (defence of Canton Roads), training vessel Tung-Chi, 1700 tons—all launched 1885–88.

DENMARK.—Armoured Ships.

Class.	NAME.	Material of Hull.	Displacement.	Length.		Beam.		Draught.		Propellers.	Indicated Horse-power.	Where Built.	Date of Launch.	Cost.	Armour.			Armament.		Speed.	Normal Coal Supply.	Complement.
															Belt.	Gun Position.	Deck Plating.	Guns.	Torpedo Tubes.			
			metric tons.	ft.	in.	ft.	in.	ft.	in.	no.				£	inches.	inches.	inches.			knots.	tons.	
c.d.s., t.	Gorm . . .	I.	2344	231	0	40	0	14	0	2	1670	Copenhagen	1870	104,000	7	8	..	2 10-in. (Armstrong) M.L.R., 8 3·4-in. (Krupp), 4 M.	..	12·25	115	158
t.	Helgoland . .	I.	5347	257	6	59	2	18	8	2	4000	Copenhagen	1878	275,000	12	10	4	1 12-in. (Krupp), 4 10·2-in., 5 4·7-in., 10 M.	4	12·0	230	350
c.d.s., t.	Herluf Trolle .	S.	3470	271	0	50	0	16	2	2	4200	Copenhagen	1899	..	..	..	..	2 9·4-in., 4 5·9-in. Q.F., 10 2·2-in., 8 smaller.	3 (sub.)	15·0	..	..
b.	Iver Hvitfeldt .	S.	3260	242	0	49	6	18	0	2	5100	Copenhagen	1886	200,000	12	8	2	2 10·2-in. (Krupp), 4 4·7-in., 12 M.	4	15·6	250	298
c.d.s., t.	Lindormen. .	I.	2076	216	0	39	5	13	9	2	1560	Copenhagen	1868	93,000	5	5	..	2 9-in. (Armstrong) M.L.R., 3 3·4-in. (Krupp), 4 M.	..	12·0	120	140
c.b.	Odin† . . .	I.	3083	237	0	50	0	15	6	1	2260	Copenhagen	1872	147,000	8	8	..	4 10-in. (Armstrong) M.L.R., 4 3·4-in. (Krupp), 7 M.	..	12·4	180	236
t.	Skjold . . .	S.	2150	226	6	38	0	13	5	2	2200	Copenhagen	1896	..	9	8–4½	2	1 9·4-in., 3 4·7-in. (Krupp), 4 1·8-in. Q.F., 1 M.	4	13·0	..	..
T. S.	Tordenskjold † .	S.	2400	221	6	43	3	15	6	2	2600	Copenhagen	1880	138,900	..	8	4–2	1 14-in. (Krupp), 4 4·7-in., 8 M.	4	14·0	170	220
c.d.s., t.	Unnamed* . . . (Herluf Trolle type)	S.	5317	..		..		..		..	..	Coponhagen	Bldg.	..	..	..	..	..	..	..	..	..

Esbern Snare (torpedo school-ship), 530 tons, 2-in. belt.

* Estimates of 1899–1900.

† To be reconstructed.

DENMARK.—Cruising Ships, &c.

Class.	NAME.	Material of Hull.	Displacement.	Length.		Beam.		Draught.		Propellers.	Indicated Horse-power.	Where Built.	Date of Launch.	Cost.	Armour.		Armament.		Speed.	Normal Coal Supply.	Complement.
															Gun Position.	Deck.	Guns.	Torpedo Tubes.			
			metric tons.	ft.	in.	ft.	in.	ft.	in.	no.				£	inches.	inches.			knots.	tons.	
g. v.	Absalon	I.	527	150	0	26	0	10	2	1	500	Blackwall	1862	..	2½	2½	4 3·4-in. (Krupp), 4 M.	..	10·0	65	70
”	Falster	I.	356	111	0	28	10	7	6	2	510	Copenhagen	1873	33,000	..	..	1 10-in. (Armstrong) M.L.R., 2 3·4-in. (Krupp), 2 M.	..	9·8	20	35
cr.	Fyen	S. shd.	2596	226	6	45	6	18	1	1	2700	Copenhagen	1882	170,000	..	1½	18 5·9-in. (Krupp), 8 M.	2	13·0	290	407
3rd cl. *cr.*	Geiser	S.	1280	257	6	27	6	11	4	2	3000	Copenhagen	1892	..	..	1½	2 4·7-in. Q.F., 4 3·4-in., 6 M.	4	17·1	..	..
”	Heimdal	S.	1280	257	6	27	6	11	4	2	3000	Copenhagen	1894	..	..	1½	2 4·7-in. Q.F., 4 3-pr., 6 M.	4	17·5	..	..
”	Hekla	S.	1280	233	0	32	10	11	2	2	3000	Copenhagen	1890	..	..	1½	2 6-in. Q.F., 4 2·2-in., 6 M.	4	17·0	..	..
g. v.	Ingolf	I.	870	192	0	28	0	12	6	1	600	Copenhagen	1876	44,000	..	..	2 5·9-in. (Krupp), 4 3·4-in., 2 M.	..	10·5	130	117
g. v.	Möen	I.	356	111	0	28	10	7	6	1	523	Copenhagen	1875	..	..	..	1 10-in. (Armstrong) M.L.R., 2 3·4-in. (Krupp), 2 M.	..	9·2	20	35
corv.	Saint Thomas	W.	1572	224	0	33	0	17	0	1	1870	Copenhagen	1871	..	..	..	8 4·7-in. (Krupp), 6 M.	..	13·0	190	182
cr.	Valkyrien	S.	2900	268	0	43	6	18	0	2	5300	Copenhagen	1887 1890	..	..	2½	2 8·2-in. (Krupp), 6 5·9-in., 4 Q.F., 10 M.	5	17·0	450	300

Gunboats.—Five in number (*Lille Belt, Öresund, Store Belt, Grönsund, Guldborgsund*), of 150 to 240 tons, 200 to 400 I.H.P. *Dagmar* (training-ship), corvette, 1200 tons; *Hjaelperen* (mining), 280 tons; *Sleipnir* (ice-breaker), 1260 tons, 3000 I.H.P.

FRANCE.—Armoured Ships.

Class.	NAME.	Material of Hull.	Displacement.	Length.		Beam.		Draught.		Propellers.	Indicated Horse-power.	Where Built	Date of Launch.	Cost.	Armour. Belt.	Armour. Gun Position.	Armour. Deck Plating.	Armament. Guns.	Armament. Torpedo Tubes.	Speed.	Normal Coal Supply.	Complement.
			metric tons.	ft.	in.	ft.	in.	ft.	in.	no.				£	inches.	inches.	inches.			knots.	tons.	
a.g.b.	Achéron .	I. & S.	1721	181	0	40	4	11	10	2	1700	Cherbourg .	1885	100,000	8	8	2½	1 10·8-in., 3 3·9-in. Q.F., 2 1·8-in., 4 M.	..	13·0	100	101
a.c.	Amiral Aube .	S.	10,014	45[illegible]	0	66	3	24	7	3	20,500 W.T.	St. Nazaire .	Bldg.	942,940	6 H.S.	5–7¾	2	2 7·6-in., 8 6·4-in. Q.F., 4 3·9-in., 26 small Q.F. and M.	4 (2 sub.)	21·0	970 / 1590	..
b.	Amiral Baudin	I. & S.	11,911	321	6	69	10	26	2	2	8320	Brest .	1883 1899	600,000	21½	16½	4	2 14·6-in., 4 6·4-in. Q.F., 8 5·5-in., 25 small Q.F. and M.*	4	15·0	800	630
a.c.	Amiral Charner	S.	4792	348	0	46	0	19	2	2	8300 B	Rochefort .	1893	353,200	3½	2	2	2 7·6-in., 6 5·5-in. Q.F., 14 small Q.F. and M.	5	18·2	413	375
b.	Amiral Duperré	I. & S.	11,209	311	0	66	11	26	9	2	8120	La Seyne .	1879	570,000	21½	15½	2¼	4 13·3-in., 1 5·5-in. Q.F., 14 4-in., 42 small Q.F. and M.	4	14·22	850	664
a.c.	Amiral de Gueydon	S.	9517	459	0	63	8	24	7	3	20,200 Nic.	Lorient .	1899	808,600	6–3¾ H.S.	8, 3¾ H.S.	2	2 7·6-in., 8 6·4-in. Q.F., 4 3·9-in., 16 1·8-in., 6 1·4-in.	2 (sub.)	21·0	1020 / 1600	612
a.c.	Amiral Pothuau	S.	5360	370	6	50	2	21	0	2	10,398 B	Havre .	1895	384,000	3⅓–2	9½	3⅜	2 7·6-in., 10 5·5-in. Q.F., 16 1·8-in., 8 1·4-in.		19·2 *t*	538	461
t.	Amiral Tré-houart	S.	6629	293	2	58	4	23	2	2	8500 B	Lorient .	1893	593,100	17¾	14½	4	2 12-in., 8 3·9-in. Q.F., 4 1·8-in. Q.F., 4 1·4-in., 8 M.	2	15·76 *t*	300	337
a.c.	Bayard .	W. & I.	6011	265	9	57	2	24	11	2	4538	Brest .	1880	..	10	8	2	4 9·4-in., 2 7·6 in., 6 5·5-in., 8 1·8-in., 12 M.	2	14·0	500	450
t.	Bouvet .	S.	12,200	401	2	70	3	27	6	3	14,000 B	Lorient .	1896	1,100,770	15¾–8 H.S.	14½ H.S.	3½ H.S.	2 12-in., 2 10·8-in., 8 5·5-in. Q.F., 8 3·9-in., 19 small Q.F. and M.	4 (2 sub.)	18·2 *t*	621	621
t.	Bouvines .	S.	6610	293	9	58	3	23	3	2	8400 D'A	La Seyne .	1892	594,640	17¾	14½	4	2 12-in., 8 3·9-in. Q.F., 4 1·8-in., 10 1·4-in. M.	2	16·05	300	323
t.	Brennus .	S.	11,395	361	0	67	0	26	3	2	14,000 B	Lorient .	1891	991,767	15¾ comp.	15¾	4¾	3 13·4-in., 10 6·4-in. Q.F., 23 small Q.F. and M.	4	17·1	800	696

* New armament; Amiral Duperré to receive the same.

a.c.	**Bruix**	S.	4754	365	3	46	0	19	10	2	9049 B	Rochefort	1894	409,622	3¾	3¾	2	2 7·6-in., 6 5·5-in. Q.F., 4 2·5-in., 4 1·8-in., 6 1·4-in., M.	4	18·3 *t*	406	391
b.	**Caïman ‡**	I. & S.	7239	278	3	59	0	24	7	2	6000	Toulon	1885	..	19½	9¾	3	2 10·8-in., 6 3·9-in. Q.F., 6 1·8-in., 2 M., 6 1·4-in.	4	14·5	400	332
t.	**Carnot**	S.	12,008	382	2	70	6	27	3	2	16,300 *t* D'A	Toulon	1894	1,070,088	17¾ 10¾	14½	2¾	2 12-in., 2 10·8-in., 8 5·5-in. Q.F., 4 2·5-in., 16 1·8-in., 10 1·4-in.	4 (2 sub.)	17·86 *t*	705	625
a.c.	**Chanzy**	S.	4933	348	0	46	0	19	2	2	8300 B	Bordeaux	1894	360,000	3¾	2	2	2 7·6-in., 6 5·5-in. Q.F., 4 2·5-in., 6 1·8-in., 6 1·4-in., M.	4	19·0	413	375
t.	**Charlemagne**	S.	11,275	385	6	66	6	27	6	3	14,500 B	Brest	1895	1,096,432	15¾	15¾	3½	4 12-in., 10 5·5-in. Q.F., 8 3·9-in., 16 1·8-in., 10 1.4-in., 8 M.	4 (2 sub.)	18·1 *t*	680 / 1100	631
t.	**Charles Martel**	S.	11,880	392	6	71	0	27	6	2	14,996 *t* D'A	Brest	1893	1,092,830	17¾	15¾	3½	2 12-in., 2 10·8-in., 8 5·5-in. Q.F., 4 2·5-in., 14 1·8-in., 5 1·4-in.	6	18·1 *t*	677	632
a.g.b.	**Cocyte**	I. & S.	1714	181	10	40	4	11	10	2	1700	Cherbourg	1887	100,000	8	8	2½	1 10·8-in., 2 3·9-in. Q.F., 2 1·8-in., 4 M.	..	13·0	100	101
c.b.	**Colbert**	W.	8924	317	9	56	6	28	5	1	4652	Brest	1875	..	8½	6¼	..	8 10·8-in., 2 9·4-in., 6 5·5-in., 2 Q.F., 14 M.	4	14 47	700	706
a.c.	**Condé**	S.	10,000	453	0	63	9	24	7	3	20,000 W.T.	Cherbourg	Bldg.	808,600	6–5 H.S.	8, 5 H.S.	2	2 7·6-in., 8 6·4-in. Q.F., 6 3·9-in., 16 1·8-in., 6 1·4-in.	5 (2 sub.)	21·0	..	..
c.b.&b.	**Courbet**	I. & S.	10,808	312	0	67	0	25	0	2	8100†	Toulon	1881 1899	800,000	15	9½	2½	4 10·8-in., 4 9·4-in., 6 5·5-in., 2 Q.F., 18 M.*	5	15·4	1000	669
a.c.	**C 11**											Cherbourg										
a.c.	**C 12**	S.	12,416	474	0	71	2	26	3	3	24,000 W.T.	Brest	Pro.	1,170,000	6 H.S.	5–7¾	2	4 7·6-in., 16 6·4-in. Q.F., 20 1·8-in., 4 1·4-in.	5 (2 sub.)	21·0	1350 / 1950	..
a.c.	**C 13**											Toulon										
a.c.	**Desaix**	S. shd.	7700	426	6	58	4	24	4	3	17,100 B	St. Nazaire	Bldg.	620,000	4	3½	2¾	10 6·4-in. Q.F., 10 1·8-in., 6 1·4-in.	2	21·0	880 / 1200	531
c.b.&b.	**Dévastation**	I. & S.	10,704	312	0	67	0	25	0	2	8320	Lorient	1879	..	15	9½	2½	4 10·8-in., 4 9·4-in., 6 5·5-in., 2 Q.F., 18 M.*	4	15·17	950	685
a.c.b.	**Duguesclin**	I. shd.	6210	266	0	57	0	25	8	2	3300	Rochefort	1883	220,000	9	8 *comp.*	2	4 9·4-in., 1 7·6-in., 6 5·5-in., 1 3·5-in. Q.F., 10 M.	2	14·0	400	430
a.c.	**DupetitThouars**	S.	9517	452	9	63	8	24	7	3	19,600 B	Toulon	Bldg.	804,600	6–3¾ H.S.	8, 3¾ H.S.	2	2 7·6-in., 8 6·4-in. Q.F., 4 3·9-in., 16 1·8-in., 6 1·4-in.	(2 sub.)	21·0	§ 1020	612

* New armament. † Has received new boilers. § Including liquid fuel.

Reconstruction of Amiral Duperré to be completed 1901, Caïman 1900, Dévastation 1900.

FRANCE.—Armoured Ships—*continued.*

Class.	NAME.	Material of Hull.	Displacement.	Length.		Beam.		Draught.		Propellers.	Indicated Horse-power.	Where Built.	Date of Launch.	Cost.	Armour. Belt.	Armour. Gun Position.	Armour. Deck Plating.	Armament. Guns.	Armament. Torpedo Tubes.	Speed.	Normal Coal Supply.	Complement.
			metric tons.	ft.	in.	ft.	in.	ft.	in.	no.				£	inches.	inches.	inches.			knots.	tons.	
a.c.	**Dupleix** . .	S. shd.	7700	426	6	58	4	24	4	3	17,100 B	Rochefort .	1900	620,000	4	3½	2¾	10 6·4-in. Q.F., 10 1·8-in., 6 1·4-in.	2	21·0	880	531
”	**Dupuy de Lôme**	S.	6406	374	0	51	6	23	6	3	14,000	Brest . .	1890	416,000	4	4	2	2 7·6-in., 6 6·4-in. Q.F., 12 2·5-in., 1·8-in., 8 M.	4	20·0	900	515
a.g.b.	**Flamme** . .	S.	1128	165	0	32	7	10	4	2	1500	Cherbourg .	1885	68,000	10	4	2	1 9·4-in., 1 3·5-in., 4 M.	1	13·0	120	84
b.	**Formidable** .	S.	12,165	321	6	69	6	26	2	2	9700*	Lorient .	1885 1899	467,520	21½	17¾	3	2 14·6-in., 8 6·4-in. Q.F., 8 5·5-in., 9 1·8-in., 14 M.**	6	16·0	900	640
c.b. & b.	**Friedland** .	I.	8994	317	0	58	0	29	11	1	4428	Lorient .	1873	..	8	7	..	8 10·8-in., 8 5·5-in., 20 M. .	4	13·3	800	676
c.d.s., t.	**Fulminant** .	I. & S.	5965	248	0	57	9	21	4	1	4500	Cherbourg .	1877	..	13	12	2	2 10·8-in., 4 1·8-in. Q.F., 6 M.	2	13·8	400	248
c.d.s., b.	**Furieux** . .	I. & S.	6019	247	10	59	0	21	9	2	5033 Nic.	Cherbourg .	1883	264,640	20	17¾	3½	2 9·4-in., 5 Q.F., 10 M.** .	2	14·0	290	248
a.g.b.	**Fusée** . .	S. shd.	1142	165	0	32	7	10	4	2	1500	Lorient .	1884	68,000	10	4	2	1 9·4-in., 1 3·5-in., 4 M. .	1	13·0	120	84
t.	**Gaulois** . .	S.	11,275	385	6	66	6	27	6	3	14,500 B	Brest . .	1896	1,093,925	15¾	15¾	3½–1½	4 12-in., 10 5·5-in. Q.F., 8 3·9-in., 16 1·8-in., 10 1·4-in., 8 M.	6 (2 sub.)	18·0	680 / 1100	632
a.c.	**Gloire** . .	S.	10,000	453	0	63	9	24	7	3	20,000 W.T.	Lorient .	Bldg.	808,600	6–5 H.S.	8, 5 H.S.	2	2 7·6-in., 8 6·4-in. Q.F., 6 3·9-in., 16 1·8 in., 6 1·4 in.	5 (2 sub.)	21·0	..	..
a.g.b.	**Grenade** . .	S. shd.	1089	165	0	32	7	10	4	2	1500	Lorient .	1888	68,000	10	4	2	1 9·4-in., 1 3·5-in., 4 M. .	1	13·0	120	84
t.	**Henri IV.** .	S.	8948	354	4	72	3	22	11	3	12,000 Nic.	Cherbourg .	1899	800,000	11–4¾ H.S.	11¾ H.S.	3	2 10·8-in., 7 5·5-in. Q.F., 12 1·8-in., 2 M.	2 (sub.)	17	725 / 1100	464
t. & b.	**Hoche** . .	I. & S.	10,997	333	0	65	7	27	3	2	11,300 B	Lorient .	1886	700,000	18	16	3	2 13·4-in., 2 10·8-in., 8 5·5-in., Q.F., 8 Q.F., 12 M.	5	16·0	800	660
t.	**Iéna** . . .	S.	12,052	400	9	68	2	27	6	3	15,500 B	Brest . .	1898	1,136,385	13¾–4¾ H.S.	..	2½	4 12-in., 8 6·4-in. Q.F., 8 3·9-in., 16 1·8-in., 5 1·4-in., 13 1·4-in. M.	4 (2 sub.)	18·0	820 / 1100	631

b.	**Indomptable** † .	I. & S.	7234	279	10	59	0	23	6	2	6605 Nic.	Lorient .	1883	..	20	9¾	3	2 10·8-in., 6 3·9-in. Q.F., 6 1·8-in., 6 1·4-in., 2 M.	4	14·8	400	332
t.	**Jauréguiberry** .	S.	11,824	364	0	72	10	27	9	2	15,800 D'A.	La Seyne .	1893	1,069,536	17¾	14½	2¾	2 12-in., 2 10·8-in., 8 5·5-in. Q.F., 4 2·5-in., 12 1·8-in., 8 1·4-in., 8 M.	6	18·07	700	625
a.c.	**Jeanne d'Arc** .	S.	11,329	477	2	63	8	26	7	3	28,000 Guyot	Toulon .	1899	882,955	6–3 H.S.	6 H.S.	2·2 H.S.	2 7·6-in., 8 5·5-in. Q.F., 10 3·9-in., 16 1·8-in., 8 1·4-in., 2 M.	2 (sub.)	23	1400 / 2100	626
c.d.s.,t.	**Jemmapes** .	S.	6592	284	0	57	8	22	0	2	9250 D'A.	St. Nazaire .	1892	525,000	17¾–10	17¾	4–2¾	2 13·4-in., 4 3·9-in. Q.F., 4 1·8-in. Q.F., 10 1·4-in. M.	2	16·7	300	334
a.c.	**Kléber** . .	S. shd.	7700	426	6	58	4	24	4	3	18,000 Nic.	Bordeaux .	Bldg.	620,000	4	3½	2¾	10 6·4-in. Q.F., 10 1·8-in., 6 1·4-in.	2	21·0	880	531
a.c.	**Latouche - Tréville**	S.	4756	348	0	46	0	19	2	2	8300 B	Havre . .	1892	360,000	3⅝	3¾	2–1½	2 7·6-in., 6 5·5-in. Q.F., 4 2·5-in., 4 1·8-in., 6 1·4-in. M.	4	18·2	406	375
b.	**Magenta** . .	I. & S.	10,851	330	0	65	7	27	3	2	12,000	Toulon .	1890	760,960	18	16	3	4 13·4-in., 17 5·5-in. Q.F., 4 2·5-in., 12 1·8-in., 8 M.	3	16·25	800	660
b.	**Marceau** . .	I. & S.	10,850	330	0	65	7	27	3	2	14,000 Nic.	La Seyne .	1887	769,080	18	16	3	4 13·4-in., 17 5·5-in. Q.F., 4 2·5-in. and 12 1·8-in., 8 M.	6	16·4	800	660
a.c.	**Marseillaise** .	S.	10,014	453	0	63	9	24	7	3	20,500 W.T.	Brest . .	Bldg.	868,626	6 H.S.	5–7¾	2	2 7·6-in., 8 6 4-in. Q.F., 6 3·9-in., 2 2·5-in., 18 1·8-in., 6 1·4-in.	4 (2 sub.)	21·0	970 / 1590	..
t.	**Masséna** . .	S.	11,924	384	10	66	0	27	0	3	13,500 D'A.	St. Nazaire .	1895	1,100,400	17¾–9¾	15¼ 15¾ H. S.	3½ H. S.	2 12-in., 2 10·8-in., 8 5·5-in. Q.F., 8 3·9-in., 12 1·8-in. and 12 1·4-in.	6 (2 sub.)	17·1 *t*	630	642
a.g.b.	**Mitraille** . .	S. shd.	1128	165	0	32	7	10	4	2	1500	Rochefort .	1886	70,000	10	4	2	1 9·4-in., 1 3·5-in., 4 M.	..	13·0	120	84
a.c.	**Montcalm**	S.	9517	452	9	63	8	24	7	3	19,600 N. S.	La Seyne .	1900	..	6–3¾ H.S.	8, 3¾ H.S.	2	2 7·6-in., 8 6·4-in. Q.-F., 4 3·9-in., 16 1·8-in., 6 1·4-in.	2 (sub.)	21·0	‡ 1020 / 1600	612
b.	**Neptune** .	I. & S.	10,983	330	0	65	7	27	3	2	12,000 B	Brest . .	1887	780,000	18	16	3	4 13·4-in., 17 5·5-in. Q.F., 4 2·5-in. and 12 1·8-in., 8 M.	5	16·02	800	660
c.d.s.	**Onondaga** . .	I.	2593	226	6	49	3	16	0	2	642	New York .	1863	..	5½	11¾	1	4 9·4-in., 4 M.	..	6·5	200	100
a.g.b.	**Phlégéton** .	I. & S.	1796	187	0	40	4	11	10	2	1700	Cherbourg .	1890	142,000	9	8	2	1 10·8-in., 1 5·5-in. Q.F., 4 1·8-in., 4 M.		12·4	72	101
c.b. & b.	**Redoutable** .	I. & S.	9437	318	2	64	8	25	6	2	6071	Lorient .	1876 1896	..	14	9½	2½	8 9·4-in., 6 3·9-in. Q.F., 2 Q.F., 12 M.**	4	14·66	1000	700

* Has received new boilers. † New armament, armour, and approximately reduced displacement given. ** New armament. ‡ Including liquid fuel.

Reconstruction of Furieux to be completed 1901, Indomptable 1900, Magenta, Marceau and Neptune 1901, Requin 1900.

FRANCE.—Armoured Ships—*continued.*

Class.	NAME.	Material.	Displacement.	Length.	Beam.	Draught.	Propellers.	Indicated Horse-power.	Where Built.	Date of Launch.	Cost.	Armour. Belt.	Armour. Gun Position.	Armour. Deck Plating.	Armament. Guns.	Armament. Torpedo Tubes.	Speed.	Normal Coal Supply.	Complement.
			metric tons.	ft. in.	ft. in.	ft. in.	no.				£	inches.	inches.	inches.			knots.	tons.	
b.	**Requin**	I. & S.	7822	279 10	59 0	24 7	2	7000 Nic.	Bordeaux	1885 1898	..	19½	9¾	3	2 10·8-in., 6 3·9-in. Q.F., 2 1·8-in., 6 1·4 in., 12 M.*	4	15·0	400	332
c.b. & b.	Richelieu	W.	9128	323 6	57 10	27 11	2	4240	Toulon	1873	..	8½	6¼	..	6 10·8-in., 5 9·4-in., 8 5·5-in. 18 M.	4	13·11	900	720
t.	**Saint Louis**	S.	11,275	385 6	66 6	27 6	3	14,500 B	Lorient	1896	1,080,997	15¾ H.S.	8–15¾ H.S.	3½	4 12-in., 10 5·5-in. Q.F., 8 3·9-in., 16 1·8-in., 10 1·4-in., 8 M.	4 (2 sub.)	18·0	820 / 1150	631
a.g.b.	**Styx**	I. & S.	1796	187 0	40 4	11 10	2	1700	Cherbourg	1892	142,000	9	8	2	1 10·8-in., 1 5·5-in. Q.F., 4 1·8-in., 4 M.	..	13·0	72	101
t.	**Suffren**	S.	12,728	411 9	70 4	27 6	3	16,200 Nic.	Brest	1899	1,180,000	11¾–5 H.S.	..	2¾	4 12-in., 10 6·4-in. Q.F., 8 3·9-in., 20 1·8-in.	4 (2 sub.)	18·0	820 / 1100	..
a.c.	Sully	S	10,000	453 0	63 9	24 7	3	20,000 W. T.	La Seyne	Bldg.	808,600	6–5 H.S.	8,5 H.S.	2	2 7·6-in., 8 6·4-in. Q.F., 6 3·9-in., 16 1·8-in., 6 1·4-in.	5 (2 sub.)	21·0	..	..
c.d.s., t.	Tempête	I. & S.	4869	248 0	57 9	16 9	1	2193	Brest	1876	..	13	12	2	2 10·8-in., 4 1·8-in. Q.F., 6 M.	2	11·7	200	197
b.	Terrible	I. & S.	7575	279 10	59 0	24 7	2	6230	Brest	1881 1899	..	19½	17¾	3	2 13·4-in., 6 3·9-in. Q.F., 2 1·8-in., 12 M.†	4	14·5	400	332
c.d.s., b.	Tonnant	I.&S.	5091	248 7	58 5	17 3	1	1935	Rochefort	1880	..	18	14½	3	2 13·4-in., 4 M.	..	11·5	200	197
c.d.s., t.	Tonnerre	I. &S.	5858	248 0	57 9	21 4	1	4165	Toulon	1875	..	13	12	2	2 10·8-in., 4 1·8-in. Q.F., 6 M.	2	14·01	400	249
c.b. & b.	Trident	W.	8857	317 9	56 4	29 1	1	5083	Toulon	1876	.	8½	6¼	½	8 10·8-in., 2 9·4-in., 6 5·5-in., 2 Q.F., 14 M.	6	14·17	650	730
a.c.	Turenne	W. & I.	6349	265 9	57 2	23 11	2	4160	Lorient	1879	..	10	8	2	4 9·4-in., 2 7·4-in., 6 5·5-in., 12 M.	2	14·14	500	450
c.d.s., t.	**Valmy**	S.	6592	293 9	57 4	23 3	2	8954	St. Nazaire	1892	578,957	17¾	17¾	4	2 13·4-in., 4 3·9-in. Q.F., 4 1·8-in. 10 M.	2	16·7	300	297
a.c.	Vauban	I.	6208	267 9	57 3	24 0	2	4560	Cherbourg	1882	..	10	8	2	4 9·4-in., 1 7·6-in., 6 5·5-in., 12 M.	2	14·32	550	440
c.d.s., t.	Vengeur	I.&S.	4709	248 0	57 9	16 9	2	2030	Cherbourg	1878	..	13	12	2	2 12·5-in., 4 1·8-in. Q.F., 6 M.	2	10·83	200	107

* New armament. † New armament. Reconstruction to be completed 1899.

FRANCE.—Cruising Ships, &c.

Class.	NAME.	Material.	Displacement.	Length.	Beam.	Draught.	Propellers.	Indicated Horse-power.	Where Built.	Date of Launch.	Cost.	Armour. Gun Position.	Armour. Deck.	Armament. Guns.	Armament. Torpedo Tubes.	Speed.	Normal Coal Supply.	Complement.
			metric tons.	ft. in.	ft. in.	ft. in.	no.				£	in.	in.			knots.	tons.	
2nd cl. cr.	Alger	S.	4382	346 0	45 3	19 6	2	8254 B	Cherbourg	1889	280,000	..	3½	4 6·4-in. Q.F., 6 5·5-in., 10 other Q.F., 10 M.	4	19·61	860	325
g. v.	Amiral Parseval	W. & I.	869	197 6	28 0	12 2	1	918	Rochefort	1879	33,772	..	..	4 5·5-in., 4 M.	..	12·33	150	116
cr.	Amiral Rigault de Genouilly.	W.	1756	236 3	35 5	17 0	1	2043 B	Brest	1876	62,796	..	..	8 5·5-in., 8 M.	..	14·49	200	198
cr.	Aréthuse	W.	3665	277 6	43 6	21 9	1	4200	Toulon	1882	..	..	..	4 6·4-in., 22 5·5-in., 8 M.	..	14·0	500	474
g. v.	Aspic	C.	476	145 4	23 10	10 6	1	453	Rochefort	1880	..	..	..	2 5·5-in., 2 3·9-in.	..	10·3	50	80
to. g. b.	Bombe	S.	420	196 10	21 7	5 11	2	2000 D'A.	Havre	1885	..	..	..	4 1·8-in. Q.F., 3 M.	2	18·0	100	63
2nd cl. cr.	Bugeaud	S.	3740	308 6	43 6	20 6	2	9000 B	Cherbourg	1893	308,650	2 shield	3	6 6·4-in. Q.F., 4 3·9-in., 8 1·8-in., 11 1·4-in.	6	18·9	587	358
g. v.	Capricorne	W.	483	148 4	23 10	10 5	1	443	Havre	1882	..	..	..	2 5·5-in., 2 3·9-in.	..	11·18	70	80
to. g. b.	Casabianca	S.	960	262 6	26 10	11 6	2	5200 D'A.	Bordeaux	1895	98,985	..	½	1 3·9-in. Q.F., 3 2·5-in. 5 1·8-in. 4 1·4-in.	..	22·0	116	143
2nd cl. cr.	Cassard	S.	3952	325 6	44 11	20 6	2	10,143 D'A.	Cherbourg	1896	318,712	2 shield	3	6 6·4-in., Q.F., 4 3·9-in., 10 1·8-in., 3 1·4-in., 2 M.	2	19·8	630	385
to. g. b.	Cassini	S.	958	262 6	27 4	11 0	2	5500 D'A.	Bordeaux	1894	98,500	..	½	1 3·9-in. Q.F., 3 2·5-in. 4 1·4-in.	2	21·2	110	118
2nd cl. cr.	Catinat	S. Shd.	4065	331 10	44 8	21 1	2	9000 B	Havre	1896	324,992	2 shield	3	4 6·4-in. Q.F., 10 3·9-in., 10 1·8-in., 4 1·4-in. M.	2	19·0	563	384
2nd cl. cr.	Cécille	I. & S.	5933	378 9	49 3	19 9	2	10,200	La Seyne	1888	299,666	..	4	8 6·4-in. Q.F., 10 5·5-in., 6 1·8-in., 14 M.	4	19·0	940	486
2nd cl. cr.	Chasseloup-Laubat	S.	3758	308 6	43 6	20 10	2	9000 D'A.	Cherbourg	1893	256,320	..	3	6 6·4-in. Q.F., 4 3·9-in., 8 1·8-in., 12 1·4-in. M.	6	19·25	587	358
1st cl. cr.	Châteaurenault	S. Shd.	8018	442 10	55 9	24 6	3	23,000 N.S.	La Seyne	1898	606,656	2 shield	2½	2 6·4-in. Q.F., 6 5.5-in., 10 1·8-in.	..	23	1400 / 2100	625

FRANCE.—Cruising Ships, &c.—*continued.*

Class.	NAME.	Material.	Displacement.	Length.	Beam.	Draught.	Propellers.	Indicated Horse-power.	Where Built.	Date of Launch.	Cost.	Armour: Gun Position.	Armour: Deck.	Armament: Guns.	Armament: Torpedo Tubes.	Speed.	Normal Coal Supply.	Complement.
			metric tons.	ft. in.	ft. in.	ft. in.	no.				£	in.	in.			knots.	tons.	
3rd cl. *cr.*	Coëtlogon	S.	1982	312 0	30 5	14 0	2	5800	St. Nazaire	1889	134,000	..	1½	4 5·5-in. Q.F., 8 other Q.F., 4 M.	5	19·3	200	190
g. v.	Comète	C.	495	151 6	24 9	10 6	1	631	Chérbourg	1884	..	..	..	2 5·5-in., 2 3·9-in., 2 M.	..	12.2	60	84
to. cr.	Condor	S.	1243	216 6	29 3	15 5	2	3800	Rochefort	1885 1896	80,000	..	1½	5 3·9-in. Q.F., 1 2·5-in., 6 M.	5	17·7	160	134
3rd cl. *cr.*	Cosmao	S.	1954	312 0	30 5	14 0	2	6000	Bordeaux	1888	133,000	..	1½	4 5·5-in. Q.F., 8 other Q.F., 4 M.	5	20·5	200	190
to. g. b.	Couleuvrine	S.	435	196 10	21 7	5 11	2	2047	Havre	1885	33,778	..	..	4 1·8-in. Q.F., 3 M.	2	18·0	100	63
to. g. b.	Dague	S.	408	196 10	21 7	5 11	2	2000	Havre	1885	36,119	..	..	4 1·8-in. Q.F., 3 M.	2	18·0	100	63
2nd cl. *cr.*	D'Assas	S.	4000	325 6	44 11	20 6	2	9500 D'A.	St. Nazaire	1896	292,682	2 shield	3	6 6·4-in. Q.F., 4 3·9-in., 10 1·8-in., 11 1·4-in.	2	19·25	630	393
2nd cl. *cr.*	Davout	S.	2291	295 4	40 0	17 6	2	8881	Toulon	1890	221,827	..	3	6 6·4-in. Q.F., 4 3·9-in., 4 2·5-in., 4 1·8., 6 M.	4	20·07 *t.*	600	336
g. v.	Décidée	S.	645	184 8	26 3	12 2	1	1000 Nic.	Lorient	1899	54,100	..	..	2 3·9-in. Q.-F., 4 2·5-in., 4 1·4-in.	..	13·0	99	99
1st cl. *cr.*	D'Entrecasteaux	S. shd.	8114	383 7	58 6	25 9	2	13,500	La Seyne	1896	667,740	10–2¾ H.S.	4	2 9·4-in., 12 5·5-in Q.F., 12 1·8-in.	6	19·2*t*	650	521
2nd cl. *cr.*	Descartes	S. shd.	3990	326 0	42 4	21 4	2	9000 B	St. Nazaire	1894	334,725	..	1½	4 6·4-in. Q.F., 10 3·7-in., 8 1·8-in., 4 1·4-in.	2	21·0*t*	552	386
cr.	D'Estaing	W. & I.	2435	262 5	37 5	18 8	1	3700	Brest	1879	84,718	..	..	15 5·5-in., 8 M.	..	15·31	300	264
3rd cl. *cr.*	D'Estrées	S. shd.	2452	311 8	39 4	17 8	2	8500 Nor.	Rochefort	1897	208,200	..	1½	2 5·5-in. Q.F., 4 3·9-in., 8 1·8-in., 2 1·4-in.	..	20·5	345/480	234
to. g. b.	D'Iberville	S.	967	262 6	27 0	11 2	2	5060 D'A.	St. Nazaire	1893	99,120	..	½	1 3·9-in. Q.F., 1 2·5-in., 4 1·4-in.	6	21·4	117	118
to. g. b.	Dragonne *	S.	410	196 10	21 7	5 11	2	2000	Havre	1885	36,074	..	..	4 5·5-in. Q.F., 3 M.	2	18·0	100	63
cr.	Dubourdieu	W.	3577	253 7	46 6	22 10	1	3300	Cherbourg	1884	154,553	..	..	4 6·4-in., 12 5·5-in., 10 M.	2	14·0	600	496
2nd cl. *cr.*	Du Chayla	S. shd.	3952	325 6	44 11	20 6	2	10,009 D'A.	Cherbourg	1895	315,835	2 shield	3	6 6·4-in. Q.F., 4 3·9-in., 10 1 8-in., 3 1·4-in., 2 M.	2	20·2 *t*	624	385

to. g. b.	**Dunois** (ex M 3)	S.	896	256	0	27	10	12	8	2	7000 N.S.	Cherbourg	1897	123,383	..	..	6 2·5-in. Q.F., 6 1·8-in.	..	23·0	137	128
cr.	**Duquesne**	I. & W.	5986	333	5	50	3	25	6	1	6589	Rochefort	1876	221,570	..	..	7 6·4-in. Q.F., 14 5·5-in., 8 M.	..	16·8	900	550
cr.	**Éclaireur**	I. & W.	1769	236	3	35	5	17	0	1	2050	Toulon	1877	16,232	..	..	8 5·5-in., 6 M.	..	15·0	200	195
to. cr.	**Epervier**	S.	1288	216	6	29	3	15	5	2	3200	Rochefort	1885	80,000	..	1½	5 3·9-in. Q.F., 1 2·5-in., 6 M.	5	17·6	160	134
g. v.	**Étoile**	C.	502	149	3	24	7	8	4	2	450	France	1885	29,782	..	..	6 3·9-in., 1 2·5-in., 2 M.	..	10·0	60	77
cr.	**Fabert**	W.	2100	294	4	36	0	18	4	1	1107	Rochefort	1874	61,967	..	..	8 5·5-in., 4 M.	..	12·42	300	218
to. cr.	**Faucon**	S.	1239	216	6	29	3	15	5	2	3200	Toulon	1887	80,000	..	1½	5 3·9-in. Q.F., 1 2·5-in., 6 M.	5	18·0	150	134
to. g. b.	**Flèche**	S.	425	196	10	21	7	5	11	2	2000	Havre	1885	37,517	..	..	4 1·8-in. Q.F., 3 M.	2	18·0	100	63
3rd cl. *cr.*	**Fleurus**	S.	1310	229	8	29	2	15	4	2	4000 Nic.	Cherbourg	1893	128,530	..	..	5 3·9-in. Q.F., 6 1·8-in., 4 M.	..	17·6 *t*	118	179
„ „	**Forbin**	S	1820	312	0	30	5	16	0	2	5700	Rochefort	1888	123 739	..	1½	4 5·5-in. Q.F., 8 other Q.F., 4 M.	5	20·6	200	190
cr.	**Forfait**	W. & I.	2464	249	4	38	0	18	0	1	2764	Toulon	1879	77,019	..	..	15 5·5-in., 8 M.	..	13·44	400	264
cr.	**Foudre** (torpedo transport)	S.	6090	370	6	52	6	23	6	2	11,900 D'A. *t*	Bordeaux	1895	407,712	..	2½	10 3·9-in. Q.F., 4 2·5-in., 4 1·4-in.	..	19·9 *t*	840	410
2nd cl. *cr.*	**Friant**	S.	3739	308	6	43	6	20	10	2	9000 Nic.	Brest	1893	308,750	..	3	6 6·4-in. Q.F., 4 3·9-in. 8 1·8-in. 6 1·4-in.	2	18·19	587	358
g. v.	**Fulton**	W.	913	199	5	28	5	12	8	1	850	Lorient	1887	37,000	..	..	2 5·5-in., 1 3·9-in., 5 M.	..	13·0	160	116
„	**Gabès**	C.	493	151	6	24	9	10	6	1	450	Rochefort	1884	28,624	..	..	2 5·5-in., 2 3·9-in.	..	11·0	60	84
3rd cl. *cr.*	**Galilée**	S.	2317	330	2	34	6	17	10	2	6600 B	Rochefort	1896	208,152	2 shields	1½	4 5·5-in. Q.F., 2 3·9-in., 8 1·8-in., 8 1·4-in.	..	20·0	226	248
1st cl. *cr.*	**Guichen**	S. shd.	8277	436	4	54	10	24	7	3	24,000 D'A	St. Nazaire	1897	611,945	2 shield	2½	2 6·4-in. Q.F., 6 5·5-in., 10 1·8-in.	..	23·0	1460	625
d. v.	**Inconstant**	W.	891	199	5	28	5	12	7	1	850	La Seyne	1886	37,000	..	..	2 5·5-in., 1 3·9-in., 5 M.	..	13·0	160	116
cr.	**Iphigénie** (Training ship)	W.	3431	244	6	46	6	22	4	1	2800	..	1881	115,323	..	..	6 6·4-in., 2 3·5-in., 10 M.	1	14·6	400	391

* Converted experimentally into a bowitzer gunboat.

FRANCE.—Cruising Ships, &c.—*continued.*

Class.	NAME.	Material of Hull.	Displacement.	Length.		Beam.		Draught.		Propellers.	Indicated Horse-power.	Where Built.	Date of Launch.	Cost.	Armour.		Armament.		Speed.	Normal Coal Supply.	Complement.
															Gun Position.	Deck.	Guns.	Torpedo Tubes			
			metric tons.	ft.	in.	ft.	in.	ft.	in.	no.				£	inches.	inches.			knots.	tons.	
3rd cl. *cr.*	**Infernet**	S shd.	2452	311	8	39	4	15	6	2	8500 Nor.	Bordeaux	1899	193,000	..	..	2 5·5-in. Q.F., 4 3·9-in. 8 1·8-in.	..	20·5	345/480	234
2nd cl. *cr.*	**Isly**	S.	4477	346	0	43	6	19	6	2	8100	Brest	1891	252,760	..	3	4 6·4-in. Q.F., 6 5·5-in., 14 2·5-in. and 1·8-in., 8 M.	5	18·3	880	332
2nd cl. *cr.*	**Jurien de la Gravière**	S. shd.	5605	440	0	48	8	22	0	3	17,000 Guyot	Lorient	1899	481,000	..	3	8 6·4-in. Q.F., 12 1·8-in.	2	23·0	600/900	511
2nd cl. *cr.*	**Jean Bart**	S.	4109	346	0	43	6	19	6	2	8000	Rochefort	1889	283,240	..	4	4 6·4-in. Q.F., 6 5·5-in., 14 2·5-in., and 1·8-in., 8 M.	5	19·0	940	332
g. v.	**Kersaint**	S. shd.	1243	226	0	34	5	15	0	1	2200	Rochefort	1897	107,933	..	..	1 5·5-in. Q.F., 5 3·9-in., 7 1·4-in.	..	15·0	199	110
to. g. b.	**La Hire**	S.	896	256	0	27	10	12	9	2	7000 N.S.	Cherbourg	1898	123,383	..	..	6 2·5-in. Q.F., 6 1·8-in.	..	23·0	137	128
3rd cl. *cr.*	**Lalande**	S.	1926	311	6	31	2	14	0	2	6000	Bordeaux	1888	133,800	..	1½	6 5·5-in. Q.F., 8 other Q.F., 4 M.	5	22·0	200	190
to. g. b.	**Lance**	S.	402	196	10	21	7	5	11	2	2000 Du T.	Havre	1886	39,964	..	..	4 1·8-in., Q.F., 3 M.	2	18·0	100	63
cr.	**Lavoisier**	S	2317	330	2	34	6	17	10	2	6400 B	Rochefort	1897	202,024	2 shield	1½	4 5·5-in. Q.F., 2 3·9-in., 8 1·8-in., 2 1·4-in., 4 M.	2	20·0	226	248
to. g. b.	**Léger**	S.	517	197	0	23	0	10	6	2	2360 B	Lorient	1891	52,000	..	..	1 3·9-in. Q.F., 3 2·5-in., 4 1·4-in.	3	18·8	130	69
to. g. b.	**Lévrier**	S.	505	197	0	23	0	10	6	2	2240 B	Lorient	1891	52,000	..	..	1 3·9-in. Q.F., 3 2·5-in., 4 1·4-in.	3	18·5	130	69

3rd cl. *cr.* .	**Linois** . .	S.	2345	321	6	34	6	17	5	2	6600	La Seyne .	1894	163,014	3·9	1½	4 5·5-in. Q.F., 2 3·9-in., 8 1·8-in., 4 1·4-in., 4 M.	4	20·5	200	248
g. v. . .	**Lion.** . . .	C.	503	151	6	24	9	10	6	1	576	Havre .	1884	23,146	..	..	2 5·5-in., 4 M. . . .	..	11·8	70	84
" . .	**Lutin** . . .	C.	493	141	9	23	11	9	4	1	373	Cherbourg .	1877	20,295	..	..	2 5·5-in., 2 3·9-in. . .	..	10·0	60	77
" . .	**Lynx** . . .	C.	485	141	9	23	11	9	4	1	427	Cherbourg .	1878	21,478	..	..	2 5·5-in., 2 3·9-in. . .	..	10·38	60	77
" . .	**Météore** . .	C.	504	151	6	24	9	10	6	1	434	Cherbourg .	1886	26,262	..	..	2 5·5-in., 3 M. . . .	..	10·0	70	77
3rd cl. *cr.* .	**Milan** . . .	S.	1733	303	2	32	10	14	7	2	3986 B	St. Nazaire .	1886	89,058	..	..	5 3·9-in. Q.F., 8 M. . .	2	18·1	400	186
cr. . .	**Naiade** . .	I. shd.	3686	246	0	47	2	22	10	1	2700	Toulon .	1881	128,275	..	..	2 6·4-in., 18 5·5-in., 10 M. .	..	13·68	500	490
cr. . .	**Nielly** . . .	W. & I.	2400	262	5	37	5	18	8	1	2921	Brest . .	1880	84,037	..	..	15 5·5-in., 8 M. . . .	..	15·23	300	264
g. v. . .	**Papin** . . .	W. & I.	891	199	6	28	6	13	7	1	855	La Seyne .	1886	37,000	..	..	2 5·5-in., 1 3·9-in., 5 M. .	..	13·0	160	116
2nd cl. *cr.* .	**Pascal** . .	S.	4015	326	0	42	4	21	4	2	9000 *t*, B	Toulon .	1895	322,321	..	1½	4 6·4-in Q.F., 10 3·9-in., 8 1·8-in., 4 1·4-in. M.	2	20·0 *t*	650	378
cr. . .	**Primauguet** .	W. & I.	2447	262	5	37	5	18	8	1	2268	Rochefort .	1882	108,592	..	..	15 5·5-in., 8 M. . . .	..	14·50	300	264
2nd cl. *cr.* .	**Protet** . . .	S. shd.	4055	331	10	44	8	21	1	2	9300 *t*, B	Bordeaux	1898	324,992	2 shield	2½	4 6·4-in. Q.F., 10 3·9-in., 10 1·8-in., 2 1·4-in.	2	20·2 *t*	563	384
cr. . .	**Roland** . . .	W.	2476	249	4	38	0	17	7	1	2294	Cherbourg .	1882	84,184	..	..	15 5·5-in., 8 M. . . .	..	14·50	350	264
to. g. b. .	**Sainte Barbe** .	S.	437	196	10	21	7	5	11	2	2000	Rouen . .	1885	43,233	..	1½	4 1·8-in. Q.F., 3 M. . .	2	18·0	100	63
" . .	**Salve** . . .	S.	413	196	10	21	7	5	11	2	2000	Rouen . .	1886	42,538	..	1½	4 1·8-in. Q.F., 3 M. . .	2	18·0	100	63
g. v. . .	**Scorpion** . .	C.	505	151	6	24	9	10	6	1	511	Havre . .	1883	23,459	..	..	2 5·5-in., 3 M. . . .	..	11·0	70	84

FRANCE.—Cruising Ships, &c.—*continued.*

Class.	NAME.	Material of Hull.	Displacement.	Length.		Beam.		Draught.		Propellers.	Indicated Horse-Power.	Where Built.	Date of Launch.	Cost.	Armour. Gun Position.	Armour. Deck.	Armament. Guns.	Armament. Torpedo Tubes.	Speed.	Normal Coal Supply.	Complement.
			metric tons.	ft.	in.	ft.	in.	ft.	in.	no.				£	inches.	inches.			knots.	tons.	
2nd cl. *cr.*	Sfax	S. & W.	4728	288	9	49	3	24	9	2	6522	Brest	1884 1898	200,000	..	1¼	6 6·4-in. Q.F., 10 5·5-in., 6 1·8-in., 6 1·4-in., 4 M.†	2	16·84	715	473
2nd cl. *cr.*	Suchet	S.	3440	318	3	43	6	17	6	2	9000	Toulon	1893	226,360	..	3	4 6·4-in. Q.F., 4 3·9-in., 4 1·8-in., 8 1·4-in., 6 M.	7	20·4 *t*	480	246
3rd cl. *cr.*	Surcouf	S.	2044	312	0	30	5	14	0	2	6000	Cherbourg	1888	131,200	..	1½	4 5·5-in. Q.F., 8 other Q.F., 4 M.	5	20·5	200	190
g. v.	Surprise	S.	627	184	8	24	7	12	8	1	853 *t*	Havre	1895	50,954	..	..	2 3·9-in. Q.F., 4 2·5-in., 4 1·4-in.	..	18·4 *t*	73	99
1st cl. *cr.*	Tage	S. shd.	7589	390	0	53	8	22	10	2	12,410	St. Nazaire	1886	93,857	..	..	8 6·4-in., 10 5·5-in., 2 2·5-in., 6 Q.F., 14 M.	7	19·0	1000	400
2nd cl. *cr.*	Tourville*	I. & W.	5576	333	5	50	3	25	4	1	7466	La Seyne	1876	271,499	..	..	7 6·4-in. Q.F., 14 5·5-in., 8 M.	..	16·89	800	550
3rd cl. *cr.*	Troude	S.	2026	311	6	31	2	14	0	2	6000	Bordeaux	1888	83,383	..	1½	4 5·5-in. Q.F., 8 other do., 4 M.	5	20·9	200	190
t. g. b.	Vautour	S.	1235	216	6	29	3	15	5	2	3391	Toulon	1886	87,733	..	1½	5 3·9-in. Q.F., 1 2·5-in. do., 6 M.	5	17·3	150	134
g. v.	Vipère	C.	486	145	4	23	10	10	6	1	441	Rochefort	1881	26,835	..	..	2 5·5-in., 2 3·9-in.	.	10·3	60	80
"	Voltigeur	W. & I.	943	199	5	28	5	12	7	1	999 B	Brest	1878	23,077	..	..	4 5·5-in., 4 M.	..	12·48	150	116
t. g. b.	Wattignies	S.	1292	230	0	29	3	15	0	2	4189	Rochefort	1891	111,000	..	..	5 3·9-in. Q.F., 6 1·8-in., 7 1·4-in., M.	4	18·61	160	180
g. v.	Zelée	S.	646	185	6	26	0	10	6	1	1000 Nic.	Rochefort	1899	..	..	..	2 3·9-in. Q.F., 4 2·5-in., 4 1·4-in.	..	13·0	..	75

* New engines, 1893. † New armament.

Shallow-draught gunboats Argus and Vigilante launched at Chiswick (Thornycroft) 1900:—displacement, 122 tons; length, 145 ft.; beam, 24 ft.; draught, 2 ft.; 2 screws; 550 I.H.P.; 13 knots; 2 3·5-in., 4 1·4-in. Q.F. guns; complement, 30; coal capacity, 80.

Merchant Cruisers (Auxiliary to French Navy).

To what Company belonging.	Name.	Register Tonnage.	Length.	Beam.	Depth.	H.P. (nominal.)	Speed.	When built.
		Tons.	Feet.	Feet.	Feet.		Knots.	
Compagnie Générale Transatlantique	La Touraine	8893	520·2	56·0	34·6	1616	19	1890
	Duc de Bragance	2096	334·6	34·2	16·8	426	17½	1889
	Eugène Pereire	2078	334·6	35·1	23·9	437	17½	1888
	Général Chanzy	2299	341·2	35·7	15·5	478	17½	1891
	La Bretagne	7112	495·4	51·8	34·5	1149	17½	1886
	La Champagne	7087	493·4	51·8	34·5	1149	17½	1885
	La Gascogne	7395	495·4	52·2	34·8	1308	17½	1886
	Maréchal Bugeaud	2206	342·5	34·1	23·0	482	17½	1890
	Ville d'Alger	2211	342·7	36·1	23·0	208	17½	1890
	La Navarre	6648	471·0	50·5	36·4	983	17	1892
	La Normandie	6283	459·3	49·2	34·1	1147	16	1882
	Ville de Tunis	1966	317·3	34·6	16·8	444	15½	1884
	Moïse	1873	310·0	33·5	16·7	443	15	1880
	St. Augustin	1854	314·0	33·8	16·5	443	15	1880
	Versailles	4336	373·7	45·3	27·0	780	..	1882
	Ville de Madrid	1874	308·7	33·5	16·7	370	15	1880
	Ville de Naples	1879	311·6	34·1	16·7	506	15	1881
Messageries Maritimes	Armand Béhic	6467	486·6	50·1	36·8	821	17½	1892
	Australien	6428	482·3	49·2	34·1	818	17½	1889
	Polynésien	6506	482·3	49·2	34·1	818	17½	1890
	Ville de la Ciotat	6461	485·8	49·9	36·8	819	17½	1892
	Ernest Simons	4562	442·9	47·1	36·7	727	..	1893
	Indus	6357	446·2	50·8	36·1	417	..	1897
	Brésil	5876	463·9	46·4	32·5	743	16½	1889
	Chili	6375	462·6	47·6	36·7	719	..	1894
	Cordillère	6379	462·6	47·6	36·1	721	..	1895
	La Plata	5807	462·6	45·9	32·5	520	16½	1889

NOTE.—The armament for the larger ships is 7 5·5-in. and smaller quick-firers.

GERMANY.—Armoured Ships.

Class.	NAME.	Material.	Displacement.	Length.	Beam.	Mean Draught.	Propellers.	Indicated Horse-power.	Where Built.	Date of Launch.	Cost.	Armour. Belt.	Armour. Gun Position	Armour. Deck Plating	Armament. Guns.	Armament. Torpedo Tubes.	Speed.	Normal Coal Supply.	Complement.
			metric tons.	ft. in.	ft. in.	ft. in.	no.				£	ins.	ins.	ins.			knots.	tons.	
c. d. s.	Aegir	S.	3600	236 6	50 6	17 9	2	4800 T.	Kiel	1895	233,500	9½ H.S.	8¾ H.S.	2	3 9·4-in., 10 3·4-in. Q.F. 6 M.	3 (1 sub.)	15·0	225 †	255
a. c.	B.	S.	8868	396 0	64 3	25 3	3	15000 Dürr.		Bldg.		4 K.S.	6 K.S.	2¾	2 9·4-in., 10 5·9-in. Q.F., 10 3·4-in., 10 1·4-in., 4 M.	4 (3 sub.)	20·5	950	528
b.	Baden	I.	7441	321 6	60 0	15 8	2	6200 Dürr.	Kiel	1880 1897	444,886	16	10	3	6 10·2-in., 8 3·4-in. Q.F., 8 1·4-in., 1 l., 6 M.	5 (2 sub.)	14·0	700	376
a. g. b.	Basilisk	I.	1109	154 3	36 0	10 2	2	759	Bremen	1878	58,042	8	8	2	1 12-in., 2 3·3-in., 2 M.	2	9·0	40	76
b.	Bayern	I.	7441	321 6	60 0	19 8	2	6326 W.T.	Kiel	1878 1896	406,660	16	10	3	6 10·2-in., 8 3·4-in. Q.F., 8 1·4-in., 1 l., 6 M.	5 (2 sub.)	14·0	700	376
c. d. s.	Beowulf	S.	3500	259 2	49 3	17 9	2	4800	Bremen	1890	175,000	9½	8	1¼	3 9·4-in., 8 3·4-in. Q.F., 6 M.	4	15·0	225	225
b.	Brandenburg	S.	10,100	354 4	65 0	24 7	2	9640	Stettin (Vulcan)	1891	606,500	15¾ comp.	11¾ comp.	2½	6 11-in., 6 4·1-in. Q.F., 8 3·4-in., 12 1·4-in., 8 M., 2 l.	6	16·5	750	552
a. g. b.	Biene	I.	1109	154 3	36 0	10 2	2	759	Bremen	1876	62,853								
a. g. b.	Camäleon	I.	1109	154 3	36 0	10 2	2	759	Bremen	1878	57,564	8	8	2	1 12-in., 2 3·3-in., 2 M.	2	10·0	40	76
"	Crocodil	I.	1109	154 3	36 0	10 2	2	759	Bremen	1879	57,237								
b.	Deutschland	I.	7319	280 0	62 4	24 7	1	5360	Poplar	1874	412,022	10	8	2	8 10·2-in., 7 5·9-in., 9 3·4-in. Q.F., 12 M., 2 l.	5	14·5	710	668

t.	Friedrich der Grosse	I.	6770	307	0	53	6	24	7	1	5400	Kiel	1874	365,170	9¾	8 br. 10 tur.	..	4 10·2-in., 2 6·6-in., 10 8·4-in. Q.F., 8 M., 2 l.	4	14·0	550	537
a. c.	Fürst Bismarck	S. shd.	10,650	393	8	66	9	26	0	3	14,000 Dürr.	Kiel	1897	..	7¾ H. S.	7¾ H. S.	3	4 9·4-in., 12 5·9-in. Q.F., 10 3·4-in., 10 1·4-in., 8 M.	6 (5 sub.)	19.0	1000†	565
c. d. s.	Frithjof	S.	3500	240	0	49	3	17	9	2	4800	Bremen	1891	175,000								
c. d. s.	Hagen*	S.	3500	240	0	49	3	17	9	2	4516	Kiel	1893	..	9¾ H. S.	7¾ H. S.	1¼	3 9·4-in., 8 3·4-in. Q.F., 6 M.	4	15·0	225	225
c. d. s.	Heimdall	S.	3500	240	0	49	3	17	9	2	4393	Wilhelmshaven	1892	233,500								
c. d. s.	Hildebrand	S.	3500	240	0	49	3	17	9	2	4413	Kiel	1892	218,000								
a. g. b.	Hummel	I.	1109	143	0	36	0	10	2	2	759	Bremen	1881	56,741	8	8	2	1 12-in., 2 3·3-in., 2 M.	2	10·0	40	76
b.	Kaiser	I.	7531	292	0	62	4	24	7	1	5700	Poplar	1874	411,301	10	10	2	8 10·2-in., 1 5·9-in., 6 4-in., 9 3·4-in. Q.F., 2 M., 2 l.	5	14·6	710	668
t	Kaiser Friedrich III. §											Wilhelmshaven	1896									
t	Kaiser Wilhelm II.											Wilhelmshaven	1897									
t	Kaiser Wilhelm der Grosse (Ersatz König Wilhelm) (A)	S.	11,000	377	4	65	8	25	8	3	13,000 Schulz	Kiel (Germania)	1899	706,000	11¾ H. N. S.	9¾–6 H. N. S.	3	4 9·4-in. Q.F., 18 5·9-in. Q.F., 12 3·3-in., 12 1·4-in., 8 M.	6 (5 sub.)	18·0	650 / 1000	700
t	Kaiser Barbarossa											Danzig (Schichau)	1900									
t	Kaiser Karl der Grosse (B)											Hamburg (Blohm & Voss)	1899									
t	C. D. E. F. G.	S.	11,700	393	8	68	10	24	10	3	15,000 W. T.	Wilhelmshaven, Kiel, Stettin	Bldg.		9–4 K.S.	10–6 K.S.	3	4 9·4-in. Q.F., 18 5·9-in. Q.F., 12 3·3-in., 12 1·4 in., 8 M.	6 (5 sub.)	18·0	650 / 1000	700
b	König Wilhelm	I.	9757	355	0	60	0	26	7	1	8350	Blackwall	1868 1896	505,141	12	6	2½	20 5·9-in. Q.F., 18 3 4-in., 8 M., 4 l.	5	14·7	700	759
b.	Kurfürst Friedrich Wilhelm.	S.	10,100	354	4	65	0	24	7	2	9959	Wilhelmshaven	1891	653,000	15¾ comp.	11¾ comp.	2½	6 11-in., 6 4·1-in. Q.F., 8 3·4-in., 12 1·4-in., 8 M., 2 l.	6	16·0	750	552
a. g. b.	Mücke	I.	1109	154	3	36	0	10	2	2	759	Bremen	1877	60,960	8	8	2	1 12-in., 2 3·3-in., 2 M.	2	10·0	40	76

* The Hagen is being reconstructed, and will be lengthened 25 ft. by the addition of a new middle section. † Also liquid fuel. § Kaiser Friedrich III, 8 cylindrical and 4 Thornycroft boilers.

GERMANY.—Armoured Ships—*continued*.

Class.	NAME.	Material of Hull.	Displacement.	Length.	Beam.	Draught.	Propellers.	Indicated Horse-Power.	Where Built.	Date of Launch.	Cost.	Armour. Belt.	Armour. Gun Position.	Armour. Deck Plating.	Armament. Guns.	Armament. Torpedo Tubes.	Speed.	Normal Coal Supply.	Complement.
			metric tons.	ft. in.	ft. in.	ft. in.	no.				£	in.	in.	in.			knots.	tons.	
a. g. b.	**Natter**	I.	1109	154 3	36 0	10 2	2	759	Bremen	1880	52,822	8	8	2	1 12-in., 2 3·3-in., 2 M.	2	10·0	40	76
a. d. s. b.	**Odin**	S.	3600	236 6	50 6	17 9	2	4800	Danzig	1894	..	9½ H.S.	8¾ H.S.	2	3 9·4 in., 10 3·4-in. Q.F., 6 M.	8 (1 sub.)	15·0	225†	266
b.	**Oldenburg**	S.	5200	246 0	59 0	19 6	2	3900	Stettin	1884	235,342	13 comp.	8 comp.	1	8 9·4-in. (Krupp), 2 3·4-in. Q.F., 6 M.	4	13·5	475	356
t.	**Preussen**	I.	6770	308 6	53 6	24 7	1	4383	Stettin	1878	351,904	9	8 10	..	4 10·2-in. (Krupp), 2 6·6-in., 10 3·4-in. Q.F., 6 M., 2 1.	4	14·0	550	537
a. c.	**Prinz Heinrich (A)**	S.	8868	396 0	64 3	25 3	3	15000 Dürr.	Kiel	1900	..	4 K.S.	6 K.S.	2¾	2 9·4-in., 10 5·9-in. Q.F., 10 3 4-in., 10 1·4-in., 4 M.	4 (3 sub.)	20·5	950	528
b.	**Sachsen**	I.	7441	321 6	59 0	21 0	2	6000 W. T.	Stettin	1877 1896	422,178	15¾	15¾	3	6 10·2-in. (Krupp), 8 3·4-in. Q.F., 8 1·4-in., 1 1., 6 M.*	5 (3 sub.)	14·0	700	376
a. d. s.	**Siegfried**	S.	3500	240 0	49 3	17 9	2	4800	Kiel (Germania)	1889	175,000	9½	7¾	1¼	3 9·4-in., 6 3·4-in. Q.F., 6 M.	4	15·0	225†	225
a. g. b.	**Salamander**	I.	1109	154 3	36 0	10 2	2	759	Bremen	1880	56,914								
”	**Skorpion**	I.	1109	154 3	36 0	10 2	2	759	Bremen	1877	60,796	8	8	2	1 12-in., 2 3·3-in., 2 M.	2	10·0	40	76
”	**Viper**	I.	1109	154 3	36 0	10 2	2	759	Bremen	1876	61,463								
”	**Wespe**	I.	1109	154 3	36 0	10 2	2	759	Bremen	1876	53,771								
b.	**Weissenburg**	S.	10,100	354 4	65 0	24 7	2	9000	Stettin (Vulcan)	1891	659,475								
b.	**Wörth**	S.	10,100	354 4	65 0	24 7	2	10,224 (t)	Kiel	1892	595,250	15¾	11¾	2½	6 11-in., 6 4·1-in. Q.F., 8 3·4-in., 12 1·4-in., 8 M., 2 1.	6	16·0‡	750†	552
b.	**Württemberg**	I.	7441	321 6	60 0	19 8	2	6000 Dürr	Stettin	1878 1898	402,512	15¾	10	3	6 10·2-in. (Krupp), 8 3·4-in. Q.F., 8 1·4-in., 1 1., 6 M.*	5 (2 sub.)	14·0	700	376

* New armament. † Also liquid fuel. ‡ Wörth: trial, 17·2 knots.

The Arminius, Friedrich Carl, and Kronprinz are now used for harbour service.

GERMANY—Cruising Ships.

Class.	NAME.	Material of Hull.	Displacement.	Length.	Beam.	Draught.	Propellers.	Indicated Horse-power.	Where Built.	Date of Launch.	Cost.	Armour. Gun Position.	Armour. Deck.	Armament. Guns.	Armament. Torpedo Tubes.	Speed.	Normal Coal Supply.	Complement.
			metric tons.	ft. in.	ft. in.	ft. in.	no.				£	inches.	inches.			knots.	tons.	
" "	**Alexandrine**	I. S. & W.	2373	236 3	42 7	18 4	1	2400	Kiel	1885	102,877	..	..	10 5·9-in., 4 4·1-in., 10 M., 1 l.	1	14·0	..	267
" "	**Arcona**	I. S. & W.	2373	236 3	42 7	18 4	1	2400	Danzig	1885	109,875	..	..	10 5·9-in., 4 4·1-in., 10 M., 1 l.	1	14·0	..	267
" "	**Blitz**	S.	1382	246 0	32 10	13 5	2	2839	Kiel	1882	66,935	..	..	6 3·4-in., 4 M.	1	16·0	250	127
cr.	**Blücher**	I. & W.	2856	244 4	44 10	19 8	1	2990	Kiel	1877	136,408	..	..	2 5·9-in., 6 M.	7	14·0	400	206
g. b.	**Bremse**	S.	866	203 5	27 10	10 6	1	1500	Bremen	1884	49,308	..	2½	1 8·2-in.	..	15·0	65	73
g. b.	**Brummer**	S.	866	203 5	27 10	10 6	1	1500	Bremen	1884	52,422	..	2¼	1 8·2-in.	..	15·0	65	73
3rd cl *cr.*	**Bussard**	S.	1857	256 0	30 2	18 4	2	2900	Danzig	1890	..	..	3	8 4·1-in. Q.F., 7 M.	2	16·5	400	..
" "	**Carola***	I. & W.	2169	226 4	42 7	18 4	1	2100	Stettin	1880	109,617	..	..	6 5·9-in., 2 4·1-in. Q.F., 8 3·4-in.	..	14·0	250	267
" "	**C. D. E. F.†**	S.	2600	328 0	38 7	15 1	2	8000 Schulz	.. ..	Bldg.	167,500	..	2	10 4·1-in. Q.F., 14 1·4-in., 4 M., 2 l.	2 (sub.)	20	550	210
" "	**Falke**	S. & W.	1731	246 0	33 6	15 0	2	2900	Kiel	1891	..	..	3	8 4·1-in. Q.F., 7 M.	2	15·5	400	..
2nd cl. *cr.*	**Freya**	S.	5650	344 5	57 0	20 8	3	10,000 Nic.	Danzig	1897	..	4 N.S.	4 N.S.	2 8 2-in. Q.F., 8 6-in. Q.F., 10 3·4-in., 10 1·4-in., 4 M.	3 (sub.)	20	500†	440
3rd. cl. *cr.*	**Gazelle**	S. & bronze shd.	2650	328 0	38 7	16 9	2	6000 Nic.	Kiel (Germania)	1898	130,000	..	2	10 4·1-in. Q.F., 14 1·4-in., 4 M., 2 l.	3 (1sub.)	19·0	..	210
2nd "	**Gefion**	S.	4207	344 6	42 8	20 8	2	9000	Danzig (Schichau)	1893	..	..	1½	10 4·1-in. Q.F., 6 2·1-in., 1 l., 8 M.	2	19·0	950	312
3rd "	**Geier**	S. & W. shd.	1776	249 4	34 10	15 6	2	2960	Wilhelmshaven	1894	..	..	3	8 4·1-in. Q.F., 7 M.	2	16·2	400	..
" "	**Greif**	S.	2000	318 0	32 0	14 9	2	5400	Kiel	1886	..	..	..	2 3·4-in. Q.F., 4 M.	..	19·0	..	130
g. v.	**Habicht**	I. & W.	848	174 0	29 6	11 5	1	600	Elbing	1879	33,054	..	..	5 4·9-in., 5 M.	..	12·0	..	128
2nd cl. *cr.*	**Hansa**	S. shd.	5900	345 7	57 10	21 8	3	10,000 B	Stettin (Vulcan)	1898	..	4 N.S.	4 N.S.	2 8·2-in. Q.F., 8 6-in., 10 3·4-in., 10 1·4-in., 4 M.	3 (sub.)	19·0	500	440

* Gunnery ship for quick-firing guns.

† Of these, C is in hand at the Imperial Yard, Danzig, D and E are at the Weser Yard, Bremen; and F is at the Germania Yard, Kiel.

GERMANY.—Cruising Ships—*continued*.

Class.	NAME.	Material of Hull.	Displacement.	Length.	Beam.	Draught.	Propellers.	Indicated Horse-power.	Where Built.	Date of Launch.	Cost.	Armour. Gun Position.	Armour. Deck.	Armament. Guns.	Armament. Torpedo Tubes.	Speed.	Normal Coal Supply.	Complement.
			metric tons.	ft. in.	ft. in.	ft. in.	no.				£	in.	in.			knots	tons.	
d.v.	**Hela**	S.	2000	328 0	36 0	..	2	5860	Bremen	1895	..	..	1¼	4 3·4-in. Q.F., 6 1·9-in., 2 M.	3	20·0	347	169
2nd cl. *cr.*	**Hertha**	S.	5650	344 6	57 0	21 8	3	10,000 B	Stettin (Vulcan)	1897	..	4	4 N.S.	2 8·2-in. Q.F., 8 6-in., 10 3·4-in., 10 1·4-in., 4 M.	3 (sub.)	20·0	500†	440
g. b.	**Hyäne**	I.	489	139 8	25 1	9 10	1	340	Wilhelmshaven	1878	24,340	..	..	1 4·9-in., 1 3·4-in., 4 M.	..	9·0	..	83
g. b.	**Iltis**	S. (shd.)	895	208 6	29 10	10 8	2	1300	Danzig (Schichau)	1898	100,000	..	..	8 3·4-in. Q.F., 6 1·4-in., 2 M.	..	13·5	160	121
2nd cl. *cr.*	**Irene**	S.	4400	308 0	46 0	21 0	2	8000	Stettin	1887	220,000	..	3	4 5·9-in., 8 4·1-in. Q.F., 6 1 9-in., 1 l., 8 M.	4	19·8	900	358
to. g. b.	**Jagd**	S.	1250	275 6	31 6	13 9	2	4000	Bremen	1888	..	..	2	4 3·4-in. Q.F., 2 M.	3	20·0	..	126
g. b.	**Jaguar**	S. (shd.)	895	203 6	29 10	10 8	2	1300	Danzig (Schichau)	1898	100,000	..	..	8 3·4-in. Q.F., 6 1·4-in., 2 M.	..	13·5	160	..
1st cl. *cr.*	**Kaiserin Augusta**	S. (shd.)	6331	387 0	52 6	23 0	3	14,000	Kiel (Germania)	1892	..	..	3½	12 5·9-in., 8 3·4-in. Q.F., 2 l., 8 M.	5	21·0	800	427
to. g. b.	**Komet**	S.	946	262 6	31 2	13 9	2	5000	Stettin	1892	..	..	2	4 3·4-in. Q.F., 2 M.	1	21·0 (*t*)	..	90
3rd cl. *cr.*	**Kondor**	S. & W.	1640	246 0	33 6	15 0	2	2930	Hamburg	1892	..	..	3	8 4·1-in. Q.F., 7 M.	2	16·5	400	..
3rd cl. *cr.*	**Kormoran**	S. & W.	1640	246 0	33 6	15 0	2	2930	Danzig	1892	..	..	3	8 4·1-in. Q.F., 7 M.	2	16·0	400	..
cr.	**Marie**	I. & W.	2100	226 4	42 7	18 4	1	2100	Hamburg	1881	..	..	..	8 5·9-in., 2 3·4-in. Q.F., 1 l., 6 M.	..	13·5	..	267
d. v.	**Meteor**	S.	946	262 6	29 6	11 6	2	4500	Gaarden	1890	..	..	2	2 3·4-in. Q.F., 2 M.	3	21·0	..	90
3rd cl. *cr.*	**Niobe**	S.	2600	328 0	38 7	15 1	2	8000 T.	Bremen (Weser)	1899	167,500	..	2	10 4·1-in. Q.F., 14 1·4-in., 4 M., 2 l.	2 (1sub.)	21·5	..	250
cr.	**Nixe §**	I. & W.	1760	177 2	42 8	18 0	1	700	Danzig	1885	..	..	..	8 5·9-in.	..	10·5	..	116
3rd cl. *cr.*	**Nymphe**	S.	2600	328 0	38 7	15 1	2	8000 Schulz	Kiel (Germania)	1899	167,500	..	2	10 4·1-in. Q.F., 14 1·4-in., 4 M., 2 l.	2 (1sub.)	21·5	550	210
cr.	**Olga**	I. & W.	2100	226 4	42 7	18 4	1	2100	Stettin	1880	113,812	..	..	8 5·9-in., 2 3·4-in. Q.F., 1 l., 6 M.	..	14·0	320	267

cr.	**Pelikan** (mining ship)	S.	2360	259 0	38 0	14 5	2	3000	Kiel	1890	..	..	..	4 3·4-in. Q.F., 4 M.	..	15·4	370	183
3rd cl. *cr.*	**Pfeil**	S.	1382	246 0	32 10	13 5	2	2700	Wilhelmshaven	1882	73,605	..	..	4 3·4-in. Q.F., 4 M.	1	16·0	250	127
2nd cl. *cr.*	**Prinzess Wilhelm**	S.	4400	339 6	46 0	21 0	2	8000	Gaarden	1887	220,000	..	3	4 5·9-in., 8 4·1-in. Q.F., 6 1·9-in., 1 l., 8 M.	4	18·7	900	358
cr.	**Schwalbe**	S. & W.	1120	203 0	30 6	12 4	2	1500	Wilhelmshaven	1887	..	..	3	8 4·1-in. Krupp, 7 M.	..	13·5	300	150
3rd cl. *cr.*	**Seeadler**	S. & W.	1640	246 0	33 6	15 0	2	2800	Hamburg	1892	..	..	3	8 4·1-in., 7 M.	2	16·0	400	156
„ „	**Sophie**	I. & W.	2100	226 4	42 7	18 4	1	2100	Danzig	1892	117,155	..	..	8 5·9-in., 2 3·4-in. Q.F., 1 l., 6 M.	..	14·0	320	267
4th cl. *cr.*	**Sperber**	S. & W.	1120	236 0	29 8	12 6	2	1500	Wilhelmshaven	1888	..	..	3	8 4·1-in., 6 M.	..	13·5	300	150
g. b.	**Tiger**	S.	894	203 6	29 10	10 8	2	1300	Danzig	1899	..	..	..	8 3·4-in. Q.F., 6 1·4-in., 2 M.	..	13·5	160	..
g. b.	**"A."**								Danzig	Bldg.								
2nd cl. *cr.*	**Victoria Luise**	S.	5650	344 5	57 0	21 8	3	10,000 Dürr.	Bremen	1897	..	4 H.S.	4 H.S.	2 8·2-in. Q.F., 8 6-in. Q.F., 10 3·4-in., 10 1·4-in., 4 M.	3 sub.	19·1	500	440
2nd cl. *cr.*	**Vineta**	S. shd.	5900	345 7	57 10	21 8	3	10,000 Dürr.	Danzig							18·0	500	440
3rd cl. *cr.*	**Wacht**	S.	1250	262 0	31 6	13 9	2	4000	Bremen	1887	..	..	2	4 3·4-in. Q.F., 2 M.	3	19·6	230	126
g. b.	**Wolf**	I.	480	139 8	25 1	9 10	1	340	Wilhelmshaven	1878	24,343	..	..	1 4·9-in., 1 3·4-in., 4 M.	..	9·0	110	83
g. b.	**Wolf** (Ersatz)	S.	894	203 6	29 10	10 8	2	1300	Danzig	Bldg.	..	..	..	8 3·4-in. Q.F., 6 1·4-in., 2 M.	..	13·5	160	..
d. v.	**Zieten**	I.	975	196 10	29 6	11 6	1	2323	Blackwall	1876	81,755	..	..	4 1·9-in. Q.F., 6 M.	..	16·0	140	111
g. b.	**Zieten** (Ersatz)	S.	894	203 6	29 10	10 8	2	1300	Danzig	Bldg.	..	..	..	8 3·4-in. Q.F., 6 1·4-in., 2 M.	..	13·5	160	..

S

The Charlotte, Mars, Grille, Hay, Ulan, Gneisenau, Moltke, Stein and Stosch, in addition to others given in the list, are used as schoolships.
The Imperial Yacht Hohenzollern, 4187 tons, 9460 I.H.P., 22 knots, carries 8 1·9-in. Q.F., but provision is made for mounting 3 4·1-in., 12 1·9-in. Q.F. and 4 M.
A station vessel for Constantinople has been bought and named Loreley, the older ship having been removed from the list.

† Displacement with 950 tons of coal, 6100 tons. Provision made for liquid fuel.

§ Training Ship.

Merchant Cruisers (Auxiliaries to the German Navy).

To what Company belonging.	Name of Ship.	Displacement.	Length.		Beam.		Draught of Water.		Indicated H.P.	Ocean Speed.	When Built.	Armament of each Ship.
		tons.	ft.	in.	ft.	in.	ft.	in.		knots.		
Hamburg-American S.S. Co.	Fürst Bismarck . .	10,500	502	0	57	6	22	3	16,400	19½	1891	8 5·9-in., 4 4·7-in., 2 3·4-in. Q.F., 2 2·2-in., 14 M.
	Augusta Victoria . .	9,500	459	3	56	0	23	0	12,280	18	1889	
North German Lloyd	Spree	8,900	462	6	51	10	22	0	12,770	19	1890	
	Lahn	7,700	449	6	49	0	22	0	9,500	18½	1887	
	Kaiser Wilhelm der Grosse	20,000	625	0	66	0	27	0	27,000	22	1897	Not known.
	Kaiser Friedrich III. .	17,000	580	9	63	11	..		25,000	22	1897	
	Aller	4,965	436	6	48	0	..		1,300(a)	16	1885	
	Saale	4,965	436	6	48	0	..		1,300(a)	16	1886	
	Trave	4,965	436	6	48	0	..		1,300(a)	16	1886	

(a) Nominal horse-power.

GREECE.—Armoured Ships.

Class.	NAME.	Material of Hull.	Displacement.	Length.		Beam.		Draught.		Propellers.	Indicated Horse-power.	Where Built.	Date of Launch.	Cost.	Armour.			Armament.		Speed.	Normal Coal Supply.	Complement.
															Belt.	Battery.	Deck Plating.	Guns.	Torpedo Tubes.			
			metric tons.	ft.	in.	ft.	in.	ft.	in.	no.					inches.	inches.	ins.			knots.	tons.	
c.d.s.	**Basileos Georgios**	I.	1774	200	2	36	0	15	6	2	2100	Blackwall	1867	..	7	6	9	2 6·6-in. (Krupp), 1 5·9-in. 9 M.	1	12 0	210	120
br.	**Basilissa Olga*** .	W.	2030	230	0	59	0	18	0	1	1950	San Rocco	1869	..	6	4½	..	4 6·6-in. 5½-ton (Krupp), 2 6·6-in. 3½-ton, 4 M., 4 l. .	3	10·0	240	400
b.	**Hydra** . .	S.	4885	334	6	51	10	23	3	2	7000	St. Nazaire	1889	..	11¾	13½	2½	3 10·6-in., 5 5·9-in., 7 2·2-in., 16 M.	3	17·0	600	400
b.	**Psara** . . .	S.	4885	334	6	51	10	23	3	2	7000	Havre . .	1890	..	11¾	13½	2½	3 10·6-in. Canet, 5 5·9-in., 1 3·9-in., 8 2·5-in., 4 1·8-in., 12 1·4-in.	3	17·0	600	400
b.	**Spetsai** . .	S.	4885	334	6	51	10	23	3	2	7000	Havre . .	1889	..	11¾	13½	2½	3 10·6-in., 5 5·9-in., 7 2·2-in., 16 M.	3	17 0	600	400

* Has received two fighting masts and new machinery; similar changes in the Georgios.

The Hydra and Spetsai are intended to receive 1 3·9-in. Q.F. and 8 2·5-in. Q.F. guns (Canet), in addition to the present armament, but the transformation has been deferred.

GREECE.—Cruising Ships.

Class.	NAME.	Material of Hull.	Displacement.	Length.		Beam.		Draught.		Propellers.	Indicated Horse-power.	Where Built.	Date of Launch.	Cost.	Armour.		Armament.		Speed.	Normal Coal Supply.	Complement.
															Gun Position.	Deck.	Guns.	Torpedo Tubes.			
			Metric tons.	ft.	in.	ft.	in.	ft.	in.	no.									knots.	tons.	
g.v.	**Acheloos** . .	S.	420	130	0	24	6	11	6	1	400	Blackwall	1884	..	..	..	2 3·7-in. (Krupp), 3 M. .	..	10·0	50	..
g.v.	**Alphios** . . .	S.	420	130	0	24	6	11	6	1	400	Blackwall	1884	..	..	..	2 3·7-in. (Krupp), 3 M. .	..	10·0	50	..
g.v.	**Aphroessa** . .	I.	380	124	7	22	11	9	10	1	160	Pt. Glasgow	1858	..	..	..	1 3·4-in. (Krupp) .	..	9·0	30	..
g.v.	**Eurotas** . . .	S.	420	130	0	24	6	11	6	1	400	Dumbarton	1884	..	..	..	2 3·7-in. (Krupp), 3 M. .	..	10·0	50	..
corv.	**Hellas** (training) .	W.	1654	200	2	37	0	19	4	1	1500	Northfleet	1858 rep. 1878–90	..	..	..	6 5·9-in. (Krupp), 2 M.	..	11·0	230	..
cr.	**Mykale** (transport) .	S.	1000	210	6	32	6	18	0	2	2400	Glasgow	1880	..	..	..	2 M.	..	1·4	..	..
corv.	**Nauarchos Miaulis**	I. & W.	1800	246	0	36	0	14	5	1	2200	La Seyne	1879	..	..	..	3 6·6-in., 5½-ton (Krupp), 1 6·6-in. 3½-ton do., 2 M., 4 l.	..	15·0	220	250
g.v.	**Paralos** . . .	I.	380	123	0	23	11	9	10	1	204	Pt. Glasgow	1858	..	..	..	1 3·4-in. (Krupp), 1 M. .	..	8·0	60	..
g.v.	**Pinios** . . .	S.	420	130	0	24	6	12	6	1	400	Dumbarton	1884	..	..	..	2 3·7-in. (Krupp), 3 M. .	..	10·55	50	..
g.v.	**Plixaura** . .	I.	380	124	7	22	11	9	10	1	160	Pt. Glasgow	1856	..	..	..	1 3·4-in. (Krupp) . .	..	9·0	55	..
g.v.	**Salaminia** . .	I.	380	123	0	23	11	9	10	1	200	Pt. Glasgow	1858	..	..	..	1 3·4-in. (Krupp), 1 M. .	..	8·0	60	..
corv.	**Sfaktirea** . .	S.	1000	216	6	29	3	18	0	1	2400	England	1885	..	..	..	2 3·9-in. (Krupp), 2 M. .	..	14·5	100	..
g.v.	**Syros** . . .	I.	380	124	7	22	11	9	10	1	160	Pt. Glasgow	1858	..	..	..	1 3·4-in. (Krupp) . .	..	9·0	18	100

Torpedo depôt-ship.—Kanaris, 1100 tons, 500 I.H.P., 2 3·9-in. (Krupp) guns, 2 Whitehead torpedo-launching guns on broadside, 2 under-water torpedo tubes ahead; 14 knots speed.

There are also 2 gunboats, Ambrakia and Aktion, of 410 tons displacement, 380 horse-power, 10 knots speed, fitted with 1 10·2-in. Krupp gun and 2 machine guns; launched 1885; 4 gunboats, Α. Β. Γ. Δ. (52 tons, 1 4·7-in. Krupp), launched 1881; and 3 mining vessels (300 tons), launched 1881.

ITALY.—Armoured Ships.

Class.	NAME.	Material of Hull.	Displacement.	Length.		Beam.		Draught.		Propellers.	Indicated Horse-power.	Where Built.	Date of Launch.	Cost.	Armour. Belt.	Armour. Gun Position	Armour. Deck Plating.	Armament. Guns.	Armament. Torpedo Tubes.	Speed.	Normal Coal Supply.	Complement.
			Metric tons.	ft.	in.	ft.	in.	ft.	in.	no.				£	inches.	inches.	ins.			knots.	tons.	
b.	A. B.	S.	8000	400	3	63	1	22	0	..	19,000 W.T.	Spezia. / Castellamare	Pro.	..	6 H.S.	6 H.S.	1½	12 8-in. Q.F., 12 3-in.	4 (2 sub.)	23·0	600 / 2100	..
t.	Affondatore	I.	4062	290	0	40	0	20	0	1	3240	Millwall	1865	197,600	5	5	2	2 28-ton (Armstrong), 6 4·7-in. Q.F., 2 2·9-in., 4 2·2-in., 4 1·4-in., 2 M.	2	12·0	460	303
"	Ammiraglio di St. Bon	S.	9800	344	6	69	4	24	9	2	13,500	Venice	1897	..	9¾–4 H.S.	9¾ H.S.	3–1½	4 10-in., 8 6-in. Q.F., 8 4·7-in., 2 2·9-in., 8 2·2-in., 12 1·4-in., 2 M.	8	18·0	600	548
a.c.	Ancona	I.	4460	256	0	50	0	25	0	1	2548	Bordeaux	1864	172,000	4½	4½	..	6 6-in. Q.F., 6 4·7-in., 2 2·9-in., 8 2·2-in., 12 1·4-in., 2 M.	3	12·0	485	423
b.	Andrea Doria	S.	11,000	328	2	65	4	27	2	2	10,500	Spezia	1885	765,500	18 comp.	18 comp.	3	4 105-ton (Armstrong), 2 6-in. 4 4·7-in. Q.F., 2 2·9-in., 10 2·2-in., 17 1·4-in., 2 M.	5 (2 sub.)	16·1	850	509
b.	Benedetto Brin	S.	13,427	426	6	78	2	27	4	2	19,000 B.	Castellamare	Bldg.	..	6 H.S.	8 H.S.	3	4 12-in., 12 6-in. Q.F., 18 3-in., 8 1·8-in., 4 M.	4 (sub.)	20·0	1000	700
a.c.	Carlo Alberto	S.	6500	325	0	59	0	22	11	2	13,220 (t)	Spezia	1896	..	6 H.S.	6 H.S.	1½	12 6-in. Q.F., 6 4·7-in., 2 2·9-in., 10 2·2-in., 10 1·4-in., 2 M.	5	19·2 (t)	1000	460
a.c.	Castelfidardo	I.	4250	256	0	50	0	21	11	1	2125	St. Nazaire	1863	233,000	4½	4½	13¾	6 6-in. Q.F., 6 4·7-in., 2 2·9-in., 8 2·2-in., 12 1·4-in., 2 M.	3	12·0	485	423
t.	Dandolo *	I & S	11,202	340	11	64	9	26	7	2	8045	Spezia	1878 1898	872,640	21½	18	2	4 10-in. (Armstrong), 7 6-in. Q.F., 5 4·7-in., 2 2·9-in., 10 2·2-in., 14 1·4-in., 2 M.	4	15·6	1000	487
t.	Duilio	I & S	11,138	340	11	64	9	26	7	2	7710	Castellamare	1876	850,400	21½	18	2	4 100-ton M.L.R. (Armstrong), 3 4·7-in. Q.F., 2 2·9-in., 8 2·2-in., 22 1·4-in., 2 M.	4	15·0	1000	487
t.	Emanuele Filiberto	S.	9800	344	6	69	4	24	9	2	13,500	Castellamare	1897	..	9¾–4 H.S.	9¾ H.S.	3–1	4 10-in., 8 6-in. Q.F., 8 4·7-in., 2 2·9-in., 8 2·2-in., 12 1·4-in., 2 M.	8	18·0	600	536
b.	Francesco Morosini	S.	11,000	328	2	65	4	27	2	2	10,000	Venice	1885	770,680	18 comp.	18 comp.	3	4 105-ton (Armstrong), 2 6-in., 4 4·7-in. Q.F., 2 2·9-in., 10 2·2-in., 17 1·4-in., 2 M.	5 (2 sub.)	17·0	850	509
a.c.	Francesco Ferrucio										13,500 Nic.	Venice	Bldg.									
a.c.	Giuseppe Garibaldi	S.	7398	344	0	59	9	23	4	2	13,500 Nic.	Sestri-Ponente (Ansaldo)	1899	..	6 H.S.	6 H.S.	1½	1 10-in., 2 8-in. Q.F., 14 6-in., 10 2·9-in., 6 1·8-in., 2 M.	4 (sub.)	20·0	655 / 1200	540

* New armament given. The reconstruction of the Duilio is not likely to be proceeded with.

ITALY.—Armoured Ships—*continued.*

Class.	NAME.	Material of Hull.	Displacement.	Length.		Beam.		Draught.		Propellers.	Indicated Horse-power.	Where Built.	Date of Launch.	Cost.	Armour. Belt.	Armour. Gun Position.	Armour. Deck Plating.	Armament. Guns.	Armament. Torpedo Tubes.	Speed.	Normal Coal Supply.	Complement.
			metric tons.	ft.	in.	ft.	in.	ft.	in.	no.					inches.	inches.	inches.			knots.	tons.	
b.	Italia	S.	14,387	400	6	74	0	31	2	2	11,986 (*t*)*	Castellamare	1880	1,167,680	16 funnel op'nings	19 *comp.*	3	4 100-ton (Armstrong), 8 6-in., 4 4·7-in. Q.F., 12 2·2-in., 24 1·4-in., 2 M.	4	18·0	1650	748
b.	Lepanto	S.	14,400	400	6	74	0	31	2	2	15,800 (*t*)	Leghorn (Orlando)	1883	1,150,880	16 funnel op'nings	19 *comp.*	3	4 100-ton (Armstrong), 8 6-in., 4 4·7-in. Q.F., 12 2·2-in., 34 1·4-in., 2 M.	4	18·38	1650	748
a.c.	Marco Polo	S.	4583	327	0	48	3	19	6	2	10,543 (*t*)	Castellamare	1890	344,400	4	4	1	6 5·9-in. Q.F., 10 4·7-in., 2 2·9-in., 9 2·2-in., 4 1·4-in., 2 M.	5 (1 sub.)	19·0	..	315
a.c.	Maria Pia	I.	4268	256	0	49	4	22	7	1	2924	La Seyne	1863	215,000	4½	4½	..	8 5·9-in., 6 4·7-in. Q.F., 2 2·9-in. 10 2·2-in. 10 1·4-in., 2 M.	2	12·0	485	428
b.	Regina Margherita	S.	13,427	426	6	78	2	27	4	2	19,000 Nic.	Spezia	Bldg.	..	6 H.S.	8 H.S.	3	4 12-in., 12 6-in. Q.F., 18 3-in., 8 1·8-in., 4 M.	4 (sub.)	20·0	1000	700
b.	Re Umberto	S.	13,825	400	0	76	9	28	6	2	19,500	Castellamare	1888	1,058,500	4	18	3	4 67-ton (Armstrong), 8 6-in. Q.F., 16 4·7-in., 2 9-in., 15 2·2-in., 14 1·4-in., 2 M.	8	19·0	1200	785
b.	Ruggiero di Lauria	S.	11,000	328	2	65	4	27	2	2	10,600	Castellamare	1884	777,560	18 *comp.*	18 *comp.*	3	4 105-ton (Armstrong), 2 6-in. 4 4 7-in. Q.F., 2 2·9-in., 10 2·2-in., 17 1·4-in., 2 M.	5 (2 sub.)	17·0	850	509
a.c.	San Martino (training service)	I.	4268	256	0	49	4	22	7	1	2620	La Seyne	1863	213,880	4½	4½	..	8 5·9-in., 6 4·7-in. Q.F., 2 2·9-in., 10 2·2-in., 10 1·4-in., 2 M.	2	12·0	490	423
b.	Sardegna	S.	13,860	411	0	76	9	28	6	2	19,650 (*t*)	Spezia	1890	1,057,440	4	14½ *comp.*	3	4 67-ton (Armstrong), 8 5·9-in. Q.F., 16 4·7-in., 2 2·9-in., 20 2·2-in., 10 1·4-in., 2 M.	5	20·1 (*t*)	1200	785
”	Sicilia	S.	13,375	400	0	76	9	28	6	2	19,500	Venice	1891	1,050,000	4	18 *comp.*	3	4 67-ton (Armstrong), 8 5·9-in. Q.F., 16 4·7-in., 2 2·9-in., 20 2·2-in., 10 1·4-in., 2 M.	5	19·2	1200	785
a.c.	Varese	S.	7400	344	6	59	9	23	9	2	13,500 B	Leghorn (Orlando)	1899	..	6 H.S.	6 H.S.	1½	1 10-in., 2 8-in. Q.F., 14 6-in., 10 2·9-in., 6 1·8-in., 2 M.	4	20·0	650 / 1200	500
a.c.	Vettor Pisani	S.	6500	325	0	59	0	22	11	2	13,000	Castellamare	1895	..	6 H.S.	6 H.S.	1½	12 6-in. Q.F., 6 4·7-in., 2 2·9-in., 10 2·2-in., 10 1·4-in., 2 M.	5	20·0	600	..

Note.—The Palestro, Principe Amedeo, and Roma are non-effective, or only available for coast defence.

* The Italia is to have Niclausse water-tube boilers.

ITALY.—Cruising Ships.

Class.	NAME.	Material of Hull.	Displacement.	Length.		Beam.		Draught.		Propellers.	Indicated Horse-power.	Where Built.	Date of Launch.	Cost.	Armour.		Armaments.		Speed	Normal Coal Supply.	Complement.
															Gun Position.	Deck.	Guns.	Torpedo Tubes.			
			tons.	ft.	in.	ft.	in.	ft.	in.	no.				£	in.	in.			knots.	tons.	
to.cr .	Agordat . . .	S.	1313	287	6	30	6	11	1	2	8000 Bl.	Castellamare .	1899	..	..	1	4 4·7-in. Q.F., 8 2·2-in., 2 1·4-in.	2	23·0	260	158
3rd cl. *cr.*	Amerigo Vespucci (training)	S.	2795	255	11	42	7	17	0	1	3340	Venice .	1882	176,300	..	¾	6 5·9-in., 4 2·2-in., 8 1·4-in., 2 l., 4 M.	2	14·0	500	265
g.v. .	Andrea Provana .	S.	649	167	4	26	3	10	2	1	1080	Leghorn . (Orlando)	1884	39,760	..	..	4 4·7 in., 3 1·4-in., Q.F. .	..	13·0	120	103
d.v. .	Archimede . .	S.	784	230	0	26	3	10	0	1	1700	Venice .	1887	60,120	..	..	4 4·7-in., 2 2·2-in. Q.F., 2 1·4-in.	2	16·0	210	109
to.g.b. .	Aretusa . . .	S.	846	230	0	26	10	11	9	2	4420	Leghorn . (Orlando)	1891	72,920	..	1	1 4·7-in., 6 2·2-in., and 3 1·4-in.	6	20·7	180	111
3rd cl. *cr.*	Calabria . . .	S.	2442	249	4	42	0	16	7	2	4094 *t.*	Spezia .	1894	133,120	..	2	4 5·9-in. Q.F., 6 4·7-in., 1 2·9-in., 8 2·2-in., 8 1·4-in., 2 M.	2	16·4 *t.*	500	257
to.g.b. .	Calatafimi . .	S.	840	229	6	27	0	10	2	2	4000	Castellamare .	1893	72,920	..	1	1 4.7-in. Q.F., 6 2·2-in., 3 1·4-in.	6	20·0	120	111
" .	Caprera . . .	S.	853	230	0	27	4	10	2	2	4800	Leghorn . (Orlando)	1894	72,920	..	1	2 4·7-in. Q.F., 4 2·2-in., 2 1·4-in. Q.F.	5	21·0	180	111
g.v. .	Cariddi . . .	W.	1050	177	2	28	6	12	5	1	956	Castellamare .	1875	65,480	..	..	2 4·7-in., 4 2·2-in. Q.F. .	..	10·0	164	..
to.cr. .	Coatit . . .	S.	1313	287	6	30	6	11	1	2	8000 Bl.	Castellamare .	1899	..	..	1	4 4·7-in. Q.F., 8 2·2-in., 2 1·4-in.	2	23·0	260	158
to.g.b. .	Confienza . .	S.	768	230	0	25	6	9	0	2	1887	Spezia .	1887	61,480	..	1	1 4·7-in. Q.F., 6 2·2-in., 2 1·4-in.	4	17·0	180	111

ITALY.—Cruising Ships—*continued.*

Class.	NAME.	Material of Hull.	Displacement.	Length.		Beam.		Draught.		Propellers.	Indicated Horse-power.	Where Built.	Date of Launch.	Cost.	Armour.		Armament.		Speed.	Normal Coal Supply.	Complement.
															Gun Position.	Deck.	Guns.	Torpedo Tubes.			
			tons.	ft.	in.	ft.	in.	ft.	in.	no.				£	in.	in.			knots.	tons.	
3rd cl. *cr.*	**Cristoforo Colombo**	S.	2675	249	0	36	0	17	6	..	3800	Venice	1892 Reblt	157,240	..	..	6 4 7-in., 4 2·2-in., Q.F., 4 1·4-in.	..	16·0	500	203
g.v	**Curtatone**	S.	1040	177	3	32	8	13	6	1	1100	Venice	1887	58,440	..	..	6 4·7-in. Q.F., 4 2·2-in., 2 1·4-in., 2 M.	..	12·0	197	131
3rd cl. *cr.*	**Dogali**	S.	2088	250	0	37	0	14	6	2	7600	Elswick	1887	156,040	4½	2	6 6-in. (Armstrong), 1 2·9-in., 9 2·2-in. Q.F., 2 1·4-in., 2 M.	4	19·66	480	257
"	**Elba**	S.	2730	272	6	40	8	16	7	2	7471 (*t*)	Castellamare	1893	200,000	4½	2	4 5·9-in. Q.F., 6 4·7-in., 1 2·9-in., 8 2·2-in., 8 1·4-in., 2 M.	4	17·9*t*	..	257
2nd cl. *cr.*	**Etna**	S.	3530	282	2	42	7	19	0	2	7480	Castellamare	1885	226,720	5	1½	2 9·8-in. (Armstrong), 6 5·9-in., 1 2·9-in., 5 2·2-in. Q.F., 8 1·4-in., 2 M.	4	17·8	630	315
to.g.b.	**Euridice**	S.	840	229	6	27	0	10	2	2	4000	Castellamare	1891	72,920	..	1	1 4·7-in. Q.F., 6 2·2-in., 3 1·4-in.	5	19·8*t*	120	111
3rd cl. *cr.*	**Etruria**	S.	2280	262	6	39	6	16	7	2	7585 (*t*)	Leghorn (Orlando)	1891	183,120	4½	2	4 5·9-in. Q.F., 6 4·7-in., 1 2·9-in., 8 2·2-in., 10 1·4-in., 2 M.	2	19·84	400	257
2nd cl. *cr.*	**Fieramosca**	S.	3600	290	0	43	6	19	4	2	7700	Leghorn (Orlando)	1888	240,120	5	1½	2 9·8-in., 6 6-in. Q.F., 1 2·9-in., 5 2·2-in., 8 1·4-in., 2 M.	4	17·5	590	315
cr.	**Flavio Gioja** (training)	S.	2533	255	11	42	7	17	0	1	4150	Castellamare	1881	193,920	..	1½	6 5·9-in., 4 2·2-in. Q.F., 8 1·4-in., 2 l., 4 M.	2	15·0	500	265
to.g.b	**Folgore**	S.	370	187	0	19	8	6	7	2	2040	Castellamare	1886	39,840	..	..	2 6-pr. and 4 3-pr. Q.F.	4	20·0	60	45
d.v.	**Galileo**	S.	770	230	0	26	3	8	2	1	1700	Venice	1887	56,720	..	..	4 4·7-in., 2 2·2-in. Q.F., 2 1·4-in.	2	15·0	210	109

3rd cl. *cr.*	**Giovanni Bausan**	S.	3068	275	7	42	7	18	4	2	6500	Elswick	1883	179,120	5	1½	2 9·8-in. (Armstrong), 6 5·9-in., 1 2·9-in., 4 2·2-in. Q.F., 8 1·4-in., 2 M.	3 (1sub.)	17·5	600	267
to.g.b.	**Goito**	S.	812	230	0	25	6	11	9	2	2620	Castellamare	1887	70,680	..	1	4 2·2-in. Q.F., 5 1·4-in.	5	19·0	180	111
g.v.	**Governolo**	S.	1255	185	0	33	9	13	9	1	1100	Venice	1894	58,440	..	..	4 4·7-in. Q.F., 4 2·2-in., 2 1·4-in., 2 M.	..	13·0	200	131
to.g.b.	**Iride**	S.	840	229	6	27	0	10	2	2	4000	Castellamare	1891	72,920	..	1	1 4·7-in. Q.F., 6 2·2-in., 3 1·4-in.	6	19·6	120	111
3rd cl. *cr.*	**Liguria**	S.	2280	262	6	39	4	16	9	2	7677 (*t*)	Sestri (Ansaldo)	1893	183,120	4½	2	4 5·9-in. Q.F., 6 4·7-in., 1 2·9-in., 8 2·2-in., 10 1·4-in., 2 M. (2 Maxims).	2	19·6*t*	430	257
" "	**Lombardia**	S.	2380	262	6	39	6	16	7	2	6843 (*t*)	Castellamare	1890	183,120	4½	2	4 5·9-in. Q.F., 6 4·7-in., 1 2·9-in., 8 2·2-in., 8 1·4-in., 2 M.	2	17·0*t*	430	257
d.v.	**Marcantonio Colonna**	S.	656	216	6	23	11	10	10	1	1700	..	1879	51,480	..	..	5 2·2-in. Q.F., 2 M.	..	15·4	197	100
to.g.b.	**Minerva**	S.	846	246	0	27	6	11	9	2	4800 W.T.	Sestri (Ansaldo)	1892	72,720	..	1	1 4·7-in. Q.F., 6 2·2-in., 3 1·4-in.	5	21·0	120	111
"	**Montebello**	S.	814	230	0	25	6	11	9	3	2776	Spezia	1888	74,120	..	1	6 2·2-in. Q.F., 2 1·4-in.	4	19·0	100	111
"	**Monzambano**	S.	840	230	0	25	6	11	9	3	1953	Spezia	1888	70,680	..	1	6 2·2-in. Q.F., 2 1·4-in.	4	17·0	100	111
"	**Partenope**	S.	840	246	0	27	6	11	9	2	4200	Castellamare	1890	71,000	..	1	1 4·7-in. Q.F., 6 2·2-in., 3 1·4-in.	5	19·0	100	111
3rd cl. *cr.*	**Piemonte**	S.	2500	300	0	38	0	15	0		12,000	Elswick	1888	220,000	3	3	6 6·6-in. Q.F., 6 4·7-in., 10 2·2-in., 6 1·4-in., 4 M.	3	21·0	560	296
3rd cl. *cr.*	**Puglia**	S.	2550	269	0	41	0	16	9	2	7000	Taranto	1898	200,000	4½	1	4 5·9-in. Q.F., 6 4·7-in., 1 2·9-in., 8 2·2-in., 8 1·4-in., 2 M.	2	20·0	650	257
d.v.	**Rapido**	I.	1568	262	5	30	6	12	6	1	1920	Leghorn (Orlando)	1876	77,400	..	..	5 2·2 in. Q.F., 2 M.	..	13·4	300	135
to.g.b.	**Saetta**	S.	400	187	0	19	8	6	7	2	2400	Castellamare	1887	38,880	..	..	2 2·2-in. Q.F., 4 1·4-in.	3	20·0	90	58

ITALY.—Cruising Ships—*continued.*

Class.	NAME.	Material of Hull.	Displacement.	Length.	Beam.	Draught.	Propellers.	Indicated Horse-power.	Where Built.	Date of Launch.	Cost.	Armour. Gun Position.	Armour. Deck.	Armament. Guns.	Armament. Torpedo Tubes.	Speed.	Normal Coal Supply.	Complement.
			tons.	ft. in.	ft. in.	ft. in.	no.				£	inches.	inches.			knots.	tons.	
cr.	Savoia (used as the Royal Yacht)	S.	2850	275 6	42 7	17 0	1	3340	Castellamare	1883	176,160	..	1·5	6 2·2-in. Q.F., 6 1·4-in., 4 l., 2 M.	2	14·0	600	216
g.v.	Scilla	W.	1076	177 2	28 6	12 5	1	826	Castellamare	1874	65,520	..	..	4 2·2-in. Q.F., 2 M.	..	10·0	140	111
g.b.	Sebastiano Veniero	S.	629	170 0	26 3	10 6	1	1160	Leghorn (Orlando)	1884	36,160	..	..	4 4·7-in., 3 1·4-in. Q.F.	..	13·0	150	103
d.v.	Staffetta	I.	1388	252 7	30 10	13 2	1	1800	S. Pierdarena (Ansaldo)	1876	82,600	..	..	4 4·7-in., 7 1·4-in. Q.F.	1	13·5	300	135
2nd cl. *cr.*	Stromboli	S.	3475	282 2	42 7	19 0	2	7394	Venice	1886	220,080	5	1·5	2 9·8-in. (Armstrong), 6 5·9-in., 1 2·9-in., 5 2·2-in. Q.F., 8 1·4-in., 2 M.	4	17·0	630	315
to. g.b.	Tripoli	S.	848	230 0	25 10	11 9	3	2543	Castellamare	1886	72,080	..	1	4 2·2-in. Q.F., 4 1·4-in.	5	18·0	130	111
3rd cl. *cr.*	Umbria	S.	2280	262 6	39 6	16 7	2	7104 (*t*)	Leghorn (Orlando)	1891	183,120	4½	2	4 5·9-in. Q.F., 6 4·7-in., 8 2·2-in., 10 1·4-in., 1 l., 2 M.	4	18·83 (*t*)	430	257
to. g.b.	Urania	S.	846	230 0	27 0	11 2	2	4000	Sestri (Odero)	1891	72,920	..	1	1 4·7-in. Q.F., 6 2·2-in., 3 1·4-in.	6	20·0	120	111
d.v.	Vedetta	I.	827	183 9	26 11	11 5	1	670	Genoa	1866	32,400	..	..	4 4·7-in., 6 1·4-in. Q.F.	..	11·0	137	40
2nd cl. *cr.*	Vesuvio	S.	3427	282 2	42 7	19 0	2	6820	Leghorn (Orlando)	1886	218,320	5	1·5	2 9·8-in., 6 5·9-in., 1 2·9-in., 5 2·2-in. Q.F., 8 1·4-in., 2 M.	4	17·0	600	315
g.v.	Volturno	S.	1040	177 3	32 8	14 4	1	1100	Venice	1887	58,960	..	..	4 4·7-in., 4 2·2-in. Q.F., 2 1·4-in., 2 M.	..	13·0	206	131

Subsidised auxiliary cruisers and despatch vessels.—Nord America, Vittoria, Duca de Galliera, and Duchessa di Genova (La Veloce S.S. Co.), Regina Margherita, Elettrico, Candia, Malta, Perseo and Orione (Navigazione Generale). The armament of these vessels is 2 2·2-in. Q.F., and 4 1·4-in. M. The gun vessels Castore and Polluce (530 tons) have been converted into tank-ships, and their guns landed and placed in the forts at Taranto.

JAPAN.—Armoured Ships.

Class.	NAME.	Material of Hull.	Displacement.	Length.	Beam.	Draught.	Propellers.	Indicated Horse-power.	Where Built.	Date of Launch.	Cost.	Armour. Belt.	Armour. Gun Position	Armour. Deck Plating.	Armament. Guns.†	Armament. Torpedo Tubes.	Speed.	Normal Coal Supply.	Complement.
			tons.	ft. in.	ft. in.	ft. in.	no.					inches.	inches	inches.			knots.	tons.	
t.	Asahi . . .	S.	15,200	400 0	75 2	27 3	2	15,000 B	Clydebank .	1899	..	9–4 H. S.	14–10 H. S.	4–2½	4 12-in., 14 6-in Q.F., 20 12-pr., 8 3-pr., 4 2½-pr.	4 (sub.)	18	1400	..
a.c.	Asama . . .	S.	9750	408 0	67 0	24 § 3	2	19,000 t	Elswick .	1898	..	7–3½ H. S.	6	2	4 8-in. Q.F., 14 6-in. (Armstrong), 12 12-pr., 7 2½-pr.	5 (4 sub.)	22·1 t	600 / 1200	..
a.c.	Azuma . . .	S.	9436	445 10	59 6	28 0	2	17,000 B	St. Nazaire .	Bldg.	..	6	..	3	4 8-in., 12 6-in. Q.F., 12 3-in., 12 1·8-in.	5 (4 sub.)	..	..	482
b.	Chin-Yen . . (Ex. Chen Yuen)	S.	7400	308 5	59 0	20 0	2	6200	Stettin .	1882	..	14	12	3	4 12-in. (Krupp), 4 6-in. Q.F., 8 l., 8 M.	3	14	1000	250
a.c.	Chiyoda . .	S.	2450	308 0	42 6	14 0	2	5700 B	Clydebank .	1889	..	4½	..	1–2	10 4·7-in. Q.F., 14 3-pr., 8 M.	3	17·5	420	300
b.	Fuji . . .	S.	12,320	374 0	73 0	26 6	2	14,000	Thames .	1896	..	18—6 H. S.	14 H. S.	2½ H. S.	4 12-in., 10 6-in. Q.F., 20 3-pr., 4 4½-pr.	5 (4 sub.)	19·2 18·5 t	1100	600
b.	Hatsuse . .	S.	15,000	400 0	76 6	27 0	2	14,500 B	Elswick .	1899	..	9–4 H. N. S.	6 H. N. S.	4	4 12-in., 14 6-in. Q.F., 20 12-pr., 8 3-pr., 4 2½-pr.	4 (sub.)	19·0	700 / 1400	741
c.d.s.	Hei - Yen (Ex. Ping-Yuen-Go)	S.	2000	200 0	40 0	16 0	2	2400	Foo Chow .	1890	..	8	5	2	1 10·2-in. (Krupp), 2 5·9-in., 6 M.	4	11·0	350	250
a.c.	Hi-yei * . .	C.	2200	231 0	40 9	17 4	1	2490	Milford .	1878	..	4½	..	..	3 6·6-in. (Krupp), 6 5·9-in., 4 M., 1 l.	..	13·0	280	308
a.c. / a.c.	Idzumo . . / Iwate . .	S.	9750	400 0	68 6	24 § 3	2	14,500 B	Elswick .	1899 / Bldg.	..	7–3½ H. N. S.	6 H. N. S.	2½	4 8-in., 14 6-in. Q.F., 17 12-pr., 7 2½-pr.	4 (sub.)	20·7	600 / 1400	482
"	Kon-go * . .	C.	2200	231 0	40 9	17 4	1	2450	Hull . .	1877	..	4½	..	..	3 6·6-in. (Krupp), 6 5·9-in., 4 M., 1 l.	..	13·7	280	308
b.	Mikasa . .	S.	15,200	400 0	76 0	27 3	2	15,000 B	Barrow .	Bldg.	..	12 4 H. N. S.	14–6 H. N. S.	2	4 12-in., 14 6-in. Q.F., 20 12-pr., 8 3-pr., 4 2½-pr.	(4 sub.)	18·0	1400	730
b.	Shikishima .	S.	14,850	400 0	75 6	27 § 3	2	14,500 B	Thames .	1898	..	9–4 H. N. S.	14–6	3–5	4 12-in., 14 6-in. Q.F., 20 12-pr., 8 M.	5 (4 sub.)	18·5	700	741
a.c.	Tokiwa . .	S.	9750	408 0	67 0	24 § 3	2	20,000	Elswick .	1898	..	7–3½ H. S.	6 H. S.	2	4 8-in. Q.F., 14 6-in. (Armstrong), 12 12-pr., 7 2½-pr.	5 (4 sub.)	23·0	600 / 1200	500
a.c.	Yakumo . .	S.	9850	407 9	64 4	23 9	2	16,000 B	Stettin .	1899	..	7–3½	6	2½	4 8-in. (Armstrong) Q.F., 6 6-in., 12 12-pr. (Armstrong), 7 2½-pr.	5 (4 sub.)	20·0	600 (1100)	500
b.	Yashima . .	S.	12,320	374 0	73 0	26 6	2	14,000	Elswick .	1896	..	18–6 N. S.	14 N. S.	2½	4 12-in., 10 6-in. Q.F., 20 3-pr., 4 4½-pr.	5 (4 sub.)	19·2 18·5 t	1100	600

* These are now used as training ships; they have no armour as against end-on fire, and no armoured deck; also the older Riojo (2450 tons)—gunnery—now without engines.
† All Q.F. guns and 12-in. for new ships are Armstrong. § Mean draught.
The old central battery ironclad *Fu-So* (3718 tons) built on the Thames, 1877, and sunk off Shikoku Island, 1897, has been refloated, and is being repaired at Kure dockyard.

JAPAN.—Cruising Ships, &c.

Class.	NAME.	Material of Hull.	Displacement.	Length.	Beam.	Draught.	Propellers.	Indicated Horse-power.	Where Built.	Date of Launch.	Cost.	Armour. Gun position.	Armour. Deck.	Armament. Guns.	Armament. Torpedo Tubes.	Speed.	Normal Coal Supply.	Complement.
			tons.	ft. in.	ft. in.	ft. in.	no.				£	in.	in.			knots.	tons.	
g.v.	**Akagi**	S.	615	164 0	27 0	10 0	1	700	Japan	1889	..	..	..	1 8·2-in., 1 5·9-in., 2 l., 2 M.	..	13·0	..	113
cr.	**Akashi**	S.	2700	306 9	40 0	16 4	2	8500	Japan (Yokosuka)	1895	327,000	..	2–1	2 6-in. Q.F. (Armstrong), 6 4·7-in., 12 3-pr., 4 M.	2	20·0	200	..
,,	**Akitsushima**	S.	3150	302 0	42 7	18 5	2	8400	Japan	1892	..	4½	3	4 6-in. Q.F., 6 4·7-in., 10 3-pr.	4	19·0	..	330
g.v.	**Atago**	S.	615	154 0	27 0	10 6	2	700	Japan	1887	..	..	..	1 8·2-in., 1 4·7-in., 2 M.	..	12·0	60	113
t.g b.	**Chihaya**	S.	875	240 0	27 6	13 0	2	5500	Yokosuka	Bldg.	..	..	..	2 4·7-in. Q.F., 4 3-pr.	5	21·0	200	..
cr.	**Chitose***	S.	4760	396 0	49 0	17 7	2	15,500	San Fr'cisco	1898	205,200	4½	4½–1¾	2 8-in. Q.F., 10 4·7-in., 12 12-pr., 2 6-pr., 2 2½-pr.	5	22·5	350/1000	405
,,	**Hashidate**	S.	4277	295 0	50 10	21 2	2	5400	Japan	1891	..	12 (Hashidate and Itsukushima)	2 (Hashidate and Itsukushima)	1 12·5-in. (Canet), 11 4·7-in. Q.F., 5 6-pr., 11 3-pr., 6 M. (Hashidate and Itsukushima)	4	17·0	400	350
,,	**Itsukushima**	S.	4277	295 0	50 10	21 2	2	5400	La Seyne	1891	..							
g.v.	**Iwaki**	W.	700	147 0	25 0	11 0	1	700	Japan	1883	..	..	..	1 5·9-in., 2 4·7-in.	..	10·0	600	115
,,	**Idzumi** (ex Esmeralda)	S.	2950	270 0	40 0	18 3	2	6500	Elswick	1878	..	22	5–1	2 10·2-in. (Armstrong), 6 4·7-in. Q.F., 2 l. 6 M.	3	18·6	400	300
cr.	**Kasagi**	S.	5416	393 6	48 9	19 0	2	15,797	Philadelphia	1897	205,200	4½	4½–1¾	2 8-in., 10 4·7-in. Q.F., 12 12-pr., 6 1·8-in.	5	22·5	350/1000	405
t.c.	**Katsuraki** / **Musashi**	S.	1476	206 9	36 0	15 0	2	1600	Japan	1885	..	..	..	2 6-in. (Krupp), 5 4·7-in., 2 M.	2	13·0	..	242
g.v.	**Maya**	S.	615	154 0	27 0	10 0	2	700	Japan	1886	..	..	..	1 8·2-in., 1 4·7-in., 2 M.	..	13·0	60	113
cr.	**Matsushima**	S.	4277	295 0	50 10	21 2	2	5400	La Seyne	1890	..	12	2	1 12·5-in. (Canet), 11 4·7-in. Q.F., 5 6-pr., 11 3-pr., 6 M.	4	17·5	400	350
,,	**Miyako**	S.	1800	314 9	36 0	13 2	2	6130	Kure	Bldg.	..	..	..	2 4·7-in. Q.F., 10 1·8-in.	2	20·0	..	..

cr.	Naniwa	S.	3650	300	0	46	0	18	6	2	7235	Elswick	1885	..	1½	3–2	2 10·2-in. (Armstrong), 6 5·9-in. Q.F., 2 3-pr., 10 M.	4	18·72	800	350
g.v.	Oshima	S.	630	164	0	27	0	10	0	1	700	Japan	1890	..	..	..	4 4·7-in., Q.F., 8 l.	..	13·0	..	..
cr.	Sai yen (ex Tsi Yuen)	S.	2300	263	3	33	0	15	9	2	2800	Stettin	1883	..	9	3	2 8·2-in., 1 5·9-in., 4 l. 10 M.	4	14·5	230	200
cr.	Suma	S.	2700	306	9	40	0	16	4	2	8500	Japan (Yokosuka)	1896	237,000	..	2–1	2 6-in. Q.F., 6 4·7-in., 12 3-pr., 4 M.	2	20·0	200	..
„	Takao	S.&W.	1774	230	0	33	0	13	0	2	2330	Japan	1888	..	..	..	4 6-in. Q.F., 1 4¾-in. do., 6 M.		15·0	300	255
„	Takachiho	S.	3700	300	0	46	0	18	6	2	7500	Elswick	1885	..	1½	3–2	2 10·2-in. (Armstrong), 6 5·9-in., 2 3-pr., 10 M.	4	18·7	800	365
„ „	Takasago Unnamed	S.	4160	360	0	46	6	17	0	2	15,500 B	Elswick	1897 Bldg.	..	4½	4½	2 8-in. Q.F., 10 4·7-in., 12 12-pr., 6 2½-pr.	5	23·0	800 / 1000	..
to.g.b.	Tatsuta	S.	875	240	0	27	6	13	0	2	5500	Elswick	1894	..	..	..	2 4·7-in. Q.F., 4 3-pdr.	5	21·0	200	..
„	Ten-riu	W.	1500	200	0	32	0	16	5	1	1250	Japan	1882	..	..	..	1 6·6-in. (Krupp), 6 4·7-in., 2 l.	..	12·0	256	222
cr.	Tsukushi (ex Arturo Prat)	S.	1350	210	0	32	0	15	0	2	2887	Elswick	1882	..	4½	..	2 10-in. (Armstrong), 4 4·7-in. Q.F., 2 l., 4 M.	2	16·5	250	190
„	Yayeyama	S.	1600	315	0	34	6	15	0	2	5400	Japan	1889	..	..	..	3 4·7-in. Q.F., 6 M.	2	20·0	..	200
„	Yamato	..	1476	206	9	36	0	15	0	1	1600	Japan	1885	..	..	..	2 6·6-in. (Krupp), 5 4·7-in., 4 M.	2	13·0	..	242
„	Yoshino	S.	4180	350	0	46	6	17	0	2	15,000	Elswick	1892	..	4½	4½–1¾	4 6-in. Q.F., 8 4·7-in., 23 3-pr.	5	23·0	1000*	300

The gunboats Chen-Pei, Chen Pien, Chen Nan, Chen Hsi, Chen Chung and Chen Tung (440 tons) were captured from the Chinese.

* Dimensions doubtful.

NETHERLANDS.—Armoured Ships.

Class.	NAME.	Material of Hull.	Displacement.	Length. ft.	Length. in.	Beam. ft.	Beam. in.	Draught. ft.	Draught. in.	Propellers.	Indicated Horse-power.	Where Built.	Date of Launch.	Cost.	Armour. Belt.	Armour. Turret.	Armour. Deck Plating.	Armament. Guns.	Armament. Torpedo Tubes.	Speed.	Normal Coal Supply.	Complement.
			metric tons.	ft.	in.	ft.	in.	ft.	in.	no.					inches.	inches.	inches.			knots.	tons.	
c.d.s.t.	Bloedhond	I.	1683	182	8	46	3	10	6	2	680	Amsterdam	1869	..	5½	9½	1	1 11-in. 28-ton (Krupp), 1 2·9-in., 2 3-pr. Q.F., 2 M.	..	7·7	104	118
”	Cerberus	I.	1584	185	8	44	0	9	9	2	534	Amsterdam	1869	..	5½	9½	1	1 11-in. 28-ton (Krupp), 1 2·9-in., 2 3-pr. Q.F., 2 M.	..	7·0	80	118
”	Draak	I.	2234	213	6	49	2	12	0	2	807	Amsterdam	1877	..	8	11	1	2 11-in. (Krupp), 1 2·9-in., 2 3-pr. Q.F., 2 M.	..	8·0	100	133
”	Evertsen	S.	3400	282	9	47	0	16	9	2	4735	Flushing	1894	..	6 H.S.	9½ H.S.	2	3 8·2-in., 2 5·9-in., 6 2·9-in. Q.F., 8 1·4-in.	3	16·0	280	260
c.d.s.t.	Haai	I.	1580	195	2	44	0	9	6	2	672	Rotterdam	1871	..	5½	9½	1	1 11-in. 28-ton (Krupp), 1 2·9-in., 2 3-pr. Q.F., 2 M.	..	8·0	76	118
”	Heiligerlee	I.	1543	187	0	44	0	9	6	2	630	Birkenhead	1868	..	5½	9½	1	1 11-in. 28-ton (Krupp), 1 2·9-in., 2 3-pr. Q.F., 2 M.	..	9·0	120	118
”	Hyena	I.	1580	192	2	44	0	9	6	2	654	Amsterdam	1870	..	5½	9½	1	1 11-in. 28-ton (Krupp), 1 2·9-in., 2 3-pr. Q.F., 2 M.	..	7·0	76	118
a.g.b.	Isala	I.	383	159	4	25	0	4	5	2	306	Amsterdam	1876	..	5	5	1	2 4·7-in. (Krupp)	..	7·7	28	44
t.	Koning der Nederlanden (I)	I.	5400	279	9	49	9	19	8	2	4500	Amsterdam	1874	..	7½	9½	¾	4 11-in., 4 4·7-in., 2 2·9-in., 6 1·4-in., 4 1·4-in. Q.F.	..	12·0	520	308
c.d.s.t. c.d.s.t.	Koningin Regentes New Ship	S.	4950	316	10	51	6	21	8	2	5300 Y	Amsterdam	Bldg. Pro.	347,500	6 N.S.	10 N.S.	2	2 9·4-in., 4 5·9 Q.F., 4 2·9-in., 8 1·4-in.	3 2 sub.	16·0	680	..
t. & b.	Koningin Wilhelmina der Nederlanden * (I)	S. shd.	4600	327	5	48	10	20	0	2	5900	Amsterdam	1892	..	..	11	3	1 11-in., 1 8·2-in., 2 6·6-in., 2 2·9-in., 4 2·9-in. Q.F., 4 1·4-in., 6 1·4-in. Q.F., 2 M.	4	16·5	448	274
c.d.s.t.	Kortenaer	S.	3400	282	9	46	11	16	9	2	4658	Amsterdam	1894	..	6 H.S.	9½ H.S.	2	3 8·2-in., 2 5·9-in., 6 2·9-in. Q.F., 8 1·4-in.	3	16·0	280	260
”	Krokodil	I.	1547	187	0	44	0	9	8	2	630	Birkenhead	1868	..	5½	9½	1	1 11-in. 28-ton (Krupp), 1 2·9-in., 2 3-pr. Q.F., 2 M.	..	9·0	120	118
”	Luipaard	I.	1610	194	9	44	0	9	7	2	680	Rotterdam	1876	..	5½	9½	1	1 11-in. 28-ton (Krupp), 1 2·9-in., 2 3-pr. Q.F., 2 M.	..	9.0	120	118

* Has received new engines and boilers.

c.d.s.t.	**Matador**	I.	2000	209	6	47	3	10	6	2	691	Rotterdam .	1878	..	5½	11	..	2 11-in. 28-ton (Krupp), 1 2·9-in., 2 3-pr. Q.F., 2 M.	..	7·5	100	130
a.g.b.	**Merva**	I.	383	159	4	24	11	4	5	2	395	Amsterdam .	1879	..	5	5	1	2 4·7-in. (Krupp) . . .	..	8·0	28	44
,,	**Mosa**	I.	373	159	6	24	11	4	5	2	400	Amsterdam .	1878	..	5	5	1	2 4·7-in. (Krupp) . . .	..	8·0	28	44
c.d.s.t.	**Panter**	I.	1580	159	3	44	0	9	6	2	560	Amsterdam .	1870	..	5½	9½	..	1 11-in. 28-ton (Krupp), 1 2·9-in., 2 3-pr. Q.F., 2 M.	..	7·0	76	118
c.d.s.t.	**Piet-Hein** . . .	S.	3400	282	9	46	11	16	9	2	4736 (*t*)	Rotterdam .	1894	..	6 H.S.	9½ H.S.	2	3 8·2-in., 2 5·9-in., 6 2·9-in., Q.F., 8 1.4-in.	3	16·2 (*t*)	280	260
t.	**Prins Hendrik der Nederlanden**	I.	3375	240	0	44	0	17	10	2	2000	Birkenhead .	1866	..	4½	10	⅝	4 9-in. 13-ton M.L.R. (Armstrong), 4 4·7-in. (Krupp), 2 2·9-in., 4 1·4-in. Q.F., 6 M.	..	11·0	380	228
t. & b.	**Reinier Claeszen** . .	S.	2479	229	5	44	4	15	0	2	350	Amsteram .	1891	..	4¾–2 *comp.*	11	3	1 8·2-in. (Krupp), 1 6·6-in., 1 2·9-in., 4 1·9-in. Q.F., 3 1·4-in.,	2	12·5	88	160
a.g.b.	**Rhenus**	I.	388	160	5	25	0	4	6	2	310	Amsterdam .	1877	..	5	5	1	2 4·7-in. (Krupp) . . .	..	7·5	28	44
c.d s.t.	**Schorpioen** . .	I.	2235	205	0	38	0	16	6	2	2225	La Seyne .	1868	..	6	11	1	1 11-in. 28-ton (Krupp), 2 2·9-in., 5 3-pr. Q.F., 2 M.	..	13·0	200	160
,,	**Stier**	I.	2112	205	0	38	0	16	4	2	2250	Birkenhead .	1868	..	11	8	1	1 11-in. 28-ton (Krupp), 2 2·9-in., 5 3-pr. Q.F., 2 M.	..	12·4	160	154
a.g.b	**Vahalis**	I.	365	126	0	27	9	5	3	2	240	Rotterdam .	1870	..	4	1	½	2 3-pr. Q.F.	..	7·5	24	34
c.d.s.t.	**Wesp**	I.	1580	195	2	44	0	9	6	2	740	Amsterdam .	1871	..	5½	9½	1	1 11-in. 28-ton (Krupp), 1 2·9-in., 2 3-pr. Q.F., 2 M.	..	8·0	76	118

* Three armoured ships of the Koningin Regentes class and two of the Reinier Claeszen class are projected.

NETHERLANDS.—Cruising Ships.

((I) denotes vessels of the Dutch Indian Navy.)

Class.	NAME.	Material of Hull.	Displacement.	Length.	Beam.	Draught.	Propellers.	Indicated Horse-power.	Where Built.	Date of Launch.	Cost.	Armour: Gun Position.	Armour: Deck.	Armament: Guns.	Armament: Torpedo Tubes.	Speed.	Normal Coal Supply.	Complement.
			metric tons.	ft. in.	ft. in.	ft. in.	no.				£	inches.	inches.			knots.	tons.	
cr.	Alkmaar	Cp. shd.	1068	178 9	30 2	15 7	1	686	Amsterdam	1874	..	..	..	1 5·9-in. (Krupp), 6 4·7-in., 1 2·9-in., 4 1·4-in. Q.F., 2 M.	..	10·0	130	112
cr	Atjeh	I. & W.	3440	301 0	41 0	21 4	1	2700	Amsterdam	1876	..	..	..	6 6·6-in. 6-ton, 8 4·7-in. (Krupp), 2 2·9-in., 8 3-pr. Q.F., 8 smaller.	..	13·5	440	301
g. v	Bali (I)	I. & W. shd.	853	175 9	29 6	21 1	1	400	Rotterdam	1878	..	..	..	1 5·9-in., 3 4·7-in. (Krupp), 1 2·9-in., 2 1·4-in. Q.F.	..	9·0	80	104
"	Batavia (I)	I. & W. shd.	850	175 6	29 6	11 10	1	400	Amsterdam	1876	..	..	..	1 7-in. 7-ton M.L.R. (Armstrong), 2 4·7-in. (Krupp), 1 2·9-in., 2 1·4-in. Q.F.	..	9·0	95	104
"	Bellona	I.	920	178 5	32 9	12 6	1	310	Amsterdam	1892	..	..	..	1 5·9-in., 7 4·7-in., 3 2·9-in., 13 smaller.	..	8·5	50	87
"	Benkoelen (I)	I. & W. shd.	853	147 7	29 6	11 10	1	446	Rotterdam	1879	..	..	..	1 5·9-in., 3 4·7-in. (Krupp), 1 2·9-in., 2 1·4-in. Q.F.	..	9·5	100	99
"	Bonaire	I. & W. shd.	853	175 6	29 6	12 7	1	412	Rotterdam	1877	..	..	..	1 5·9-in., 2 4·7-in. (Krupp), 1 2·9-in., 2 1·4-in. Q.F., 2 M.	..	9·0	104	84
"	Borneo (I)	Cp. shd.	800	176 6	31 0	13 4	1	1040	Glasgow	1892	..	..	..	6 4·1-in., 1 2·9-in., 2 1·4-in. Q.F., 2 M.	..	13·0	124	106
"	Ceram (I)	S. & W. shd.	550	173 2	25 7	10 3	1	800	Flushing	1887	..	..	..	3 4·7-in. (Krupp), 1 2·9-in., 2 1·4-in. Q.F.	..	12·5	70	82
"	Condor (I)	Cp. shd.	350	126 0	20 0	10 0	1	300	Amsterdam	1885	..	..	..	1 2·3-in., 2 2-in.	..	10·0	26	40
cr.	De Ruyter	I. & W. shd.	3517	302 1	41 0	22 1	1	3300	Amsterdam	1880	..	..	..	6 6·6-in. 6-ton, 8 4·7-in. (Krupp), 2 2·9-in., 8 3-pr. Q.F., 8 smaller.	..	14·5	360	30
g. v.	Edi (I)	S.	810	166 0	30 9	11 9	2	1100	Flushing	Bldg.	..	..	..	3 4·7-in. Q.F., 2 2·9-in., 4 1·4-in.	..	13·0	113	95
"	Flores (I)	S. & W. shd.	550	173 2	25 7	11 4	1	650	Amsterdam	1887	..	..	..	3 4·7-in, 1 2·9-in., 2 1·4-in. Q.F.	..	11·7	75	82
cr.	Friesland	S.	3900	294 0	48 6	17 8	2	10418 Y	Rotterdam	1896	285,700	..	2	2 5·9-in. Q.F., 6 4·7-in., 4 2·9-in., 8 1·4-in., 4 smaller.	4	19·8 (*t*)	400	306

cr.	Gelderland	S.	4033	310	8	48	6	17	8	2	9750 Y	Feijenoord	1898	..	..	2¼	2 5·9-in. Q.F., 6 4·7-in., 4 2·9-in., 4 1·4-in., 4 M.	4	20·0	850	..
cr.	Holland	S.	3900	294	0	48	6	17	8	2	10548 Y	Amsterdam	1896	285,700	..	2	2 5·9-in. Q.F., 6 4·7-in., 4 2·9-in., 8 1·4-in., 4 M.	4	19·6 (t)	400	306
g.v.	Java (I)	I. & W. shd.	1300	205	4	31	2	14	1	1	1050	Rotterdam	1885	..	..	..	1 5·9-in., 3 4·7-in., 1 2·9-in., 2 1·4-in. Q.F.	..	12·5	160	114
cr.	Koningin Emma der Nederlanden	I. & W. shd.	3528	301	0	41	0	21	4	1	2730	Amsterdam	1879	..	..	..	6 6·6-in. 6-ton, 8 4·7-in. (Krupp), 2 2·9-in., 8 3-pr. Q.F., 8 M.	..	14·0	470	301
g.v.	Lombok (I)	S. & W. shd.	600	172	0	27	3	11	0	1	990	Amsterdam	1891	..	..	..	3 4·7-in., 1 2·9-in., 2 3-pr. Q.F.	..	12·0	55	87
,,	Makasser (I)	I. & W. shd.	850	177	0	29	6	11	10	1	320	Amsterdam	1877	..	..	..	1 6.3-in. 7-ton M.L.R. (Armstrong), 2 4·7-in. (Krupp), 1 2·9-in., 2 1·4-in. Q.F.	..	8·5	96	104
,,	Madura (I)	I. & W. shd.	853	177	0	29	6	11	10	1	400	Amsterdam	1880	..	..	..	1 5·9-in., 3 4·7-in. (Krupp), 1 2·9-in., 2 1·4-in. Q.F.	..	9·0	85	104
,,	Mataram (I)	S.	810	166	0	30	9	11	9	2	1100	Amsterdam	1896	..	..	..	3 4·7-in. Q.F., 2 3-in., 2 1·4-in.	..	13·0	113	95
,,	Nias (I)	S.	810	166	0	30	9	11	9	2	1227	Amsterdam (Huygens)	1895	..	..	..	3 4·7-in. Q.F., 2 2·9-in., 4 1·4-in.	..	13·0	120	95
cr.	Noord-Brabant	S.	4033	299	0	48	6	17	9	2	9750 Y	Flushing	1899	..	..	2¼	2 5·9-in. Q.F., 6 4·7-in., 4 2·9-in., 4 1·4-in., 4 M.	4	20·0	850	..
,,	Padang (I)	I. & W. shd.	853	176	6	29	6	11	10	1	400	Rotterdam	1878	..	..	..	1 5·9-in., 3 4·7-in. (Krupp), 1 2·9-in., 2 1·4-in. Q.F.	..	9·0	80	104
,,	Pelikaan (I)	S. & W. shd.	400	131	2	24	1	8	8	..	485	Rotterdam	1891	..	..	..	3 4·7-in. Q.F., 1 3-in., 2 3-pr. do.	..	11·35	43	40
,,	Pontianak (I)	C. shd.	730	163	8	30	3	11	9	1	360	Amsterdam	1873	..	..	..	1 6·3-in. 7-ton M.L.R. (Armstrong), 2 4·7-in. (Krupp), 1 2·9-in., 2 1·4-in. Q.F.	..	9·5	90	104
,,	Sambas (I)	C.	654	137	10	28	10	11	10	1	374	Rotterdam	1874	..	..	..	1 7-in. 7-ton M.L.R. (Armstrong), 2 4·7-in. (Krupp), 1 2·9-in., 1 1·4-in. Q.F.	..	8·5	85	100
,,	Serdang (I)	S	810	166	0	30	9	11	9	2	1100	Flushing	1897	..	..	..	3 4·7-in. Q.F., 2 2·9-in., 4 1·4-in.	..	13·0	113	95
sl.	Sommelsdijk	I. & W. shd.	1013	178	5	31	0	14	0	1	700	Amsterdam	1881	..	..	..	1 5·9-in., 3 4·7-in. (Krupp), 1 2·9-in.	..	10·0	150	88
cr.	Sumatra (I)	S.	1720	229	6	37	0	14	0	2	8750	Amsterdam	1890	..	..	1½	1 8·2-in., 1 5·9-in., 2 4·7-in., 1 2·9-in., 4 8-pr. Q.F., 2 M.		17·0	225	183

NETHERLANDS.—Cruising Ships—*continued.*

((I) denotes vessels of the Dutch Indian Navy.)

Class.	NAME.	Material of Hull.	Displacement.	Length.	Beam.	Draught.	Propellers.	Indicated Horse-power.	Where Built.	Date of Launch.	Cost.	Armour. Gun Position.	Armour. Deck.	Armament. Guns.	Armament. Torpedo Tubes.	Speed.	Normal Coal Supply.	Complement.
			metric tons.	ft. in.	ft. in.	ft. in.	no.				£	inches.	inches.			knots.	tons.	
g.v.	**Sumbawa (I)**	S.	600	174 6	26 6	11 4	1	930	Flushing	1891	..	..	..	3 4·7-in., 1 2·9-in., 2 3-pr. Q.F.	..	12·5	60	87
"	**Suriname**	I. & W. shd.	884	177 0	29 6	11 5	1	440	Amsterdam	1877	..	..	..	1 5·9-in. (Krupp), 2 4·7-in., 1 2·9-in., 2 1·4-in. Q.F., 3 M.	..	9·0	105	84
cr.	**Tromp**	I. & W. shd.	3512	301 0	41 0	21 4	1	2772	Amsterdam	1877	..	..	..	6 6·6-in. 6-ton, 8 4·7-in. (Krupp), 2 2·9-in., 6 3-pr. Q.F., 2 M.	..	14·0	470	301
cr.	**Utrecht**	S.	4033	310 8	48 6	17 8	2	9750 Y	Amsterdam	1898	..	..	2½	2 5·9-in. Q.F., 6 4·7-in., 4 2·9-in., 4 1·4-in., 4 M.	4	20·0	850	..
cr.	**Van Speyk**	I. & W. shd.	3728	302 1	41 0	23 0	1	2891	Amsterdam	1880	..	..	..	6 6·6-in. 6-ton, 8 4·7-in. (Krupp), 2 2·9-in., 6 3-pr. Q.F., 2 M.	..	14·0	360	301
cr.	**Zeeland**	S.	3900	294 0	48 6	17 8	2	10589 Y *t.*	Flushing	1897	285,700	..	2	2 5·9-in. Q.F., 6 4·7-in., 4 2·9-in., 8 1·4-in., 4 M.	4	19·4 *t*	400	306
g.v.	**Zwaluw (I)**	I. & W. shd.	340	126 0	20 0	10 0	1	240	Flushing	1882	..	..	..	2 3-in., 2 2-in.	..	10·0	26	40

Gun-vessels of the Indian Navy, Arend, Flamingo, Raaf, Reiger, Valk, Zeeduif, and Zwaan (400 tons), launched between 1880 and 1891; Glatik (417 tons), 1894; Argus and Cycloop (438 tons), 1893; Sindoro and Soembing (642 tons), built at Soerabaia, 1877–78.

Sixteen Gunboats (*Staunch class*) of 268 tons, and of 100 to 171 H.P.; also *five small gunboats*, of 210 tons, and 124 to 174 H.P., and *one steel gunboat* of 108 tons and 172 I.H.P. The new programme contemplates the building of 3 unarmoured monitors, 14 gunboats and 3 schooners (see Chap. II.).

NORWAY.—Armoured Ships.

Class.	NAME	Material of Hull.	Displacement.	Length.		Beam.		Draught.		Propellers.	Indicated Horse-power.	Where Built.	Date of Launch.	Cost.	Armour.			Armament.		Speed.	Normal Coal Supply.	Complement.
															Belt.	Gun Position	Deck Plating.	Guns.	Torpedo Tubes.			
			metric tons.	ft.	in.	ft.	in.	ft.	in.	no.				£	in.	in.	in.			knots.	tons.	
c.d.s.t.	Eidsvold .	S.	3847	290	0	50	6	16	6	2	4500 Y.	Elswick	Bldg.	..	6	5	2	2 8·2-in., 6 5·9-in. Q.F., 8 12-pr., 6 3-pr.	2 (sub)	16·5	250 / 400	248
"	Mjölner .	I.	1515	203	5	45	11	11	10	1	450	Norrkoping	1868	56,800	H.S. 5	H.S. 12	1	2 4·7-in., 2 2·5-in. Q.F., 3 M., 1 l.	..	8·0	138	80
"	Norge .	S.	3847	290	0	50	6	16	6	2	4500	Elswick	1900	..	6	5	2	2 8·2-in., 6 5·9-in. Q.F., 8 12-pr., 6 3-pr.	2 (sub)	16·5	250 / 400	248
" / "	Harald Haarfagre . / Torkenskjold	S.	3556	280	0	48	6	16	6	2	Y. 3700	Low Walker	1896 / 1897	190,000	H.S. 7	H.S. 8	2	2 8-in. Q.F., 6 4·7-in., 6 12-pr., 6 1½-pr.	2 (sub)	17·2† t.	200 / 540	248
"	Skorpionen .	I.	1447	200	2	45	11	11	6	1	850	Horton	1866	..	H.S. 5	12	1	2 4·7-in., 2 2·5-in. Q.F., 3 M., 1 l.	..	6·0	138	80
"	Thor .	I.	2003	203	5	49	3	13	2	1	600	Horten	1872	..	7	14¼	1	2 4·7-in., 2 2·5-in. Q.F., 3 M., 1 l.	..	8·0	200	90
"	Thrudvang .	I.	1515	200	2	45	11	11	10	1	500	Horten	1867	..	5	12	1	2 4·7-in., 2 2·5-in. Q.F., 3 M., 1 l.	..	8·0	138	80

† Natural draught.

Cruising Ships.

Class.	NAME.	Material of Hull.	Displacement.	Length.		Beam.		Draught.		Propellers.	Indicated Horse-power.	Where Built.	Date of Launch.	Cost.	Armour.		Armament.		Speed.	Normal Coal Supply.	Complement.
															Gun Position.	Deck.	Guns.	Torpedo Tubes.			
			metric tons.	ft.	in.	ft.	in.	ft.	in.	no.				£	in.	in.			knots.	tons.	
g.b.	Æger .	S.	393	108	6	29	6	8	0	2	450	Horten	1892	..	..	1½	1 8·2-in., 1 2·7-in. Q.F., 2 1·9-in.	..	9·0	..	..
g.v.	Ellida .	W.	1000	187	0	32	8	14	4	2	900	Horten	1880	..	..	..	5 5·9-in. 4-ton (Krupp), 1 4·7-in., 1 l., 2 M.	1	12·0	97	128
"	Frithjof .	S.	1371	216	6	32	10	13	3	2	300	Horten	1896	..	..	..	2 4·7-in. 4 2·9-in. Q.F., 4 1·4-in., 2 l.	3 (1 sub.)	..	..	..
"	Heimdal .	S.	630	167	3	26	9	11	8	1	700	Christiania	1892	..	..	..	4 2·5-in. Q.F.	..	12·0	22	..
corv.	Nord Stjernen .	W.	1609	216	6	39	4	17	9	1	800	Horten	1862	..	..	..	6 6·2-in. 3-ton M.L.R., 10 8-in. smooth-bore, 3 l.	..	9·0	195	216
g.v.	Sleipner .	I.	580	173	10	25	11	9	6	2	800	Horten	1877	..	..	..	1 10·2-in. 22-ton (Krupp), 1 5·9-in. 4-ton do., 1 M.	1	12·0	80	87
to.g.b.	Valkyrien .	S.	380	190	0	24	2	9	2	2	3300	Elbing .	1896	..	..	..	2 2·7-in. Q.F., 1 M.	2	23·2 t	..	..
g.v.	Viking .	S.	1113	203	6	30	6	13	0	2	2000	Horten	1891	..	..	1½	2 5·9-in. (Arms.), 4 2·5-in Q.F., 4 1·4-in., 2 M.	3	15·0	140	..

Eleven Gunboats, of 189 to 280 tons, and of 180 to 450 I.H.P., armed with one large gun and machine guns in each.
Sixteen smaller Gunboats, of 60 tons, 70 I.H.P., and 7½ knots speed; each armed with one 5¾-inch gun. Also several smaller gunboats.
A first-class gunboat, No. 4, of 395 tons, in hand.

PORTUGAL.—Armoured Ships.

Class.	NAME.	Material of Hull.	Displacement.	Length.	Beam.	Draught.	Propellers.	Indicated Horse-power.	Where Built.	Date of Launch.	Cost.	Armour: Belt.	Armour: Battery.	Armour: Deck Plating.	Armament: Guns.	Armament: Torpedo Tubes.	Speed.	Normal Coal Supply.	Complement.
			metric tons.	ft. in.	ft. in.	ft. in.	no.				£	inches.	inches.	inches.			knots	tons.	
c.b.	**Vasco da Gama**	I.	2422	200 0	40 0	18 0	2	3600	Blackwall	1876	132,000	9	10	3	2 10·2-in. 18-ton (Krupp), 1 5·9-in., 2 2·5-in., Q.F., 2 M.	..	13·2	280	218
c.d.s.	2 Unnamed	S.	2500	229 8	42 8	13 6	2	3000	..	Pro.	..	7¾	7¾	2	2 9·4-in., 4 4·7-in. Q.F., 4 1·8-in., 4 M.	2	15·0	..	..

Cruising Ships.

Class.	NAME.	Material of Hull.	Displacement.	Length.	Beam.	Draught.	Propellers.	Indicated Horse-power.	Where Built.	Date of Launch.	Cost.	Armour: Gun Position.	Armour: Deck.	Armament: Guns.	Armament: Torpedo Tubes.	Speed.	Normal Coal Supply.	Complement.
			metric tons.	ft. in.	ft. in.	ft. in.	no.				£	in.	in.			knots.	tons.	
cr.	**Adamastor**	S.	1993	250 0	35 0	14 0	2	4000	Leghorn	1896	..	5	3	2 5·9-in. Q.F., 4 4·7-in., 4 2·2-in., 4 M.	3	18·0 *t.*	..	..
corv.	**Affonso de Albuquerque**	I. & W.	1111	203 0	33 0	13 6	1	1360	Blackwall	1884	56,500	..	..	2 6-in. (Armstrong), 5 4·7-in., 2 2·5-in. Q.F., 2 M.	..	13·3	140	183
,,	**Bartholomeu Dias**	W.	2377	207 0	37 5	20 6	1	400 (nom.)	Blackwall	1858	..	..	..	8 5-in.	..	10·0	360	271
g.v.	**Bengo**	I. & W.	462	125 6	24 6	9 0	1	400	Birkenhead	1879	22,500	..	..	1 6-in., 2 3·4-in.	..	10·0	80	88
,,	**Diu**	W.	729	147 0	27 6	13 0	1	700	Lisbon.	1889	..	..	..	1 5·9-in. (Krupp), 2 4·7-in., 1 3-pr. Q.F., 2 M.	..	12·0	80	114

cr.	**Dom Carlos I.**	S. shd.	4100	360	0	46	6	17	6	2	12,500 Y.	Elswick	1898	..	..	4	4 5·9-in. Q.F. (Armstrong), 8 4·7-in., 12 3-pr., 6 1-pr., 4 M.	5 (3 sub.)	22·0	1000	260
g.v.	**Dom Luiz I.**	S.	721	151	0	27	3	13	8	2	512	Lisbon	1895	..	..	..	4 4·1-in., 3 2·5-in. Q.F., 3 M.	..	9·9	100	..
„	**Douro**	W.	587	142	9	26	0	11	0	1	400	Lisbon	1873	..	..	..	1 5·9-in. 4-ton, 2 4·7-in., 1 M.	..	10·0	85	107
corv.	**Duque da Terceira** (training)	W.	1430	179	6	34	0	15	6	1	660	Lisbon	1877 1864	..	..	..	2 4·7-in., 2 2·5-in. Q.F., 1 M.	..	9·0	130	178
g.v.	**Liberal**	I. shd.	580	140	0	25	6	10	6	1	580	Birkenhead	1884	32,500	..	..	1 6-in. 4-ton (Armstrong), 3 4-in., 2 M.	..	11·0	90	109
corv.	**Mindello**	C.	1124	170	0	35	9	14	0	1	900	Blackwall	1876	74,500	..	..	2 7-in. 4-ton M.L.R. (Armstrong), 4 4·7-in. 2 M.	..	11·5	130	169
g.v.	**Mandovi**	I. shd.	462	125	6	24	6	9	0	1	400	Birkenhead	1879	22,500	..	..	1 5·9-in., 2 3·4-in., 2 M.	..	10·0	80	86
„	**Quanza**	W.	587	142	9	26	0	11	0	1	500	Lisbon	1877	..	..	..	1 5·9-in. 4-ton, 2 4·7-in., 1 M.	..	10·0	85	107
corv.	**Rainha de Portugal**	C.	1124	170	0	35	9	14	0	1	900	Blackwall	1876	74,500	..	..	2 7-in. M.L.R. (Armstrong), 4 4·7-in., 2 M.	..	11·5	130	169
cr.	**Rainha Amelia**	S.	1660	246	0	36	0	14	8	2	5000 Nor.	Lisbon	1899	..	..	1	4 5·9-in. Q.F., 2 3·9-in., 2 3-pr., 4 M.	2	17·5	..	250
g.v.	**Rio Ave**	W.	378	120	0	22	0	10	6	1	180	Lisbon	1880	..	..	..	1 4·7-in. (Armstrong), 2 3-in.	..	8·0	60	90
„	**Rio Lima**	I. shd.	638	148	6	27	6	10	6	1	500	Birkenhead	1875	33,000	..	..	1 7-in. 4-ton (Armstrong), 4 4-in., 2 M.	..	11·0	100	109
„	**Sado**	C.	645	148	6	28	0	10	6	1	500	Birkenhead	1875	35,500	..	..	1 7-in. 4-ton (Armstrong), 4 4·7-in., 1 M.	..	11·0	100	109
cr.	**São Gabriel** / **São Rafael**	S. shd.	1800	246	0	35	6	14*	3	2	4000 N.S.	Havre	1898	..	..	1½	2 5·9-in. Q.F. (Canet), 4 4·7-in., 8 1·8-in., 2 M.	1	17·5 *t.*	500	..
g.b.	**São Salvador**	S.	721	151	0	27	3	13	8	2	..	Lisbon	Bldg.	..	..	..	4 4·1-in., 8 5·2-in. Q.F., 3 M.	..	11·0	100	..
„	**Tamega**	C.	645	148	6	28	0	10	6	1	500	Birkenhead	1875	35,500	..	..	1 7-in. 4-ton (Armstrong), 4 4·7-in., 1 M.	..	11·0	100	109
„	**Tejo**	W.	587	142	9	26	0	11	0	1	400	Lisbon	1869	..	..	..	1 5·9-in. 4-ton, 2 4·7-in.	..	10·0	85	107
„	**Vouga**	W.	730	160	9	27	6	12	0	1	600	Lisbon	1882	..	..	..	4 4-in., 2 1·8-in. Q.F., 2 M.	..	10·0	100	109
„	**Zaire**	I. shd.	580	140	0	25	6	10	6	1	580	Birkenhead	1884	32,500	..	..	1 6-in. (Armstrong), 3 4-in. 2 M.	..	11·0	90	109
„	**Zambeze**	W.	641	143	0	25	9	12	0	1	500	Lisbon	1886	..	..	..	1 6-in. (Armsg.), 2 4-in., 2 M.	..	10·0	85	107
g.v.	One unnamed	S.	300	..	..	..	..	..	..	..	..	Lisbon	Bldg.	..	..	..	.. .. ..	..	..	..	..

* Mean draught. † Bunker capacity.

Fifteen small Gunboats and about 29 light draught steel river-gunboats. Two gunboats of 220 tons, the Al. Baptista de Andrade and Thomaz Andrea, are building for Mozambique and Timor.

RUSSIA.—Armoured Ships.

(B.S., Black Sea Fleet.)

Class.	NAME.	Material of Hull.	Displacement.	Length.	Beam.	Draught.	Propellers.	Indicated Horse-power.	Where Built.	Date of Launch.	Cost.	Armour. Belt.	Armour. Gun Position.	Armour. Deck Plating.	Armament. Guns. B.L.R. are of Russian Krupp pattern.	Armament. Torpedo Tubes.	Speed.	Normal Coal Supply.	Complement.
			tons.	ft. in.	ft. in.	ft. in.	no.				£	in.	in.	in.			knots.	tons.	
c.d.s., t.	Adm. Chichagoff	I.	3505	254 0	42 7	18 0	1	2060	St. Petersburg	1868	..		6	..	2 11-in. 28-ton, 4 4-pr., 6 Q.F., 4 l.	..	10·5	300	264
„	Adm. Greig	I.	3462	254 0	42 7	17 6	1	2031	St. Petersburg	1868	..	4½	6	..	3 11-in. 28-ton, 6 Q.F., 2 l.	..	10·0	300	280
„	Adm. Lazareff	I.	3462	254 0	43 0	17 6	1	2004	St. Petersburg	1867	..	4½	6	..	3 11-in. 28-ton, 6 Q.F., 4 l.	..	10·25	300	280
a.c.	Adm. Nahimoff	S. shd.	8524	333 0	61 0	25 0	2	8000	St. Petersburg	1885	572,000	10 comp.	8 comp.	3	8 8-in., 10 6-in., 10 Q.F., 4 3-pr., 6 M.	4	16·7	1200	567
„	Adm. Boutakoff*	S.	4126	265 0	52 6	17 0	2	5000	St. Petersburg (New Admiralty)	Bldg.	..	10	7–8	3	3 10-in., 4 6-in. Q.F., 6 1·8 in., 8 M.	4	16·0	400	318
„	Adm. Oushakoff } Adm. Seniavin }	S.	4126	265 0	52 6	17 0	2	5000	St. Petersburg	1893 / 1894	410,000	10	7–8	3	4 9-in., 4 6-in. Q.F., 6 1·8-in. Q.F., 8 M.	4	16·0	400	318
c.d.s., t.	Adm. Spiridoff	I.	3505	254 3	42 7	19 1	1	2007	St. Petersburg	1868	..	6	6	..	2 11-in. 28-ton, 4 4-pr., 6 Q.F., 4 l.	..	10·5	300	260
b.	Alexander II.	S. shd.	9927	326 0	67 0	23 0	2	8000	St. Petersburg	1887	..	14 comp.	10 comp.	2½	2 12-in. 50-ton, 4 9-in. 19-ton, 8 6-in., 4 6-pr. Q.F., 4 3-pr., 6 M.	5	16·5	1200	604
b.	Alexander III.	S.	13,600	397 0	76 0	26 0	2	16,000 W T	St. Petersburg (Baltic)	Bldg.	..	9 H.S.	5 H.S.	4	4 12-in., 12 6-in. Q.F., 20 3-in., 20 3-pr., 6 1-pr.	6 5 sub.	18	2000	..
a.c.	Bayan	S.	7800	445 0	57 0	21 2	2	16,500 B	La Soyne	Bldg.	..	7¾	..	..	2 8-in., 8 6-in Q.F., 20 2·9-in., 7 1·8-in.	..	21	..	..
b.	Borodino	S.	13,600	397 0	76 0	26 0	2	16,000 B	St. Petersburg (New Admiralty)	Bldg.	..	9 H.S.	5 H.S.	4	4 12-in., 12 6-in. Q.F., 20 3-in., 20 3-pr., 6 1-pr.	6 5 sub.	18	2000	..
b.	Catherine II. (B.S.)	I. & S.	10,180	331 0	69 0	26 6	2	10,600 B	Nicolaieff	1886	900,000	16 comp.	14 comp.	3	6 12-in. (56-ton), 7 6-in., 8 6-pr. Q.F., 6 M.	7	15·5	886	325
c.d.s., t.	Charodeika	I.	1881	206 9	42 7	10 6	2	786	St. Petersburg	1867	..	4½	6	1	4 9-in., 2 Q.F. and 2 M.	..	8·0	250	171
a.c.	Dmitri Donskoi	S. shd.	5882	296 5	52 0	24 4	2	7000	St. Petersburg	1883	..	6	unard.	2½	6 6-in. Q.F., 10 4 7-in. Q.F., 16 Q.F. and M., 4 l.	4	16·5	400	510

* Particulars doubtful. It is said the Boutakoff will be larger, better protected, and more heavily armed than the Oushakoff and others of the class.

t.	**Dvenadzat Apostoloff** (Twelve Apostles) B.S.	S.	8076	330	0	60	0	25	6	2	11,500	Nicolaieff	1890	..	14 *comp.*	12 *comp.*	2½	4 12-in. 52-ton, 4 6-in., 8 3-pr. Q.F., 10 M.	6	16·6	800	500
t.	**Gangoot***	S.	6592	278	0	62	0	21	0	2	8300	St. Petersburg	1890	..	16 *comp.*	7·8	2½	1 12-in., 4 9-in., 4 6-in., 10 Q.F.	5	14·7	500	58
a.c.	**General Admiral**	I. shd.	4722	285	5	49	3	21	0	1	4472	St. Petersburg	1873	..	6	..	..	6 8-in., 2 6-in., 10 Q.F. and M., 5 l.	4	14·2	1000	312
c.d.s.	**General Admiral Apraxine**	S.	4200	277	6	52	6	17	6	2	5757 (*t*)	St. Petersburg (New Admiralty)	1896	..	10 H.S.	7·8	3	3 10-in., 4 6-in. Q.F., 6 1·8-in., 8 1·4-in.	4	15·0 (*t*)	215	318
a.c.	**Gertzog Edinburgski**	I. & W.	5050	285	5	49	3	21	0	1	5222	St. Petersburg	1875	..	6	6	..	4 8-in., 5 6-in., 12 Q.F., 6 l.	2	15·2	1000	500
b.	**Georgi Pobiedonosetz** (George the Victorious) B.S.	S.	10,280	320	0	69	0	26	7	2	10,600 13,468 (*t*)	Sebastopol	1892	431,000	16	12	..	6 12-in. 56-ton, 7 6-in., 8 3·9-in. Q.F., 6 M.	7	16·5	700	500
a.g.b.	**Gremiastchy**	S.	1500	225	0	41	0	11	0	2	2500 B	St. Petersburg	1892	..	5	..	1½	1 9-in., 1 6-in., 10 Q.F.	2	15·0	100	142
a.c.	**Gromoboi**	S. shd.	12,336	473	0	68	6	26	0	3	14,500	St. Petersburg (Baltic)	1899	..	6 H.S.	6 H.S.	3	4 8-in. Q.F., 16 6-in., 6 4·7-in., 20 3-in., 36 small Q.F and M.	5 4 sub.	20	2500	..
a.g.b.	**Grozjastchy**	S.	1492	229	0	41	8	11	0	2	2000 B	St. Petersburg	1890	..	5	..	1½	1 9-in., 1 6-in., 8 Q.F.	2	15·0	100	120
”	**Khrabry**	S.	1492	229	0	41	8	11	0	2	3000 Nic.	St. Petersburg (New Admiralty)	1895	..	5	..	1½	1 9-in., 1 6-in., 8 Q.F.	2	15·0	100	120
c.b.	**Kniaz Pojarski**	I.	5138	272	4	49	3	23	11	1	2835	St. Petersburg	1867	..	4½	4½	..	2 8-in., 2 6-in., 10 Q.F. and M., 4 l.	..	11·0	600	452
t.	**Kniaz Potemkine Tavritchesky, B.S.**	S.	12,480	358	0	73	0	27	0	2	10,600 B	Nicolaieff	Bldg.	..	9	10	2½	4 12-in., 10 6-in. Q.F., 16 3-in., 16 1·8-in., 14 1·4-in., 2 l.	5	17·0	900†	636
c.d.s., br.	**Kreml**	I.	3480	219	10	52	5	15	0	1	2822	St. Petersburg	1864	..	4½	4½	..	8 8-in., 6 6-in., 5 Q.F., 6 l.	..	9·0	..	63
a.c.	**Minin**	I.	6136	298	6	49	3	25	3	1	6000 B	St. Petersburg	1878 altd.	..	7	8	..	4 8-in., 12 6-in., 16 Q.F., 4 l.	..	14·0	1200	450
c.d.s., br.	**Netron-Menya**	I.	3494	219	10	53	0	15	6	1	1600	St. Petersburg	1864	..	4½	4½	..	14 8-in., 4 Q.F., 2 M., 4 l.	1	9·0	500	63
t.	**Navarin**	S.	10,206	338	0	67	0	25	0	2	9000	St. Petersburg	1891	772,995	16	12	..	4 12-in., 8 6-in., 14 Q.F., 4 l.	6	16·0	1200	..
t.	**Nicolai I.**	S. shd.	9672	326	0	67	0	23	0	2	8000 B	St. Petersburg	1888	453,000	14 *comp.*	10 tur. 6-in. b. *comp.*	2½	2 12-in. 52-ton, 4 9-in. 19-ton, 8 6-in., 12 Q.F., 8 M., 4 l.	6	14·8	..	604
cir. c.d.s.	**Novgorod, B.S.**	I. shd.	2706	101	0	101	0	13	0	6	2000	Nicolaieff	1873	..	9–7	9	..	2 11-in., 8 Q.F., 2 M., 2 l.	..	6·0	200	150
b.	**Orel**	S.	13,600	397	0	76	0	26	0	2	16,000 W T	St. Petersburg (Galerny)	Bldg.	..	9 H.S.	5 H.S.	4	4 12-in., 12 6-in. Q.F., 20 3-in., 20 3-pr., 6 1-pr.	6 5 sub.	18	2000	..

* Foundered near Viborg, June 24th, 1897. Efforts to refloat the vessel appear not to have been abandoned.

† And liquid fuel.

RUSSIA.—Armoured Ships—*continued.*

(B.S., Black Sea Fleet.)

Class.	NAME.	Material of Hull.	Displacement.	Length.	Beam.	Draught.	Propellers.	Indicated Horse-power.	Where Built.	Date of Launch.	Cost.	Armour. Belt.	Armour. Gun Positio	Armour. Deck Plating.	Armament. Guns. B.L.R. are of Russian Krupp pattern	Armament. Torpedo Tubes.	Speed.	Normal Coal Supply.	Complement.
			tons.	ft. in.	ft. in.	ft. in.					£	inches.	inches.	inches.			knots.	tons.	
t.	**Oslabya** . . .	S.	12,674	434 6	71 6	26 0	3	14,500	St. Petersburg (New Admiralty)	1898	..	9½ H. S.	9 H. S.	2¾	4 10-in., 11 6-in. Q.F., 16 3-in., 10 1·8-in., 17 1·4-in., 2 l.	6	18·0	1063 2056	732
a.g.b.	**Otvazny** . . .	S.	1500	225 0	41 0	11 0	2	2000 2589 (*t*) B	St. Petersburg	1892	..	5	..	1½	1 9-in., 1 6-in., 10 Q.F.	2	15·0 15.5 (*t*)	100	142
a.c.	**Pamyat Azova** . .	S. shd.	6675	377 0	51 0	23 0	2	8000	St. Petersburg	1888	350,000	9 *comp.*	8 *comp.*	2½	2 8-in., 13 6-in., 14 Q.F., and 3 M.	7	18·8	1000	525
t.	**Peresviet** . . .	S.	12,674	434 0	71 9	26 0	3	14,500	St. Petersburg (Baltic)	1898	..	9½ H. S	9 H. S.	2¾	4 10-in., 11 6-in. Q.F., 16 3-in., 10 1 8-in., 17 1·4-in., 2 l.	6	18 0	1063 2056	732
c.d.s., br.	**Pervenetz** . . .	I.	3279	219 10	52 5	14 9	1	1067	Blackwall .	1863	..	4½	4½	..	6 8-in., 9 6-in., 7 Q.F., 8 l. .	..	9·0	..	63
t.	**Peter Veliky** . .	I.	9891	328 2	62 4	23 9	2	8258	St. Petersburg	1872	..	14·8	8·6	3	4 12-in. 40-ton, 13 Q.F., 4 l. .	1	14·5	1200	436
t.	**Petropavlovsk** . .	S.	10,960	367 6	69 0	26 0	2	14,213 (*t*)	St. Petersburg	1894	1,098,000	15¾	10 H. S.	3½	4 12-in., 12 5·9-in. Q.F., 34 smaller.	6	16·3 *t*	900	..
t.	**Poltava** . . .	S.	10,960	367 6	69 0	26 0	2	11,255 (*t*)	St. Petersburg	1894	1,098,000	15¾	10 H. S.	3½	4 12-in., 12 5·9-in. Q.F., 34 smaller.	6	16 2	900	..
b.	**Pobieda** (Victory) .	S.	12,674	434 0	71 9	26* 0	3	16,000 W.T.	St. Petersburg (Baltic)	Bldg.	..	9½ H. S.	9 H. S.	2¾	4 10-in, 11 6-in. Q.F., 16 3-in., 10 1·8-in., 17 1·4-in., 2 l.	6	18·0	1063 2056	732
b.	**Retwisan** . . .	S.	12,700	374 0	72 2	25 0	2	16,000 Nic.	Philadelphia .	B.dg.	..	9 K. S.	10 K. S.	4	4 12-in., 12 6-in. Q.F., 20 3-in., 20 3-pr., 6 1-pr.	..	18·0	1616 2000	..
a. c.	**Rossia** . . .	S. shd.	12,130	480 0	68 6	26 0	3	14,500 B	St. Petersburg	1896	..	10 H.S.	2 H.S.	2½	4 8-in., 16 6-in. Q.F., 12 3-in., 36 small Q.F. & M.	5	20·0	2500 †	725
t.	**Rostislav,** B.S. . .	S.	8880	341 0	66 6	24 0	2	8,500	Nicolaieff .	1896	..	15¾ *comp.*	15¾ *comp.*	2–3	4 10-in., 8 5·9-in. (Canet), 12 1·8-in. Q.F., 4 1·5-in., 2 M.	2–3	16·0	550 † 800	..

a. c.	**Rurik**	S.	10,923	396	6	67	0	26	0	2	13250	St. Petersburg .	1894	..	10 *comp.*	..	2½	4 8-in., 16 6-in., 6 4·7-in. Q.F., 18 small Q.F. & M.	5	18·0	2000	768
t.	**Sevastopol** . . .	S.	10,960	367	6	69	0	26	0	2	13600	St. Petersburg .	1895	1,098,000	15¾	10 H. S.	3½	4 12-in., 12 5.9-in. Q.F., 34 smaller.	6	17·5	900	..
b.	**Sinope**, B.S. . .	I. & S.	10,180	331	0	69	0	26	6	2	13000 B	Sebastopol .	1887	900,000	16 *comp.*	14 *comp.*	3	6 12-in. 50-ton, 7 6-in., 8 Q.F., 6 M.	7	16·75	886	325
t.	**Sissoi Veliky** (Sissoi the Great)	S.	8880	341	0	66	6	24	0	2	8500	St Petersburg .	1894	796,333	15¾ *comp.*	15¾ *comp.*	3	4 12-in., 6 6-in. Q.F., 12 1·8-in., 4 1·4-in., 2 M.	6	16·0	550	..
b.	**Tchesmé**, B.S. . .	I. & S.	10,180	331	0	69	0	26	6	2	11000	Sebastopol .	1886	900,000	16 *comp.*	14 *comp.*	3	6 12-in. 50 ton, 7 6-in., 8 Q.F., 6 M.	7	15·0	886	325
„	**Tria Sviatitelia**, B.S. . (Three Saints.)	S.	12,480	357	6	72	2	27	0	2	10600	Nicolaieff .	1893	..	18–16	16	3	4 12-in., 12 6-in. Q.F., 4 4-in., 4 7-in., 56 smaller Q.F. & M.	6 2 sub.	16·0 18·0 (*t*)	1000	582
b.	**Tsarevitch** . . .	S.	12,900	388	8	75	5	26	*0	2	16300 B	La Seyne .	Bldg.	..	9 K. S.	10·5 K. S.	3	4 12-in., 12 6-in. Q.F., 20 3-in., 2 2·5-in., 20 1·8-in., 6 1·4-in.	4	18·0	1016 2000†	..
circular c.d.s.	**Vice-Admiral Popoff**, B.S.	I. shd.	3590	120	0	120	0	13	5	6	3066	Nicolaieff .	1875	..	16	16	..	2 12-in. 40-ton, 2 Q.F., 6 l. .	..	8·0	250	453
a.c.	**Vladimir Monomach**	S. shd.	6061	296	6	52	0	24	0	2	7000	St. Petersburg .	1882	..	6 *comp.*	..	2	5 8-in., 12 6-in., 18 Q.F. & M., 4 l.	2	15·2	400	550

* Mean draught. † And liquid fuel.

Ten old Monitors of 1566 tons have been removed from this list:—Uragan, Tifon, Streletz, Edinorog, Koldun, Lava, Bronenosetz, Latnik, Perun, and Vieschun; and one of 1461 tons—Smerch.

RUSSIA.—Cruising Ships, &c.

(B.S., Black Sea Fleet.)

Class.	NAME.	Material of Hull.	Displacement.	Length.		Beam.		Draught.		Propellers.	Indicated Horse-power.	Where Built.	Date of Launch.	Cost.	Armour. Gun Position.	Armour. Deck.	Armament. Guns.	Armament. Torpe in Tubes.	Speed.	Normal Coal Supply.	Complement.
			tons.	ft.	in.	ft.	in.	ft.	in.	no.				£		ins.			knots.	tons.	
to.g.b.	Abrek	S.	535	212	2	24	10	9	0	2	4506 (t.)	Abo	1896	53,600	..	½	.. ..	2	21·2 (t.)	..	..
2nd cl. cr.	Admiral Korniloff	S.& W.	5000	351	0	48	6	20	0	2	9000	St. Nazaire	1887 1895	296,000	..	2½	2 8-in., 14 6-in., 6 1·8-in. Q.F., 6 1·4-in., 5 l.	6	17·5	1100	425
cr.	Afrika	I.	2590	285	5	39	4	17	1	1	1350	Chester, U.S.	1877	..	..	..	3 6-in., 6 Q.F., 4 M., 4 l.	..	13·0	975	257
cr.	Asia	I.	2500	269	0	36	0	16	5	1	1100	Philadelphia	1878	..	..	..	2 6-in., 5 Q.F., 6 M., 5 l.	..	13·0	750	260
cr.	Askold	S.	6000	426	6	49	3	20	4	3	19000 Schulz	Kiel (Germania)	1900	..	..	3	12 6-in. Q.F., 12 3-in., 8 1·8, 2 1·4 in., 2 M.	6 (2sub.)	23·0	710	500
cr.	Aurora	S.	6630	413	4	55	9	21	0	3	11610 B	St. Petersburg (Galerny)	Bldg.	..	..	2½	8 6-in. Q.F., 20 3-in., 8 1·4-in.	4	20·0	1100 ..	..
to.g.b.	Bakan (Mining)	S. shd.	840	180	6	15	7	11	6	2	3800 B	St. Petersburg (Baltic)	1896	..	..	..	4 Q.F.	..	12·0	..	..
g.v.	Bobr	S.	950	187	0	35	2	9	6	2	1150	Kretona	1884	43,000	..	..	1 9-in., 1 6-in., 5 Q.F., M., & 6 l.	..	12·0	..	..
cr.	Bogatyr	S.	6600	416	8	54	6	20	8	2	19500 Nor.	Stettin (Vulcan)	Bldg.	..	5 N.S.	2¾	12 6-in. Q.F., 12 3-in., 6 1·4 Hotchkiss.	6 (2sub.)	23·0	720 1000	580
cr.	Boyarin											Copenhagen	Bldg.								
cr.	A	S.	6375	413	4	52	5	20	8	2	20000 B	Windau	Pro.	..	..	2	12 6-in. Q.F., 12 3-in., 8 1·8-in.	5 (2sub.)	23·0	..	..
cr.	B											St.Petersburg	Pro.								
to.g.b.	Captain Sacken, B.S.	S.	742	210	0	24	0	8	10	2	3400	Nicolaieff	1888	40,700	..	..	7 4·7-in. Q.F., 7 M.	6	18·5	97	120
g.v.	Chernomoretz, B.S.	S.	1224	210	0	35	0	11	1	1	2000	Nicolaieff	1889	40,000	..	..	2 8-in., 1 6-in., 7 Q.F. & M.	2	13·5	250	161
,,	Coreetz	S.	1213	206	0	35	0	10	6	2	1500	Stockholm	1886	..	..	1½	2 8-in., 1 6-in., 2 Q.F., 4 l.	2	13·5	..	..
cr.	Diana	S.	6630	413	4	55	9	21	0	3	11610 B	St. Petersburg (Galerny)	1899	..	..	2½	6 6-in. Q.F., 20 3-in., 8 1·4 in.	4	20·0	..	..
corv.	Djigit	I. & W.	1456	206	9	32	10	16	1	1	1700	St. Petersburg	1876	..	..	..	3 6-in., 8 Q.F. & M., & 4 l.	..	13·0	250	172
,,	Donetz, B.S.	S.	1224	210	0	35	0	11	0	1	2000	Nicolaieff	1887	40,000	..	..	2 8-in., 1 6-in., 7 Q.F. & M.	2	13·5	250	161

to.g.b.	Gaidamak	S.	500	192	6	24	2	7	6	2	3000	Abo	1893	..	..	..	2 1·8-in. Q.F., 7 1·4-in., 10 M.	3	22·0	90	87
g.v.	Gilyak	S.	963	200	0	37	0	9	3	2	1000 B	St. Petersburg (New Admiralty)	1897	..	2	..	1 4·7-in. Q.F., 5 3-in., 2 2·6-in., 4 1·8-in.	1	12·0	..	..
to.g.b.	Griden, B.S.	S.	400	192	6	24	2	7	6	2	3500	Nicolaieff	1893	66,600	..	..	2 1·8-in. Q.F., 7 1·4-in., 10 M.	3	22·0	90	60
g.v.	Jermak	I.	706	154	3	26	3	11	2	1	125	St. Petersburg	1870	..	..	..	2 guns	..	..	..	..
to.g.b.	Kazarsky, B.S.	S.	400	190	0	24	0	8	6	2	3500	Elbing	1890	32,500	..	..	9 1·8-in. Q.F. (Hotchkiss)	2	23·0	90	60
corv.	Kreisser	I. & W.	1653	206	9	32	10	16	1	1	1800	St. Petersburg	1875	..	..	..	2 6-in., 7 Q.F., 1 M., 4 l.	..	13·0	..	..
g.v.	Kubanetz, B.S.	S.	1224	210	0	35	0	11	0	1	1500 B	Sebastopol	1888	40,000	..	..	2 8-in., 1 6-in., 7 Q.F.	2	13·8	250	161
to.g.b.	Lieutenant Ilyin	S.	714	230	0	24	0	8	10	2	3500	St. Petersburg	1887	40,150	..	..	7 3-pr. Q.F., 10 M.	7	20·1	97	120
g.v.	Mandjur	S.	1416	210	0	35	0	11	0	2	1400	Copenhagen	1886	..	..	1½	2 8-in., 1 6-in., 7 Q.F., M., & 4 l.	2	14·0	..	..
corv.	Nayezdnik	I. & W.	1334	206	9	32	10	14	0	1	1719	St. Petersburg	1878	..	..	..	3 6-in., 7 Q.F. & M., 4 l.	..	13·0	250	172
cr.	Novik	S.	3000	347	10	40	0	..		3	170 0 W T	Elbing	Bldg.	..	..	2	6 6-in. Q.F. & smaller guns	2	25·0	..	..
corv.	Oprichnik	S. & W.	1426	206	9	32	10	14	0	1	1268	St. Petersburg	1880	..	..	..	3 6-in., 7 Q.F. & M., 4 l.	..	13·0	230	172
cr.	Pallada	S.	6630	413	4	55	8	21	0	3	11610 B	St. Petersburg (Galerny)	1899	..	..	2½	6 6-in. Q.F., 20 3-in., 8 1·4-in.	4	20·0	..	..
3rd cl. cr.	Pamyat Merkuriya, B.S.	I. & S.	3050	295	0	41	0	17	0	1	3000	Toulon	1880	..	..	..	6 6-in., 8 Q.F. & M., 4 l.	2	16·0	1100	200
sl.	Plastun	I. & W.	1255	206	9	32	10	14	0	1	1268	St. Petersburg	1879	..	..	..	3 6-in., 7 Q.F. & M., & 4 l.	..	13·0	250	172
to.g.b.	Posadnik	S.	462	192	6	24	2	7	6	2	3600	Elbing	1892	111,000	..	..	2 1·8-in. Q.F., 7 1·4-in., 3 M.	3	22·0	90	87
corv.	Razboynik	I. & W.	1329	206	9	32	10	14	0	1	1786	St. Petersburg	1878	125,000	..	..	3 6-in., 7 Q.F. & M., & 4 l.	..	13·0	250	172

RUSSIA.—Cruising Ships, &c.—*continued.*

Class.	NAME.	Material of Hull.	Displacement.	Length.	Beam.	Draught.	Propellers.	Indicated Horse-power.	Where Built.	Date of Launch.	Cost.	Armour: Gun Position.	Armour: Deck.	Armament: Guns.	Armament: Torpedo Tubes.	Speed.	Normal Coal Supply.	Complement.
			tons.	ft. in.	ft. in.	ft. in.	no.				£	in.	in.			knots.	tons.	
3rd cl. *cr.*	**Rynda**	S. shd.	3508	265 9	45 11	16 1	2	3000	St. Petersburg	1885	..	..	$1\frac{1}{2}$	10 6-in., 9 Q.F., M., & 4 l.	4	14·8	710	322
g.v.	**Sivootch**	S.	950	187 0	35 0	9 6	2	1125	Stockholm	1884	43,000	..	..	1 9-in., 1 6-in., 5 Q.F., M., & 6 l.	..	12·5	..	..
corv.	**Strjelok**	I. & W.	1343	206 9	32 10	14 0	1	1528	St. Petersburg	1880	..	..	..	3 6-in., 7 Q.F., M., & 4 l.	..	13·0	250	172
cr.	**Svietlana**	S.	3828	331 3	42 8	18 9	2	3828 B	Havre	1896	..	4	2	6 5·9 Q.F. (Canet), 10 1·8-in.	4	20·2 (t.)	1000	..
g.v.	**Teretz, B.S.**	S.	1224	210 0	35 0	11 0	..	1500 B	Sebastopol	1888	40,000	..	..	2 8-in., 1 6-in., 7 Q.F. & M.	2	13·8	250	161
,,	**Uraletz, B.S.**	S.	1224	210 0	35 0	11 0	..	1500 B	Sebastopol	1888	40,000	..	..	2 8-in., 1 6-in., 7 Q.F. & M.	2	13·8	250	161
to.g.b.	**Voevoda**	S.	400	192 6	24 2	7 6	2	3600	Elbing	1892	111,000	..	..	2 1·8-in. Q.F., 7 1·4-in., 3 M.	3	22·0	90	87
sl.	**Vjestnik**	I. & W.	1255	206 9	32 10	14 2	1	1268	St. Petersburg	1879	..	..	..	3 6-in., 7 Q.F. & M., & 4 l.	..	13·0	250	172
to.g.b.	**Vzadnik**	S.	400	192 6	24 2	7 6	2	3000	Abo	1893	..	..	..	4 1 8-in. Q.F., 7 1 4-in. 10 M. & l.	3	22·0	250	172
cr.	**Waryag**	S.	6500	420 0	52 0	20 8	2	20000 Nic.	Philadelphia	1899	..	..	3 N S.	12 6-in. Q.F., 12 3-in., 6 1·4 Hotchkiss.	6 (2 sub.)	23·0	770/1250	571
sl.	**Zabiyaka**	I.	1234	219 10	29 6	14 9	1	1194	Philadelphia	1878	..	..	..	6 Q.F., 4 M., 5 l.	..	14·5	90	87
g.v.	**Zaporojetz**	S.	1224	210 0	35 0	10 0	2	1500	Nicolaieff	1887	40,000	..	..	2 8-in., 1 6-in., 7 Q.F. & M.	2	13·5	250	161
cr.	Unnamed, C	S.	6500	..	..	..	..	..	Nicolaieff	Pro.	..	..	..	..	..	..	..	..
cr.	,, D.	S.	6500	..	..	..	..	..										

Baltic:—*Ten Gunboats, Staunch Class,* of 270 to 402 tons, 195 to 445 I.H.P., with 1 11-inch breech-loader, and 9 knots speed, and *two Gunboats* of about 180 tons and 7 knots speed. *Training Ships,* Bajan, Voin, Vierny, and Moriak. Ermach, very powerful ice-breaker. Black Sea:—*Twelve Steamers (Gun-vessels, Despatch-vessels, &c.)* 90 to 298 tons. *Imperial Yachts,* Standart, Polarnaia Svezda, Tsarevna, &c.

Auxiliary Steamers.

Class.	NAME.	Material of Hull.	Displacement.	Length.	Beam.	Draught.	Propellers.	Indicated Horse-power.	Where Built.	Date of Launch.	Complement.
	BLACK SEA CO.		tons.	ft. in.	ft. in.	ft. in.					
Auxiliary Cruiser	Czar	S.	2340	319 0	37 0	23 6	1	350 nom.	Newcastle	1883	14
,,	Czarevna	,,	2340	319 0	37 0	23 6	1	350 nom.	,,	1883	14
,,	Czaritza	,,	2340	319 0	37 0	23 6	1	350 nom.	,,	1883	14
,,	Grand Duke Alexis	,,	2350	284 0	37 0	14 9	1	3500	Hebburn	1890	16
,,	Grand Duke Constantine	,,	2400	284 0	37 0	15 0	1	3500	,,	1891	16
,,	Grand Duke No. 1	,,	2400	288 0	37 0	15 0	1	2500	,,	Bldg.	14½
,,	Grand Duke No. 2	,,	2400	288 0	37 0	15 0	1	2500	,,	,,	14½
,,	Emperor Nicolas II.	,,	..	..	..	..	..	..	,,	1895	..
,,	Roumantzeff	,,	760	212 0	28 0	7 6	2	1000	,,	1894	13
	VOLUNTEER FLEET.										
,,	Ekaterinoslav	,,	10,500	440 0	49 6	24 0	2	3200	,,	1896	12
,,	Khabarovsk	I.	2700	265 0	36 0	14 6	2	1800	,,	1894	13
,,	Kherson*	S.	10,225 B	493 0	54 3	24 0	2	{ 12,500 B 13,150 (*t.*) }	,,	1895	19½
,,	Kiev	,,	10,500	440 0	49 6	24 0	2	3200	Clydebank	1895	13
,,	Kostroma	I.	7975	360 0	42 0	23 6	1	2700	Hebburn	1888	14
,,	Moskva*	S.	11,700 B	508 0	58 0	25 0	2	12,500 B	Clydebank	1898	20
,,	Nijni Novgorod	I.	7876	325 0	40 0	23 6	1	2000	Elswick	1891	11½
,,	Orel	,,	7990	445 0	48 0	23 6	2	10,000	Hebburn	1889	19
,,	Petersburg	,,	9252	460 0	52 0	24 0	2	11,000	,,	1894	19
,,	Poltava	S.	10,225 B	493 0	54 3	24 0	2	12,500	Dumbarton	Bldg.	11½
,,	Saratoff	,,	8556	462 0	50 0	24 0	2	10,000	Glasgow	1892	19
,,	Tamboff	,,	8640	385 0	45 0	24 6	1	2,500	Dumbarton	1893	12¾
,,	Vladimir	,,	10,500	440 0	49 6	24 0	2	3,200	,,	1895	12
,,	Voronej	,,	10,500	440 0	49 6	24 0	2	3,200	,,	1895	12
,,	Yaroslav	,,	8640	325 0	45 0	24 6	1	2,500	,,	1893	12¾

* Armament, 3 4·7-in. Q.F., 20 smaller. Three other ships of 10,000 tons, 20 knots, in hand.

SPAIN.—Armoured Ships.

Class.	NAME.	Material of Hull.	Displacement.	Length.		Beam.		Draught.		Propellers.	Indicated Horse-power.	Where Built.	Date of Launch.	Cost.	Armour. Belt.	Armour. Gun Position	Armour. Deck Plating.	Armament. Guns.	Armament. Torpedo Tubes.	Speed.	Normal Coal Supply.	Complement.
			metric tons.	ft.	in.	ft.	in.	ft.	in.	no.				£	in.	ins.	ins.			knots.	tons.	
a.c.b.	**Cardenal Cisneros** .	S.	7000	347	10	61	0	21	10	2	15,000	Ferrol .	1896	600,000	12	10½	2	2 11-in., 10 5·5-in. Q.F., 2 2·7-in., 4 2·2-in., 4 1·4-in., 2 M.	8	20·0	1200	500
”	**Cataluña** . . .	S.	7000	347	10	61	0	21	10	2	15,000	Carthagena .	Bldg.	600,000	12	10½	2	2 11-in., 10 5·5-in. Q.F., 2 2·7-in., 4 2·2-in., 4 1·4-in., 2 M.	8	20·0	1200	484
a.c.t.	**Emperador Carlos V.**	S.	9235	380	0	67	0	25	0	2	13,500	Cadiz (Vea Murguia)	1895	734,000	2	10	6⅛	2 11-in. (Hontoria), 8 5·5-in. Q.F., 4 3·9-in., 2 2·7-in., 4 2·2-in., 6 M.	6	20·0	1200	535
br.	**Numancia** . . .	I.	7305	314	10	55	9	25	3	1	8708	La Seyne .	1863 1897	315,600	5½	5	..	8 10-in. M.L.R. (Armstrong), 6 6·2-in. Q.F., 6 4·7 in. (Hontoria), 8 M., 3 l.	2	8·0	1100	600
b.	**Pelayo** . . .	S.	9900	330	0	66	0	24	11	2	9000 Nic.	La Seyne .	1887 1897	..	17¾	19¼	4	2 12·5-in. 48-ton, 2 11-in. 38-ton, 9 5·5-in. Q.F., 6 smaller, 12 M.	7	16·0	800	600
a.c.b.	**Princesa de Asturias**	S.	7000	347	10	61	0	21	10	2	15,000	Carraca .	1896	600,000	12	10½	2	2 11-in., 10 5·5-in. Q.F., 2 2·7-in., 4 2·2-in., 4 1·4-in., 2 M.	8	20·0	1200	500
c.s., t.	**Puig-cerda** . (Monitor) (torpedo training)	I.	553	127	11	29	6	6	7	2	328	La Seyne .	1874	..	4	4	9	1 6·2-in. (Palliser), 2 4·7-in. bronze smooth bores.	..	8·0	23	..
br.	**Vitoria** (training) * .	I.	7250	318	3	55	10	25	3	1	4500	Blackwall .	1865	..	5¼	5	..	8 9-in. M.L.R. (Armstrong), 3 8-in., 1 7·8-in. (Hontoria), 8 M., 2 l.	2	11·0	875	561

* Furnished with fighting-masts and 5·5-in. Q.-F. guns at La Seyne.

SPAIN.—Cruising Ships.

Class.	NAME.	Material.	Displacement.	Length.	Beam.	Draught.	Propellers.	Indicated Horse-power.	Where Built.	Date of Launch.	Cost.	Armour. Gun Position.	Armour. Deck.	Armament. Guns.	Armament. Torpedo Tubes.	Speed.	Normal Coal Supply.	Complement.
			metric tons.	ft. in.	ft. in.	ft. in.	no.				£		ins.			knots.	tons.	
cr.	**Alfonso XII.**	S.	3090	278 10	42 7	16 5	1	4800	Ferrol	1887	..	..	..	6 6·2-in. (Hontoria), 2 2·7-in. 6 6-pr. Q.F., 4 3-pr., 5 M.	5	17·5	600	300
cr.	**Alfonso XIII.**	S.	5000	318 6	50 6	20 0	2	11,000	Ferrol	1891	..	..	4½	4 7·8-in. (Hontoria), 6 4·7-in., 6 2·2-in. Q.F., 6 1·4-in., 3 M.	5	20·0	1200	276
cr.	**Aragon**	W.	3342	246 0	45 11	20 11	1	4400	Carthagena	1879	..	..	..	6 6·2-in. (Hontoria), 2 3·3-in. (Krupp), 4 2·9-in., 2 M.	2	14·0	470	300
g.b.	**General Concha**	I.	524	157 5	25 7	8 7	2	600	Ferrol	1883	..	..	..	3 4·7-in. (Hontoria), 2 Q.F., 1 M.	1	11·5	80	93
cr.	**Conde de Venadito**	I.	1130	210 0	32 0	12 6	1	1600	Carthagena	1888	..	..	..	4 4·7-in. (Hontoria), 2 2·7-in., 2 Q.F., 5 M.	2	14·0	220	130
to.g.b.	**Don Alvaro de Bazan**	S.	823	233 0	26 9	22 0	2	4600	Ferrol	1897	..	..	..	2 4·7-in. (Hontoria) Q.F., 4 1·6-in., 2 M.	4	20·0	..	110
to.g.b.	**Doña Maria de Molina**	S.	823	233 0	26 9	22 0	2	4600	Ferrol	1896	..	..	..	2 4·7-in. (Hontoria), Q.F., 4 1·6-in., 2 M.	4	20·0	..	110
to.g.b.	**Destructor**	S.	458	192 6	25 0	7 0	2	3800	Clydebank	1887	..	..	..	1 3·5-in., 4 6-pr. Q.F., 4 M.	3	22·56	104	55
d.v.	**Fernando el Catolico** (Torpedo training)	I.	500	157 5	25 7	8 5	2	550	La Seyne	1875	..	..	..	1 6·2-in. M.L.R. (Palliser), 2 4·7-in., smooth-bores, 1 M.	..	10·0	20	98
to.g.b.	**Filipinas**	S.	750	213 0	27 0	8 6	2	4600	Cadiz	1892	..	..	..	2 4·7-in. Q.F., 4 1·5-in., 4 M.	4 sub.	20·0	120	110
to.g.b.	**Galicia**	S.	571	190 0	23 0	10 4	2	2600	Le Graña	1891	..	..	..	2 4·7-in. (Hontoria), 4 2·2-in. Q.F., 1 M.	2	19·0	106	80

SPAIN.—Cruising Ships—*continued.*

Class.	NAME.	Material of Hull.	Displacement.	Length.		Beam.		Draught.		Propellers.	Indicated Horse-power.	Where Built.	Date of Launch.	Cost.	Armour. Gun Position.	Armour. Deck.	Armament. Guns.	Armament. Torpedo Tubes.	Speed.	Normal Coal Supply.	Compl ment.
			metric tons	ft.	in.	ft.	in.	ft.	in.	no.				£	ins.	ins.			knots.	tons.	
g. v.	General Lezo	I.	524	157	5	25	7	8	6	2	600	Carthagena	1885	..	..	..	2 4·7-in. (Hontoria), 1 3·5-in., 2 Q.F., 1 M.	2	11·0	80	97
cr.	Infanta Isabel	I.	1130	210	11	32	2	12	5	1	1500	Cadiz	1885	..	..	..	4 4·7-in. (Hontoria), 2 2·7-in., 3 Q.F., 4 M.	2	14·0	220	130
cr.	Isabel II.	I.	1130	210	11	32	2	12	5	1	1500	Ferrol	1886	..	..	..	4 4·7-in. (Hontoria), 2 7·7-in., 4 Q.F., 3 M.	2	14·0	220	130
cr.	Isabel la Catolica	S.	3500	..		..		..		..	..	Carraca	Bldg.	..	..	..	.. .. ..	..	..	..	..
cr.	Lepanto	S.	4826	318	6	50	6	20	0	2	12,000	Carthagena	1892	..	..	4¾	4 7·8-in. (Hontoria), 6 4·7-in. Q.F., 6 6-pr., 4 3-pr. 5 M.	5	20·0	1100	276
g.v.	Magellanes	I.	524	157	5	25	7	8	6	2	600	Cadiz	1885	..	..	..	3 4·7-in. (Hontoria), 3 M.	1	11·0	80	97
to.g.b.	Marques de la Victoria	S.	823	233	0	26	9	22	0	2	4600	Ferrol	1897	..	..	..	2 4·7-in. (Hontoria) Q.F., 4 1·6-in., 2 M.	4	20·0	..	110
cr.	Marques de la Enseñada	S.	1030	185	0	30	0	11	6	2	1600	Carraca	1890	..	..	2½	4 4·7-in. (Hontoria), 5 Q.F., 4 M.	4	15·0	160	164
d.v.	Marques del Duero	I.	500	157	5	25	7	8	5	2	550	La Seyne	1875	..	..	..	1 6·2-in. M.L.R. (Palliser), 2 4·7-in. smooth-bores, 1 M.	..	10·0	90	98
to.g.b.	Marques de Molins	S.	571	190	0	23	0	10	4	2	2600	La Graña	1891	..	..	..	2 4·7in. (Hontoria), 4 2·2-in. Q.F., 1 M.	2	19·0	106	80
"	Martin Alonzo Pinzon	S.	571	190	0	23	0	10	4	2	2600	La Graña	1892	..							
cr.	Navarra	W.	3342	232	11	42	7	20	4	1	4400 Nic.	Ferrol	1881	..	..	..	4 5·9-in., 2 4·7-in., 2 3·4-in., 4 2 9-in., 4 M.	2	14·0	470	300

cr.	Nueva España	S.	630	190	0 23	0 11	9	2	2600	Carraca	1889	.	..	..	2 4·7-in. (Hontoria), 4 2·2-in. Q.F.,	2	18·0	106	91
,,	Quiros	S.	315	155	0 23	0 11	0	..	500	Hong Kong	1895	..	..	..	2 2·2-in. Q.F., 2 M. 1 M.	..	11·5	..	..
to.g.b.	Rapido	S.	570	190	0 23	0 10	4	2	2600	Carraca	1891	..	..	..	2 4·7-in. (Hontoria), 4 2·2-in. Q.F., 1 M.	2	18·0	106	80
cr.	Reina Regente	S.	5000	..	..	..	..	..	..	Ferrol	Bldg.	..	..	..	..	..	20·0	..	..
,,	Rio de la Plata	S. shd.	1800	246	0 35	4 15	0	2	7100	Havre	1898	..	..	1	2 5·5-in. Q.F., 4 3·9-in., 4 2·2-in., 6 M.	2	20·0	270	..
to.g.b.	Temerario	S.	570	190	0 23	0 10	4	2	2600	..	1889	..	..	..	2 4·7-in. (Hontoria), 4 2·2-in. Q.F., 1 M.	2	20·5	106	82
,,	Veloz	S.	750	213	0 27	0 8	6	2	4500	..	Bldg.	..	..	..	2 5·9-in. (Hontoria), 4 2·2-in. Q.F., 4 M.	2	20·0	106	..
,,	Vincente Yanez Pinzon	S.	571	190	0 23	0 10	4	2	2600	La Graña	1891	..	..	..	2 4·7-in. (Hontoria), 4 2·2-in. Q.F., 1 M.	2	19·0	106	80
,,	Villalobos	S.	315	155	0 23	0 11	0	..	500	Hong Kong	1896	..	..	..	2 2·2-in. Q.F., 2 M.	..	11·5	..	..

In the war with the United States a number of Spanish gunboats, including nearly all of the first class, were captured (see the United States Tables), while many were destroyed. After the war a number of gunboats were sold to the United States Government. For a list of these see under "Spain" in Chapter II. The cruisers lost at Manila and Santiago have been removed from the list. A cruiser of 2000 tons, subscribed for by Spaniards in Mexico, is understood to be in hand at Cadiz.

SWEDEN.—Armoured Ships.

Class.	NAME.	Material of Hull.	Displacement.	Length.		Beam.		Draught.		Propellers.	Indicated Horse-power.	Where Built.	Date of Launch.	Cost.	Armour.			Armament.		Speed.	Normal Coal Supply.	Complement.
															Belt.	Gun Position.	Backing, Deck Plating.	Guns.	Torpedo Tubes.			
			metric tons.	ft.	in.	ft.	in.	ft.	in.	no.				£	inches.	inches.	inches.			knots.	tons.	
c.d.s., t.	A. B. C.	S.	3670	287	0	49	3	16	5	2	5500 W. T.	Sweden	Bldg.	..	$7\frac{7}{8}$	$5\frac{3}{4}$–$7\frac{7}{8}$	$1\frac{7}{8}$	2 8·2-in., 6 5·9-in. Q.F., 10 2·2-in., 2 1·4-in., 2 M.	2 sub.	16·5	300	..
a.g.b.	Berserk	I.	460	130	3	26	3	8	3	2	155	Norrköping	1874	..	3	$16\frac{1}{2}$	$\frac{3}{4}$	1 4·7-in. Q.F., 2 2·2-in.	..	8·0	19	45
„	Björn	I.	460	130	3	26	3	8	3	2	155	Norrköping	1874	..	$3\frac{3}{4}$	$16\frac{1}{2}$	$\frac{3}{4}$	1 9·4-in., 2 M.	..	8·0	19	45
c.d.s., t.	Dristigheten	S.	3500	285	1	48	7	16	1	2	5400 W. T.	Gothenburg	Bldg.	..	$7\frac{7}{8}$ N.S.	$5\frac{3}{4}$–$7\frac{7}{8}$ N.S.	$1\frac{7}{8}$	2 8·2-in., 6 5·9-in. Q.F., 10 2·2-in., 2 M.	2 sub.	16·5	300	..
„	Folke	I.	460	130	3	26	3	8	3	2	155	Norrköping	1875	..	3	$16\frac{1}{2}$	$\frac{3}{4}$	1 9·4-in., 2 M.	..	8·0	19	45
„	Gerda	I.	460	130	3	26	3	8	3	2	133	Stockholm	1873	..	3	$16\frac{1}{2}$	$\frac{3}{4}$	1 9·4-in., 2 M.	..	7·6	19	45
c.d.s., t.	Göta	S.	3290	258	6	47	11	16	9	2	4750	Gothenburg	1890	..	$11\frac{9}{16}$–$7\frac{13}{16}$	$11\frac{5}{8}$–$9\frac{5}{8}$	2	2 10-in., 4 6-in., 5 2·2 Q.F., 8 M.	3	16·0	240	150
a.g.b.	Hildur	I.	460	130	3	26	3	8	3	2	133	Stockholm	1872	..	3	$16\frac{1}{2}$	$\frac{3}{4}$	1 9·4-in., 2 M.	..	7·6	19	45
c.d.s., t.	John Ericsson	I.	1500	199	10	45	3	12	2	1	380	Norrköping	1865	..	$4\frac{7}{8}$	$10\frac{5}{8}$	$1\frac{1}{13}$	2 6-in., 2 2·2-in. Q.F., 2 M.	..	7·5	112	80
„	Loke	I.	1600	204	9	44	11	11	10	1	430	Norrköping	1871	..	$4\frac{7}{8}$	$17\frac{5}{8}$	1	2 9·4-in., 2 M.	..	8·5	112	75
„	Njord	S.	3500	278	3	48	7	17	5	2	5350	Gothenburg	1898	..	9·5	$7\frac{7}{8}$	$1\frac{7}{8}$	2 9·8-in., 6 4·7-in. Q.F., 10 2·2-in., 4 M.	1	16·5	275	200
„	Oden	S.	3500	278	3	48	7	17	6	2	5330	Stockholm	1896	..	9·5	$9\frac{7}{8}$–$7\frac{7}{8}$	$1\frac{7}{8}$	2 9·8-in., 4 4·7-in. Q.F., 10 2·2-in., 4 M.	1	16·5	275	200
a.g.b.	Sölve	I.	460	130	3	26	3	8	3	2	155	Norrköping	1875	..	3	$16\frac{1}{2}$	$\frac{3}{4}$	1 9·4-in., 2 M.	..	8·0	19	45
c.d.s., t.	Svea	S.	3100	248	4	49	3	17	1	2	3640	Gothenburg	1886	..	$11\frac{1}{2}$–$7\frac{3}{4}$	$11\frac{1}{2}$–$9\frac{1}{2}$	$1\frac{15}{16}$	2 10-in. (Armstrong), 4 4·7-in., 6 2·2-in., 8 M.	1	14·7	220	268
„	Thor	S.	3500	278	3	48	7	17	6	2	5350	Stockholm	1898	..	9·5	$7\frac{7}{8}$	$1\frac{7}{8}$	2 9·8-in., 6 4·7-in. Q.F., 10 2·2-in., 4 M.	1	16·5	275	200
t.	Thordön	I.	1500	199	10	45	3	11	10	1	380	Norrköping	1866	..	$4\frac{5}{8}$	$10\frac{1}{4}$	1	2 9·4-in., 2 M.	..	6·7	112	80
c.d.s., t.	Thule	S.	3300	260	10	47	11	16	9	2	4740	Stockholm	1892	..	$11\frac{1}{2}$–$7\frac{3}{4}$	$11\frac{1}{2}$–$9\frac{1}{2}$	$1\frac{7}{8}$	2 10-in. (Armstrong), 4 6-in., 5 2·2 Q.F., 8 M.	2	16·2	250	165
„	Tirfing	I.	1500	199	10	45	3	11	10	1	380	Norrköping	1867	..	$4\frac{5}{8}$	$10\frac{1}{4}$	1	2 9·4-in., 2 M.	..	6·8	112	80
a.g.b.	Ulf	I.	460	130	3	26	3	8	3	2	155	Norrköping	1873	..	$3\frac{3}{4}$	$16\frac{1}{2}$	$\frac{3}{4}$	1 4·7-in. Q.F., 2 2·2-in. Q.F.	..	8·0	20	45

SWEDEN.—Cruising Ships, &c.

Class.	NAME.	Material of Hull.	Displacement.	Length. ft.	Length. in.	Beam. ft.	Beam. in.	Draught. ft.	Draught. in.	Propellers.	Indicated Horse-power.	Where Built.	Date of Launch.	Cost.	Armour. Gun Position.	Armour. Deck.	Armament. Guns.	Armament. Torpedo Tubes.	Speed.	Normal Coal Supply.	Complement.
			metric tons.	ft.	in.	ft.	in.	ft.	in.	no.									knots.	tons.	
corv.	**Balder**	W.	1886	202	9	36	5	18	4	1	1380	Carlskrona	1870	..	..	..	1 6-in. (Armstrong), 8 4·7-in., 2 1·5-in., 1 2·5-in., 5 M.	..	12·1	200	218
g.v.	**Blenda**	I.	500	167	8	25	11	9	10	2	590	Gothenburg	1874	..	..	..	1 10·6-in., 1 4·7-in., 2 M.	..	11·5	98	71
to.g.b.	**Psilander**	S.	800	232	0	27	3	8	10	2	4500 W.T.	Stockholm	Bldg.	..	..	..	2 4·7-in. Q.F., 4 2·2-in.	1 (sub.)	20·5	..	..
to.g.b.	**Claes Uggla**																				
g.v.	**Disa**	I.	500	167	8	25	11	9	2	2	590	Carlskrona	1877	..	..	..	1 6-in., 1 4·7-in., 2 2·2., 2 M.	..	11·6	98	72
tor. ship.	Drott (ex Ran)	I.	690	175	2	25	11	9	10	2	960	Stockholm	1877	..	..	..	4 Engström, Q.F.	1	13·0	100	..
g.v.	**Edda**	I.	640	183	9	26	11	10	2	2	960	Carlskrona	1885	..	..	..	1 10·6-in., 1 6-in., 2 1·5., 2 M.	..	13·6	80	76
corv.	**Freja**	S.&W.	2000	215	11	40	0	19	8	1	1750	Malmö	1885	..	..	..	4 6-in., 8 4·7-in., 4 1·5-in., 2 2·5-in., 5 M.	..	14·1	180	250
to.g.b.	Jacob Bagge	S.	800	222	1	26	11	10	2	2	3970	Malmö	1898	..	..	..	2 4·7-in. Q.F., 4 2·2-in.	1 (sub.)	19·5	..	..
”	Örnen										4100	Gothenburg	1896						19·5		
”	Claes Horn										3600	Stockholm	1899						20·0		
g.v.	**Rota**	I.	536	171	3	25	3	10	10	2	780	Stockholm	1878	..	..	..	1 10·6-in., 1 4·7-in., 2 M.	..	13·0	80	72
corv.	**Saga**	W.	1530	200	6	34	5	17	1	1	900	Carlskrona	1878	..	..	..	1 6-in. (Armstrong), 6 4·7-in., 1 M., 3 l.	..	11·2	170	189
g.v.	**Skäggald**	I.	536	171	3	25	11	9	10	2	780	Stockholm	1879	..	..	..	1 10·6-in., 1 4·7-in., 2 M.	..	13·2	80	72
”	**Skagul**	I.	536	171	7	25	7	10	2	2	780	Stockholm	1878	..	..	..	1 6-in., 1 4·7-in., 2 2·2-in., 2 M.	..	13·1	80	72
”	**Skuld**	I.	536	171	3	25	3	10	10	2	780	Carlskrona	1879	..	..	..	1 10·6-in., 1 4·7-in., 2 M.	..	13·0	80	72
”	Urd	I.	536	172	3	25	7	10	6	2	780	Malmö	1877	..	..	..	1 6-in., 1 4·7-in., 2 2·2 Q.F., 2 M.	..	13·5	80	71
”	Verdande	I.	536	171	7	25	7	10	2	2	780	Carlskrona	1879	..	..	..	1 10·6-in., 1 4·7-in., 2 M.	..	13·2	80	72

Four gunboats of 190 to 200 tons, and about 180 I.H.P. each, and carrying 1 5-in B.L.R. and 2 M.; also one vessel of 280 tons and 440 H.P., armed with 4 quick-firing guns —the Svenskund, used as a mining and torpedo-ship and ice-breaker.

TURKEY.—Armoured Ships.

A number of ships have been struck out of these lists owing to information obtained from Constantinople. Of the remainder few have any fighting value.

Class.	NAME.	Material.	Displacement.	Length.	Beam.	Draught.	Propellers.	Indicated Horse-power.	Where Built.	Date of Launch.	Cost.	Armour. Belt.	Armour. Gun Position	Armour. Deck.	Armament. Guns.	Armament. Torpedo Tubes.	Speed.	Normal Coal Supply	Complement.
			tons.	ft. in.	ft. in.	ft. in.	no.					inches.	inches.	inches.			knots.	tons.	
c.	Assar-i-Shefket	I.	2080	203 5	42 7	16 5	2	1750	La Seyne	1868	..	6	5	..	1 9-in. (Armstrong), 4 7-in., 4 M., 4 l.	..	11·0	375	220
c.b.	Assar-i-Tewfik	I.	4687	272 4	52 6	24 11	1	3560	La Seyne	1868	..	8	6	..	2 9·2-in., 6 6-in. Q.F., 10 12-pr., 12 6-pr.*	..	13·0	400	..
c.	Avni-Illah	I.	2400	226 4	36 0	16 5	1	2200	Thames	1869	..	6	6	1½	4 9-in. M.L.R. (Armstrong), 4 M., 4 l.	1	12·0	220	225
b.	Azizieh	I.	6400	292 0	55 9	25 7	1	3735	Clyde	1864	..	5½	4½	..	2 9·2-in. (Krupp), 8 8·2-in., 6 3·9-in., 7 M., 2 l.	2	13·0	750	600
c.b.	Feth-i-Bulend	I.	2806	236 3	39 4	18 1	1	3250	Thames	1869	..	9	9	5	4 9-in. M.L.R. (Armstrong), 4 M., 4 l.	1	13·0	300	250
a.g.b.	Feth-el-Islam	I.	335	101 9	24 7	5 11	1	290	Gironde	1864	..	3	3	..	2 7-in. (Armstrong), 2 l.	..	8·0	20	..
c.b.	Hamidieh	I.	6700	292 0	55 9	24 10	1	4500	Turkey	1885	..	9	5	3	10 10·2-in. (Krupp), 2 6·6-in., 6 l., 2 M.	2	13·0	600	..
b.	Mahmoudieh	I.	6400	292 0	55 9	25 7	1	3735	Thames	1864	..	5½	4½	..	2 9·2-in. (Krupp), 8 8·2-in., 6 3·9-in., 7 M., 2 l.	2	12·0	750	600
c.b.	Messoudieh	I.	9120	331 5	59 0	25 11	1	11,000 W.T.	Thames	1874 1900	..	12	6-9	1	2 9·2-in., 12 6-in. Q.F., 14 3-in., 10 6-pr., 2 3-pr.*	..	15·0	600	..
,,	Muin-i-Zaffer	I.	2400	230 0	36 0	16 5	2	2200	Thames	1869	..	6	6	1½	4 10-in. M.L.R. (Armstrong), 1 4·7-in. (Krupp), 4 M., 4 l.	1	12·0	220	..
,,	Mukadim-i-Hair	I.	2806	236 3	39 4	18 1	1	3000	Turkey	1872	..	9	9	5	4 10-in. M.L.R. (Armstrong), 1 4·7-in. (Krupp), 4 M., 4 l.	1	12·0	300	250
,,	Nedjim-i-Schefhet	I.	2050	203 5	42 7	16 5	2	1900	La Seyne	1868	..	6	5	..	1 9-in., 4 7-in. (Armstrong), 4 M., 4 l.	1	11·0	300	220
b.	Orkanieh	I.	6400	292 0	55 9	25 7	1	3735	Clyde	1865	..	5½	4½	..	2 9·2-in. (Krupp), 8 8·2-in., 6 3·9-in., 7 M., 2 l.	2	12·0	750	600
,,	Osmanieh	I.	6400	292 0	55 9	25 7	1	3735	Clyde	1864	..	5½	4½	..	2 9·2-in. (Krupp), 8 8·2-in., 6 3·9-in., 7 M., 2 l.	2	12·0	750	600

* The Assar-i-Tewfik and Messoudieh are now at Genoa, and the latter is in hand at Messrs. Ansaldo's yard, receiving the new armament and machinery indicated. Nothing has as yet been decided in regard to the Assar-i-Tewfik.

TURKEY.—Cruising Ships, &c.

NAME.	Material of Hull.	Displacement.	Length.		Beam.		Draught.		Propellers.	Indicated Horse-power.	Where Built.	Cost.	Date of Launch.	Armour.		Armament.		Speed.	Normal Coal Supply.	Complement.
														Gun Position.	Deck.	Guns.	Torpedo Tubes.			
		tons.	ft.	in.	ft.	in.	ft.	in.				£		in.	in.			knots.	tons.	
Fezibahri	S.	1815	226	0	35	0	14	0	2	2500	Turkey	..	Bldg.	..	½	6 6-in. (Krupp)	7	17·0	..	..
Heibetnuma	S. &W.	1960	226	0	37	0	14	0	1	2500 ind.	Turkey	..	1890	..	..	3 6·6-in. (Krupp), 6 4·7-in., 6 Q.F.	2	14·0	..	..
Hundavendikiar	S.	4050	279	0	49	3	21	0	2	..	Turkey	..	Bldg.	..	2	2 8·9-in. (Krupp), 6 5·9-in., 4 3·9-in.	5	..	..	300
Lutfi-hamayoun	C.	1313	210	0	35	0	14	0	1	2800	Turkey	..	1892	..	..	4 6-in. (Krupp), 6 4·7-in. 6 Q.F.	2	13·0	..	..
Namet	S.	900	230	0	31	0	16	6	2	4500	Gaarden	..	1890	¼	..	2 4-in. (Krupp), 16 M.	2	19·0	..	111
Pelenk-i-deria	S.	840	236	3	31	0	16	6	2	5000	Gaarden	..	1890	¼	..	2 4-in. (Krupp), 16 M.	2	20·0	..	111
Sedul Bahr	W.	800	173	6	26	7	11	6	1	160	Turkey	..	1894	..	2	4 4·7-in. (Krupp), 6 M.	2	12·7	120	..
Selimieh	S.	4050	279	0	49	3	21	0	2	..	Turkey	..	Bldg.	..	½	2 8·2-in. (Krupp), 6 5·9-in., 4-in., 6 M.	..	..	..	300
Shadie	S.	1815	226	0	35	0	14	0	2	2500	Turkey	..	Bldg.	..	..	6 5·9-in. (Krupp)	7	17·0	..	..
Shahani-deria	S.	450	200	0	23	0	9	0	2	8000	Turkey	..	1892	..	..	2 4·7-in. Q.F. (Krupp), 6 M.	4	22·0	..	..
Zuhaf	W.	800	173	6	26	7	11	6	1	160	Turkey	..	1894	..	..	4 4·7-in. (Krupp), 6 M.	2	12·7	120	..

UNITED STATES.—Armoured Ships.

Class.	NAME.	Material of Hull.	Displacement.	Length.		Beam.		Draught.		Propellers.	Indicated Horse-power.	Where Built.	Date of Launch.	Cost.	Armour. Belt.	Armour. Gun Position	Armour. Deck Plating.	Armament. Guns.	Armament. Torpedo Tubes.	Speed.	Normal *Coal Supply.	Complement.
			tons.	ft.	in.	ft.	in.	ft.	in.	no.				£	inches.	inches.	inches.			knots.	tons.	
c.d.s., t. (1 t.)	Ajax	I.	2100	225	0	43	8	13	6	1	340	Pittsburg	1864	141,083	5	10	..	2 15-in. smooth-bores	..	6·0	160	105
t.	Alabama	S.	11,525	368	0	72	3	25	6	2	10,000	Philadelphia	1898	544,539	9½–16½	10–15	2¾–4	4 13-in., 14 6-in. Q.F., 16 6-pr., 6 1-pr., 4 M., 2 l.	4	16·0	800 / 1200	489
c.d.s., t. (2 t.)	Amphitrite	I.	3990	259	6	55	6	14	8	2	1600	Wilmington	1883	..	5–9	7½–11½	1¾	4 10-in., 2 4-in. Q.F., 2 6-pr., 2 3-pr., 6 1-pr., 1 M., 1 l.	..	10·5	250	182
c.d.s., t. (1 t.)	Arkansas	S.	3000	252	0	50	0	12	6	2	2400	Newport News	Bldg.	197,267	11	10–11	1½	2 12-in., 4 4-in. Q.F., 3 6-pr., 4 1-pr.	..	11·5	200 / 400	131
a. c.	Brooklyn	S.	9215	400	6	64	8	26	2	2	18,769	Philadelphia	1895	613,583	3	5½–8	6–3	8 8-in., 12 5-in. Q.F., 12 6-pr., 4 1-pr., 4 M., 2 l.	4	21·9	900 / 1461	516
a.c.	California	S.	12,000	..		..		24	0	2	22,000	..	Bldg.	..	5–6	6	4	4 8-in., 14 6-in. Q.F., 18 14-prs. & 28 smaller.	..	23·0	2000	500
c.d.s., t. (1 t.)	Canonicus	I.	2100	225	0	43	8	13	9	1	340	Boston	1864	128,011	5	10	..	2 15-in. smooth-bores, 2 12-pr. H.	..	6·0	160	106
c.d.s., t. (1 t.)	Catskill	I.	1875	200	0	46	0	11	6	1	340	Brooklyn	1863	87,900	5	11	..	2 15-in. smooth-bores	..	6·0	150	105
c.d.s., t. (1 t.)	Comanche	I.	1875	200	0	46	0	11	6	1	340	Jersey City	1863	125,997	5	11	..	2 15-in. smooth-bores	..	6·0	160	105
c.d.s., t. (1 t.)	Connecticut	S.	3000	252	0	50	0	12	6	2	2,400	Bath, Me.	Bldg.	197,714	11	10–11	1½	2 12-in., 4 4-in. Q.F., 3 6-pr., 4 1-pr.	..	11·5	200 / 400	131
c.d.s., t. (1 t.)	Florida											Elizabeth Port	Bldg.	190,075								
t.	Georgia	S.	13,500	420	0	75	0	24	0	3	18,400 W. T.	..	Bldg.	..	6–9 K.	10–11 K.	2¾–3½	4 12-in., 4 8-in., 12 6-in. Q.F., 12 14-prs., 12 3-prs., 8 1-pr., 8 M, 2 l.	2 sub.	18.5	2200	
t.	Illinois	S.	11,525	368	0	72	3	25	0	2	10,000	Newport News	1898	533,237	9½–16½	10–15	2¾–4	4 13-in., 14 6-in. Q.F., 16 6-pr., 6 1-pr., 4 M., 2 l.	4	16·0	800 / 1200	489
b.	Indiana	S.	10,288	348	0	69	3	27	2	2	9,738	Philadelphia	1893	620,569	18	6–17	2¾	4 13-in., 8 8-in., 4 6-in. Q.F., 20 6-pr. Q.F., 7 1-pr., 2 l.	2	15·5	400 / 1597	473

b.	**Iowa**	S.	11,340	360	0	72	3	26	10	2	12,105	Philadelphia	1896	618,514	5–14 H. S.	6–15	2¾	4 12-in., 8 8-in., 6 4-in. Q.F., 20 6-pr., 4 1-pr., 4 M., 2 l.	4	17·1 (*t*)	625 / 1795	505
c.d.s., t. (1 t.)	**Jason**	I.	1875	200	0	46	0	11	6	1	340	Chester	1864	86,872	5	11	..	2 15-in. smooth-bores, 2 12-pr. H.	..	6·0	150	105
ram	**Katahdin**	S.	2155	250	9	43	5	16	0	2	5,068	Bath, Me.	1893	191,102	3–6	18	2–6	4 6-pr.	..	16·1	175 / 192	97
super-posed turrets	**Kearsarge** / **Kentucky**	S.	11,525	368	0	72	2	25	1	2	10,000	Newport News	1898	462,345 each	9½–16½ H. S.	9–17 H. S.	2¾–5	4 13-in., 4 8-in., 14 5-in. Q.F., 20 6-pr., 6 1-pr., 4 M., 2 l.	4	16·0	410 / 1645	511
c.d.s., t. (1 t.)	**Lehigh**	I.	1875	200	0	46	0	11	6	1	340	Chester	1864	86,864	5	11	..	2 15-in. smooth-bores, 2 12-pr. H.	..	6·0	150	105
c.d.s., t. (1 t.)	**Mahopac**	I.	2100	225	0	43	8	13	6	1	340	Jersey City	1865	130,560	5	10	..	2 15-in. smooth-bores, 2 12-pr. H.	..	6·0	160	105
t.	**Maine**	S.	12,500	388	0	72	2	25	6	2	16,000 (Nic.)	Philadelphia	Bldg.	592,828	8½–12 K.	9–12 K.	2¾–4	4 12-in., 16 6-in. Q.F., 20 6-pr., 6 1-pr., 4 M., 1 l.	2 sub.	18·0	1000 / 2000	515
c. d. s., t. (1 t.)	**Manhattan**	I.	2100	225	0	43	8	13	6	1	340	Jersey City	1865	129,247	5	10	..	2 15-in. smooth-bores, 2 12-pr. H.	..	6·0	160	105
b.	**Massachusetts**	S.	10,288	348	0	69	3	27	2	2	10,403	Philadelphia	1893	620,569	18	6–17	2¾	4 13-in., 8 8-in., 4 6-in., 20 6-pr., 6 1-pr., 2 M., 2 l.	3	16·2	400 / 1597	473
c.d.s., t. (2 t.)	**Miantonomoh**	I.	3990	259	6	55	6	15	0	2	1,426	Chester	1876 reblt.	..	7	11½	1¾	4 10-in., 2 6-pr. Q.F., 2 3-pr., 6 1-pr., 2 M., 1 l.	..	10·5	250	149
t.	**Missouri**	S.	12,500	388	0	72	3	25	6	2	16,000 W. T.	Newport News	Bldg.	592,828	8½–12 K.	9–12 K.	2¾–4	4 12-in., 16 6-in. Q.F., 20 6-pr., 6 1-pr., 4 M., 1 l.	2 sub.	18·0	1000 / 2000	515
c.d.s., t. (2 t.)	**Monadnock**	I.	3990	259	6	55	6	14	7	2	3,000	Vallejo, Cal.	1883 reblt.	..	5–9	7½–11½	1¾	4 10-in., 2 4-in. Q.F., 2 6-pr., 2 3-pr., 2 1·4-in., 2 1-pr., 2 M., 1 l.	..	12·0	250	183
c.d.s., t. (1 t.)	**Montauk**	I.	1875	200	0	46	0	11	6	1	340	Brooklyn	1864	86,903	5	11	..	2 15-in. smooth-bores, 2 12-pr. H.	..	5·5	160	105
c.d.s., t.	**Monterey**	S.	4084	256	0	59	0	15	4	2	5244	S. Francisco	1891	345,731	6–13	7½–13	3	2 12-in., 2 10-in., 6 6-pr., 4 1-pr., 2 M., 1 l.	..	13·6	200 / 236	195
c.d.s., t. (1 t.)	**Nahant**	I.	1875	200	0	46	0	11	6	1	340	Boston	1863	84,910	5	11	..	2 15-in. smooth-bores, 2 12-pr. H.	..	6·0	160	105

* The figures below the line in this column are bunker capacity.

UNITED STATES.—Armoured Ships—*continued.*

Class.	NAME.	Material of Hull.	Displacement.	Length.	Beam.	Draught.	Propellers.	Indicated Horse-power.	Where Built.	Date of Launch.	Cost.	Armour. Belt.	Armour. Gun Position.	Armour. Deck Plating.	Armament. Guns.	Armament. Torpedo Tubes.	Speed.	Normal Coal Supply.	Complement.
			tons.	ft. in.	ft. in.	ft. in.	no.				£	in.	in.	in.			knots.	tons.	
c.d.s.,t. (1 t.)	**Nantucket**	I.	1875	200 0	46 0	11 6	1	340	Boston	1863	83,857	5	11	..	2 15-in. smooth-bores, 2 12-pr. H.	..	5·6	160	105
a.c.	**Nebraska**	S.	12,000	..	..	24 0	3	22,000	..	Bldg.	..	5–6	6	4	4 8-in., 14 6-in. Q.F., 18 14-pr., and 28 smaller.	..	23·0	2000	500
t.	**New Jersey**	S.	13,500	420 0	75 0	24 0	3	18,400 W. T.	..	Bldg.	..	6–9 K.	10–11 K.	2¾–3½	4 12-in., 4 8-in., 12 6-in. Q.F., 12 14-pr., 12 3-pr., 8 1-pr., 8 M., 2 l.	2 sub.	18·5	2200	
a.c.	**New York**	S.	8200	380 6	64 10	26 8	2	17,401	Philadelphia	1891	613,377	4	5½–10	6–3	6 8-in., 12 4-in. Q.F., 8 6-pr., 2 1-pr., 2 M., 2 l.	2	21·0	750	556
t.	**Ohio**	S.	12,500	388 0	72 3	25 6	2	16,000 W. T.	S. Francisco.	Bldg.	595,705	8½–12 K.	9–12 K.	2¾–4	4 12-in., 16 6-in. Q.F., 20 6-pr., 6 1-pr., 4 M., 1 l.	2 sub.	18·0	1290 1000	515
b.	**Oregon**	S.	10,288	348 0	69 3	27 2	2	11,111	S. Francisco.	1893	653,447	18	6–17	2¾	4 13-in., 8 8-in., 4 6-in. Q.F., 20 6-pr., 6 1-pr., 2 M., 1 l.	3	16·7	2000 400	473
c.d.s.,t. (1 t.)	**Passaic**	I.	1875	200 0	46 0	11 6	1	340	Brooklyn	1863	86,956	5	11	..	2 15-in. smooth-bores, 2 12-pr. H.	..	6·0	1594 160	105
t.	**Pennsylvania**	S.	13,500	420 0	75 0	24 0	3	18,400 W. T.	..	Bldg.	..	6–9 K.	10–11 K.	2¾–3½	4 12-in., 4 8-in., 12 6-in. Q.F., 12 14-pr., 12 3-pr., 8 1-pr., 8 M., 2 l.	2 sub.	18·5	2200	
c.d.s.,t. (2 t.)	**Puritan**	I.	6060	289 6	60 1½	18 0	2	3,700	Chester	1882 reblt.	..	6–14	8–14	2	4 12-in., 6 4-in. Q.F., 6 6-pr., 2 1·4-in., 2 1-pr., 1 l.	..	12·4	320	230
c.d.s.,t. (2 t.)	**Terror**	I.	3990	259 6	55 6	15 4	2	1,600	Philadelphia	1883 reblt.	..	7	11½	1¾	4 10-in., 2 6-pr., 2 3-pr., 4 1-pr., 2 M., 1 l.	..	10·5	250	176
a.c.	**Texas**	S.	6315	301 4	64 1	24 6	2	8,610	Norfolk	1892	513,716	12	12	2	2 12-in., 6 6-in., 12 6-pr. Q.F., 10 1-pr., 2 M., 1 l.	6	17·8 (*t*)	500 850	389
a.c.	**West Virginia**	S.	12,000	..	..	24 0	2	22,000	..	Bldg.	..	5–6	6	4	4 8-in., 14 6-in. Q.F., 18 14-pr., and 28 smaller.	..	23·0	2000	500
t.	**Wisconsin**	S.	11,525	368 0	72 3	25 5	2	10,000	S. Francisco.	1898	549,666	9½–16½	10–15	2¾–4	4 13-in., 14 6-in. Q.F., 16 6-pr., 4 1-pr., 4 M., 2 l.	4	16·0	800 1200	489
c.d.s.,t. (1 t.)	**Wyandotte**	I.	2100	225 0	43 8	13 6	1	340	Cincinnati	1864	131,401	5	10	..	2 15-in. smooth-bores, 2 12-pr. H.	..	6·0	160	105
c.d.s.,t. (1 t.)	**Wyoming**	S.	3,000	252 0	50 0	12 6	2	2,400	S. Francisco.	Bldg.	200,850	11	10–11	1½	2 12-in., 4 4-in. Q.F., 3 6-pr., 4 1-pr.	..	11·5	200	131

The new programme (April, 1900) includes 2 battleships (13,500 tons), 3 armoured cruisers (13,000 tons), and 3 protected cruisers (8,000 tons).

UNITED STATES.—Cruising Ships, &c.

Class.	NAME.	Material of Hull.	Displacement.	Length.		Beam.		Draught.		Propellers.	Indicated Horse-power.	Where Built.	Date of Launch.	Cost.	Armour. Gun Position.	Armour. Deck.	Armament. Guns.	Armament. Torpedo Tubes.	Speed.	Normal Coal Supply.	Complement.
			tons.	ft.	in.	ft.	in.	ft.	in.	no.				£	in.	in.			knots.	tons.	
cr.	Albany (ex Abreu)	S. sh.	3440	330	0	43	9	16	10	2	7500	Elswick	1899	247,611	..	3½	6 6-in. Q.F., 4 4·7-in., 10 6-pr., 4 1-pr., 4 M., 2 l.	..	20·0	400/800	260
g.b.	Annapolis	S.	1000	168	0	36	0	12	7	1	1227 B.&W.	Elizabeth Pt.	1896	46,789	..	..	6 4-in. Q.F., 4 6-pr., 2 1-pr., 1 M.	..	13·1 (*t*)	100/225	135
cr.	Atlanta	S.	3000	271	3	42	2	21	1	1	4030 W. T.	Chester	1884	126,785	..	1½	2 8-in., 6 6-in. Q.F., 6 6-pr., 4 1-pr., 2 M., 1 l.	..	15·6	490	278
cr.	Baltimore	S.	4413	327	6	48	7	23	11	2	10,064	Philadelphia	1888	272,270	4½	4–2½	4 8-in., 6 6-in., 4 6-pr., 2 3-pr., 2 1-pr., 6 M., 1 l.	5	20·1	400/1143	386
g.v.	Bancroft	S.	839	187	6	32	0	12	11	2	1213	Elizabeth Pt.	1892	51,371	..	½	4 4-in. Q.F., 6 6-pr., 8 3-pr., 1 1-pr., 1 M.	1	14·37	100/136	195
g.v.	Bennington	S.	1710	230	0	36	0	16	7	2	3436	Chester	1890	100,894	..	½	6 6-in., 2 6-pr., 2 3-pr., 4 M., 1 l.	..	17·5	200/403	195
cr.	Boston	S.	3000	271	3	42	2	21	1	1	4300	Chester	1884	127,196	..	1½	2 8-in., 6 6-in., 2 6-pr., 2 3-pr., 2 1-pr., 2 1·8-in., 2 1·4-in., 2 M., 1 l.	..	15·6	495	278
g.v.	Castine	S.	1177	204	0	32	1	14	4	2	2199	Bath, Me.	1892	65,450	..	½	8 4-in. Q.F., 4 6-pr., 2 1-pr., 1 M., 1 l.	.	16·0	125/292	151
cr.	Charleston*	S.	3730	312	7	46	2	21	8	2	6666	S. Francisco	1888	209,103	2	2–3	2 8-in., 6 6-in., 4 6-pr., 2 3-pr., 6 1-pr., 2 M., 1 l.	4	18·2	328/757	306
cr. *cr.*	Chattanooga Cleveland	S. shd.	3100	292	0	43	0	16	8	2	4500 W. T.	..	Bldg.	..	..	..	10 5-in. Q.F., 8 6-pr., 2 1-pr., 4 M.	..	16·5	470/700	293
cr.	Chicago	S.	4500	325	0	48	2	22	6	2	9000 B.&W. (½)	Chester	1885	182,677	4	1½	4 8-in. Q.F., 14 5-in. Q.F., 7 6-pr., 2 1-pr., 2 M., 1 l.	..	18·0	831	409
cr.	Cincinnati	S.	3213	300	0	42	0	20	2	2	10,000	Brooklyn	1892	226,055	..	2½	1 6-in., 10 5-in. Q.F., 8 6-pr., 2 1-pr., 2 M., 1 l.	2	19·0	350/468	314
cr.	Columbia	S.	7375	412	0	58	2	25	7	3	18,509	Philadelphia	1892	559,950	4	4–2½	1 8-in., 2 6-in., 8 4-in. Q.F., 12 6-pr., 2 1-pr., 2 M., 1 l.	4	22·8 (*t*)	750/1670	477
g.v.	Concord	S.	1710	230	0	36	0	16	7	2	3405	Chester	1890	100,894	..	½	6 6-in., 2 6-pr., 2 3-pr., 2 1·4-in., 2 M., 1 l.	..	16·8	200/401	194

* The Charleston was wrecked on the north-east coast of Luzon, November, 1899.

UNITED STATES.—Cruising Ships, &c.—*continued.*

Class.	NAME.	Material of Hull.	Displacement.	Length.	Beam.	Draught.	Propellers.	Indicated Horse-power.	Where Built.	Date of Launch.	Cost.	Armour: Gun Position.	Armour: D.ck.	Armament: Guns.	Armament: Torpedo Tub's.	Speed.	Normal Coal Supply.	Complement.
			tons.	ft. in.	ft. in.	ft. in.	no.				£	in.	in.			knots.	tons.	
cr.	Detroit	S.	2089	257 0	37 0	16 8	2	5277	Baltimore	1891	125,860	..	½	10 5-in. Q.F., 6 6-pr., 2 1-pr., 2 M., 1 l.	2	18·71	200/340	256
cr. / cr.	Denver / Des Moines	S. shd.	3100	292 0	43 0	16 8	2	4500 W.T.	..	Bldg.	..	..	..	10 5-in. Q.F., 8 6-pr., 2 1-pr., 4 M.	..	16·5	470/700	293
g.v.	Dolphin	S.	1486	240 0	32 0	17 1	1	2253	Chester	1884	64,728	..	..	3 4-in. Q.F., 2 14-pr., 2 6-pr., 2 3-pr., 2 M.	..	15·5	173	117
cr.	Don Juan de Austria*	I.	1130	210 0	32 0	12 6	1	1600	Carthagena	1889	..	..	..	4 5-in. Q.F., 4 6-pr., 4 M.†	3	14 0		130
cr.	Galveston	S. shd.	3100	292 0	43 0	16 8	2	4500 W.T.	..	Bldg.	..	..	..	10 5-in. Q.F., 8 6-pr., 2 1-pr., 4 M.	..	16·5	470/700	293
g.v.	Helena	S.	1397	250 9	40 1	10 0	2	1988	Newport News	1896	57,536	..	½	8 4-in. Q.F., 4 6-pr., 4 1-pr., 2 M., 1 l.	..	15·5 (t)	100/300	256
cr / cr.	Isla de Cuba* / Isla de Luzon*	S.	1030	185 0	30 0	11 6	2	2200	Elswick	1887	..	..	2½	4 4-in. Q.F., 4 6-pr., 4 M.†	3	16·0	160	160
g.v.	Machias	S.	1177	204 0	32 1	14 4	2	2046	Bath, Me.	1891	65,450	..	½	8 4-in. Q.F., 4 6-pr., 2 1-pr., 1 M., 1 l.	..	15·46	125/292	151
cr.	Marblehead	S.	2089	257 0	37 0	16 8	2	5451	Boston	1892	138,498	..	½	10 5-in. Q.F., 6 6-pr., 2 1-pr., 2 M., 1 l	2	18·4	200/340	248
g.b.	Marietta	S.	1000	174 0	34 0	13 5	2	1054 B.&W.	S. Francisco	1896	45,823	..	..	6 4-in. Q.F., 4 6-pr., 2 1-pr., 1 M., 1 l.	..	13·2 (t)	120/2 6	140
cr.	Minneapolis	S.	7875	412 0	58 2	25 7	3	20,862	Philadelphia	1893	552,754	4	4–2½	1 8-in., 2 6-in., 8 4-in. Q.F., 12 6-pr., 2 1-pr., 2 M., 1 l.	4	23·0 (t)	750/1891	477
cr.	Montgomery	S.	2089	257 0	37 0	16 8	2	5580	Baltimore	1891	125,860	..	½	10 5-in. Q.F., 6 6-pr., 2 1-pr., 2 M., 1 l.	2	19·0	200/340	257
g.v.	Nashville	S.	1371	220 0	38 1	12 0	2	2536 W.T.	Newport News	1895	57,536	..	½	8 4-in. Q.F., 4 6-pr., 2 1-pr., 2 M., 1 l.	1	16·7 (t)	150/400	176
cr.	Newark	S.	4098	311 7	49 2	22 7	2	8869	Philadelphia	1890	256,437	3	3–2	12 6-in. Q.F., 8 6-pr., 4 1-pr., 2 M., 1 l.	..	19·0	400/809	384
cr.	New Orleans (ex Amazonas).	S. shd.	3437	346 0	43 9	20 2	2	7500	Elswick	1896	293,684	..	3–1¼	6 6-in. Q.F., 4 4·7-in., 10 6-pr., 4 3-pr., 4 M., 2 l.	8	20·0	700/800	407

g.b.	**Newport** . . .	S.	1000	168	0	36	0	12	7	1	1008	Bath, Me. .	1896	47,406	..	..	6 4-in. Q.F., 4 6-pr. 2 1-pr., 1 M. . .	..	12·3 (*t*)	100 / 239	135
cr.	**Olympia** . . .	S.	5870	340	0	53	1	24	10	2	17,313	S. Francisco.	1892	369,054	4½–3½	4–2½	4 8-in., 10 5-in. Q.F., 14 6-pr., 7 1-pr., 2 M., 1 l.	6	21·69	400 / 1169	450
g.v.	**Petrel** . . .	S.	892	176	3	31	0	13	5	1	1095	Baltimore .	1888	50,755	..	½	4 6-in., 2 3-pr., 2 1-pr., 2 1·4-in., 2 M. .	..	11·8	100 / 200	122
cr.	**Philadelphia** . .	S.	4324	327	6	48	7	23	5	2	8815	Philadelphia	1889	277,405	..	4–2½	12 6-in., 4 6-pr., 4 1-pr., 2 M., 1 l. . .	..	19·68 (*t*)	400 / 1085	384
g.b.	**Princeton** . . .	S.	1000	168	0	36	0	12	7	1	800	Camden .	1897	47,262	..	..	6 4-in. Q.F., 4 6-pr., 2 1-pr., 1 M. . .	..	12·0	100 / 238	135
cr.	**Raleigh** . . .	S.	3213	300	0	42	0	20	2	2	10,000	Norfolk .	1892	226,055	..	2½	1 6-in., 10 5-in. Q.F., 8 6-pr., 4 1-pr., 2 M., 1 l.	2	19·0	350 / 460	313
cr.	**Reina Mercedes**‡ .	S.	3090	282	2	42	7	16	5	1	3970	Ferrol . .	1886	..	..	..	6 6·2-in., 2 2·7-in., 3 2·2-in. Q.F., 8 smaller	5	17·5	600	370
cr.	**San Francisco** . .	S.	4098	310	0	49	2	22	3	2	9913	S. Francisco.	1889	293,435	..	3–2	12 6-in. Q.F., 4 6-pr., 4 3-pr., 2 1-pr., 6 M., 1 l.	4	19·5	350 / 628	383
cr.	**Tacoma** . . .	S. shd.	3100	292	0	43	0	16	8	2	4500 W. T.	..	Bldg.	..	..	..	10 5-in. Q.F., 8 6-pr., 2 1-pr., 4 M.	..	16·5	470 / 700	293
cr.	**Vesuvius** . . . (Dynamite Gun Cruiser)	S.	929	252	4	26	6	11	2	2	3795	Philadelphia	1888	71,963	..	¼	3 15-in. dynamite guns, 5 3-pr., 1 M. .	..	21·4	152	69
g.v.	**Vicksburg** . . .	S.	1000	168	0	36	0	12	7	1	1118	Bath, Me. .	1896	47,406	..	..	6 4-in. Q.F., 4 6-pr., 2 1-pr., 1 M. . .	..	12·7 (*t*)	100 / 239	135
g.v.	**Wheeling** . . .	S.	1000	174	0	34	0	12	7	2	1081	S. Francisco.	1897	65,540	..	..	6 4-in. Q.F., 4 6-pr., 2 1-pr., 1 M., 1 l. .	..	12·9 (*t*)	120 / 226	140
g.v.	**Wilmington** . .	S.	1397	250	9	40	1	10	0	2	1894	Newport News	1895	57,536	..	¼	8 4-in. Q.F., 4 6-pr., 4 1-pr., 4 M., 1 l. .	..	15·0 (*t*)	100 / 300	175
g.v.	**Yorktown** . . .	S.	1710	230	0	36	0	16	7	2	3392	Philadelphia	1888	93,496	..	¼	6 5-in. Q.F., 2 6-pr. Q.F., 2 3-pr., 4 1-pr., 2 M., 1 l.	2	16·1	200 / 380	195

* Captured at Manila after the battle of May 1, 1898, and under repair at Hong Kong. The following gunboats and some others were also captured at Cuban ports and in the Philippines: first-class, Hernan Cortez, Pizarro, Vasco Nuñez de Balboa, Diego Velasquez, Alvarado and Sandoval; second and third-class, Alerta, Ardilla, Fradera, Flecha, Ligera, Satelite, Margarit, Vigia, General Blanco, Intrepida, Cauto, and Leyte.
† New armament of the captured cruisers.
‡ Sunk at the mouth of Santiago Harbour, July 5, 1898, and refloated by the Americans.

Enrolled Auxiliary Cruisers of the United States Navy.

	Class.	NAME.	Material of Hull.	Gross Tonnage.	Length.	Beam.	Depth.	Propellers.	Indicated Horse-Power.	Where Built.	When Built.	Owners.	Armament, all Q.F.	Speed.
Atlantic Coast.	1st	St. Louis	S.	11,629	535 5	63 0	26 8	2	18,000	Philadelphia	1895	International Navigation Company.	8 5·5-in., 4 6-pr., 4 M.	22·2
	1st	St. Paul	S.	11,629	535 5	63 0	26 8	2	18,000	,,	1895	,,	8 5·5-in., 4 6-pr., 4 M.	22·5
	1st	Paris.	S.	10,794	517 0	63 3	22 0	2	20,000	Clydebank, Scotland	1889	,,	12 5·5-in., 6 6-pr., 6 M.	20·7
	1st	New York	S.	10,802	517 0	63 3	22 0	2	20,000	,,	1888	,,	12 5·5-in., 6 6-pr., 6 M.	20·6
	3rd	Newport	I.	2735	326 0	38 2	23 9	1	..	Chester, Pa.	1880	Pacific Mail.	8 4-in., 8 M.	16·0
	3rd	City of Para	I.	3532	345 0	38 6	19 9	1	2250	,,	1878	,,	8 4-in., 8 M.	12·0
	3rd	Caracas	I.	2584	283 6	40 2	20 5	1	..	Philadelphia	1889	Red D Line.	8 4-in., 6 M.	13·0
	3rd	Philadelphia	I.	2520	300 0	35 0	20 6	1	..	,,	1885	,,	8 4-in., 6 M.	12·0
	3rd	Venezuela	I.	2843	303 6	40 2	20 5	1	..	,,	1889	,,	8 4-in., 8 M.	..
	3rd	Orizaba	S.	3497	336 2	43 2	22 0	1	..	Chester, Pa.	1889	Cuba Mail.	8 4-in., 6 M.	14·0
	3rd	Yumuri	S.	3497	336 2	43 2	22 3	1	..	,,	1889	,,	8 4-in., 6 M.	14·0
	3rd	City of Washington	I.	2684	300 5	38 4	19 2	1	2500	,,	1877	,,	8 4-in., 8 M.	15·0
	3rd	Saratoga	I.	2820	298 0	38 6	23 5	1	..	,,	1878	,,	8 4-in., 8 M	14·0
	3rd	Seneca	I.	2729	271 5	40 0	16 4	1	..	,,	1884	,,	8 4-in., 6 M.	..
	3rd	Yucatan	I. & S.	3525	336 2	43 2	22 3	1	..	,,	1890	,,	8 4-in., 6 M.	14·0
	3rd	Segurança	S.	4033	321 3	45 3	17 4	1	..	,,	1890	,,	6 5-in., 4 4-in., 4 1-pr., 3 M.	14·0
	3rd	Vigilancia	S.	4115	321 8	45 2	16 5	1	..	,,	1890	,,	6 5-in., 4 4-in., 4 1-pr., 3 M.	14·0
	4th	Advance	I.	2605	295 0	38 0	23 4	1	1350	,,	1883	Panama R. R. Co.	Small Q.F.	15·0
	4th	Alliança	I.	2985	303 0	42 0	15 0	1	2250	,,	1886	,,	,,	14·0

Pacific Coast.	3rd	City of Sydney	I.	3017	339 0	40 2	20 5	1	1950	,,	1875	Pacific Mail.	6 6-in., 10 6-pr., 2 M.	15·0
	3rd	City of Peking	I.	5079	408 0	47 0	19 5	1	4500	,,	1874	,,	6 5-in., 12 6-pr.	13·0
	3rd	City of Rio de Janeiro	I.	3548	345 0	38 6	19 9	1	2000	,,	1878	,,	8 4-in., 8 M.	14·0
	3rd	Peru	S.	3528	336 0	45 9	27 6	1	2800	San Francisco	1892	,,	9 5-in., 12 6-pr.	14·0
	4th	City of Panama	I.	1490	248 0	36 0	20 2	1	650	Chester, Pa.	1873	,,	Small Q.F.	..
	4th	Colon	I.	2686	280 0	40 0	28 6	1	1350	,,	1872	,,	,,	..
	4th	San Jose	I.	2081	283 0	37 0	21 0	1	1400	,,	1882	,,	,,	..
	4th	San Blas	I.	2075	283 0	37 0	21 0	1	1400	,,	1882	,,	,,	..
	4th	San Juan	I.	2076	283 0	37 0	21 0	1	1400	,,	1882	,,	,,	..
	4th	Acapulco	I.	2572	290 0	40 0	20 0	1	1350	Wilmington	1873	,,	,,	..

Converted Merchant Vessels Retained.

Class.	NAME.	Material of Hull.	Displacement.	Length.		Beam.		Draught.		Propellers.	Indicated Horse-power.	Where Built.	Date of Launch.	Cost.	Armour.		Armament.		Speed.	Normal Coal Supply.	Complement.
															Gun Position.	Deck.	Guns.	Torpedo Tubes.			
			tons.	ft.	in.	ft.	in.	ft.	in.	No.				£	in.	in.			knots.	tons.	
cr.	Badger	S.	4784	326	6	42	0	18	6	1	3200	Chester	1889	75,412	..	..	6 5-in. Q.F., 6 3-pr.	..	16·0	836	235
cr.	Buffalo	S.	6888	380	6	48	0	22	0	1	3600	Newport News	1893	117,949	..	..	2 5-in. Q.F., 4 4-in., 6 6-pr., 2 M.	..	14·5	100	297
cr.	Dixie	S.	6114	389	2	48	0	19	11	1	1371	Newport News	1893	117,949	..	..	10 6-in. Q.F., 6 6-pr., 2 M.	..	16·0	1371	181
cr.	Panther	I.	4260	310	0	40	0	18	3	1	..	Philadelphia	1889	77,055	..	..	6 5-in. Q.F., 2 4-in., 6 3-pr., 1 M., 1 1.	..	13·0	475	198
cr.	Prairie	I.	6872	390	6	46	10	22	0	1	3800	Philadelphia	1890	117,949	..	..	10 6-in. Q.F., 6 6-pr., 2 M.	..	14·5	1000	295
cr.	Yankee	I.	6888	380	6	48	0	22	0	1	3800	Newport News	1892	117,949	..	..	10 5-in. Q.F., 6 6-pr., 2 M.	..	14·5	1000	282
cr.	Yosemite	I.	6179	389	2	48	0	20	1	1	3800	Newport News	1892	117,949	..	..	10 6-in. Q.F., 6 6-pr., 2 M.	..	16·0	1371	285
cr.	Mayflower(yacht)	S.	2690	275	0	36	0	17	3	2	4700	Clydebank	1896	88,359	..	..	2 5-in. Q.F., 12 6-pr., 2 M.	..	16·8	584	160

SHIPS BELONGING TO POWERS WHOSE NAVIES ARE OF LITTLE OR NO IMPORTANCE.

Belgium.—Twelve steam vessels, between 419 and 684 tons net, launched between 1870 and 1888, principally employed as packets, which are under the orders of the Government.

Bulgaria.—Eleven steamers of small size, of which one is used as the Prince's Yacht. Two armoured gunboats, for the defence of the Danube, building at Leghorn. Other ships are to be laid down. The Nadiezda, a despatch vessel (715 tons) of the French Casabianca type; length, 219 ft. 6 in.; beam, 27 ft. 6 in.; draught, 12 ft. 6 in.; launched at Bordeaux in 1898, steamed at 18·85 knots at her trials; engines, 2600 I.H.P.; Lagrafel and d'Allest boilers; armament, 2 3·9-in., 3 1·8-in. Q.F., and 2 torpedo-tubes.

Egypt.—This Power has now no efficient warships.

Hayti.—Steel gun vessel—Crête à Pierrot—940 tons, length 210 ft., beam 30 ft.; 1 6·2-in., 1 4·7-in., and 4 3·9-in. Q.F., 6 M. Steel gunboat—Capois la Mort—260 tons, 1 3·9-in., and 4 1-pr. Q.F. Iron corvette—Dessalines—1200 tons, armed with 1 3·9-in. Q.F., 2 3·9-in. B.L., 2 l., 2 M. Three iron or steel sloops:—St. Michael, 1804, and Toussaint L'Ouverture, of from 500 to 900 tons, all of 12 to 14 knots speed, and armed with one large and four to eight small guns. Gun vessel, 22nd of December, of 900 tons, 9 knots speed, armed with four 40-pr. Armstrongs.

Liberia.—The Gorronommah gunboat, of 150 tons displacement, completed 1892; and another one, the Rocktown, completed at Rotterdam in 1896 (12 knots on trial).

Mexico.—The Zaragoza, built of steel, 1200 tons, 1300 horse-power, 15 knots speed, and armed with four 4·7-in. guns and 4 rapid-firing guns. Two gun vessels, Democrata and Mexico, of 450 tons and 11 knots speed, armed with two 6½-inch muzzle loaders and two small guns. Two small gunboats of 10 knots speed. A gunboat is in hand at New Orleans. Five torpedo-boats.

Morocco.—A torpedo cruiser, of 1200 tons displacement, 2500 HP., 18 knots speed, and carrying two guns, 4·7-in. B.L., and 4 Q.F. guns, built in 1892. A gunboat of 450 tons, 1200 I.H.P., 14·5 knots, is completing at Sampierdarena (Maclaren & Wilson) and another has been laid down.

Persia.—Despatch vessel—the Persepolis—of 1200 tons and 10 knots speed. She is armed with 5 small breech-loading guns.

Peru—Lima, built in 1881, of 1700 tons displacement, 1800 horse-power, and 16 knots speed; armed with two 6-in. B.L.R. guns. Screw steamer Santa Rosa, of about 400 tons.

Roumania.—Elizabetha, protected cruiser (deck 3 in. thick), built in 1887 at Elswick; 230 ft. long, 32 ft. 10 in. beam, 1320 tons, 4500 I.H.P.; 4 5·9-in. B.L.R., 4 Q.F., 2 M., 4 torpedo tubes. Composite gunboat Mircea, 350 tons; Grivitza, 180 tons. Six gunboats of 45 to 110 tons, 7 to 9 knots speed. Six coast-guard vessels—Oltul, Siretul, Bistritza, Olteano, Smeo, and Monteano—95 tons, 100 ft. long, 13·6 in. beam, 6 ft. draught; speed, natural draught 11 knots, forced draught 13½ knots; 1 Q.F., 2 M. Screw steamer—Romania—240 tons, repaired 1890. Six first-class torpedo-boats (120 ft. 6 in., 21 knots); 2 second class (63 ft., 16·5 knots), built 1882–1888. The shipbuilding programme contemplates the building of 8 monitors of 500 tons, 12 torpedo-boats and 8 vedettes for the Danube, and 6 coast-defence vessels of 3500 tons, 4 destroyers of 300 tons, and 12 torpedo-boats for the Black Sea.

Saint Domingo.—The Independencia, built in England 1894, 170 ft. long, 25 ft. broad, displacement 322 tons, and armed with seven Hotchkiss quick-firing guns. Restauracion, steel gun-vessel, 1000 tons, launched at Glasgow in 1896. The 14-knot cruiser Presidente has been reconstructed, and carries seven guns.

Sarawak.—Two gunboats, of 175 and 118 tons respectively of low speed, each armed with two guns.

Siam.—Two corvettes (800 tons, 8 guns); six gunboats. One deck-protected cruiser, the Maha Chakrkri, 290 ft. long, 39 ft. 4 in. broad, of 2500 tons displacement and 17 to 18 knots speed; armament, four 4·7-in. quick-firing guns, and ten 6-pr. quick-firing guns. Cruiser Makut-Rajakamar, 650 tons.

Uruguay.—Gunboats: General Artigas, 274 tons, 12½ knots speed, 2 4·7-in. (Krupp), 2 M.; General Rivera, 300 tons, 12 knots speed, armed with 1 5·9-in. and 1 2·3-in. gun; and the General Jaurez.

Venezuela.—Gun-vessel, Libertador, 832 tons. Four river gunboats building.

BRITISH AND FOREIGN TORPEDO-BOAT FLOTILLAS.

Great Britain and Dependencies.

Name or Number.	Where Built.	Launched.	Dimensions.			Number of Screws.	Displacement.	Indicated Horse-Power.	Mean Speed on Trial, or expected.	Armament.	Torpedo Tubes.	Complement.	Coal Capacity.
			Length.	Beam.	Draught.								
Great Britain.			Feet	Feet.	Feet.		Tons.		Knots.				Tons
TORPEDO-BOAT DESTROYERS													
Ardent	Chiswick ..	1894	201·6	19	7·3	2	247	4,500	27·97	1-12 pr. 5-6 prs.	2	45	60
Banshee	Birkenhead ..	1894	210	19·5	..	2	290	4,400	27·97	1-12 pr. 5-6 prs.	2	50	
Boxer	Chiswick ..	1894	201·6	19	7·3	2	247	4,500	27·17	1-12 pr. 5-6 prs.	2	45	60
Bruiser	Chiswick ..	1895	201·6	19	7·3	2	247	4,500	27·97	1-12 pr. 5-6 prs.	2	45	60
Charger	Poplar	1894	190	18·5	5·25	2	250	3,100	27·98	1-12 pr. 5-6 prs.	2	45	60
Conflict	East Cowes ..	1894	205·6	20	..	2	270	4,370	27·21	1-12 pr. 5-6 prs.	2	50	
Contest	Birkenhead ..	1894	210	19·5	..	2	290	4,400	27·4	1-12 pr. 5-6 prs.	2	50	
Daring	Chiswick ..	1893	185	19	7	2	237	4,300	27·70	1-12 pr. 3-6 prs.	3	45	50
Dasher	Poplar	1895	190	18·5	5·25	2	250	3,182	26·21	1-12 pr. 5-6 prs.	2	45	60
Decoy	Chiswick ..	1894	185	19	7	2	237	4,300	27·76	1-12 pr. 3-6 prs.	3	45	50
Dragon	Birkenhead ..	1894	210	19·5	..	2	290	4,500	27·14	1-12 pr. 5-6 prs.	2	50	
Ferret	Birkenhead ..	1893	194	19·25	5	2	280	4,810	27·62	1-12 pr. 3-6 prs.	3	50	70
Fervent	Paisley	1895	200	19	7·8	2	270	3,800	[27]	1-12 pr. 5-6 prs.	2	50	70
Handy	Govan	1895	200	19	7·8	2	261	3,800	27·04	1-12 pr. 5-6 prs.	2	50	70
Hardy	Sunderland ..	1895	196	19	5	2	245	4,200	26·8	1-12 pr. 5-6 prs.	2	50	70
Hart..	Govan	1895	185	19	7	2	260	4 010	27·07	1-12 pr. 5-6 prs.	2	60	70
Hasty	Poplar	1894	190	18·5	5·25	2	250	3,250	26·03	1-12 pr. 5-6 prs.	2	45	60
Haughty..	Sunderland ..	1895	196	19	5	2	265	4,000	27·1	1-12 pr. 5-6 prs.	2	50	60
Havock	Poplar	1893	180	18·5	5·25	2	240	3,500	26·77	1-12 pr. 3-6 prs.	3	43	57
Hornet	Poplar	1893	180	18·5	5·25	2	240	4,000	27·31	1-12 pr. 3-6 prs.	3	43	57
Hunter	Govan	1895	..	..	..	2	260	4,000	27·2	1-12 pr. 5-6 prs.	2		
Janus	Jarrow	1895	200	19·7	†6·5	2	252	3,789	27·8	1-12 pr. 5-6 prs.	2	50	60
Lightning	Jarrow	1895	200	19·7	6·5	2	252	4,007	27·94	1-12 pr. 5-6 prs.	2	50	60
Lynx	Birkenhead ..	1894	194	19·25	5	2	280	4,000	27·00	1-12 pr. 3-6 prs.	3	50	70
Opossum	Hebburn ..	1895	200	19	5·2	2	290	4,052	28·24	1-12 pr. 5-6 prs.	2	50	60
Porcupine	Jarrow	1895	200	19·7	6·5	2	288	3,866	27·91	1-12 pr. 5-6 prs.	2	50	60
Ranger	Hebburn ..	1895	200	19	5·2	2	264	3,900	27·13	1-12 pr. 5-6 prs.	2	50	60
Rocket	Clydebank ..	1894	205·6	19·5	5·25	2	280	4,200	27·37	1-12 pr. 5-6 prs.	2	50	60
Salmon	Hull	1895	200	19·5	5·4	2	264	3,580	27·61	1-12 pr. 5-6 prs.	2	50	60
Shark	Clydebank ..	1894	205·6	19·5	5·25	2	280	4,250	27·59	1-12 pr. 5-6 prs.	2	50	60
Skate	Barrow	1895	195	20·5	..	2	265	4,100	27·10	1-12 pr. 5-6 prs.	2	50	60
Snapper	Hull	1895	200	19·5	5·5	2	270	4,500	27·9	1-12 pr. 5-6 prs.	2	50	60
Spitfire	Elswick ..	1895	200	19	5·3	2	300	3,780	27·5	1-12 pr. 5-6 prs.	2	45	60
Starfish	Barrow	1894	195	20·5	..	2	265	4,000	27·97	1-12 pr. 5-6 prs.	2	45	60
Sturgeon	Barrow	1894	195	20·5	..	2	265	4,010	27·16	1-12 pr. 5-6 prs.	2	45	60
Sunfish	Hebburn ..	1895	200	19	5·2	2	290	4,292	27·62	1-12 pr. 5-6 prs.	2	50	60
Surly	Clydebank ..	1894	205·6	19·5	5·25	2	280	4,400	28·05	1-12 pr. 5-6 prs.	2	50	50
Swordfish	Elswick ..	1895	200	19	5·3	2	300	4,100	[27]	1-12 pr. 5-6 prs.	2	45	60
Teazer	East Cowes ..	1895	200	19·5	5·6	2	270	4,500	[27]	1-12 pr. 5-6 prs.	2	50	60
Wizard	East Cowes ..	1895	200	19·5	5·2	2	270	4,400	[27]	1-12 pr. 5-6 prs.	2	45	60
Zebra	Blackwall ..	1895	200	20	6	2	300	3,850	27·00	1-12 pr. 5-6 prs.	2	50	60
Zephyr	Paisley	1895	200	19	5·3	2	270	3,850	[27]	1-12 pr. 5-6 prs.	2	50	60
Albatross	Chiswick ..	1898	227·6	21·25	8·5	2	360	7,900	32	1-12 pr. 5-6 prs.	2	68	100
Angler	Chiswick ..	1896	210	19·6	7·1	2	278	5,800	30·87	1-12 pr. 5-6 prs.	2	60	80
Arab	Clydebank ..	bldg.			Design not complete					1-12 pr. 5-6 prs.	2	60	80
Ariel	Chiswick ..	1897	210	19·6	7·1	2	278	5,800	30·59	1-12 pr. 5-6 prs.	2	60	80
Avon	Barrow	1896	210·6	21·6	5·6	2	300	6,000	30	1-12 pr. 5-6 prs.	2	60	80
Bat	Jarrow	1806	215	20·75	6·8	2	326	6,185	30·1	1-12 pr. 5-6 prs.	2	60	91
Bittern	Barrow	1897	210·6	21·6	5·6	2	300	6,000	30	1-12 pr. 5-6 prs.	2	60	80
Brazen	Clydebank ..	1896	218	20·0	5·6	2	300	6,000	30	1-12 pr. 5-6 prs.	2	60	80
Bullfinch..	Hull	bldg.	210	20·6	5·8	2	300	5,800	30	1-12 pr. 5-6 prs.	2	60	80
Chamois	Jarrow	1896	215	20·75	6·8	2	325	6,333	30·2	1-12 pr. 5-6 prs.	2	60	91
Cheerful	Hebburn ..	1897	210	21·0	8	2	308	6,000	30	1-12 pr. 5-6 prs.	2	62	82
Coquette	Chiswick ..	1898	210	19·5	7·2	2	285	5,800	30·31	1-12 pr. 5-6 prs.	2	60	80
Crane	Jarrow	1896	215	20·7	6·8	2	324	6,336	30·3	1-12 pr. 5-6 prs.	2	60	80
Cygnet	Chiswick ..	1898	210	19·5	7·2	2	285	5,800	30·35	1-12 pr. 5-6 prs.	2	60	80
Cynthia	Chiswick ..	1898	210	19·5	7·2	2	285	5,800	30·2	1-12 pr. 5-6 prs.	2	60	80
Desperate	Chiswick ..	1895	210	19·6	7·2	2	275	5,800	30	1-12 pr. 5-6 prs.	2	60	80
Dove	Hull	bldg.	210·0	20·6	5·8	2	300	5,800	30	1-12 pr. 5-6 prs.	2	60	80
Earnest	Birkenhead ..	1896	210·6	21·7	6·3	2	300	6,000	30·13	1-12 pr. 5-6 prs.	2	58	80
Electra	Clydebank ..	bldg.	218	20·0	5·6	2	300	6,000	30	1-12 pr. 5-6 prs.	2	58	80
Express	Birkenhead ..	1897	227·6	22·0	9	2	300	7,700	33	1-12 pr. 5-6 prs.	2	60	80
Fairy	Govan	1897	227·6	22·0	9	2	300	7,700	32	1-12 pr. 5-6 prs.	2	60	80
Falcon	Fairfield ..	bldg.	220	21·3				6,000	30				
Fame	Chiswick ..	1896	210·6	19·6	7·1	2	275	5,800	30·16	1-12 pr. 5-6 prs.	2	60	80

* Built by Yarrow, fitted with Thornycroft W. T. boilers at Earle's. All Jarrow-built destroyers have Reed's boilers. Vessels marked † have Thornycroft W. T. boilers of "Daring" type.

Great Britain and Dependencies—*continued.*

Name or Number.	Where Built.	Launched.	Dimensions. Length.	Beam.	Draught.	Number of Screws.	Displacement.	Indicated Horse-Power.	Mean Speed on Trial, or expected.	Armament.	Torpedo Tubes.	Complement.	Coal Capacity.
TORPEDO BOAT DESTROYERS			Feet	Feet.	Feet.		Tons.		Knots.				Tons
Fawn	Jarrow	1897	215	20·7	6·8	2	325	6,581	30·5	1–12 pr. 5–6 prs.	2	60	91
Flirt	Jarrow	1897	215	20·7	6·8	2	328	6,682	30	1–12 pr. 5–6 prs.	2	60	91
Flyingfish	Jarrow	1897	215	20·7	6·8	2	323	6,416	30·4	1–12 pr. 5–6 prs.	2	58	91
Foam	Chiswick	1896	210	19·6	7·1	2	275	5,800	30·18	1–12 pr. 5–6 prs.	2	58	80
Gipsy	Govan	1897	227·6	22·0	9	2	300	7,700	32	1–12 pr. 5–6 prs.	2	60	80
Greyhound													
Griffon	Birkenhead	1896	210·0	20	5·3	2	300	6,000	30·11	1–12 pr. 5–6 prs.	2	58	80
Kestrel	Clydebank	bldg.	218	20·0	5 6	2	300	6,000	30	1–12 pr. 5–6 prs.	2	60	80
Leo	Sunderland	bldg.	210 0	19·9	7·6	2	283	5,400	30	1–12 pr. 5–6 prs.	2	58	88
Leopard	Barrow	1897	210	20·0	5·6	2	300	6,000	30	1–12 pr. 5–6 prs.	2	60	80
Leven	Glasgow	bldg.	218 0	20.0	5·6	2	300	6,000	30	1–12 pr. 5–6 prs.	2	58	80
Lively													
Locust	Birkenhead	1896	210	21·7	5·3	2	300	6,000	30·16	1–12 pr. 5–6 prs.	2	58	80
Mallard	Chiswick	1896	210·6	19·6	7·1	2	275	5,800	30·11	1–12 pr. 5–6 prs.	2	60	80
Mermaid	Hebburn	1898	210	21 0	8	2	308	6,000	30	1–12 pr. 5–6 prs.	2	62	82
Myrmidon	Jarrow	bldg.	215	20·75	6·8	2	335	6,500	30	1–12 pr. 5–6 prs.	2		91
Orwell	Birkenhead	bldg.	218·0	20·0	5·6	2	300	6,000	30	1–12 pr. 5–6 prs.	2	58	80
Osprey	Govan	bldg.	227·6	22·0	9	2	300	7,700	32	1–12 pr. 5–6 prs.	2	60	80
Ostrich	Fairfield	bldg.	210	21·0				6,000	30				
Otter	Barrow	1896	210	20·0	5·6	2	300	6,000	30	1–12 pr. 5–6 prs.	2	60	80
Panther	Birkenhead	1897	210·6	21·7	5·3	2	300	6,000	30·14	1–12 pr. 5–6 prs.	2	58	80
Peterel	Jarrow	1899	215	20·75	6·8	2	334	6,500	30	1–12 pr. 5–6 prs.	2		91
Quail	Birkenhead	1895	213 6	21·6	5·3	2	300	6,000	30·38	1–12 pr. 5–6 prs.	2	58	90
Racehorse													
Recruit	Glasgow	bldg.	218·0	20·0	5·6	2	300	6,000	30	1–12 pr. 5–6 prs.	2	58	80
Roebuck													
Seal	Birkenhead	1897	218·0	20·0	5·6	2	300	6,000	30·15	1–12 pr. 5–6 prs.	2	58	80
Sparrowhawk	Birkenhead	1896	210·6	21·7	5·3	2	300	6,000	30·13	1–12 pr. 5–6 prs.	2	58	80
Spiteful	Jarrow	1898	215	20·75	6·8	2	334	6,500	30·1	1–12 pr. 5–6 prs.	2		91
Sprightly													
*b*Stag	Chiswick	1899	210	19·75	7·2	2	285	5,800	30	1–12 pr. 5–6 prs.	2	60	80
Star	Jarrow	1896	215	20·75	6·88	2	328	6,266	30·7	1–12 pr. 5–6 prs.	2	58	91
*c*Success	Sunderland		210·0	21·0	9·24	2	350	6,000	30	1–12 pr. 5–6 prs.	2	62	83
Sylvia	Sunderland	bldg.	210	19·9	7·6	2	283	5,400	30	1–12 pr. 5–6 prs.	2	58	80
Syren	Jarrow	bldg.	215	20·75	6·8	2	335	6,500	30	1–12 pr. 5–6 prs.	2		91
Thrasher	Birkenhead	1896	210·6	21·7	5·3	2	300	6,000	30·13	1–12 pr. 5 6 prs.	2	58	80
Violet	Sunderland	bldg.	210	20·75	6·88	2	283	5,400	30	1–12 pr. 5–6 prs.	2	58	80
*a*Viper	Hebburn	1899	210	21	7	8	312½	6,500	31	1–12 pr. 5–6 prs.	2	62	88
								10,000	35				
Virago	Birkenhead	1896	210·6	21·7	5·3	2	300	6,000	30·13	1–12 pr. 5 6 prs.	2	58	80
*d*Vixen	Barrow	bldg.	210·0	20·0	5·8	2	327	6,000	30	1–12 pr. 5–6 prs.	2	62	80
Vulture	Clydebank	bldg.	218	20	5·6	2	300	6,000	30	1–12 pr. 5–6 prs.	2	58	80
Whiting	Jarrow	1896	215	20·75	6·88	2	330	6,239	30·2	1– 2 pr. 5·6 prs.	2	58	91
Wolf	Birkenhead	1897	218	20	5·6	2	300	6,000	30	1–12 pr. 5·6 prs.	2	58	80
TORPEDO BOATS— FIRST CLASS—													
1 (*ex* Lightning)	Chiswick	1877	84·6	10·9	5	1	27	460	19	..	1	..	
2–9 (8 boats)	Chiswick	1878–9	87	10·9	4	1	28	450	20	..	1	15	
10	Chiswick	1880	90·5	10·9	4	1	28	450	21·7	..	1	15	
11, 12 (2 boats)	Chiswick	1880	87	10·9	4	1	28	450	20	..	1	15	7
13	Lambeth	1878	87	10·9	4	1	28	460	21	..	2	15	7
14	Poplar	1878	87	11	4·5	1	33	550	22	..	2	15	7
15	..	..	87	10·9	4	1	28	450	21	..	2	15	7
17, 18 (2 boats)	Poplar	1877	86	11	4·5	1	33	450	21	..	2	15	7
19	East Cowes	1878	87	10·9	4	1	28	460	21	..	2	15	7
20	..	1880	87	10	4	..	28	360	16·9	..	2	15	
21, 22 (2 boats)	Chiswick	1885	113	12·5	5·7	1	63	730	20	..	3	..	10
23, 24 (2 boats)	Poplar	1885–6	113	12·5	5·5	1	67	600	19·5	2–3 prs.	3		
25–29 (5 boats)	Chiswick	1886	127·5	12·5	6·2	1	60	600	21	..	4	15	
30–33 (4 boats)	Poplar	1886	125	13	5·5	1	60–66	670	19·5	2–3 prs.	5	15	20
34–38 (5 boats)	East Cowes	1886	125	14·6	4	1	60–66	950	18–19	..	5	15	
39, 40 (2 boats)	Poplar	1885	100	12·5	..	..	40	500	..	..	1	15	
41–60 (20 boats)	Chiswick	1886	127·5	12·5	6·2	1	60	700	21	2–3 prs.	4	15	
61, 63–74, 76–78 (16 boats)	Poplar	1886	125	13	5·5	1	75	700	19–20	2–3 prs.	5	15	20
79	Poplar	1886	125	13	5·5	..	75	1,000	22·4	2–3 prs.	..	15	20
80	Poplar	1887	135	14	6	1	105	1,540	23	4–3 prs.	5	21	30
81 (*ex* Swift)	East Cowes	1885	150	17·5	..	1	125	..	..	6–3 prs.	3	25	35
82–87 (6 boats)	Poplar	1889	130	13·5	5·5	1	85	1,100	23	3–3 prs.	3	19	20
88, 89 (2 boats)	Poplar	1894	142	14·75	4 5	1	112	1,600	..	3–3 prs.	3	18	20
90	Poplar	1895	140	14·25	3·7	1	100	1,430	..	3–3 prs.	3	18	18
91, 92 (2 boats)	Chiswick	1894	140	15·5	7·5	1	130	2,400	23–24	3–3 prs.	3	18	25
93	Chiswick	1893	140	15·5	5·4	2	130	2,200	23·5	3–3 prs.	3	18	25
94–96 (3 boats)	East Cowes	1894	140	15·5	..	1	130	2,000	23·2	3–3 prs.	3	18	25
97	Birkenhead	1893	140	15·5	..	1	130	2,690	23·35	3–3 prs.	3	18	25
98 and 99	Chiswick	bldg.	160	17	8·4	1	178	2,850	25	3–3 prs.	3	32	40
*e*2 Unnumbered		Design not complete.											

a This vessel is fitted with the Parsons compound steam turbine. *b* Have Thornycroft W. T. boilers of "Daring" type. *c* Has four Thornycroft W. T. boilers. *d* Has four Express W. T. boilers. *e* Programme 1900–1901. Nos. 51, 87 fitted with Earle's W. T. boilers.

Great Britain and Dependencies—*continued.*

Name or Number.	Where Built.	Launched.	Dimensions.			Number of Screws.	Displacement.	Indicated Horse-Power.	Maximum Trial Speed.	Armament.	Torpedo Tubes.	Complement.	Coal Capacity.
			Length.	Beam.	Draught.								
TORPEDO BOATS.—*cont.*			Feet.	Feet.	Feet.		Tons.		Knots.				Tons.
SECOND CLASS—													
38–48 (10 boats)	Poplar	1889	60	9·2	3·7	1	16·5	230	16·5	1 mach.	1	9	11
49, 50 (2 boats)	Poplar	1887	60	8·5	3	1	15	200	17	1 mach.	1	9	
51–62 (12 boats)	Chiswick ..	1878–9	60·5	7·5	3·5	1	..	..	16·5	..	2	7	
63	..	1879	60	..	..	1	..	..	15	..	2		
64–73 (10 boats)	Chiswick ..	1880–1	60·5	7·5	3·5	1	..	..	16–17	..	2	7	
74, 75, 96, 97 (4 boats)..	Poplar	1883	62	7·6	3·6	1	12	..	16	1 mach.	2	7	
76–95 (20 boats)	Chiswick ..	1882–3	63	7·5	3·5	1	..	..	16·5–17	..	2	7	
98	Chiswick ..	1883	66·8	7·5	2·5	hyd.	..	120	12·6	..	2	7	
99, 100 (2 boats)	Chiswick ..	1886	64	8	3·6	1	..	..	16–16·8	..	2	7	
101	..	..	64	..	..	..	..	..	..	..	2	7	
1–9 (9 boats)	East Cowes ..	..	56	..	..	1	12	..	14·5	2 mach.	sp	..	·7
COLONIAL, ETC.—													
Victoria.													
Childers	Chiswick ..	1883	113	12·5	5·9	1	65	730	20	2–1 prs.	..	12	10
One boat	Poplar	1891	130	13·5	5·7	1	82	1,150	23	3–3 prs.	3	19	20
Nepean, Lonsdale (2 boats)	Chiswick ..	1884	63	7·5	3·2	1	12	150	17·5	..	1	7	
New South Wales.													
Acheron, Avernus (2 boats)	..	1879	..	..	..	1	16	300	16				
Queensland.													
Mosquito	Chiswick ..	1884	63	7·5	3·2	1	12	..	17	..	1	7	
Wasp	..	..	..	..	..	..	12	..	..	..	..	7	
Tasmania.													
One boat	Chiswick ..	1884	63	7·5	3·2	1	12	..	17	..	1	7	
New Zealand.													
Nos. 1–4 (4 boats)	Chiswick ..	1884	63	7·5	3	1	12	170	17	1 mach.	Sp.		
India.													
Nos. 1–3 (3 boats)	Chiswick ..	1888	134·5	14·8	7·1	1	96	1,270	23·2	2 Q.F.	5		
Nos. 4–6 (3 boats)	East Cowes ..	1889	130	14·6	..	..	95	1,030	20				
No. 7	Paisley	1888	130·4	14	..	..	92	1,060	21				

Argentine Republic.

Name or Number.	Where Built.	Launched.	Dimensions. Length.	Beam.	Draught.	Number of Screws.	Displacement.	Indicated Horse-Power.	Maximum Trial Speed.	Armament.	Torpedo Tubes.	Complement.	Coal Capacity.
DESTROYERS—			Feet.	Feet.	Feet.		Tons.		Knots.				Tons.
Santa Fé†	Poplar.. ..	1896	190	19·6	7·4	2	280	4,000	26·5 *t.*	*1 14-pr., 3 6-pr. Q.F., 2 M.	3	54	80
Corrientes	Poplar.. ..	1896	190	19·6	7·4	2	280	4,000	27·4 *t.*		3	54	80
Missiones..	Poplar.. ..	1896	190	19·6	7·4	2	280	4,000	26·0 *t.*		3	54	80
Entre Rios	Poplar.. ..	1896	190	19·6	7·4	2	280	4,000	26·7 *t.*		3	54	80
FIRST CLASS—													
2 boats	Chiswick ..	1890–1	150	14·5	5·2	2	110	1,500	24·52	3 3-prs.	3	27	22
6 boats	Poplar.. ..	1890	130	13·5	6	1	85	1,200	23–24	2 3-pr. Q.F.	2	15	15
4 boats	Poplar.. ..	1880–2	100	12·5	6	1	52	600	20	2 mach.	3	14	10
SECOND CLASS—													
Nos. 1–8 (8 boats) ..	Poplar.. ..	1890	60	9·2	3	1	16	230	17	1 Q.F.	1	10	1·25
Nos. 9–10 (2 boats)	Chiswick ..	1881	60	7·5	3·5	1	16	230	17	..	1		
VEDETTE BOATS—													
Nos. 1–4 (4 boats) ..	..	1875	55	7	..	..	..	..	..	..	sp.		

The two 150-ft. boats are named Comodoro Py and Murature.

The six 130-ft. boats are named Bathurst, Buchardo, Jorge, King, Pinedo, and Thorne. They have locomotive boilers.

The four 100-ft. boats are named Alerta, Centelia, Ferre, and Py.

* ¼-in. plating over entire engine and boiler space.

† The Santa Fé was lost in 1897, and Messrs. Yarrow are building a new destroyer to replace her, in which the Santa Fé's engines, recovered from the wreck, will be placed.

Austria-Hungary.

Name or Number.	Where Built.	Launched.	Dimensions. Length.	Beam.	Draught.	Number of Screws.	Displacement.	Indicated Horse-Power.	Maximum Trial Speed.	Armament.	Torpedo Tubes.	Complement.	Coal Capacity.
FIRST CLASS—			Feet.	Feet.	Feet.		Tons.		Knots.				Tons.
Adler, Falke	Poplar.. ..	1886	135	13·7	5·6	1	95	900	22·4	2 Nord.	2	16	28
22 boats	Elbing, Trieste, &c.	1886-9	128	15·9	6·9	1	83	900 / 1,000	17·5 to 21·5	2 mach.	2	15	28
Boa .. .. / Cobra .. .. .. / Kigyo .. .. .. / Python	Poplar.. ..	1898–9	152·6	15·3	7·6	1	133	2,000	24·3	2 3-pr. Q.F.	3	24	30
Viper	Poplar.. ..	1896	147·6	14·9	7·6	1	130	2,000	26·5	2 3-pr. Q.F.	2	26	30
Natter	Elbing ..	1896	150	17·5	8·8	2	152	2,300	26·5	2 3-pr. Q.F.	3	..	30
SECOND CLASS—													
Nos. 9, 10 (2 boats)	Chiswick, Poplar, Pola and Elbing	1881	98·5	10·8	2·9	1	37	450	17	1 Q.F.	1		
Nos. 11–32 (22 boats)		1883–7	107	11·6	3·1	1	47	600	17				
Nos. 33–39 (7 boats)		1887–91	118·1	14·4	3·3	1	64	700	18	2 Q.F.	1		
Nos. 2–8 (7 boats) ..	Pola and Poplar ..	1878–81	87·4	9·6	2·8	1	27	300	15	..	1		

Brazil.

Name or Number.	Where Built.	Launched.	Dimensions. Length.	Beam.	Draught.	Number of Screws.	Displacement.	Indicated Horse-Power.	Maximum Trial Speed.	Armament.	Torpedo Tubes.	Complement.	Coal Capacity.
			Feet.	Feet.	Feet.		Tons.		Knots.				Tons
First Class—													
Nos. 1–5 (5 boats)	Poplar.. ..	1882	100	12·5	5·5	1	52	600	20	2 mach.	2	16	20
Araguary	Chiswick ..	1891	150	14·5	5·2	2	150	1,550	25·1	2 Q.F.	4	27	22
Iguatemi	Chiswick ..	1891	150	14·5	5·2	2	150	1,550	25·4	2 Q.F.	4	27	22
Marcillo Diaz	Chiswick ..	1891	150	14·5	5·2	2	150	1,550	25·8	2 Q.F.	4	27	22
5 boats	Elbing ..	1892–3	152	17·2	7·9	2	130	2,200	28	2–1 prs.	3	24	30
Piratiny	..	..	130	12	..	..	..	..	10	2–1 pr.	1		
Poty	..	..	126	12	3	..	30	..	18	1–1 pr.	1		
Second Class—													
Inbanhuay (wood)..	New York..	1893	90	10	3	..	17	..	25	1–1 pr.	1	10	
4 boats	..	1883–4	..	..	..	1	17	..	17				
1 boat	Chiswick ..	1885	63	75	3·2	1	..	..	17	..	..	..	2
1 boat	Poplar.. ..	1886	60	8	3	1	14	200	17	..	1		
Third Class—													
Moxoto	Poplar.. ..	1883	60	9·3	..	..	..	..	16	1–1 pr.	1		
5 boats	Chiswick ..	1883	45	6	1·2	1	3·5	..	12-13	1 mach.	sp.		

Chili.

Name or Number.	Where Built.	Launched.	Dimensions. Length.	Beam.	Draught.	Number of Screws.	Displacement.	Indicated Horse-Power.	Maximum Trial Speed.	Armament.	Torpedo Tubes.	Complement.	Coal Capacity.
			Feet.	Feet.	Feet.		Tons.		Knots.				Tons.
Destroyers—													
Capitan Orella ..	Birkenhead.	1896	210	21·6	..	2	300	6000	30·17	1-12 pr. Q.F. 5-6 pr.	2	65	90
Capitan Munoz Gamero.. ..	Birkenhead.	1896	210	21·6	..	2	300	6000	30·42	1-12 pr. Q.F. 5–6 pr.	2	65	90
Teniente Serrano ..	Birkenhead.	1896	210	21·6	..	2	300	6000	30·35	1–12 pr. Q.F. 5-6 pr.	2	65	90
Guardia-Marina Riquelme	Birkenhead.	1896	210	21·6	..	2	300	6000	30·09	1-12 pr. Q.F. 5-6 pr.	2	65	90
First Class—													
3 boats	Poplar.. ..	1881	86	12·5	..	1	25	400	19-20	..	4	15	
5 boats	Poplar.. ..	1881	100	12·5	..	1	35	400	18–19	4 mach.	4	15	9
Sarjento Aldea ..	Poplar.. ..	1886	125	13·5	5·5	1	70	800	20	2 Q.F.	4	18	15
Injeniero Hyatt, Cirujano Videla, Injeniero Mutilla, Guardia-Marina Contreras, Capitan Thompson, and Teniente Rodriguez (Viper type) ..	Poplar.. ..	1896 1898	152·6	15·3	7·9	1	140	2200	27·5 27·2	3–3 pr. Q.F.	3	28	40
Janequeo Guale, Rucumilla, and Guacolda	Poplar.. ..	1881	100	12·5	..	1	..	450	..	..	..	..	
Tegualda, Quidora, and Fresia	Poplar.. ..	..	87	10·9	..	1	..	400	..	..	..	..	
Second Class—													
1 boat	East Cowes	1887	50	..	..	..	..	..	16	..	..	..	
1 boat	East Cowes	1892	60	9·6	5	1	15	270	19	..	1	..	
1 boat	La Seyne ..	1895	42	8·6	..	1	..	..	..	..	1	..	

The Thompson and Rodriguez were sent out in sections, and put together at Talcahuano and Valparaiso.

China.

Name or Number.	Where Built.	Launched.	Dimensions. Length.	Dimensions. Beam.	Dimensions. Draught.	Number of Screws.	Displacement.	Indicated Horse-Power.	Maximum Trial Speed.	Armament.	Torpedo Tubes.	Complement.	Coal Capacity.
DESTROYERS—			Feet.	Feet.	Feet.		Tons.		Knots.				Tons.
Hai Lung	Elbing	1898	193 7	21·0	..	2	280	6,000	35	6 3-pr. Q.F.	2	..	67
Hai Nin	Elbing	1898											
Hai Ching	Elbing	1898											
Hai Hoha	Elbing	1899											
FIRST CLASS—													
1 boat	Elbing	1886	144·3	16·4	7·5	1	128	1,400	24·2	4 1-pr. revs.	2	20	15
1 boat	Poplar	1887	128	13	5	1	69	1,000	23·9	3 Q.F., 4 Gatlings	3	28	15
25 boats	Stettin, &c.	1886–87	110	13	4·9	1	65	1,000	19·5	1-pr. revs.	3	16	10
2 boats	Stettin	1883	86	10·4	3·4	1	28	650	18·2	1-pr. revs.	2	16	12
1 boat	Stettin	1884	123·5	21·7	..	..	..	..	19	..	5	16	
2 boats	Elbing	1895	128	15·8	..	..	120	1,250	24·5	Q.F.	2		
SECOND CLASS—													
11 boats	Elbing	1885–86	85	11·9	4·8	1	27	400	19	..	1	..	5
1 boat	Foochow	Bldg.	88·6	6·7	3·3	1	30	550	20·5				

About twenty boats only are said to be serviceable.

Costa Rica.

Costa Rica has one 62-ft., 15-knot boat.

Denmark.

Name or Number.	Where Built.	Launched.	Dimensions. Length.	Dimensions. Beam.	Dimensions. Draught.	Number of Screws.	Displacement.	Indicated Horse-Power.	Maximum Trial Speed.	Armament.	Torpedo Tubes.	Complement.	Coal Capacity.
FIRST CLASS—			Feet.	Feet.	Feet.		Tons.		Knots.				Tons.
Hajen	Copenhagen	1896	154·3	15·4	7·9	2	142	2,317	22·9	1 4·7-in. 1 1-pr.	3	..	..
Havörnen	Copenhagen	1897											
Söbjörnen	Copenhagen	1898											
Delfinen	Chiswick	1883	111·5	12·6	6	1	59	620	20	1 mach.	2	14	9
Havhesten	Chiswick	1888	137·9	14	7	1	94	1,200	22·8	2 1-pr. revs.	4	20	15
Hvalrossen	Chiswick	1884	114	12·6	6·5	1	64	660	18·7	1 mach.	2	14	10
Makrelen	Copenhagen	1893	140	14·2	7	2	112	1,200	..	..	..	..	16
Narhvalen	Chiswick	1888	137·9	14	7	1	94	1,200	22·3	2 1-pr. revs.	4	20	16
Nord Kaperen	Copenhagen	1893	140	14·2	7	2	112	1,200	..	2 1-pr. revs.	4	..	16
Sölöven	Chiswick	1887	131	14·8	6·8	1	89	1,200	23·3	2 mach.	4	20	14
Söulven	Havre	1880	94·8	10·9	3·9	1	37	450	18·1	..	2	12	5
Springeren	Copenhagen	1891	119	13	4·9	1	81	800	18·3	2 1-pr. revs.	2	20	14
Stören	Chiswick	1887	131	14·8	6·8	1	89	1,200	23	2 mach.	4	20	14
Sværdfisken	Chiswick	1881	110	12	6	1	49	600	20·7	1 mach.	2	14	8
SECOND CLASS—													
Nos. 4, 5 (2 boats)	Chiswick	1882	63	7·5	2·5	1	15	150	16·9	1 mach.	2	6	1
Nos. 6, 7 (2 boats)	Chiswick	1884	66·8	8	4·2	1	16	170	15·4	1 mach.	2	6	1·5
Nos. 8, 9 (2 boats)	Chiswick	1886	69·5	8·1	3·8	1	17	170	15·7	1 mach.	2	6	1
Nos. 10, 11 (2 boats)	Chiswick	1888	70·2	8	4	1	18	180	15·8	1 mach.	2	6	1
Nos. 12, 13 (2 boats)	Chiswick	1889	78·3	9	4·9	1	24	350	18	1 mach.	2	8	3
1 boat	Chiswick	1875	58	7·5	3	1	..	..	16	..	sp.		

Four destroyers and two boats are provided for.

France.

Name or Number.	Where Built.	Launched.	Dimensions. Length.	Beam.	Draught.	Number of Screws.	Displacement.	Indicated Horse-Power.	Maximum Trial Speed.	Armament.	Torpedo Tubes.	Complement.	Coal Capacity.
DESTROYERS—			Feet.	Feet.	Feet.		Tons.		Knots.				Tons
Durandal	Havre	1898	180·5	19·5	10·6	2	300	4800	27·4	1-9pr. 6-3prs.	2	45	48
Epée	Havre	Bldg.	185·9	19·6	9·11	2	319	5700	26.0	1–12 prs.	2	48	33
Escopette	Rochefort	Bldg.	183·9	19 6	9·11	2	303	5700	26	1-9pr. 6-3prs.	2	45	48
Espingole	Havre	Bldg.	183·9	19·6	10·6	2	303	4800	26	1-9pr. 6-3prs.	2	45	48
Fauconneau	Havre	1900	183·9	19·6	10·6	2	303	4800	26	1-9pr. 6-3prs.	2	45	48
Flamberge..	Rochefort	Bldg.	183·9	19 6	9·11	2	303	5700	26	1-9pr. 6-3prs.	2	45	48
Framée	Bordeaux	1900	185·9	19·6	9·11	2	319	5700	26·0	4–6 prs.	2	48	33
Hallebarde	Havre	1899	180·5	19·5	10·6	2	300	4800	27	1-9pr. 6-3prs.	2	45	48
Pertuisane..	Rochefort	Bldg.	183·9	19·6	9·11	2	303	5700	26	1-9pr. 6-3prs.	2	45	48
Pique	Havre	1900	185·9	19·6	9.11	2	319	5700	26 0	1–12 prs.	2	48	33
Rapière	Rochefort	Bldg.	183·9	19·6	9·11	2	303	5700	26	1 9pr. 6-3prs.	2	45	48
Yatagan	Bordeaux	1900	185·9	19·6	9·11	2	319	5700	26·0	4–6 prs.	2	48	33
M 12, M 13	Rochefort	Pro.											
M 14 to M 21	..	Pro.											
SEA-GOING—													
Agile	La Seyne ..	1889	139	14·7	7·7	2	121	1,100	20·4	3–3 prs.	2	26	14
Alarme	St. Nazaire	1889	151	15·7	8·3	2	169	1,400	20·5	2–3 prs.	4	30	40
Aquilon	Normand ..	1895	137·8	14·6	7·9	2	127	2,000	25·17	2–3 prs.	2	34	17
Archer	Normand ..	1893	138	14·7	6·5	2	131	1,250	21	2–3 prs.	2	26	17
Argonaute..	St. Denis ..	1893	141	16·4	9·3	2	131	1,500	25·1	2–3 prs.	2	34	16
Audacieux..	Nantes ..	1899	144·2	15·2	10·0	2	152	4,200	30	2–3 prs.	2	..	18
Aventurier	St. Nazaire..	1889	151	15·7	8·3	2	174	1,400	20·5	2–3 prs.	4	34	40
Averne	Havre ..	1894	141	16·4	9·3	2	133	1,500	24·4	2–3 prs.	2	27	16
Borée	Bordeaux ..	Bldg.	144·2	15·2	10·0	2	152	4,200	30	2–3 prs.	2	..	18
Bourrasque	Havre	Bldg.	144·2	15·2	10·0	2	152	4,200	30	2–3 prs.	2	..	18
Cerbère	Normand ..	1899	137 8	14·6	7·9	2	127	2,000	25·0	2–3 prs.	2	34	17
Chevalier	Normand ..	1893	144·3	15·7	6·8	2	131	2,700	27·2	2–1 prs.	2	32	17
Corsaire	St. Denis ..	1893	160 5	15	5·4	2	171	2,500	25·5	4–1 prs.	2	32	15
Coureur	Chiswick ..	1888	147·5	14·5	4·6	2	129	1,550	23·28	4 Nords.	2	27	22
Cyclone (ex-Tenare) ..	Havre	1898	144·2	15·2	10·0	2	152	4,200	30	2–3 prs.	2	..	18
Dauphin	Havre ..	1894	141	16·4	9·3	2	137	1,500	25·22	2–3 prs.	2	34	16
Défi	St. Nazaire..	1889	151	15·7	8·3	2	173	1,400	21	2–3 prs.	4	30	40
Dragon	Normand ..	1892	138	14·7	8·2	2	129	1,400	25	2–3 prs.	2	26	15·5
Eclair	La Seyne ..	1891	144·3	14·7	7·7	2	128	1,100	21·5	3–3 prs.	2	26	17
Flibustier	Normand ..	1894	143	16·4	9·3	2	132	1,500	23·5	2–3 prs.	2	34	16
Forban	Normand ..	1895	144·2	15·2	10	2	135	3,200	31·2	2–1 prs.	2	..	..
Grenadier	Normand ..	1892	138	14·7	8·2	2	129	1,400	25·25	2–3 prs.	2	26	15·5
Grondeur	Havre	1892	147·5	14·5	5	2	130	1,550	24	2–3 prs.	2	27	20
Kabyle	La Seyne ..	1891	144·3	14·7	7·7	2	128	1,100	21·6	3–3 prs.	2	27	17
Lancier	Normand ..	1893	138	14·7	8·2	2	128	1,400	25·79	2–3 prs.	2	26	15·5
Lansquenet	Nantes ..	1893	165·4	15·8	4·2	2	151	2,800	..	2–3 prs.	..	..	..
Mangini	Nantes ..	1895	147·6	14·8	7·9	2	129	2,100	27·5	2–3 prs.	2	34	17
Mistral	Havre	Bldg.	144·2	15·2	10·0	2	152	4,200	30	2–3 prs.	2	..	18
Mousquetaire	Havre ..	1892	154	15·7	7	2	150	2,100	24·77	2–1 prs.	2	32	18
Orage	La Seyne ..	1891	144·3	14·7	7·7	2	128	1,100	21·7	3–3 prs.	2	26	17
Ouragan	Nantes ..	1887	151	15·7	8·3	2	174	1,400	20	2–3 prs.	4	30	40
Rafale..	Havre	Bldg.	144·2	15·2	10·0	2	152	4,200	30	2–3 prs.	2	..	18
Sarrasin	Bordeaux..	1893	139	14·7	7·7	2	131	1,100	20·5	3–3 prs.	2	26	14
Simoun	Havre	Bldg.	144 2	15·2	10·0	2	152	4,200	30	2 3 prs.	2	..	18
Sirocco	Havre	Bldg.	144·2	15·2	10·0	2	152	4,200	30	2–3 prs.	2	..	18
Teméraire	St. Nazaire..	1889	151	15·7	8·3	2	174	1,400	21	2–3 prs.	4	30	40
Tourbillon	Bourdeaux..	1892	139	14·7	7·7	2	131	1,100	20·5	3–3 prs.	2	26	14
Tourmente	St. Denis ..	1893	141	16·4	9 3	..	132	1,500	21·6	2–3 prs.	2	25	15
Tramontane	Bordeaux ..	Bldg.	144·2	15·2	10·0	2	152	4,200	30	2–3 prs.	2	..	18
Trombe	Nantes ..	Bldg.	144·2	15·2	10·0	2	152	4,200	30	2–3 prs.	2	..	18
Turco	St. Denis ..	1892	138	14·7	8·2	2	124	1,400	21·3	2–3 prs.	2	26	15·5
Typhon	Havre	Bldg.	144·2	15·2	10·0	2	152	4,200	30	2–3 prs.	2	..	18
Véloce..	Havre	1892	147·5	14·5	5	2	130	1,550	23·6	2–3 prs.	2	27	20
Zouave	St. Denis .	1892	138	14·7	8·2	2	124	1,400	21·3	2–3 prs.	2	26	15·5
N 18 to N 21	..	Pro.											
FIRST CLASS—													
Balny	Normand ..	1886	134·5	11	7·2	1	66	700	20	2–1 pr. rev.	2	21	12
Bouët-Willaumez	St. Denis ..	1888	134·5	11	7·2	1	66	700	20	2–1 pr. rev.	2	21	12
Capt. Cuny	..	1886	134·5	11	7·2	1	66	700	20	2–1 pr. rev.	2	21	12
Capt. Mehl	..	1886	134·5	11	7·2	1	66	700	20	2–1 pr. rev.	2	21	12
Challier	..	1886	134·5	11	7·2	1	66	700	20	2–1 pr. rev.	2	21	12
Dehorter	St. Denis ..	1886	134·5	11	7·2	1	66	700	20	2–1 pr. rev.	2	21	12
Deroulède	Normand .	1886	134·5	11	7·2	1	66	700	20	2–1 pr. rev.	2	21	12
Doudart de Lagrée	Normand ..	1886	134·5	11	7·2	1	66	700	20	2–1 pr. rev.	2	21	12
Edmond Fontaine	St. Denis ..	1888	134·5	11	7·2	1	66	700	20	2–1 pr. rev.	2	21	12
161 (ex G. Charmes) ..	La Seyne ..	1886	132·5	12·5	6·	1	80	560	18·8	2–1 prs.	..	23	12
126–129 (4 boats)	Normand ..	1888–9	118	13·2	8·6	2	79	1,250	21	2–1 prs.	2	21	10
145–149 (5 boats)	Normand ..	1891–3	118	13·2	8·7	2	80	1,300	23·9	2–1 prs.	2	21	10
152–154 (3 boats)	Normand ..	1892–	118	13·2	8·7	2	80	1,300	24·6	2–1 prs.	2	21	10
155–157 (3 boats)	Bordeaux ..	1893	118	13 2	8·7	2	80	1,300	23	2–1 prs.	2	21	10
158–160 (3 boats)	Cail	1893	118	13·2	8 7	2	80	1,300	23	2–1 prs.	2	21	10

France—*continued.*

Name or Number.	Where Built.	Launched.	Dimensions. Length.	Beam.	Draught.	Number of Screws.	Displacement.	Indicated Horse-Power.	Maximum Trial Speed.	Armament.	Torpedo Tubes.	Complement.	Coal Capacity.
			Feet.	Feet.	Feet.		Tons.		Knots.				Tons.
FIRST CLASS—*continued.*													
161–163 (3 boats).. ..	St. Nazaire..	1892	118	13·2	8·7	2	8)	1,300	23	2–1 prs.	2	21	10
164–166 (3 boats).. ..	La Seyne ..	1892	118	13·2	8·7	2	79	1,300	23	2–1 prs.	2	21	10
167–169 (3 boats).. ..	Creusot ..	1892	118	13·2	8·7	2	81	1,300	23	2–1 prs.	2	21	10
170, 171 (2 boats).. ..	Normand ..	1893–4	118	13·2	8·7	2	80	1,300	23–2	2–1 prs.	2	21	10
172, 173 (2 boats).. ..	Normand ..	1893–4	118	13·2	8·7	2	89	1,390	23–24	2–1 prs.	2	21	10
174–176 (3 boats).. ..	Normand ..	1893–5	118	13·2	8·7	2	94	1,390	23–24	2 1-prs.	2	21	10
177–179 (3 boats).. ..	Havre.. ..	1893	118	13·2	8·7	2	79	1,300	23–24	2–1 prs.	2	21	10
180–187 (8 boats).. ..	Creusot, etc.	1893–4	118	13·2	8·7	2	80	1,300	23 4	2–1 prs.	2	21	10
188–191 (4 boats).. ..	Normand, etc.	1893–4	118	13·2	..7	..	79	..	24–2	..	2	21	10
192–194 (3 boats).. ..	Havre, etc.	1894–5	118	13·2	8·	2	82	1,300	23·55	2–1 prs.	2	21	10
195–200 (6 boats).. ..	Havre, etc.	1894–5	319	13·2	8·7	2	80	1,300	23·5	2 1-prs.	2	21	10
201–205 (5 boats).. ..	Havre	1897–8	121·4	13·6	..7	2	84	1,500	23·5	2–1 prs.	2	23	10
206–211 (6 boats).. ..	Bordeaux ..	1897–8	121·4	13·6	..	2	86	1,500	23·5	2 1-prs.	2	23	10
212–215 (4 boats).. ..	Havre	1898–9	121·6	13·6	..	2	88	1,500	23·5	2 1-prs.	2	23	1›
216–226 (11 boats) ..	Cherbourg, Toulon, etc.	1899	121·6	13·6	..	2	88	1,500	23·5	2–1 prs.	2	23	10
227–235 (9 boats).. ..	Bordeaux,etc.	Bldg.	121·4	13·2	8·7	2	84	1,500	23·5	2-1-prs.	2	23	10
236–255 (20 boats) ..	Bordeaux,etc.	Bldg.	121.4	13·2	8·7	2	84	1,500	23·5	2-1-prs.	2	23	10
P. 55–63 (9 boats) ..	Havre	Bldg.	121·4	13·2	3·7	2	84	1,500	23·5	2–1 prs.	2	23	10
P. 64–74 (11 boats) ..	..	Pro.											
P. 75–84..	..	Pro.											
SECOND CLASS—													
26	..	1878	108	11	5·6	1	45	400	19	2–1 prs.	2	16	10
27	..	1878	104·4	10·6	6·1	1	44	400	19	2–1 prs.	2	16	10
28	..	1878	111·5	11	5·6	1	44	400	19	2–1 prs.	2	16	10
60–64 (5 boats)	..	1878–85	108·2	10·3	6·1	1	45	400	19	2–1 prs.	2	16	10
65, 66, 68 (3 boats) ..	..	1878–85	108·2	10·7	6·4	1	49	500	20	2–1 prs.	2	16	10
69–74 (6 boats)	..	1878–85	108·2	10·7	6·5	1	50	500	20	2-1 prs.	2	16	10
75 82, 84–109 (34 boats)	Cail, etc. ..	1885–92	114·7	10·6	6	1	54	525	20	2–1 prs.	2	16	10
111–125 (12 boats) ..	La Seyne, etc.	1885–90	114·7	10·6	6	1	54	525	20	2–1 prs.	2	16	10
130–132, 134–144 (14 boats)*	Normand, etc.	1889–90	111·5	11·4	6	1	52·8	520	21	2–1 prs.	2	16	10
THIRD CLASS—													
8, 10–16, 18, 19 (10 boats)	Various Firms in France and England.	1877–82	86	10·2	5	1	27	200–450	16–19	..	..	10	
20			87	10·8	5	1	33			..	..	10	
22, 23 (2 boats)			87·6	10·4	5·2	1	30			..	..	10	
24, 25 (2 boats)			88·5	10·4	6	1	30			..	..	10	
31, 32 (2 boats)			85·5	10·4	3·8	1	27			..	..	10	
33–36 (4 boats)			89	10·4	6	1	32			..	..	10	
37–40 (4 boats)			87	10·8	5	1	32			..	..	10	
41, 42 (2 boats)			87	10·8	6	1	33			..	..	10	
43, 44 (2 boats)			89	10·4	5·7	1	32			..	..	10	
47			87	10·8	5	1	33			..	..	10	
48			89	10·4	5·8	1	32			..	..	10	
49, 50, 53 (3 boats) ..			87	10·8	5	1	32			..	..	10	
54, 55 (2 boats)			91	10	6·1	1	32			..	..	10	
VEDETTE BOATS—													
(1 boat) (aluminium)..	Poplar ..	1894	62·3	9·1	..	1	14	210	20·5	..	1	8	
29, 30 (2 boats)	Chiswick ..	1876	67	8·5	3·5	1	16	..	18	..	1	8	
56, 57 (2 boats)	Chiswick ..	1879	59	7·5	3·5	1	12	50	16	..	1	8	
58, 59 (2 boats)	Chiswick ..	1881	63	7·5	3·5	1	11	50	17	..	1	8	
A, B, C	Creusot ..	1894	62·4	8·9	4·9	1	15·	210	16·5	..	1	9	
D, E, F, G, H, I † ..	Creusot ..	Bldg.											
Libellule	Havre												
SUBMARINE—													
Algérien	Cherbourg ..	Bldg.	48·8	9·2									
Farfadet	Rochefort ..	Bldg.											
Français	Cherbourg ..	Bldg.	48·8	9·2									
Gnome	Rochefort ..	Bldg.											
Gustave Zédé	Toulon ..	1893	131	..	..	1	266	720	14	..	1	8	
Gymnote	Mourillon ..	1888	59	5·9	5·9	1	30	60	4–6	..	..	4	
Korrigan	Rochefort ..	Bldg.											
Lutin	Rochefort ..	Bldg.											
Morse..	Cherbourg ..	1899	120	9·2	..	..	146	..	13	..	1	9	
Narval	Cherbourg ..	1899	111·6	12·4	5·2	1	106	250	12	..	2	11	..
Q 13–14	Cherbourg ..	Bldg.											
Sirène	Cherbourg ..	Bldg.	111·6	12·4	5·2	1	106	250	12	..	2	11	..
Triton..	Cherbourg ..	Bldg.	111·6	12·4	5·2	1	106	250	12	..	2	11	..

* Second-class boat No. 83 lost off Cape de la Chèvre, 1897, and No. 133 near Algiers, 1898.
† For the torpedo-transport Foudre.
Two other submarine boats are to be built out of the patriotic fund initiated by the *Matin* newspaper.

Germany.

Name or Number.	Where Built.	Launched.	Dimensions. Length.	Dimensions. Beam.	Dimensions. Draught.	Number of Screws.	Displacement.	Indicated Horse-Power.	Maximum Trial Speed.	Armament.	Torpedo Tubes.	Complement.	Coal Capacity.
			Feet.	Feet.	Feet.		Tons.		Knots.				Tons.
DIVISION BOATS—													
D 1, D 2 (2 boats)	Elbing ..	1887	180·6	21·6	9·8	2	250	1,800	19	6 1-pr. revs.	3	48	50
D 3, D 4 (2 boats)	Elbing ..	1888	184	21·8	9·6	2	300	2,000	20	4 6-pr. Q.F. 2 1-pr. revs.	3	48	90
D 5, D 6 (2 boats)	Elbing ..	1888–9	190·3	23	9·6	2	320	3,000	22½	4 6-pr. Q.F. 2 1-pr. revs.	3	48	90
D 7, D 8 (2 boats)	Elbing ..	1890	190·3	23	9·9	2	350	3,500	22½	6 Q.F.	3		
D 9	Elbing ..	1894	197·0	24·3	9·9	2	380	4,500	26	6 Q.F.	3		
D 10	Chiswick ..	1898	211·9	19·6	8·1	2	310	5,800	28·5	5 3-pr. Q.F.	3	52	80
D 11	..	Pro.	..	..	..	..	350						
FIRST CLASS—													
S 1—S 65 (64 boats)*	Elbing ..	1883–92	121 / 150	15·7 / 15·6	6·7 / 6·7	..	85–88	900 / 1,600	20–22½	2 1-pr. revs.	2	..	17
S 66—S 73 (10 boats)	Elbing ..	1893	154·3	16·4	..	2	110 / 145	1,600	..	..	3		
S 74—S 81 (8 boats)	Elbing ..	1894	154·3	16·4	..	2	125	1,900	25	..	3		
S 82—S 87 (6 boats)	Elbing ..	1897–8	158·2	16·9	9·0	2	140	2.300	26	2 1-pr. revs.	3	..	32
G 88—G 89 (2 boats)	Kiel (Germania)	1898	154·3	16·5	..	..	160	2,500	26	2 mach.	3	22	
G 90—G 97 (8 boats)	Elbing ..	Bldg.	157·5	16·9	8·9	2	155	..	25	1 Q.F., 1 m.	3	..	30
V 1, V 2 (2 boats)	Stettin ..	1884	124·6	..	..	..	75	550	..	..	2		
V 3, V 4 (2 boats)	Stettin ..	1884	..	..	..	..	90	1,000	..	..	2		
V 5—V 10 (6 boats)	Stettin ..	1884	..	..	..	..			19	..	2		
G 1,	Gaarden ..	1885	124·6	15·7	6·6	..	88	1,000	19	2 1-pr. revs.	2	17	
Y 1,	Poplar ..	1884	120	12·5	5·5	1	65	650	19	2 1-pr. revs.	2	15	25
T 1, T 2 (2 boats)	Chiswick, &c.	1884	117·7	12·5	6·2	1	80	..	20·2	2 1-pr. revs.	2	16	22
H 1,	Kiel (Howaldt)	1886	..	..	..	..	80	1,000	20	2 1-pr. revs.	2		
K 1,	Kiel (Dockyard)	1887	118·1	13·4	5·9	..	85	1,000	22	2 1-pr. revs.	..	18	
SECOND CLASS—													
3 boats	..	1893	..	..	..	..	88	..	22				
2 boats	..	1893	..	..	..	..	90	..	3				
VEDETTE BOATS—													
13 boats	..	..	..	..	..	..	13·5	..	18				
2 boats	..	..	..	..	..	..	..	..	16				
1 boat	Chiswick ..	1884	63	8	4·3	1	..	..	15·5	1 mach.	2		

* S 41 lost 1895.

The Estimates of 1900 provide the initial expenditure (2,400,000 marks) for the building of a division of torpedo boats.

Greece.

Name or Number.	Where Built.	Launched.	Dimensions. Length.	Dimensions. Beam.	Dimensions. Draught.	Number of Screws.	Displacement.	Indicated Horse-Power.	Maximum Trial Speed.	Armament.	Torpedo Tubes.	Complement.	Coal Capacity.
			Feet.	Feet.	Feet.		Tons.		Knots.				Tons.
6 boats	Stettin ..	1885	128	15·3	5·4	1	85	1,050	19	4 1-pr. revs.	..	20	20
6 boats	Poplar.. ..	1881	100	12	4·2	1	48	600	19	2 1-pr. revs.	2	12	9
4 boats	La Seyne ..	1880	72	13	5·5	1	52	225	..	..	..	..	10
5 boats	La Seyne ..	1881	89	11	3·1	1	35	500	17·5	..	..	..	5
2 boats	Poplar.. ..	1878	75	10·8	2·5	..	18	295	16·2	..	..	..	1·5
8 boats	..	..	..	..	..	..	21	..	16				
20 boats	Various ..	..	..	..	..	..	..	..	..	..	sp.	..	..

Italy.

Name or Number.	Where Built.	Launched.	Dimensions. Length.	Dimensions. Beam.	Dimensions. Draught.	Number of Screws.	Displacement.	Indicated Horse-Power.	Maximum Trial Speed.	Armament.	Torpedo Tubes.	Complement.	Coal Capacity.
DESTROYERS—			Feet.	Feet.	Feet.		Tons.		Knots.				Tons.
Fulmine	Sestri (Odero)	1898	200	20·4	5·4	2	298	4,800	28	1 12-pr. 3 6-pr. Q.F.	3	60	43
Lampo .. .. .. Freccia .. .. .. Dardo .. .. .. Strale	Elbing (Schichau)	1899	196·8	21·3	5·8	2	320	6,000	30	1 12-pr. Q.F., 5 6-pr.	2		
Nembo .. .. .. Turbine .. .. .. Meteora .. .. .. Tuono	Naples (Pattison)	Bldg.	208	19·4	6·3	2	350	6,000	30	1 12-pr. Q.F., 3 6-pr.	2		
FIRST CLASS—													
5 boats: Aquila .. Sparviero Nibbio .. Avvoltolo Falco ..	Elbing ..	1888	152	17·2	7·9	2	130	2,200	26·6	2 3-pr. Q.F., 1 1-pr. Q.F., 1 1-pr. rev.	3	24	40
Nos. 78, 79 (2 boats)	Venice ..	1887	135	14	5·3	2	110	1,600	24	1 1-pr. Q.F., 1 1-pr. rev.	5	20	30
Pellicano	Sestri (Odero)	1899	157·4	19	14·8	2	147	2,700	25	2 3-prs.	2	..	40
Condore	Sestri (Ansaldo)	1898	154·3	16·8	6·9	2	136	2,500	27	2 3-prs.	2	27	40
SECOND CLASS—													
Nos. 76, 77 (2 boats)	Poplar.. ..	1887	140	14	5	2	100	1,600	25	2 3-pr. Q.F., 1 1-pr. rev.	5	20	30
Nos. 84–104, 106–111 (27 boats)	Elbing and Italy ..	1887–88	127·7	15·6	6·8	1	85	1,000	22·5	2 1-pr. Q.F.	2	17	7
Nos. 112–116, 118–135 (23 boats)	Elbing and Italy ..	1889–92	127·7	15·6	6·8	1	85	1,100 1,200	23	..	2	17	17
No. 117	..	1895	131·2	16·4	..	1	85	1,000	..	2 1 pr. Q.F.	2	17	17
Nos. 136–146 (11 boats)	Italy	1893–94	131·2	16·4	..	1	85	1,000	22	2 1-pr. Q.F.	2	17	17
Nos. 147–153 (7 boats)	Italy ..	1894–5	131·2	16·4	..	1	85	1,000	22	2 1-pr. Q.F.	2	17	17
12 boats	Italy	Bldg.	131·2	16·4	7	1	85	1,000	22	2 1-pr. Q.F.	2	17	17
Nos. 56–75 (20 boats)	Elbing and Italy ..	1885–87	127·7	15·6	6·8	1	65	1,000	22·5	2 1-pr. Q.F.	2	17	17
THIRD CLASS—													
No. 22	Poplar.. ..	1882	100	12·5	5·5	1	40	620	22	1 1-pr. rev.	2	11	10
No. 25	Poplar.. ..	1882	100	12·5	5·5	1	40	620	23	1 1-pr. rev.	2	11	10
Nos. 26–55 (30 boats)	Chiswick and Italy ..	1882–86	100	11·7	5·3	1	34	430	21·3	1 1-pr rev.	2	11	7
Nos. 80–83 (4 boats)	Genoa.. ..	1888	101·6	..	..	1	34	430	21	1 1-pr. rev.	2	11	7
Nos. 23, 24 (2 boats)	Chiswick ..	1881	92	10·6	4·9	1	33	470	21·8	1 1-pr. rev.	2	11	7
No. 11	..	1883	..	..	..	1	31	250	..	..	..	10	
FOURTH CLASS.													
No. 1	Chiswick ..	1878	78·8	9·8	3	1	25	420	19	1 1-pr. rev.	2	10	
No. 2	Poplar.. ..	1879	86	11	4·5	1	25	420	21	..	2	10	7
Nos. 6 and 18	Chiswick ..	1883	62·4	7·5	2·5	1	13	170	17	1 1-pr. rev.	2	10	
No. 11	Chiswick ..	1883	75·6	9·9	3·8	1	16	250	19·2	1 1-pr. rev.	2	10	
SUBMARINE—													
Delfino	Spezia.. ..	1895	49·0	..	..	..	..	..	10·0	..	..	..	..

Japan.

Name or Number.	Where Built.	Launched.	Dimensions. Length.	Beam.	Draught.	Number of Screws.	Displacement.	Indicated Horse-Power.	Maximum Trial Speed.	Armament.	Torpedo Tubes.	Complement.	Coal Capacity.
			Feet.	Feet.	Feet.		Tons.		Knots.				Tons.
DESTROYERS—													
Murakumo	Chiswick ..	1898											
Shinonome	Chiswick ..	1898											
Yugiri	Chiswick ..	1898											
Shiranui	Ch swick ..	1899	210·0	10·5	7·2	2	285	5,800	30 to 30·55	1 12-pr., 5 6-prs.	2	54	80
Kagerou	Chiswick ..	1895											
Usugumo	Chiswick ..	1900											
Ikadsuchi	Poplar.. ..	1898											
Inadsuma	Poplar.. ..	1899											
Akebono	Poplar.. ..	1899	220·0	20·6	8·9	2	400	6,000	31·63 to 31·38	1 12-pr., 5 6-prs.	2	55	95
Sazanami	Poplar.. ..	1839											
Oboro	Pop ar.. ..	1899											
Niji	Poplar.. ..	1899											
Hayabusa..	Havre	Bldg.											
Kasasagi	Havre	Bldg.	220·0	20·6	..	2	..	6,000	31	1 12-pr., 5 6-prs.	2		
Managuru	Havre	Bldg.											
Shirabaka	Havre	Bldg.											
4 unnamed	Elbing ..	Bldg.	220·0	20·6	8·6	2	360	6,000	33	1 12-pr., 5 6-pr. Q.F.	2	56	90
FIRST CLASS—													
Kotaka	Poplar.. ..	1886	170	19·6	5		190	1,400	19	4 mach.	6	..	
14 boats*	Creusot ..	1889	114·7	10·6	6	2	56	525	20	2 1-prs.	..	16	50
7 boats	Kobe	1889	114·7	10·6	6	1	56	525	20	2 1-prs.	..	16	
4 boats	Poplar.. ..	1879	100	12·5	..	1	40	620	20	..	..	..	
1 boat	Normand ..	1891	118	13·2	8·7	1	75	1,300	23	2 1-prs.	2	21	3
2 boats	Elbing ..	1891	128	16	..	2	90	1,300	23	3 1-prs.	3		10
10 boats	Kobe	Bldg.	..	..	..	1	..	..	..	..	..	..	24
5 boats	Havre	Bldg.	..	..	..	..	150	..	24	..	..	..	..
8 boats	Elbing ..	Bldg.	..	..	..	..	150	..	24	..	..	..	..
10 boats	Poplar.. ..	Bldg.	152·6	15·3	7·0	1	..	2,200	27	1 3-pr.	3	29	30

* No. 16 lost off the Pescadores, 1895.

The ten years' programme includes 23 first-class, 31 second-class, and 35 third-class torpedo-boats, and a 6750-ton torpedo transport.

Mexico.

Mexico has five first-class boats building or projected.

Netherlands.

Name or Number.	Where Built.	Launched.	Dimensions. Length.	Beam.	Draught.	Number of Screws.	Displacement.	Indicated Horse-Power.	Maximum Trial Speed.	Armament.	Torpedo Tubes.	Complement.	Coal Capacity.
			Feet.	Feet.	Feet.		Tons.		Knots.				Tons.
FIRST CLASS—													
Ardjoeno	Poplar.. ..	1886	125	13	6	1	83	800	21	2 1-prs.	2	16	10
Batok	Amsterdam	1887	125	13	6·9	1	83	725	20	2 1-prs.	2	16	10
Cycloop	Amsterdam	1887	125	13	6·9	1	83	680	20	2 1-prs.	2	16	10
Dempo	Amsterdam	1887	125	13	6·9	1	83	780	20	2 1-prs.	2	16	10
Empong	Poplar.. ..	1888	128	13	6·2	1	91	1,100	24·1	2 1-prs.	3	16	15
Etna	Poplar.. ..	1882	100	12·6	5·6	1	46	550	21·5	2 1-prs.	2	16	7
Foka	Amsterdam	1888	128	13	6·2	1	90	1,000	22·1	2 1-prs.	3		
Goentoer	Amsterdam	1888	128	13	6·2	1	90	950	21	2 1-prs.	3		
Habang	Amsterdam	1888	128	13	6·2	1	90	930	21·7	2 1-prs.	3		
Hekla	Poplar.. ..	1882	100	12·6	5·6	1	45	550	21·5	2 1-prs.	2	16	7
Idjen	Amsterdam	1889	128	13	6·2	1	90	840	20·6	2 1-prs.	3		
Krakatau	Amsterdam	1889	128	13	6·2	1	90	750	19·1	2 1-prs.	3		
Lamongan	Amsterdam	1890	104·5	13·3	5·2	1	50	790	20·7	2 1-prs.	2		
Makjan	Amsterdam	1890	104·5	13·3	5·2	1	50	790	20·7	2 1-prs.	2		
Nobo	Amsterdam	1890	104·5	13·3	5·2	1	50	790	20·7	2 1-prs.	2		
Scylla .. .. .. / Hydra	Poplar.. ..	Bldg.	130	13·6	6·0	1	..	1,200	23	2 1-prs.	2	18	20
3 boats	Poplar.. ..	Bldg.	152·6	15·3	7·6	1	..	2,000	23	2 3-prs.	2	25	30
13 boats	..	pro.	160	..	..	..	..	..	..	..	..	..	..
4 boats	..	pro.	100	..	..	..	..	..	..	..	..	..	..
SECOND CLASS—													
Nos. 1, 2, 4–20 (19 boats)	Chiswick, etc.	1878–86	76 / 79	10·3	5·2	1	29	250	18	1 1-pr.	2 sp	..	3
Nos. 3.21,2 (3 boats)	..	1890	83·6	10·5	5·1	1	37	460	17·9	1 1-pr.	1	..	3
1 boat	East Cowes	1883	45·5	9·7	..	1	..	..	12	1 mach.	1		
INDIAN FLEET—													
Cerberus	Flushing ..	1888	125	13	6·9	1	83	912	21·2	2-1 prs.	2	16	
1 boat	..	1891											
3 boats	..	1893–94	125	..	..	..	83	..	21·5	..	2		

Norway.

Name or Number.	Where Built.	Launched.	Dimensions.			Number of Screws.	Displacement.	Indicated Horse-Power.	Maximum Trial Speed.	Armament.	Torpedo Tubes.	Complement.	Coal Capacity.
			Length.	Beam.	Draught.								
FIRST CLASS—			Feet.	Feet.	Feet.		Tons.		Knots.				Tons.
Lyn	..	1882	94·2	9·7	2·5	1	36	430	18	..	1	..	3
Oi	..	1882	97·5	11	5·6	1	40	450	18	..	1	..	3
Orm, Otter (2 boats)	..	1387	108·2	12·2	5·6	1	40	500	20	..	2	..	3
Pil, Rask (2 boats)..	..	1887	101·7	11·8	5·6	1	40	500	20	..	2	..	3
Snar	..	1887	104·9	11·8	5·6	1	40	500	20	..	2	..	3
Springer	..	1887	97·5	11·6	5·6	1	40	450	19	..	2	..	3
Varg (8), Raket (9)	Christiania..	1894	111·5	12·4	..	1	43	..	..	..	2	..	..
Hval, Delfin, Hai (3 boats)	Elbing ..	1896	128·0	15·0	..	1	84	1,100	24·5	2 1·4-in. Q.F.	2	..	..
Storm, Ovrand, Trods	Christian'a ..	Bldg.	128·0	15·0	..	1	84	1,100	23	2 1·4-in. Q.F.	2		
SECOND CLASS—													
Rasp	Chiswick ..	1873	58	7·5	3·9	1	16	..	18	..	2		
Ulven	..	1878	56	..	..	1	16	..	9	..	sp.		
2 boats	..	..	..	..	..	..	20	..	12	..	..		

Portugal.

Name or Number.	Where Built.	Launched.	Dimensions.			Number of Screws.	Displacement.	Indicated Horse-Power.	Maximum Trial Speed.	Armament.	Torpedo Tubes.	Complement.	Coal Capacity.
			Length.	Beam.	Draught.								
			Feet.	Feet.	Feet.		Tons.		Knots.				Tons.
5 boats (5–9)	Elbing ..	1890–92											
Espadarte (1)	Poplar.. ..	1881	85	11	5	1	31	450	19·7	2 mach.	2	10	10
Nos. 2, 3, 4 (3 boats)	Poplar.. ..	1886	120	12·5	5·5	1	60	700	20	2 mach.	2	16	18
Fulminante	Blackwall ..	1880	75	15	2·6	2	40	150	11·5	2 mach.	..	..	8
1 boat	..	..	..	..	..	..	25						
Mineiro	Lisbon ..	1893	..	..	..	..	..	..	12				
SUBMARINE—													
Plongeur	..	1892	72·1	11·5	..	..	..	..	6				

Roumania.

Name or Number.	Where Built.	Launched.	Dimensions.			Number of Screws.	Displacement.	Indicated Horse-Power.	Maximum Trial Speed.	Armament.	Torpedo Tubes.	Complement.	Coal Capacity.
			Length.	Beam.	Draught.								
FIRST CLASS—			Feet.	Feet.	Feet.		Tons.		Knots.				Tons.
Naluka	Havre.. ..	1888	120·7	11·3	6·9	1	55	500	21	1 1-pr. rev	2	..	12
Sborul	Havre.. ..	1888	120·7	11·3	6·9	1	..	500	21	1 1-pr. rev.	2	..	12
Smeul	Havre.. ..	1888	120·7	11·3	6·9	1	..	500	21	1 1-pr. rev.	2	..	12
SECOND CLASS—													
Szimul	Poplar.. ..	1882	63	8	3	1	15	150	16·5	..	..	8	1
Vulturul	Poplar.. ..	1882	63	8	3	1	15	150	16·5	..	..	8	1

Russia.

Name or Number.	Where Built.	Launched.	Dimensions.			Number of Screws.	Displacement.	Indicated Horse-Power.	Maximum Trial Speed.	Armament.	Torpedo Tubes.	Complement.	Coal Capacity.
			Length.	Beam.	Draught.								
BALTIC SEA.													
DESTROYERS—			Feet.	Feet.	Feet.		Tons.		Knots.				Tons.
Sokol	Poplar.. ..	1895	190	18·6	7.0	2	240	4,400	29.7	1 12-pr. 3	2		
Krechet, Korshun (2 boats)	Abo	1898	198 10	18 6	7·0	2	240	3,800	27·5	6-pr.			
Iastreb	Ishora.. ..	1898	196·9	18·4	11·5	1	240	3,800	27	1 2·8-in.	2		
Nyrok	Ishora.. ..	1898	196·9	18·4	11·5	1	240	3,800	27	3 1·8-in.	2		
Berkout	Ishora.. ..	1898	196·9	18·4	11·5	1	240	3,800	27	3 1·8-in.	2		
Condor	Ishora.. ..	1898	196·9	18·4	11·5	1	240	3,800	27	3 1·8-in.	2		
6 boats	Ishora.. ..	Bldg.											
Kit, Skat, Delphin, Kassatka (4 boats)	Elbing ..	1899	196·9	18·4	11·5	1	350	3,800	27	3 1·8-in.	2		
Ossetr, Kephal, Losos	La Seyne ..	1899	187·0	20·7	6·6	1	320	..	..				
Forel, Sterliad ..	Havre.. ..	1899	..	..	..	1	312	..	..				
Gagara, Voron, Filin, Sova	Nevsky ..	1899	..	..	..	1	312	..	..				
Som	Birkenhead	1899	213	21·5	12·9	1	370	6,000	30				
Lebed, Pelikan, Pavlin, Fasan ...	Creighton ...												
Drozd, Diatel, Baklan, Bekass, Gorlitza, Gratch, Kulik, Perepel, Skvoretz, Strige, Shtchegol ... ... 2 boats	Nevsky and Ishora ...	Bldg.											
FIRST CLASS—													
Aspen	Ishora.. ..	1895	127·9	15·7	6·9	1	98	1,250	21	..	2	..	17
Abo	Elbing ..	1886	128	15·7	7·5	1	87	900	32·2	4 1-pr. revs.	2	13	17
Bjerke	Putiloff ..	1890	136·5	13	7·8	..	81	1,100	21				
Dago	Abo	1891	152	13	8·3	..	100	1,000	19				
Domeness	Putiloff ..	1895	127 9	15·7	6·9	1	98	1,250	21	..	2	..	17
Eckness	Abo	1890	136·5	13	7·8	..	81	1,100	21				
Hapsal	Putiloff ..	1891	126	13	8·5	1	81	1,100	21	2 1-pr. revs.	2	13	
Hogland	Ishora.. ..	1894	128	16	6·9	1	85	1,200	22	2 1-prs.	2	13	17
Kotka	Abo	1891	152	13	8·3	..	100	1,000	19				
Kotlinj	St. Petersburg	1885	124·2	12·9	5·9	2	67	500	16·5	2 1-pr. revs.	2	16	15
Kronschlot	Ishora.. ..	1891	152	13	8·3	..	100	1,000	19				
Lachta	Elbing ..	1886	128	15·7	7·5	1	87	900	20	4 1-pr. revs.	2	13	17
Libawa	Elbing ..	1886	128	15·7	7·5	1	87	1,000	22	4 1-pr. revs.	2	13	17
Louga	Elbing ..	1886	128	15·7	7·5	1	87	900	20	4 1-pr. revs.	2	13	17
Moonsund	Putiloff ..	1891	126	13	8·5	1	81	1,100	21	2 1-pr. revs.	2	13	
Nargen	Ishora.. ..	1894	128	16	6·9	1	85	1,200	22	2 1-prs.	2	13	17
Narwa	Elbing ..	1886	128	15·7	7·5	1	87	900	20	4 1-pr. revs.	2	13	17
Nyrok	Ishora.. ..	1898											
Pernoff	Normand ..	1892	138	14·7	9·9	2	118	1,000	25·4	2 mach.	2	26	
Rochensalm	Putiloff ..	1890	136·5	13	7·8	..	81	1,100	21				
Seskar	Ishora.. ..	1891	152	13	8·3	..	100	1,000	19				
Sestoresk	Normand ..	1893	118	13·2	8·7	2	130	1,900	25	2 1-prs.	2	21	10
Tosna	Putiloff ..	1893	127·9	15·7	6·9	1	98	1,250	21	..	2	13	17
Transund	Ishora.. ..	1895	127·9	15·7	6·9	1	98	1,250	21	..	2	..	17
Viborg	Clydebank..	1886	144·5	17	8·1	2	126	1,400	20	2 3-pr. revs.	3	24	45*
Vindawa	Elbing ..	1886	128	15·7	7·5	1	87	900	21	4 1-pr. revs.	2	13	17
Vzriw	St. Petersburg	1877	118	16	10·9	1	160	800	14·5	4 Q.F.	1	18	16
8 boats	St. Petersburg	1894	128	16	6·9	1	85	1,200	22	2 1-prs.	2	13	17
2 boats	Putiloff ..	1894	138	14·7	9·9	2	118	..	25	2 mach.	2	26	
2 boats	St. Petersburg	1896	128	16	6·9	2	85	1,200	22	2 1-prs.	2	13	17
6 boats	St. Petersburg	1897	138	14·7	9·9	2	120	..	25	..	2	26	
†8 boats	St. Petersburg	1898	..	..	..	..	118						
3 boats	Nicolaieff ..	1898											
SECOND CLASS—													
21 boats (Galka class)	Elbing and Russia ..	1880 &c.	74·7	8·9	5	1	30	220	16	..	2	14	3
21 boats (Woronclass)	Elbing and Russia ..	..	66	11·1	..	1	..	260	17				
1 boat	Poplar.. ..	1888	60	8·5	3	1	16	240	17·5	..	2	..	1
BLACK SEA.													
FIRST CLASS—													
A. B. C. (3 boats) ..	Nicolaieff ..	1893	126	..	..	..	81	..	21				
Adler..	Elbing ..	1890	152·0	17·2	7·9	2	130	2,200	27·4	2 1-prs.	3	24	40
Anakria	Elbing ..	1890	128·0	16	6·9	1	85	1,200	22	2 1-prs.	2	13	17
Anapa	Odessa.. ..	1891	126	13	8·5	1	81	1,100	21	2 1-pr. revs.	2	13	
Altodorj	Odessa.. ..	1891	126	13	8·5	1	81	1,100	21	2 1-pr. revs.	2	13	
Batoum	Poplar.. ..	1880	100	12·5	5·5	1	40	500	22	2 1-pr. revs.	2	12	9
D. E. (2 boats) ..	Sebastopol ..	1893	128	..	..	..	85	..	22				
Gagri	Claparède ..	1883	120·6	13·3	7	1	78	600	18	2 1-pr. revs.	2	13	12
Gelendshik	La Seyne ..	1883	122·7	12·4	6·2	1	73	560	18	2 1-pr. revs.	2	13	11
Ismail	Nicolaieff ..	1886	128	15·7	7·5	1	87	900	20	2 1-pr. revs.	2	13	17
Itzvar	Odessa.. ..	1891	..	..	..	..	81	1,100					
Kodor	Elbing ..	1886	128	15·7	7·5	1	87	900	21	4 1-pr. revs.	2	13	17

* Has received liquid fuel apparatus. † Of the Pernoff type, building on the Neva.

Russia—*continued.*

Name or Number.	Where Built.	Launched.	Dimensions. Length.	Beam.	Draught.	Number of Screws.	Displacement.	Indicated Horse-Power.	Maximum Trial Speed.	Armament.	Torpedo Tubes.	Complement.	Coal Capacity.
BLACK SEA—*contd.*			Feet.	Feet.	Feet.		Tons.		Knots.				Tons.
FIRST CLASS—*contd.*													
Kilia	Elbing ..	1886	128	15·7	7·5	1	87	900	22	4 1-pr. revs.	2	13	17
Novorossisk	Elbing ..	1886	128	15·7	7·5	1	87	900	22	4 1-pr. revs.	2	13	17
Poti	Normand ..	1883	124·6	11·9	6·7	1	72	570	18·5	2 1-pr. revs.	2	13	11
Reni	Elbing ..	1886	128	15·7	7·5	1	87	900	22	4 1-pr. revs.	2	13	17
Sookboum	Chiswick ..	1883	113	12·5	6	1	64	700	19.5	2 Nords.	2	13	10
Tchardak	Elbing ..	1886	128	15·7	7·5	1	87	900	20	4 1-pr. revs.	2	13	17
Yalta	Elbing ..	1886	128	15·7	7·5	1	87	900	22	4 1-pr. revs.	2	13	17
3 boats	Elbing ..	1886	128	15·7	7·5	1	87	900	22	4 1-pr. revs.	2	13	17
4 boats	Nicolaieff ..	Bldg.	..	..	..	..	..	..	..	..	..	..	..
SECOND CLASS—													
Istcheritza	Sebastopol ..	1878	62·3	9·7	3·9	1	24	220	15	..	..	10	
Karabin	Elbing	1877	64·3	8·4	2	1	11	120	15	..	..	8	
Kefal	Chiswick ..	1880	60·5	7·5	3·5	1	..	..	16·8	..	..	8	
Scheglensk	Sebastopol ..	1878	58·3	9·5	3·9	1	24	220	15	..	..	10	
Schehouka	Sebastopol ..	1878	59·3	9·5	3·9	1	24	220	15	..	..	10	
Scoombia	Odessa	1878	61·3	10	4	1	25	220	15	..	..	10	
Soroka	St. Petersburg	1878	62.3	9·7	3·9	1	24	220	15	..	..	10	
Soulin	..	1877	60	9·7	3·9	1	24	210	15	..	..	10	
Sultanka	Odessa	1878	64·3	10	4	1	25	220	15	..	..	10	
1 boat	Poplar	1877	75	10	..	..	..	..	..	..	..	..	
50 boats (Woron Class)	Elbing, etc.	..	66	11·1	..	1	..	260	17				
SIBERIAN FLOTILLA.													
Borgo	Abo	1890	136·5	13	7·8	..	81	1,100	21				
Forel	..	..	71·5	6·5	3·3	1	23	220	16				
Jantchiche	Elbing	1887	128	15·7	11·5	..	87	970	19	4 1-pr. revs.	2	13	17
N.	..	1893	152·5	16·8	..	..	140	2,200	26·5	2 1-pr. revs.	3	24	40
N.	..	1893	152·5	16·8	..	..	140	2,200	26·5	2 1-pr. revs.	3	24	40
Podorosnik	..	..	71·5	6·5	3·3	1	23	220	16				
Revel	Normand ..	1886	152·3	12·3	8·1	1	96	780	22	2 Q.F.	2	23	30
Sisik	..	..	71·5	6·5	3·3	1	23	220	16				
Skorpion	..	..	71·5	6·5	3·3	1	23	220	16				
Sootchena	Elbing	1887	128	15·7	11·5	..	87	970	19	4 1-pr. revs.	2	13	17
Sterliad	..	..	71·5	6·5	3·3	1	23	220	16				
Strauss	..	..	71·5	6·5	3·3	1	23	220	16				
Sunguri (*ex* Hogland)	Abo	1890	152	16	7·9	2	140	1,800	22				
Sweaborg	Normand ..	1886	152·3	12·3	8·1	1	96	780	19·7	2 Q.F.	..	..	30
Ussuri (*ex* Nargen)	Abo	1890	152	16	7·9	2	140	1,800	22				
2 Unnamed	Ochtenski ..	Bldg.	152	16	7·9	2	140	1,800	22				

Spain.

Name or Number.	Where Built.	Launched.	Dimensions. Length.	Beam.	Draught.	Number of Screws.	Displacement.	Indicated Horse-Power.	Maximum Trial Speed.	Armament.	Torpedo Tubes.	Complement.	Coal Capacity.
DESTROYERS—			Feet.	Feet.	Feet.		Tons.		Knots.				Tons.
Terror	Clydebank ..	1896	220	22	5·6	2	300	6,000	28	2 12-pr. 2 6-pr. 2 1-pr.	2	67	100
Audaz .. .. .. }													
Osado .. .. .. }	Clydebank ..	1897	225	25·6	5·8	2	400	7,500	30	2 14-pr. 2 6-pr. 2 1-pr.	2	70	80
Proserpina .. .. }													
FIRST CLASS—													
Acevedo	Chiswick ..	1885	117·7	12·5	6·2	1	63	660	20·1	2 mach.	2		
Arieta	Chiswick ..	1887	147·5	14·6	4·9	2	97	1,600	26·1	4 3-pr. Q.F.	2	..	25
Azor	Poplar	1887	134·5	14	6	1	108	1,600	24	4 3-pr. Q.F.	3	23	25
Bustamente	Normand ..	1887	126	10·9	..	..	63	800	..	3 3-prs.	2		
Ejercito	Kiel	1887	111·5	13	3·3	..	60	1,000	25	2 mach.	2		
Habana	Chiswick ..	1887	127·5	12·5	6	1	59	730	21·3	1 mach.	2		
Halcon	Poplar	1887	134·5	14	..	1	108	1,600	24	4 3-pr. Q.F.	3	23	25
Julian Ordoñez ..	Chiswick ..	1885	117·7	12·5	6·2	1	65	660	20·1	2 1-in. Nord.	2		
Orion	Gaarden ..	..	125	15·5	3·5	1	85	1,000	21·5	2 1-pr. revs.	2	18	16
Rayo	Chiswick ..	1887	147·5	14·6	4·9	2	97	1,600	25·5	4 3-pr. Q.F.	2	..	25
Retamosa	Poplar	1886	118	12·5	5·5	1	70	700	20·5	2 1-in.	2	17	20
Rigel	Bremen ..	1883	105	12·3	3·3	1	57	..	19	1 1-pr. rev.	2	18	13
Seza	Ferrol	1885	126	..	..	..	85	..	14				
4 boats	..	Bldg.	147	43.0	5	..	98	1,600	25			23	25
2 boats	Clydebank ..	Bldg.	..	..	..	..	..	..	28				
SECOND CLASS—													
Aire	Spain	1883	43·4	10·2	3	2	25	175	8	1 3·1-in.	..	16	1
Castor	La Seyne ..	1878	76·2	9·7	2·3	..	23	265	19	..	..	14	1·5
Pollux	Poplar	1879	84·5	10·7	4·6	..	33	450	19·5	..	2	14	9
VEDETTE BOATS—													
3 boats	East Cowes	1892	60	9·3	..	..	..	..	18.3				
Peral	Carraca ..	1889	70	8·5	..	2	87	60	10				

Sweden.

TORPEDO BOATS.

Name or Number.	Where Built.	Launched.	Dimensions. Length.	Beam.	Draught.	Number of Screws.	Displacement.	Indicated Horse-Power.	Maximum Trial Speed.	Armament.	Torpedo Tubes.	Complement.	Coal Capacity.
FIRST CLASS—			Feet.	Feet.	Feet.		Tons.		Knots.				Tons.
Komet	Elbing ..	1896	128	15·9	6·11	1	92	1,056	23·0	2 1·9-in. Q.F.	2	16	17
Blixt..	Carlskrona..	1898	128	15·9	6·11	1	92	1,260	23·5	2 1·9-in. Q.F.	2	18	17
Meteor	Carlskrona..	1899	128	15·9	6·11	1	92	1,330	23·8	2 1·9-in. Q.F.	2	18	17
Stjerna	Carlskrona..	1899	128	15·9	6·11	1	92	1,250	23·4	2 1·9-in. Q.F.	2	18	17
Orkan	Carlskrona..	1900	128	15·9	6·11	1	92	1,250	23·5	2 1·5-in. Q.F.	2	18	17
Vind	Carlskrona..	1900	128	15·9	6·11	1	92	1,250	23·5	2 1·5-in. Q.F.	2	18	17
Bris	Carlskrona..	1900	128	15·9	6·11	1	92	1,250	23·5	2 1·5-in. Q.F.	2	18	17
2 boats (A and B) ..	Carlskrona..	Bldg.	128	15·9	6·11	1	92	1,250		2 1·5-in. Q.F.	2	18	17
No. 1..	Chiswick ..	1884	113·2	12·2	6·3	1	65	620	18·5	1 mach.	2	16	11
2 boats (3 and 5) ..	Stockholm ..	1887	114·2	12·6	6·7	1	67	620	18·5	1 mach.	2	16	15
No. 7..	Stockholm ..	1887	114·2	12·6	6·7	1	67	620	18·7	1 mach.	2	16	15
2 boats (9 and 11) ..	Carlskrona..	1894	126·8	13·11	7·7	1	86	850	19·5	2 mach.	2	16	15
SECOND CLASS—													
No. 61	Stockholm ..	1882	91·6	11·8	5·7	1	40	350	16·0	1 mach.	1	14	9
No. 63	Chiswick ..	1883	100·1	11·10	5·11	1	45	420	19·0	1 mach.	2	14	7
No. 65	Stockholm ..	1885	100·1	11·10	5·11	1	45	420	19·0	1 mach.	2	14	9
No. 67	Stockholm ..	1888	100·9	11·10	6·1	1	46	430	19.2	1 mach.	2	14	9
No. 69	Stockholm ..	1886	100·9	11·10	6·1	1	46	450	19·9	1 mach.	2	14	9
No. 71	Stockholm ..	1887	103·4	11·10	6·7	1	58	460	18·6	1 mach.	2	14	9
No. 73	Stockholm ..	1887	103·4	11·10	6·7	1	68	460	18·6	1 mach.	2	14	9
No. 75	Stockholm ..	1892	100·5	11·6	6·3	1	49	460	18·9	1 mach.	2	14	9
No. 77	Carlskrona..	1891	100·5	11·6	6·3	1	49	460	18·9	1 mach.	2	14	9
No. 79	Stockholm ..	Bldg.	104·0	12·5	6·1	1	49			1 1·5-in. Q.F.	2	14	
No. 81	Stockholm ..	Bldg.	104·0	12·5	6·1	1	49			1 1·5-in. Q.F.	2	14	
THIRD CLASS—													
Nos. 141, 143, 145, 147, 149 (5 boats) ..	Stockholm ..	1879–1890	55·0	10·7	4·1	2	21	80	10		2		1·5

Turkey.

Name or Number.	Where Built.	Launched.	Dimensions. Length.	Beam.	Draught.	Number of Screws.	Displacement.	Indicated Horse-Power.	Maximum Trial Speed.	Armament.	Torpedo Tubes.	Complement.	Coal Capacity.
DESTROYERS—			Feet.	Feet.	Feet.		Tons.		Knots.				Tons
Berk-Efshan	Gaarden ..	1894	187	21·6	..	2	270	200	25	6 1-pr. revs.	2		
Tajjar	Gaarden ..	1894	187	21.6	..	2	270	..	25	6 1-pr. revs.	2		
FIRST CLASS—													
Edjder (No. 10) ..	Gaarden ..	1890	152·7	18·9	7·4	2	150	2,200	23	5 3-prs. Q.F.	2		
1 boat	Constantinople	1889	140	16	6·9	2	120	1,800	23	5 1-pr. revs.	2		
5 boats	Gaarden ..	1889–90	126·7	15·4	8·6	1	85	1,300	22	2 1-pr. revs.	2	21	8
Timsah	London ..	1887	126	15	..	..	..	..	21·7				
5 boats	Elbing ..	1886	120·3	16·2	..	..	85	900	21	2 Nords.	2	20	10
4 boats *	Constantinople	1886–89	100·3	11·8	5·5	1	42	550	19·5	2 mach.			
Tewfik	Normand ..	1885	100·7	13	5·5	1	42	550	20				
2 boats	La Seyne and Constantinople	1885	100·7	13	5·5	1	42	550	20·3	2 Nords.			
2 boats	Teddington	1887	124	15	..	..	..	..	22				
2 boats	Kiel	1892	127	..	..	..	..	..	22				
SUBMARINE—													
Abdul Hamid	Chertsey ..	1886	100	12	..	3	160	250	10	2 mach.	1	..	8
Abdul Medjid	Chertsey ..	1886	100	12	..	3	160	250	10	2 mach.	1	..	8

United States.

Name.	Where Built.	Launched.	Dimensions.			Number of Screws.	Displacement.	Indicated Horse-Power.	Maximum Trial Speed.	Armament.		Complement.	Maximum Coal Capacity.
			Length.	Beam.	Draught.					Guns.	Torpedo Tubes.		
DESTROYERS—		Bldg.	ft. in.	ft. in.	ft. in.		Tons.		Knots.				Tons
Bainbridge ..	Philadelphia	1900	245 0	23 7	6 6	2	420	8,000	29	2 12-pr., 5 6-pr.*	2	64	139
Barry	Philadelphia	1900	245 0	23 7	6 6	2	420	8,000	29	2 12-pr., 5 6-pr.	2	64	[illegible]
Chauncey.. ..	Philadelphia	1900	245 0	23 7	6 6	2	420	8,000	29	2 12-pr., 5 6-pr.	2	64	130
Dale	Richmond ..	1900	245 0	23 7	6 6	2	420	8,000	28	2 12-pr., 5 6-pr.	2	64	139
Decatur	Richmond ..	1900	245 0	23 7	6 6	2	420	8,000	28	2 12-pr., 5 6-pr.	2	64	139
Hopkins.. ..	Wilmington	1900	244 0	24 6	6 0	2	408	7,200	29	2 12-pr., 5 6-pr.	2	64	150
Hull	Wilmington	1900	244 0	24 6	6 0	2	408	7,200	29	2 12-pr., 5 6-pr.	2	64	150
Lawrence.. ..	Weymouth, Mass. ..	1900	242 3	22 3	6 2	2	400	8,400	30	2 12-pr., 5 6-pr.	2	64	115
Macdonough ..	Weymouth, Mass. ..	1900	242 3	22 3	6 2	2	400	8,400	30	2 12-pr., 5 6-pr.	2	64	115
Paul Jones ..	San Francisco	1900	245 0	23 7	6 6	2	420	7,000	29	2 12-pr., 5 6-pr.	2	64	139
Perry	San Francisco	1900	245 0	23 7	6 6	2	420	7,000	29	2 12-pr., 5 6-pr.	2	64	139
Preble	San Francisco	1900	245 0	23 7	6 6	2	420	7,000	29	2 12-pr., 5 6-pr.	2	64	139
Stewart	Morris Heights	1900	245 0	23 7	6 6	2	420	8,000	29	2 12-pr., 5 6-pr.	2	64	139
Truxtun	Baltimore ..	1900	248 0	23 3	6 0	2	433	8,300	30	2 12-pr., 5 6-pr.	2	64	232
Whipple	Baltimore ..	1900	248 0	23 3	6 0	2	433	8,300	30	2 12-pr., 5 6-pr.	2	64	232
Worden	Baltimore ..	1900	248 0	23 3	6 0	2	433	8,300	30	2 12-pr., 5 6-pr.	2	64	232
Bagley	Bath	Bldg.	157 0	17 0	4 7	2	167	..	28	3 3-pr.	3	29	..
Bailey	Morris Heights	1899	205 0	19 0	6 0	2	235	5,000	30	4 6-pr.	2	..	20
Barney	Bath	Bldg.	157 0	17 0	4 7	2	167	..	28	3 3-pr.	3	29	..
Biddle	Bath	Bldg.	157 0	17 0	4 7	2	167	..	28	3 3-pr.	3	29	..
Blakely	Boston ..	Bldg.	175 0	17 6	4 8	2	165	3,000	26	3 3-pr.	3	29	70
De Long	Boston ..	Bldg.	175 0	17 6	4 8	2	165	3,000	26	3 3-pr.	3	29	70
Du Pont	Bristol, R.I.	1897	175 0	17 8	4 8	2	165	3,400	28·58	4 1-pr.	3	32	76
Farragut	San Francisco	1898	213 6	20 8	6 0	2	273	5,000	30	4 6-pr.	2	..	76
Foote..	Baltimore ..	1896	160 0	16 1	5 0	2	142	..	24·5	3 1-pr.	3	24	44
Goldsborough ..	Portland, Ore	1899	194 8	20 5	5 0	2	247·5	..	30	4 6-pr.	2	..	131
Nicholson ..	Elizabethport	Bldg.	174 6	17 0	4 6	2	174	..	26	3 3-pr.	3	29	..
O'Brien	Elizabethport	Bldg.	174 6	17 0	4 6	2	174	..	26	3 3-pr.	3	29	..
Porter	Bristol, R.I.	1896	175 0	17 8	4 8	2	165	..	28·63	4 1-pr.	3	32	76
Rodgers	Baltimore ..	1896	160 0	16 1	5 0	2	142	2,000	24·5	3 1-pr.	3	24	44
Rowan	Seattle, Wash.	1898	170 0	17 0	5 11	2	182	3,200	26	4 1-pr.	3	32	60
Shubrick	Richmond ..	1899	175 0	17 6	4 8	2	165	3,000	26	3 3-pr.	3	29	70
Stockton	Richmond ..	1899	175 0	17 6	4 8	2	165	3,000	26	3 3-pr.	3	29	70
Stringham ..	Wilmington	1899	225 0	22 0	6 6	2	340	7,200	30	7 6-pr.	2	..	120
Thornton	Richmond	Bldg.	175 0	17 6	4 8	2	165	3,000	26	3 3-pr.	3	29	70
Tingey	Baltimore ..	Bldg.	175 0	17 6	4 8	2	165	3,000	26	3 3-pr.	3	29	70
Wilkes	Morris Heights	Bldg	175 0	17 6	4 8	2	165	3,000	26·25	3 3-pr.	3	29	70
Winslow	Baltimore ..	1897	160 0	16 1	5 0	2	142	2,000	24·5	3 1-pr.	3	24	44
SEA-GOING—													
Cushing	Bristol, R.I.	1890	138 9	14 3	4 11	2	105	1,720	22·5	3 1-pr.	3	23	36
Davis	Portland, Ore.	1898	146 0	15 4	5 4	2	132	1,750	22.5	3 1-pr.	3	..	..
Dahlgren	Bath	1899	147 0	16 4	4 7	2	146	4,200	30·5	4 1-pr.	2	..	33
Ericsson	Dubuque, Iowa	1894	149 7	15 6	4 9	2	120	1,800	24	4 1-pr.	3	23	35
Fox	Portland, Ore.	1898	146 0	15 4	5 4	2	132	1,750	22·5	3 1-pr.	3	..	..
Manly	Yarrow ..	..	..	..	..	..	..	..	..	..	..	..	..
Morris	Bristol, R.I.	1898	138 3	15 6	4 1	2	105	1,750	24	3 1-pr.	3	..	28
Somers	Schichau, Elbing ..	..	149 3½	17 5	..	2	145	..	..	..	..	..	..
T. A. M. Craven	Bath	1899	147 0	16 4	4 7	2	146	4,200	30·5	4 1-pr.	2	..	32
THIRD CLASS—													
Gwin..	Bristol, R.I.	1897	99 6	12 6	3 3	1	46	850	20·88	1 1-pr.	2	..	8
Mackenzie ..	Philadelphia	1898	99 3	12 9	4 3	1	65	850	20	1 1-pr.	2	..	15·3
McKee	Philadelphia	1898	99 3	12 9	4 3	1	65	850	19·82	2 1-pr.	2	..	..
Talbot	Bristol, R.I.	1897	99 6	12 6	3 3	1	46	850	21·15	1 1-pr.	2	..	8·8
Stiletto (wood)	Bristol, R.I.	..	88 6	11 0	3 0	1	31	359	18·22	..	2	..	4
SUBMARINE—													
Plunger	Baltimore ..	1893	85 3	11 6	..	2	168	1,200	8	..	1	..	..
Holland	Elizabethport	1896	54 0	10 0	..	..	65	150	..	..	2	5	..

* Guns of Destroyers of this class are Driggs Semi-Automatic Quick-Firers.

The Barcelo and some other Spanish torpedo-boats were captured during the war.

The programme of 1898 included 16 destroyers and 12 sea-going boats.

PLANS

OF

BRITISH AND FOREIGN SHIPS.

SCALE FOR FULL-PAGE PLATES
50 FEET TO THE INCH.

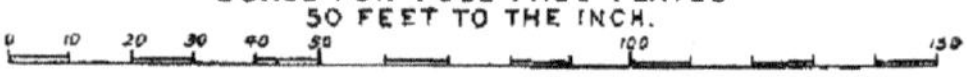

SCALE FOR HALF-PAGE PLATES.
100 FEET TO THE INCH

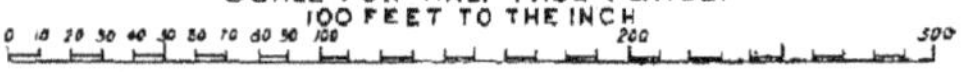

GREAT BRITAIN.

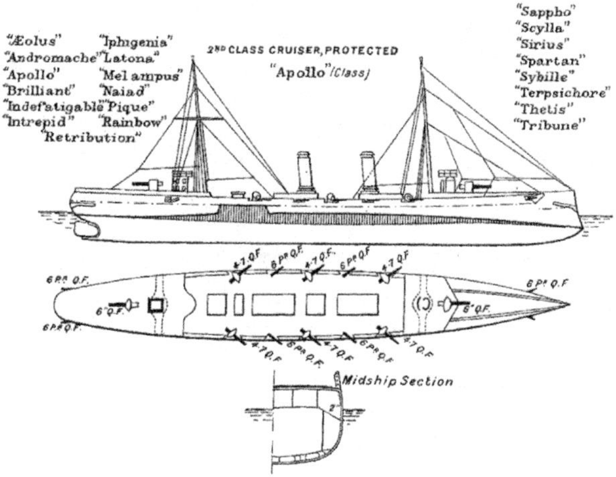

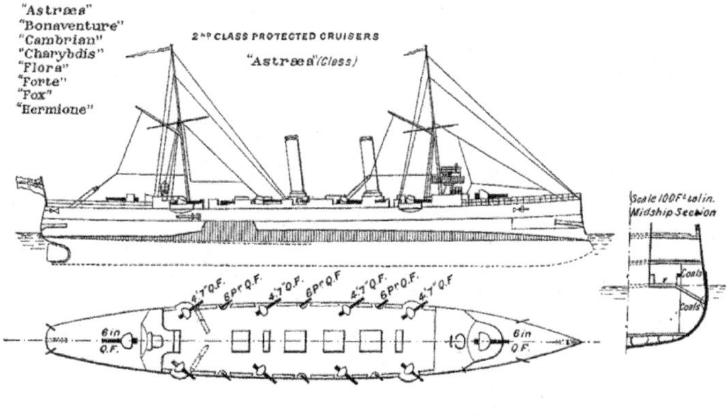

PLATE 1.

b

GREAT BRITAIN

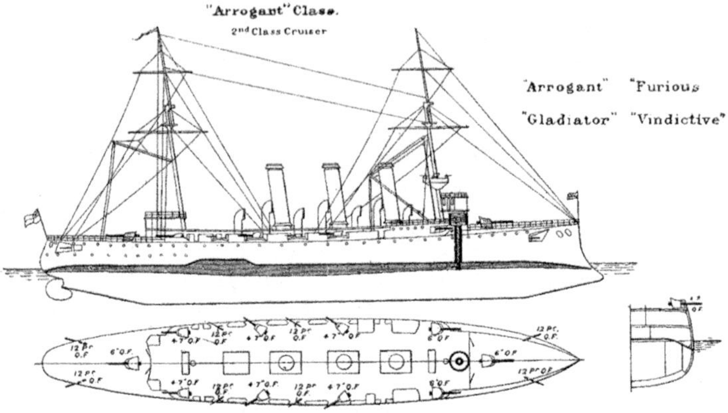

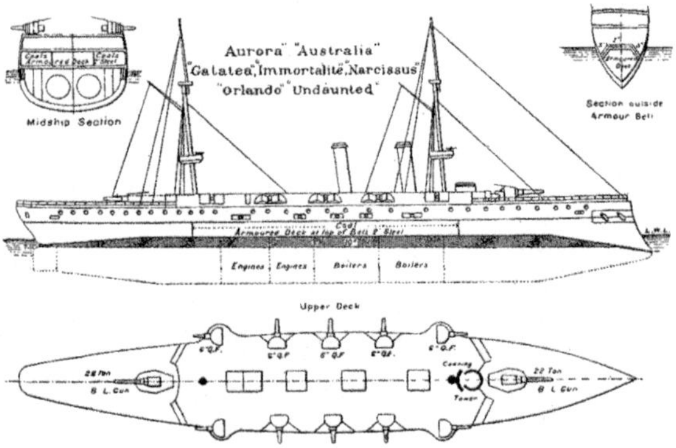

PLATE 2.

GREAT BRITAIN

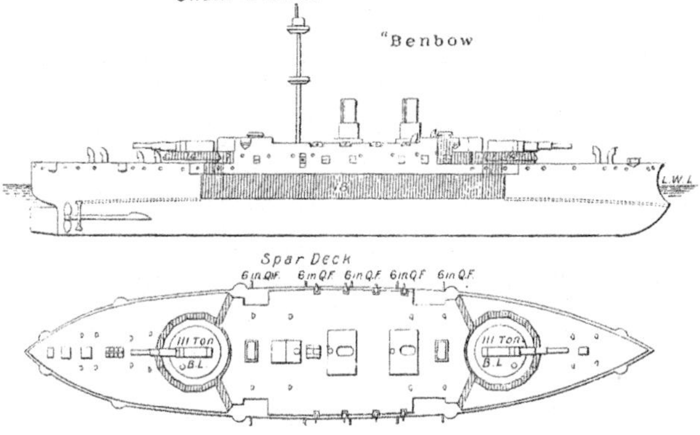

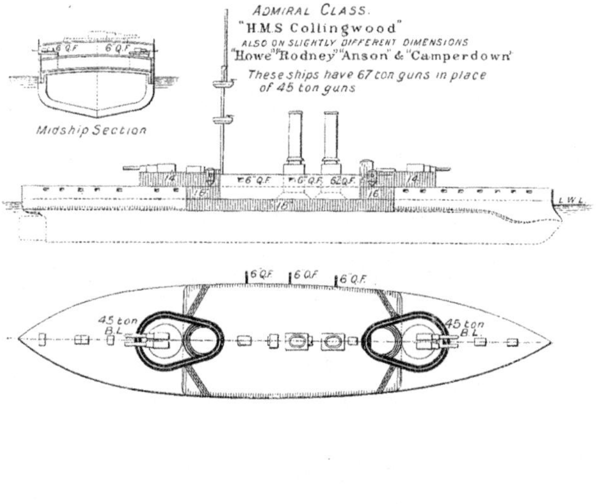

PLATE 3.

b 2

GREAT BRITAIN

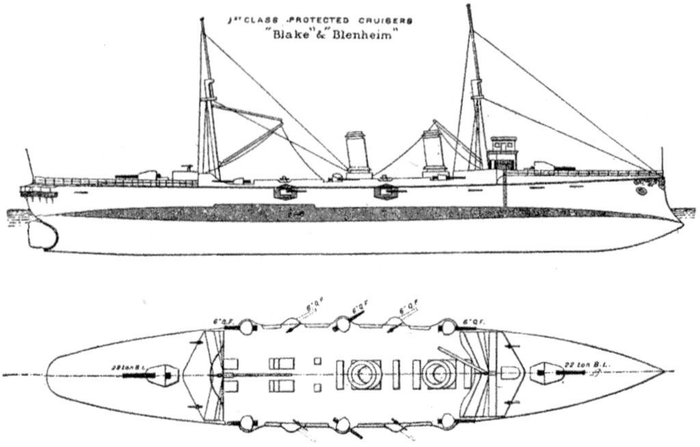

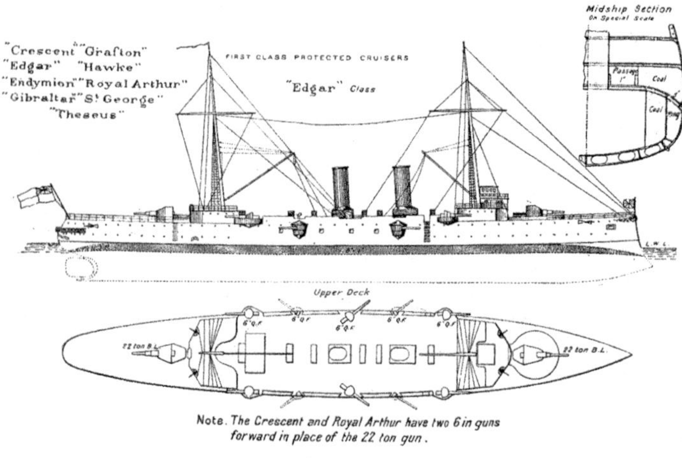

Note. The Crescent and Royal Arthur have two 6 in guns forward in place of the 22 ton gun.

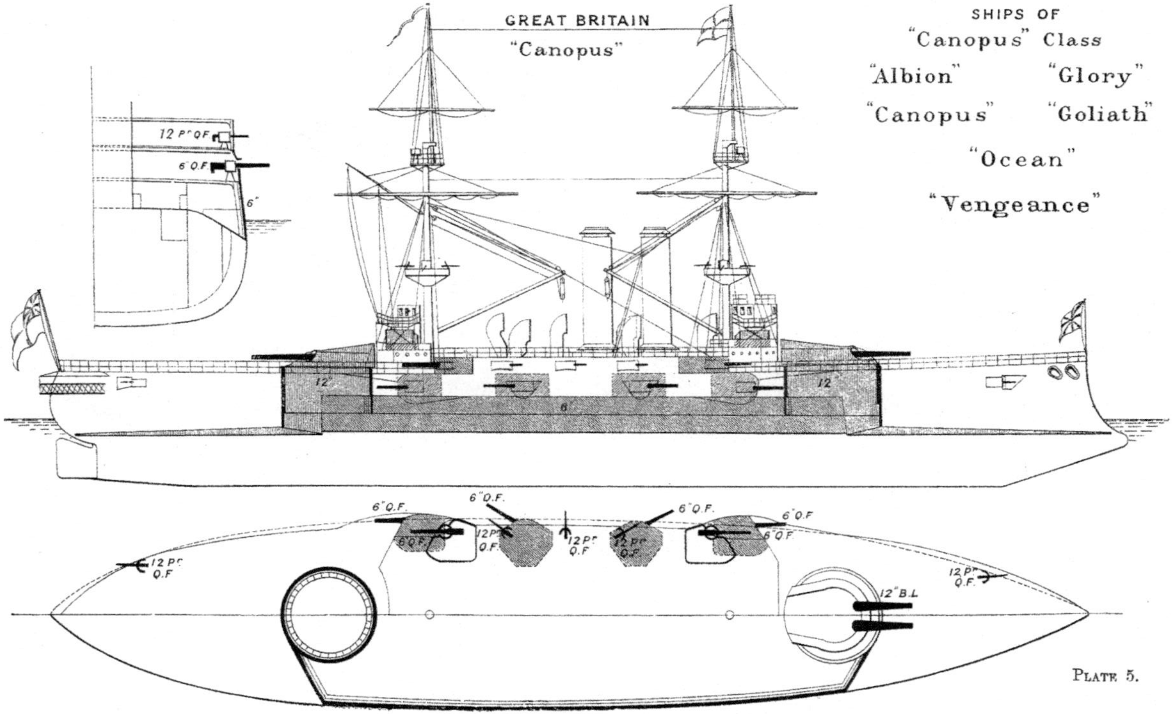

SHIPS OF
"Canopus" Class
"Albion" "Glory"
"Canopus" "Goliath"
"Ocean"
"Vengeance"

PLATE 5.

CREAT BRITAIN

FIRST CLASS BATTLE SHIPS

"Centurion" & "Barfleur"

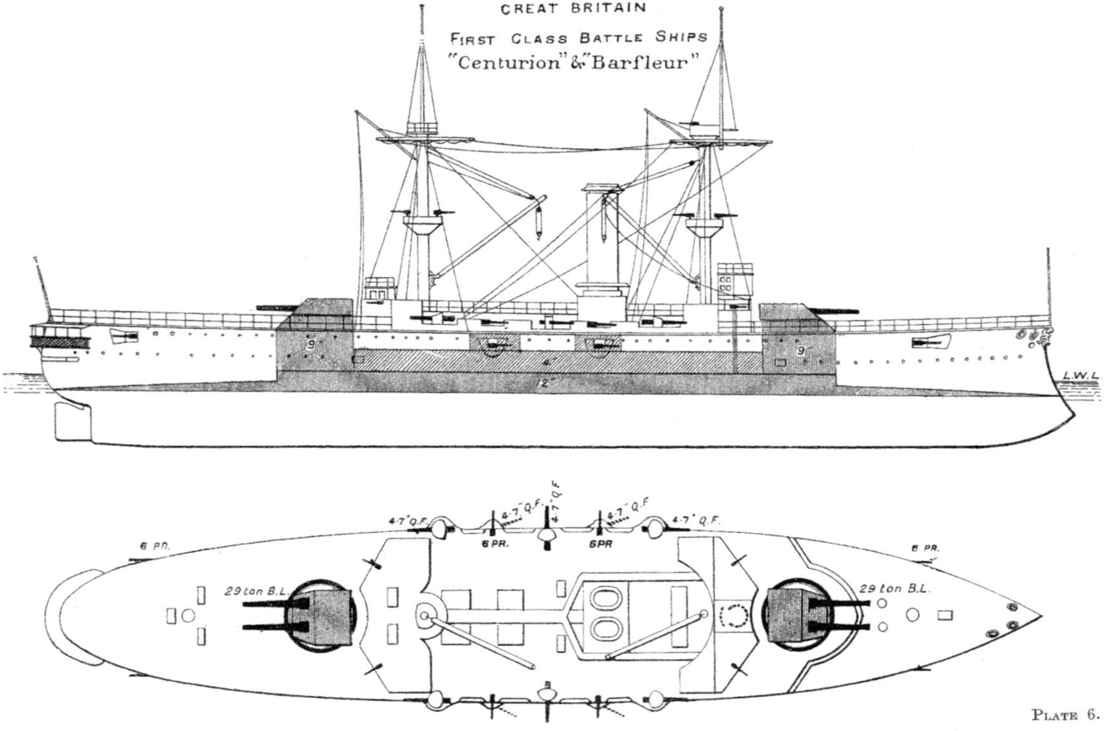

PLATE 6.

CREAT BRITAIN

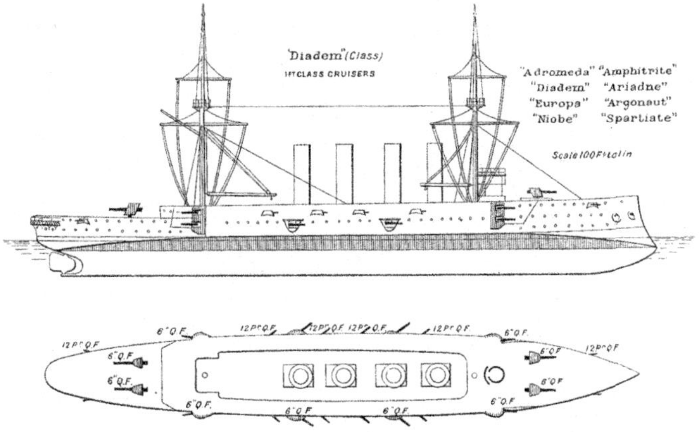

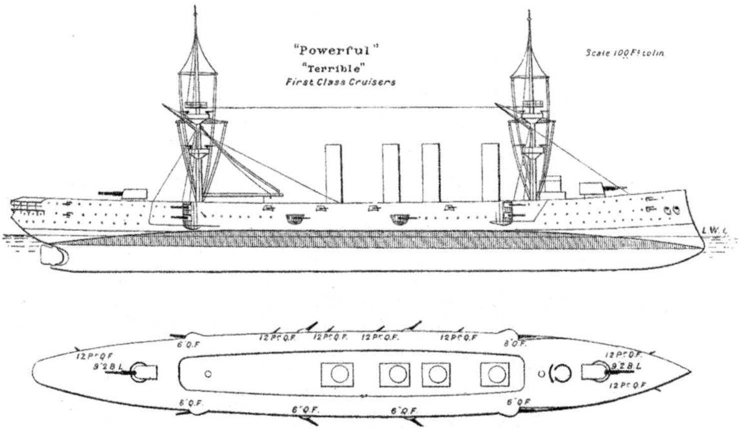

Plate 7.

GREAT BRITAIN

FIRST CLASS ARMOURED CRUISERS OF THE

"CRESSY" CLASS

"Cressy". "Aboukir". "Hogue".

"Euryalus". "Sutlej". "Bacchante".

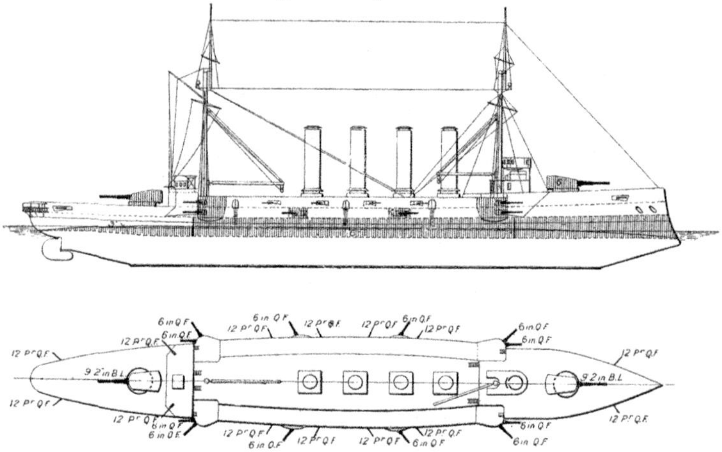

"FORMIDABLE" CLASS

"Formidable". "Irresistible". "Implacable".

"Bulwark", "London". "Venerable".

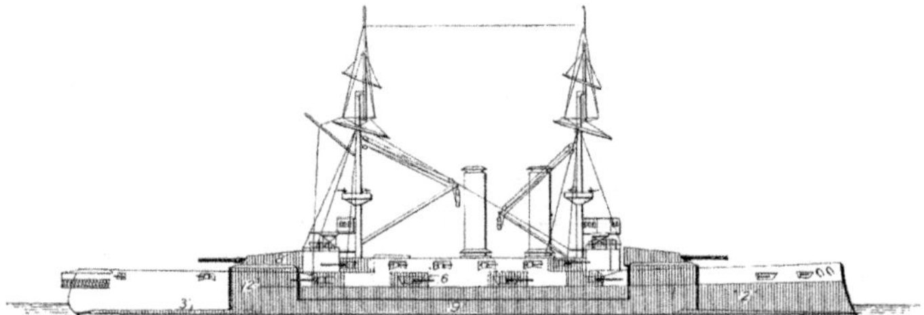

PLATE 8.

GREAT BRITAIN

"Dreadnought"

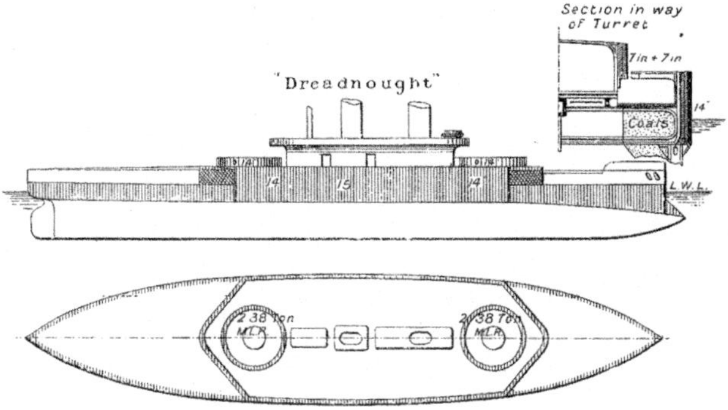

"Inflexible"

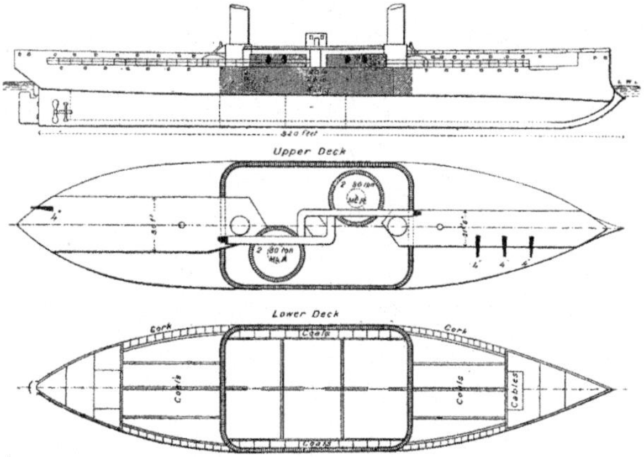

PLATE 9.

GREAT BRITAIN

FIRST CLASS BATTLE SHIPS.

"Duncan" Class

"Duncan"
"Cornwallis"
"Russell"
"Exmouth"
"Montagu"
"Albemarle"

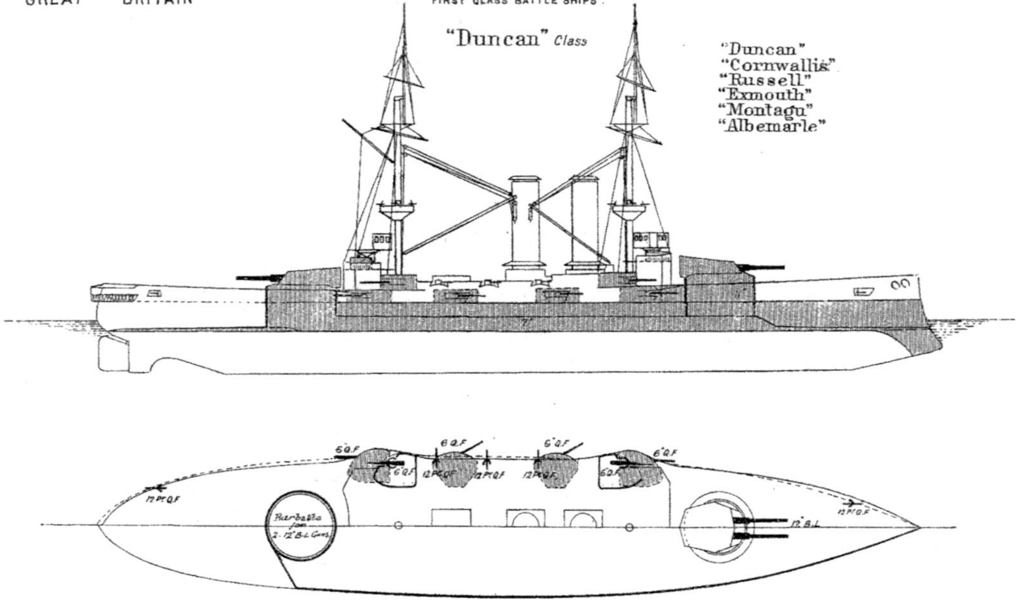

Plate 10.

GREAT BRITAIN.

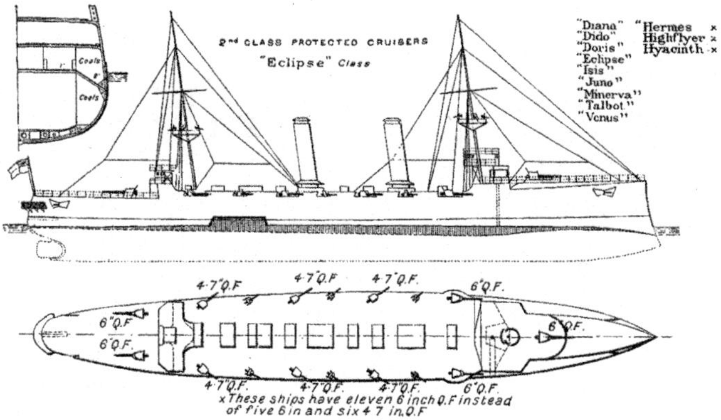

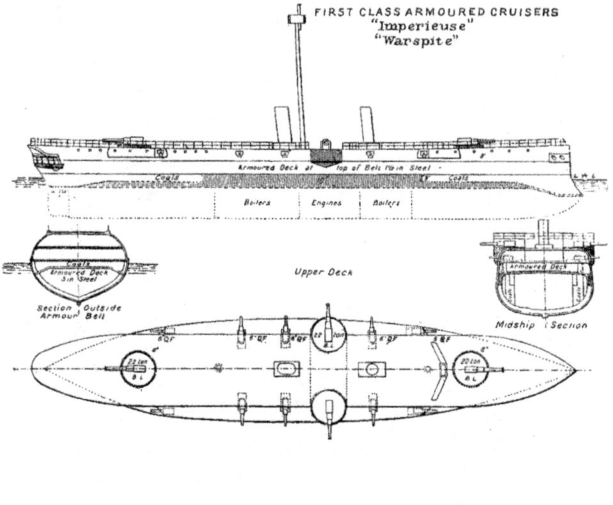

Plate 11.

GREAT BRITAIN

"Hood"

TURRET SHIP WITH REDOUBTS

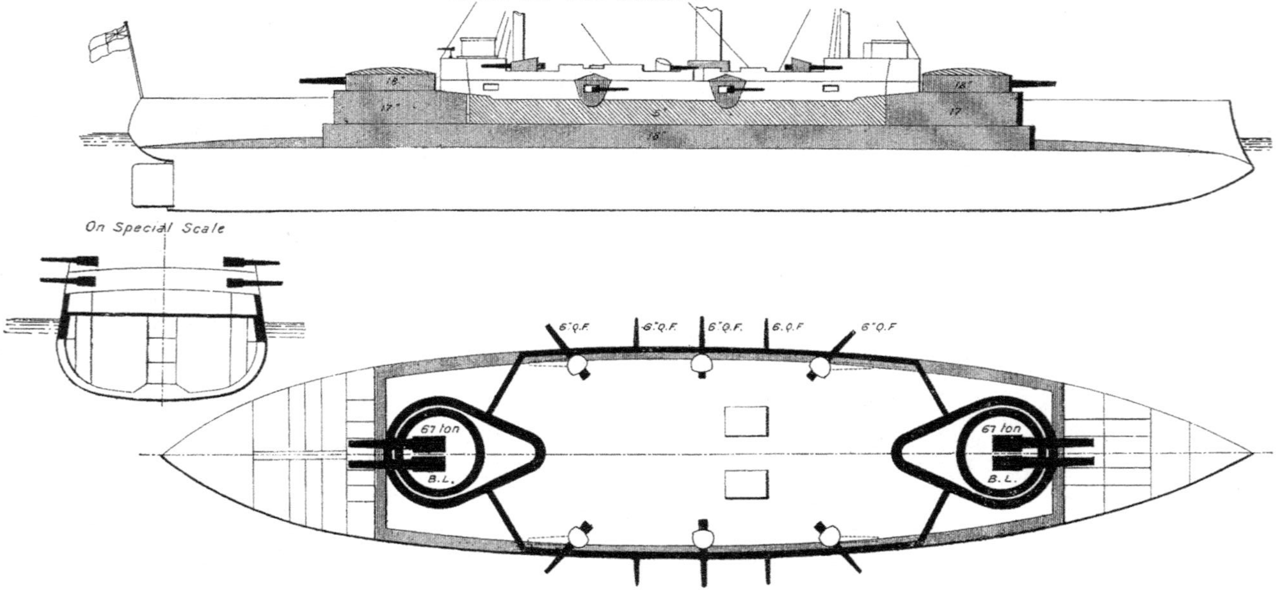

PLATE 12.

GREAT BRITAIN

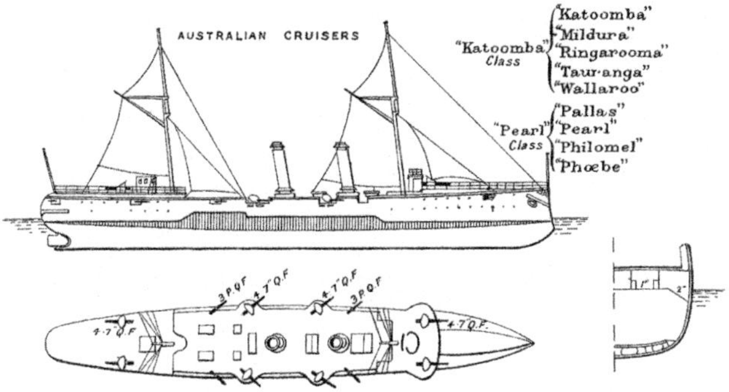

Pelorus Class

Pelorus. *Pomone.*
Pactolus. *Prometheus*
Pegasus. *Proserpine.*
Perseus *Pyramus.*
Pandora *Psyche*
Pioneer

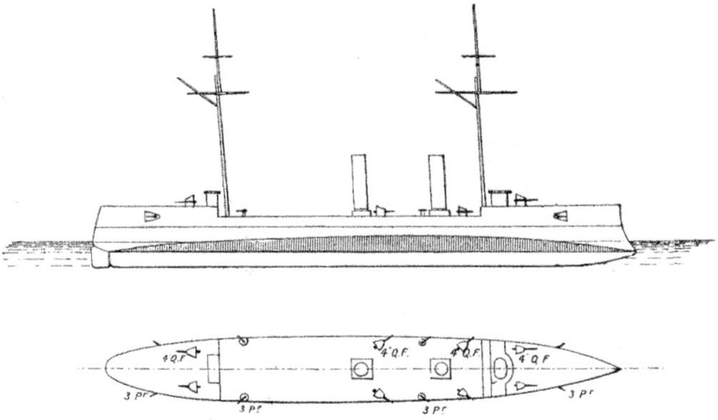

Plate 13.

GREAT BRITAIN

FIRST CLASS ARMOURED CRUISERS

"Monmouth" *Class*

"Monmouth"
"Kent"
"Essex"
"Bedford"

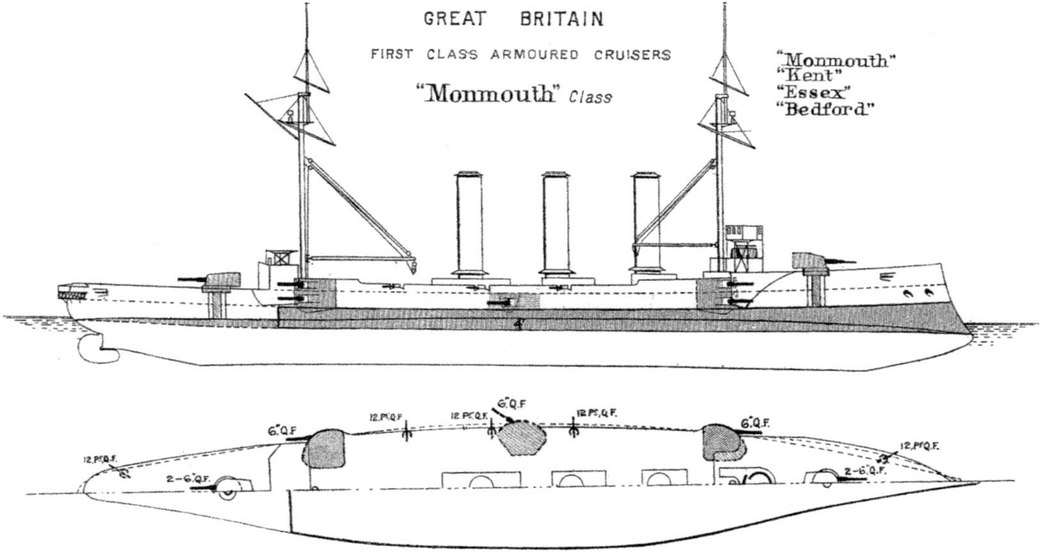

GREAT BRITAIN

"Majestic"

"Majestic" Class

"Majestic" "Victorious"
"Magnificent" "Mars"
"Hannibal" "Illustrious"
"Prince George" "Cœsar"
"Jupiter"

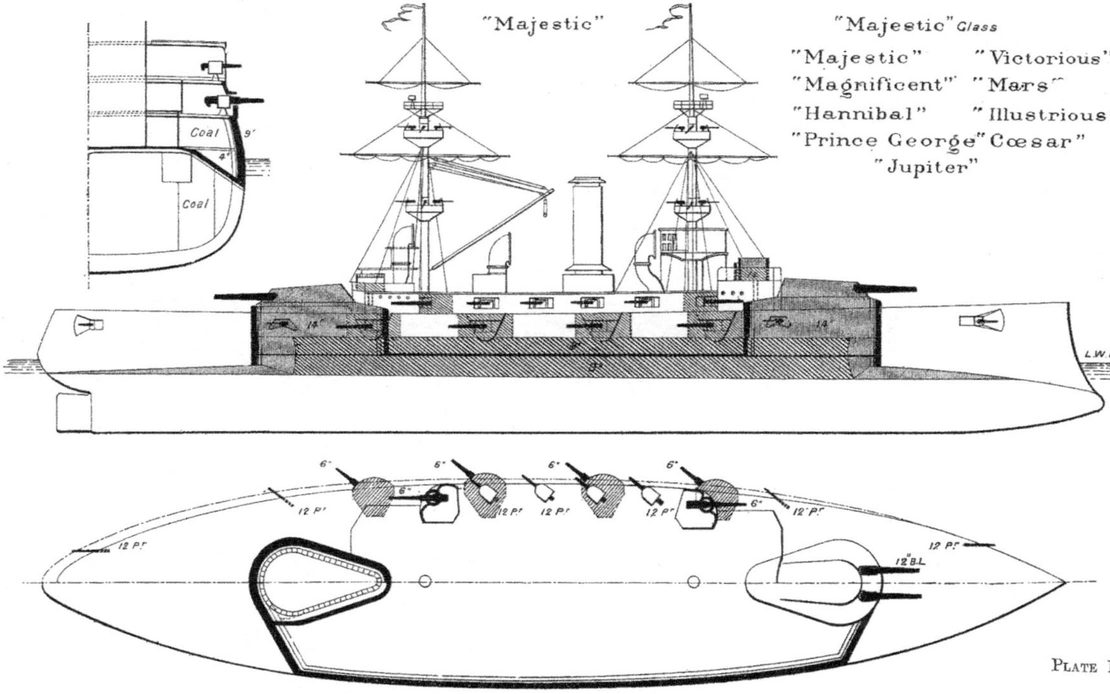

PLATE 15.

GREAT BRITAIN

FIRST CLASS BATTLESHIP

"Renown"

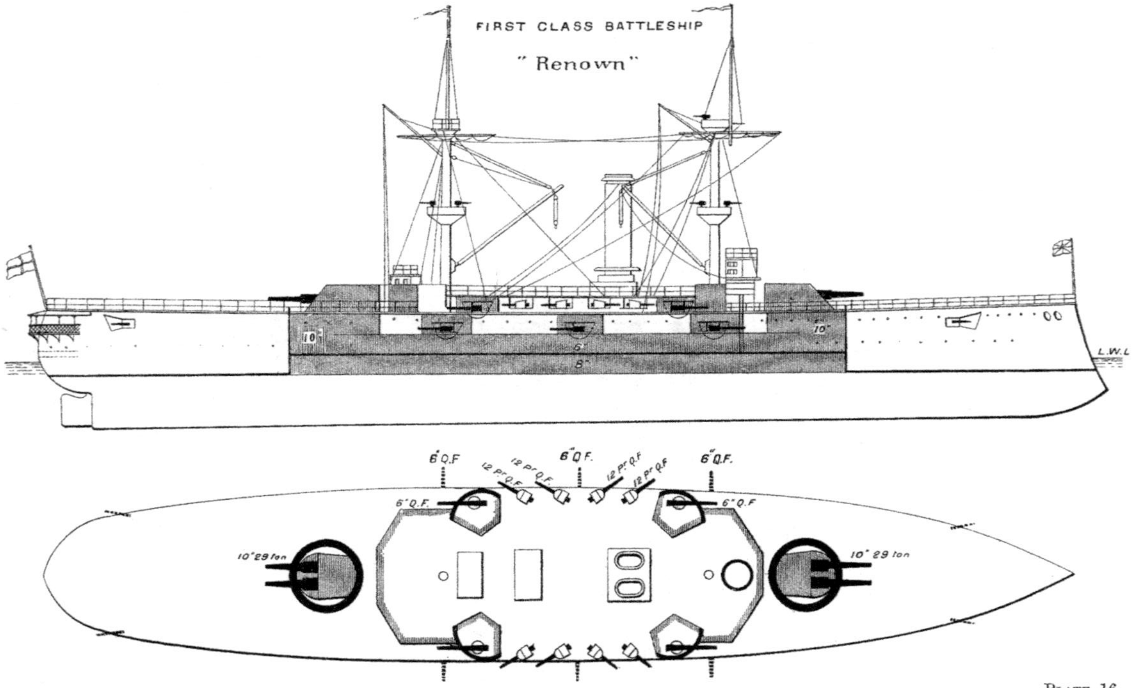

PLATE 16.

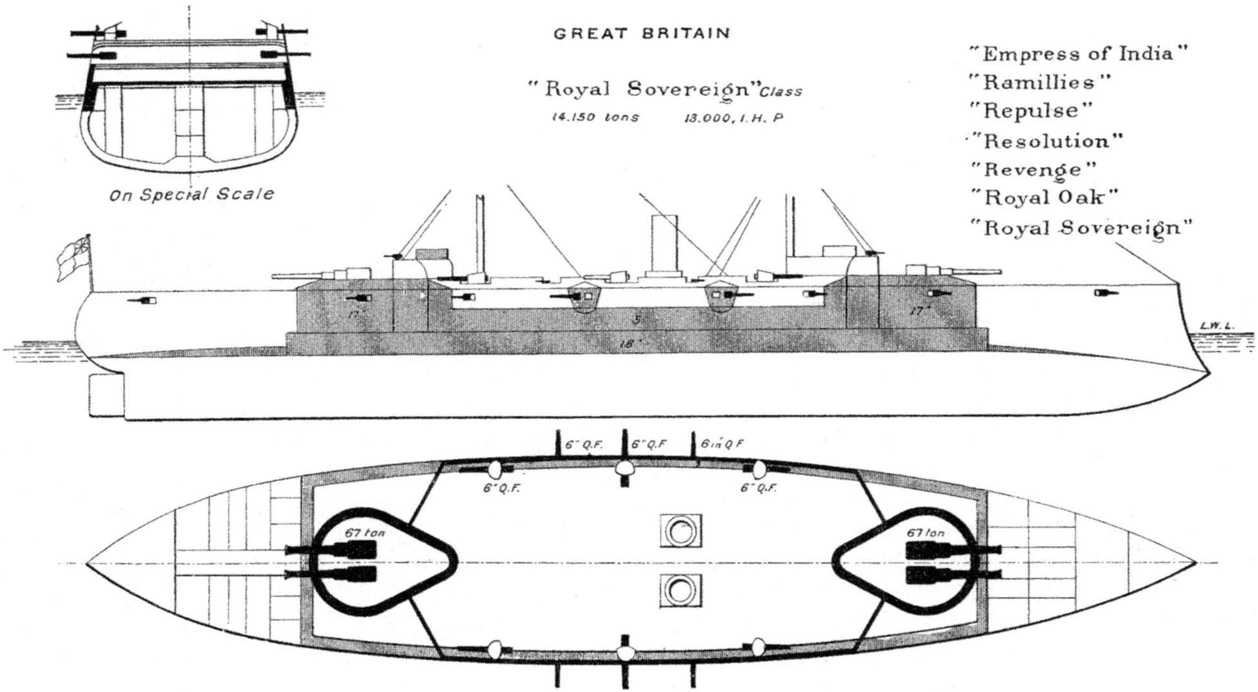
GREAT BRITAIN
"Royal Sovereign" Class
14.150 tons 13.000, I.H.P
"Empress of India"
"Ramillies"
"Repulse"
"Resolution"
"Revenge"
"Royal Oak"
"Royal Sovereign"
On Special Scale
L.W.L.
17"
5"
18"
6" Q.F.
6" Q.F
6in Q.F
67 ton

PLATE 17.

GREAT BRITAIN.

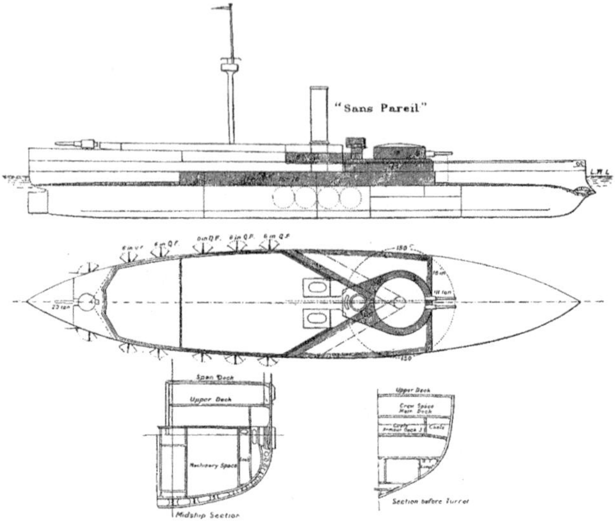

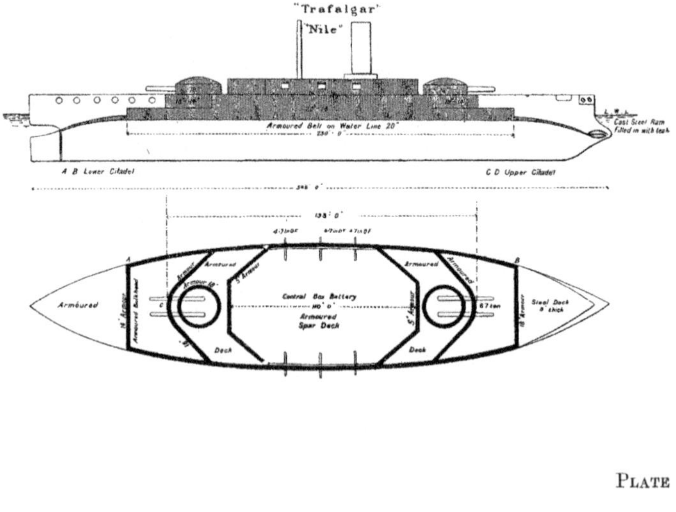

ARGENTINA

PROTECTED CRUISER

"Buenos Aires."

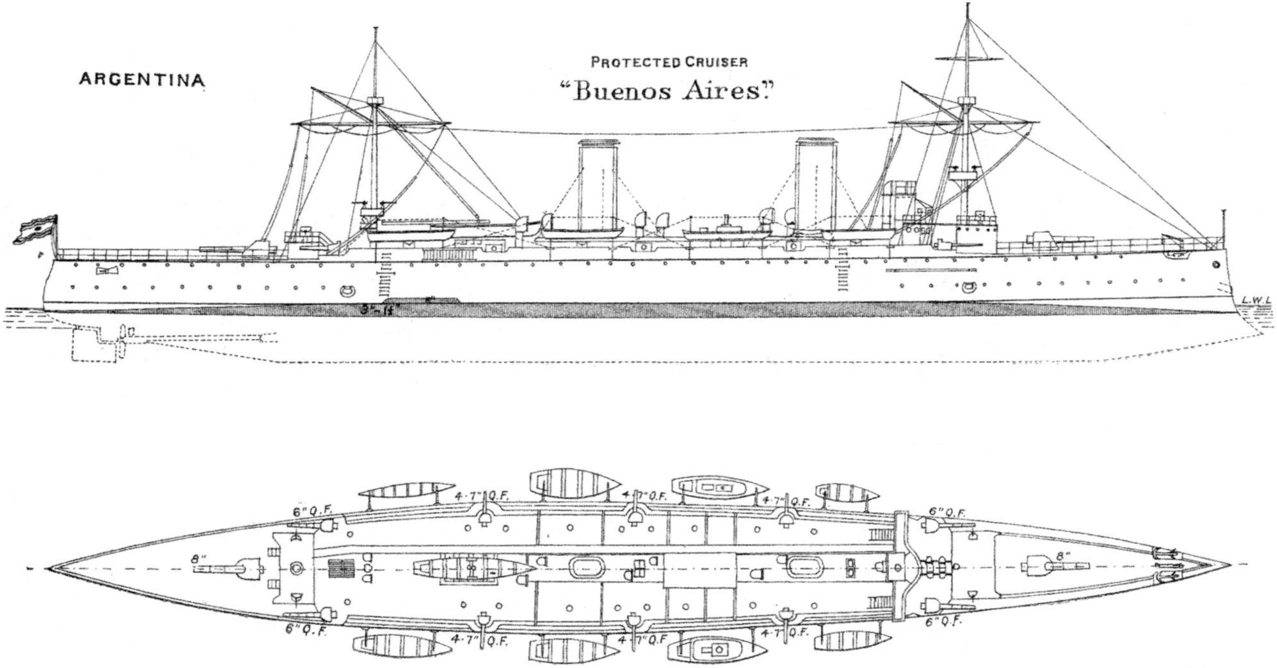

PLATE 19,

ARGENTINA.

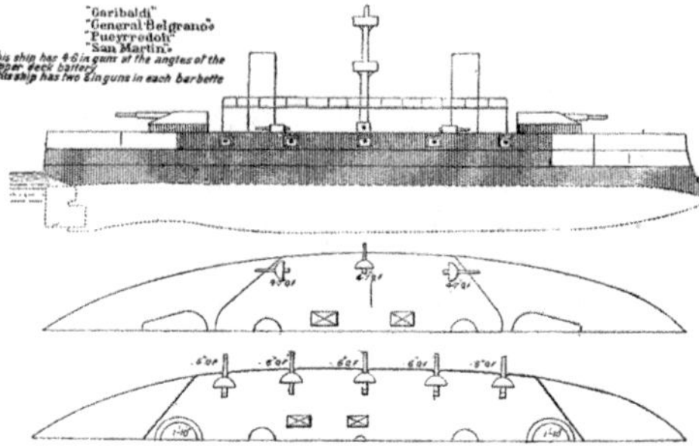

"Libertad" & "Independencia."

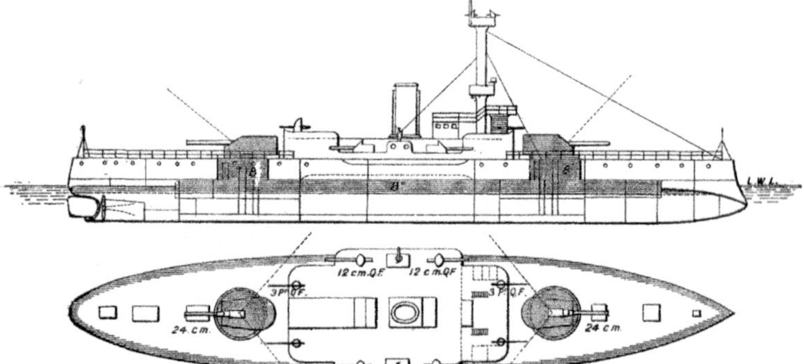

ARGENTINA.

"Nueve de Julio."

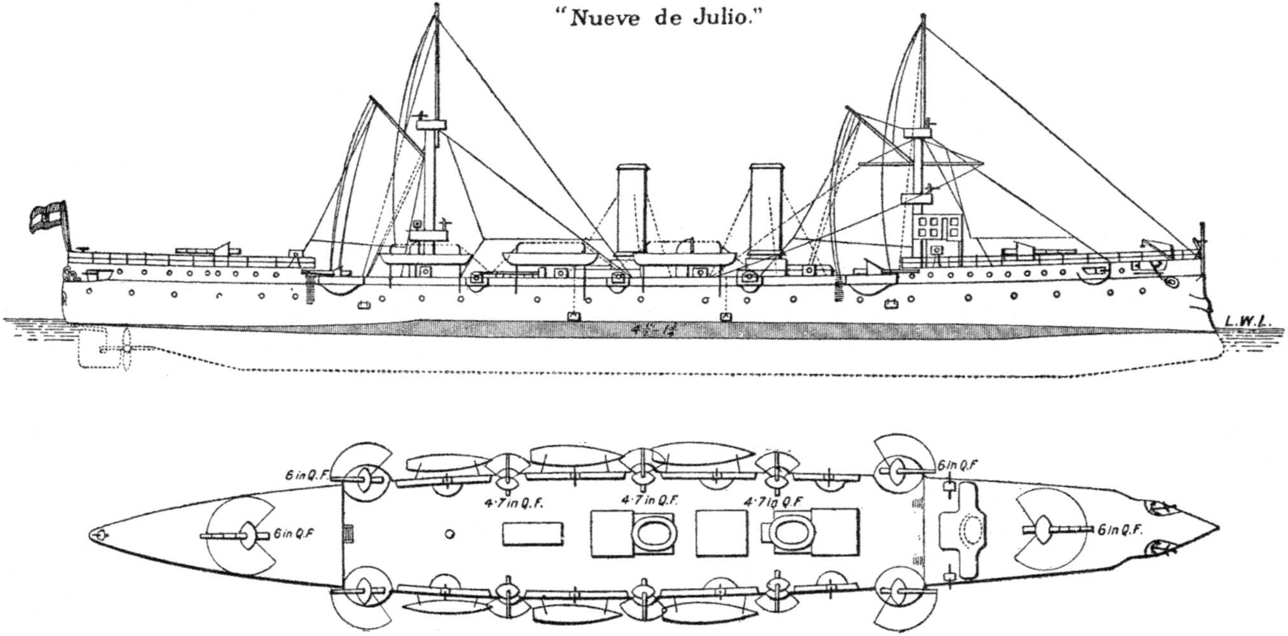

PLATE 21.

AUSTRIA-HUNGARY.

ARMOURED CRUISER.

Kaiser Karl VI.

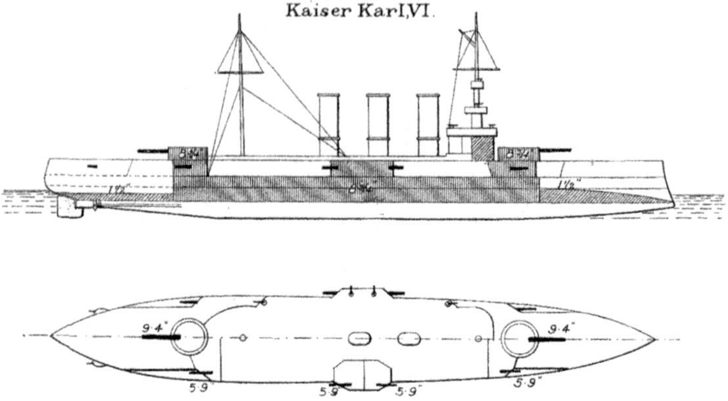

Zenta

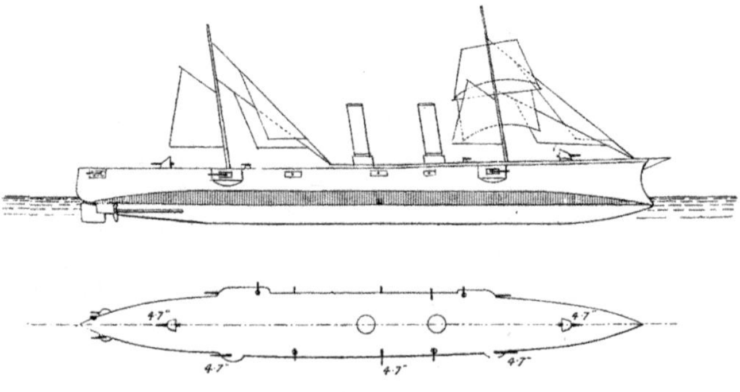

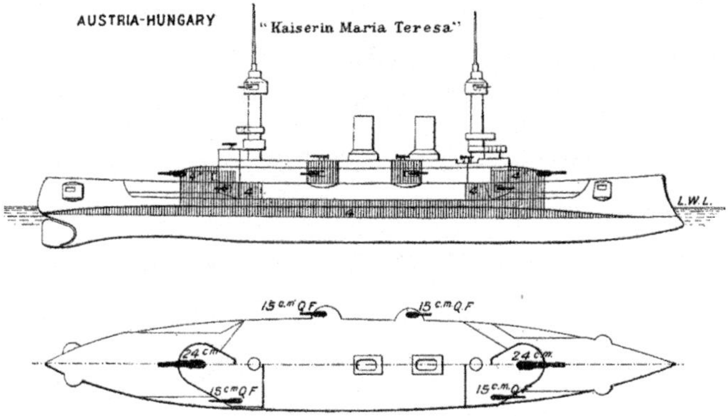
AUSTRIA-HUNGARY
"Kaiserin Maria Teresa"
L.W.L.
15 c.m Q.F
15 c.m Q.F
24 c.m
24 c.m
15 c.m Q.F
15 c.m Q.F

PLATE 23.

AUSTRIA-HUNGARY.

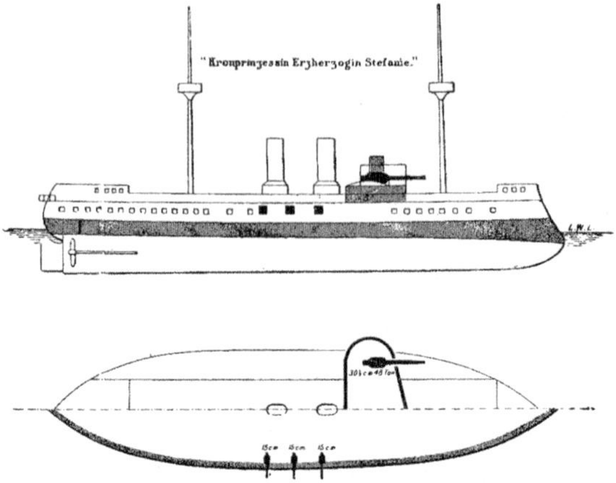

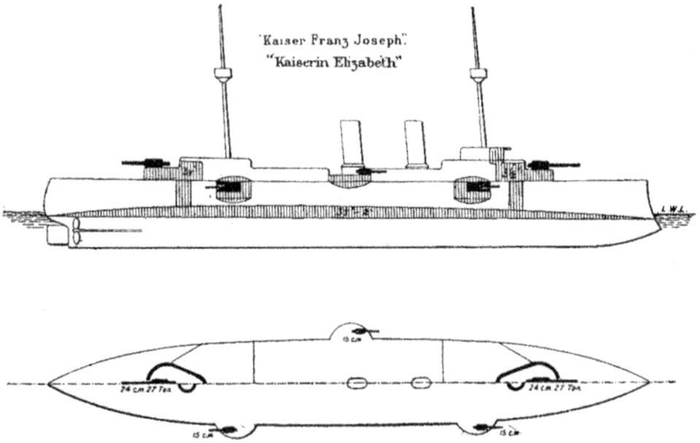

AUSTRIA HUNGARY.

"Kronprinz Erzherzog Rudolph"

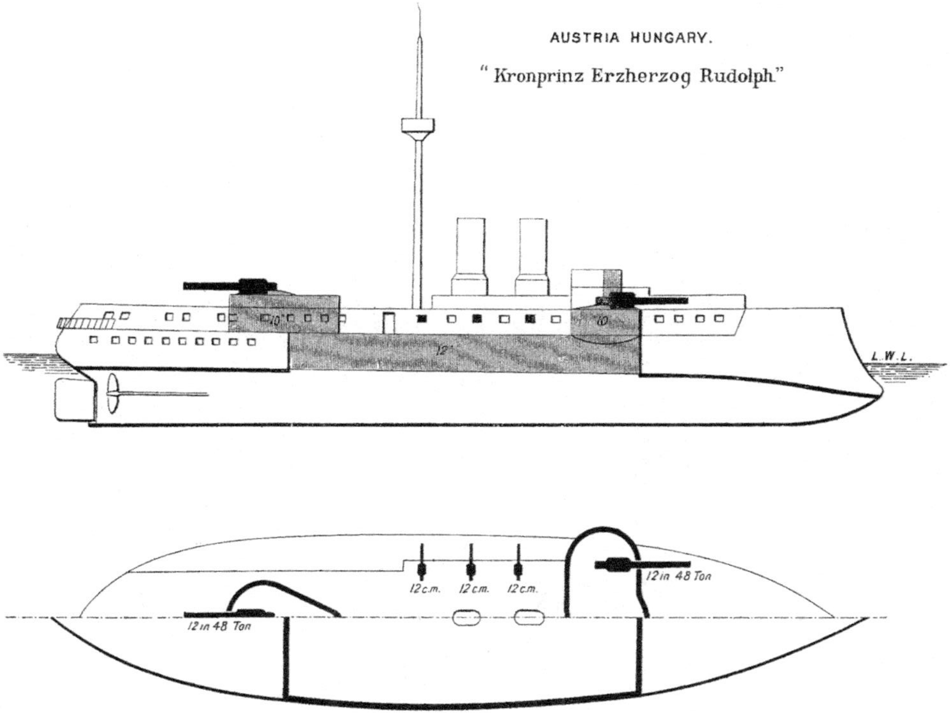

PLATE 25.

AUSTRIA-HUNGARY

"Budapest"
"Monarch"
"Wien"

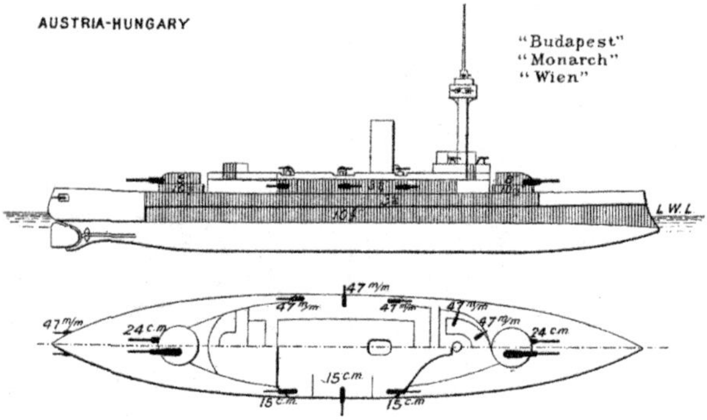

PLATE 26.

BRAZIL.

Marshal Deodoro
Marshal Floriano.

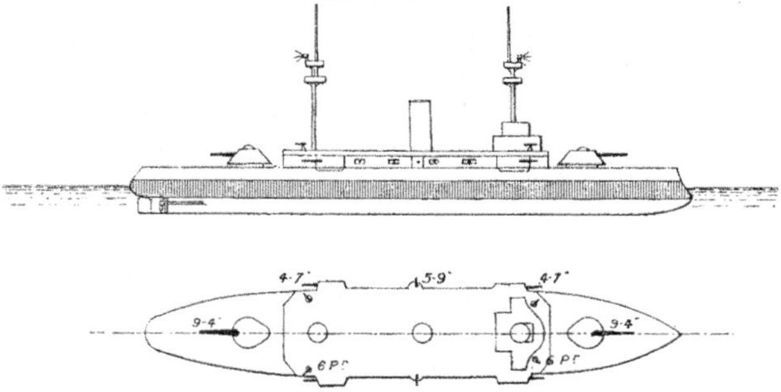

"Riachuelo"

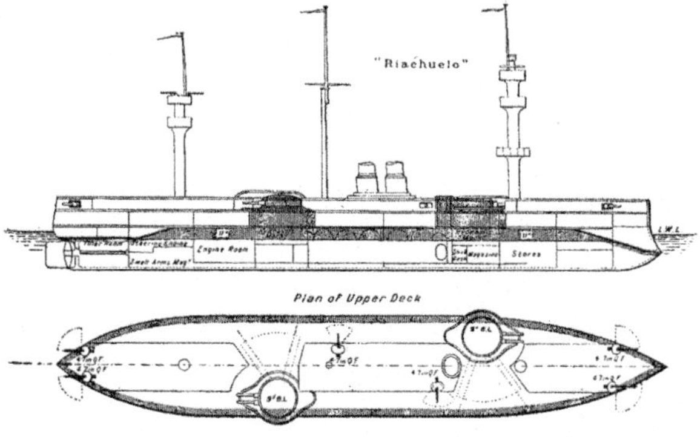

CHILI.

Torpedo Gunboat

Almirante Simpson

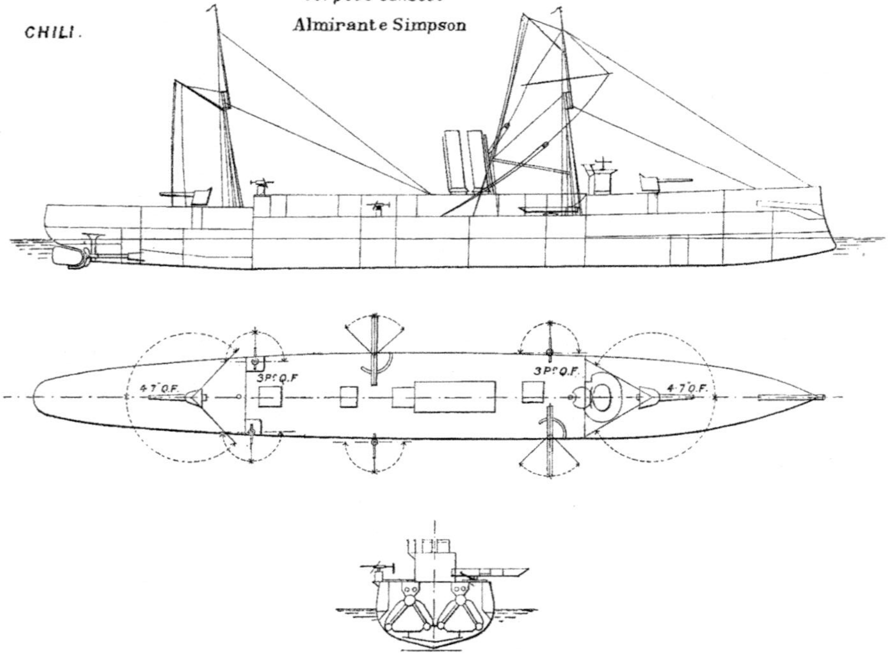

Section at Boilers

Plate 28.

CHILI.

"Blanco Encalada"

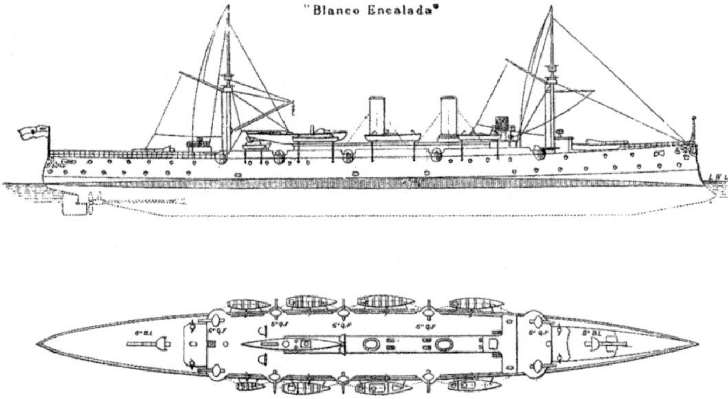

"Capitan Prat"

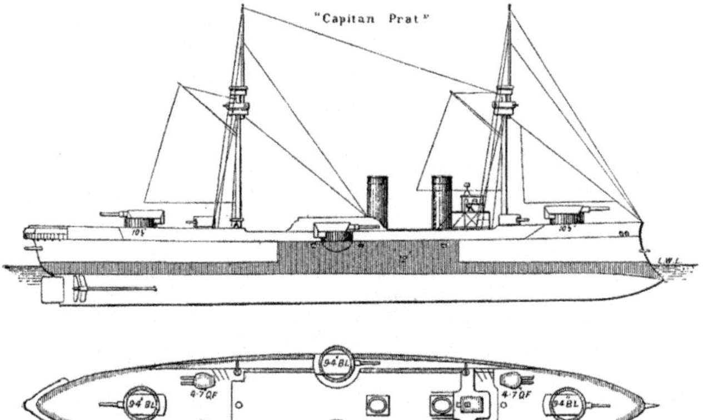

CHILI

Armoured Cruiser

"Esmeralda"

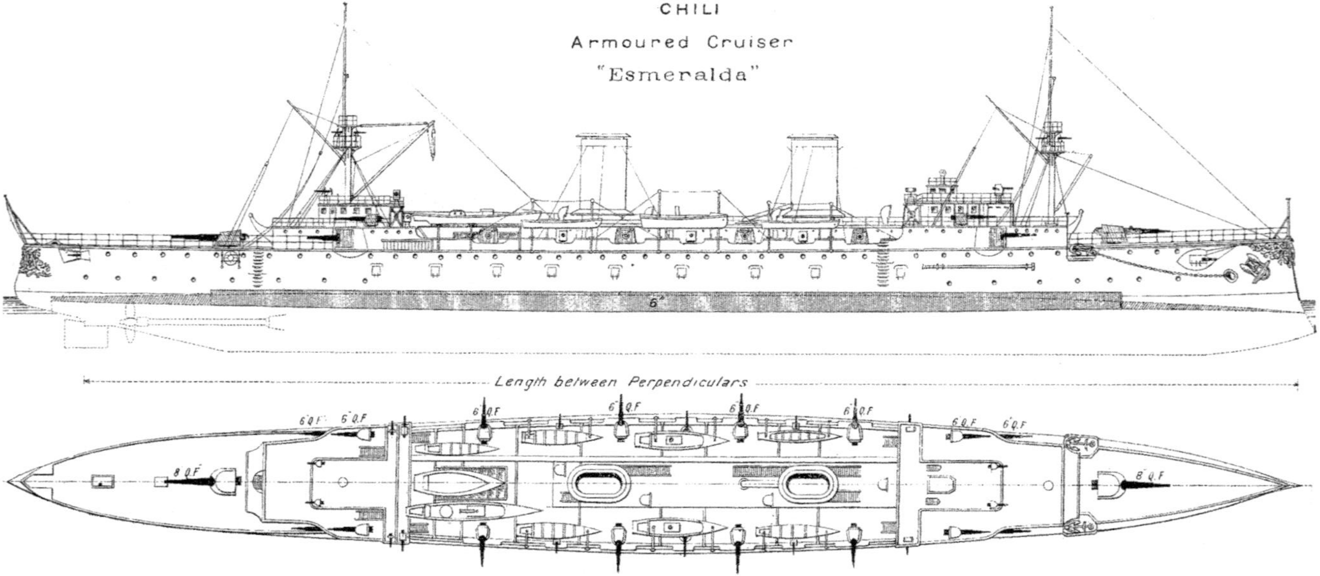

PLATE 30.

CHILI.

Protected Cruiser "Ministro Zenteno"

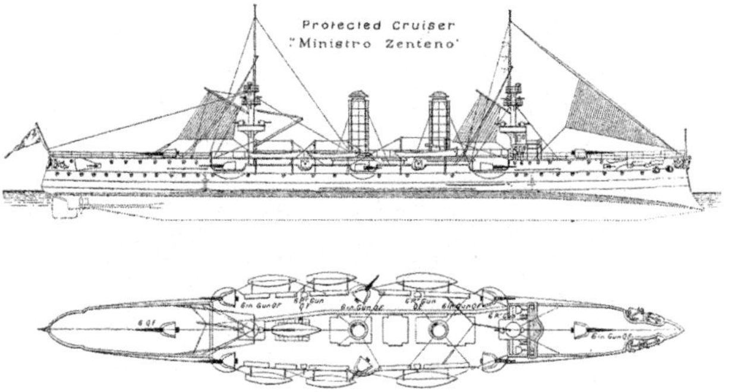

O' Higgins

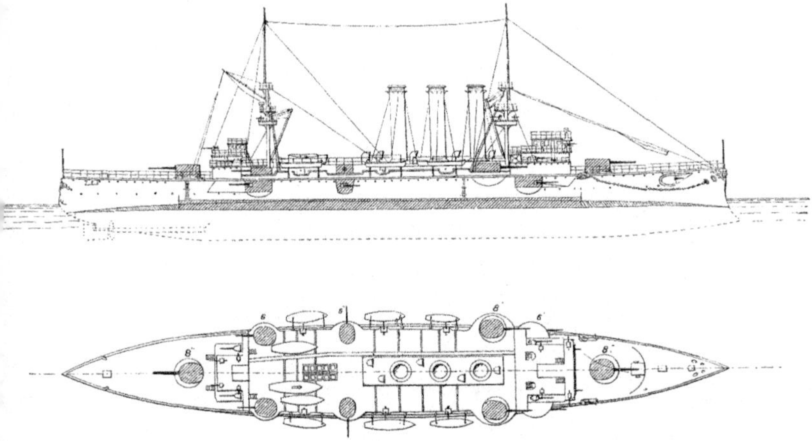

PLATE 31.

DENMARK.

"Helgoland"

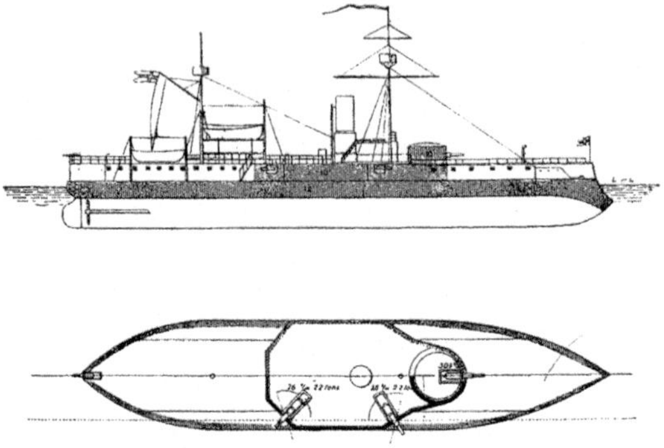

"Iver Hvitfeld"

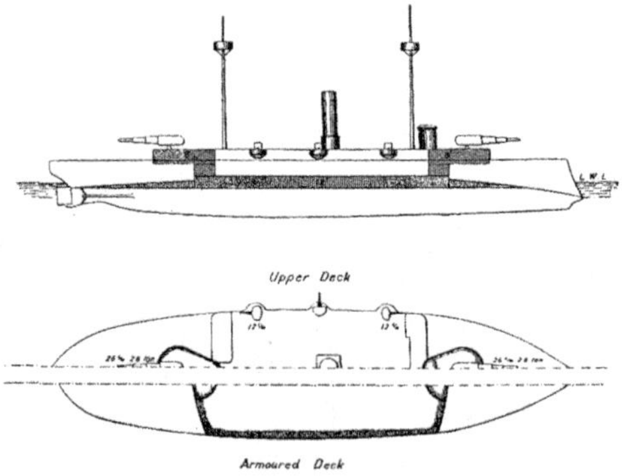

Plate 32.

FRANCE.

"Amiral Baudin."
"Formidable"

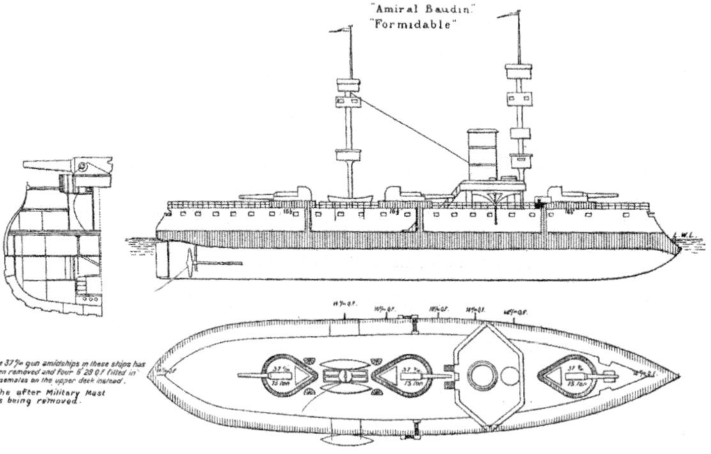

The 37 c/m gun amidships in these ships has been removed and four 5" 28 Q.F. fitted in Casemates on the upper deck instead.
The after Military Mast is being removed.

"Amiral Duperré"

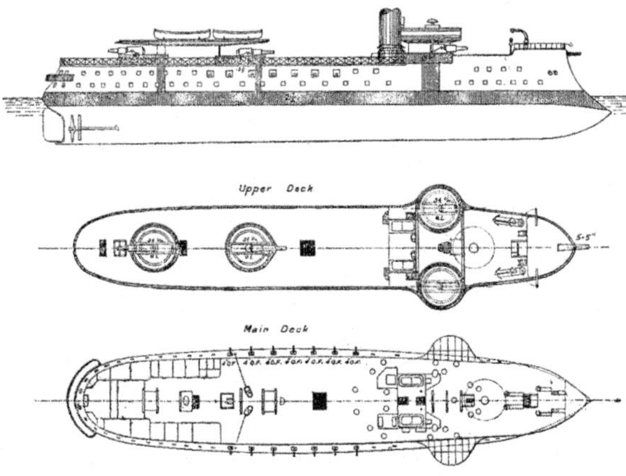

PLATE 33.

FRANCE.

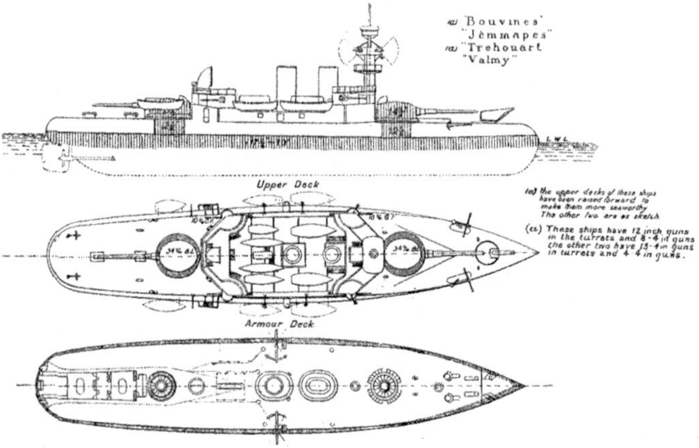

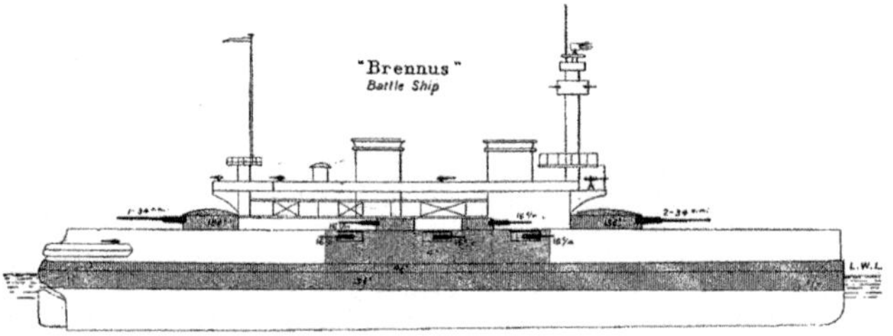

PLATE 34.

FRANCE.

"Bruix"
"Chanzy"
"Charner"
"Latouche Tréville"

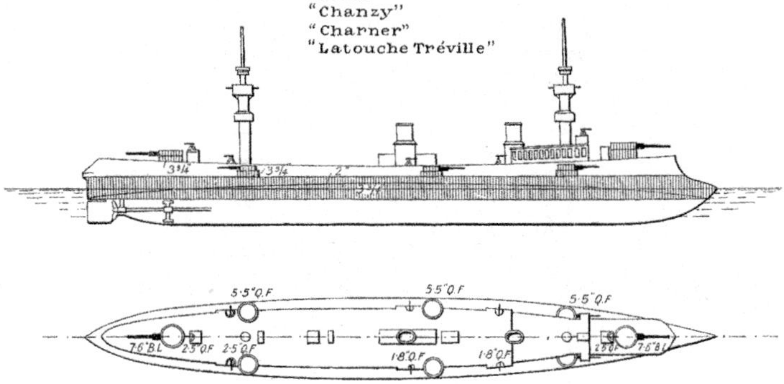

"Caiman"
"Indomptable"
"Requin"
"Terrible"

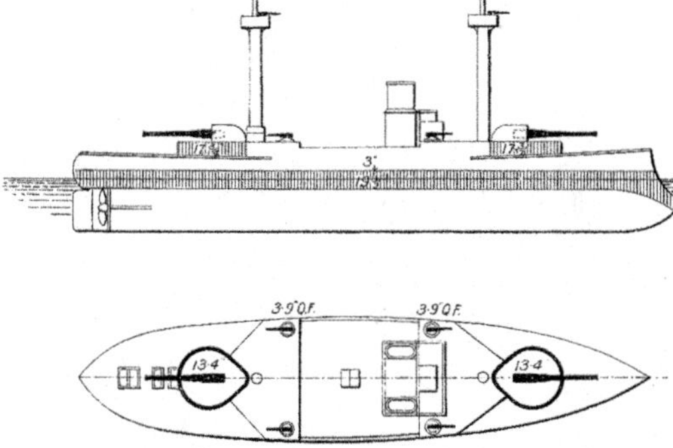

FRANCE.

"Carnot"

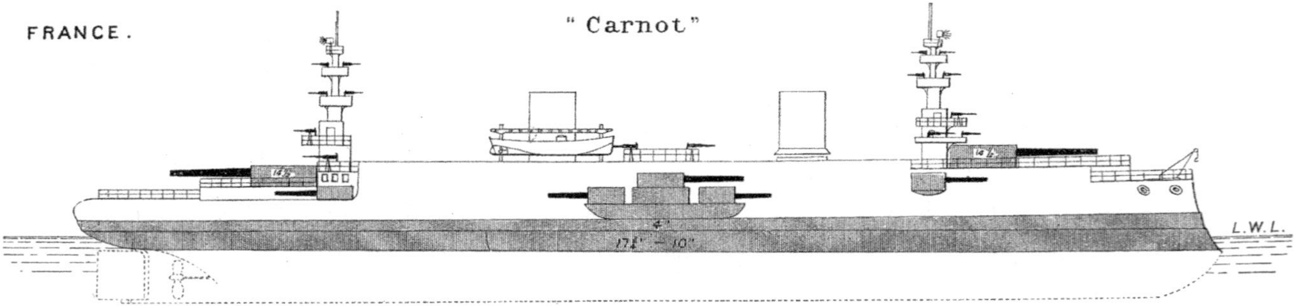

Note To reduce weight the after Military Mast has been taken out and the bridges cut down

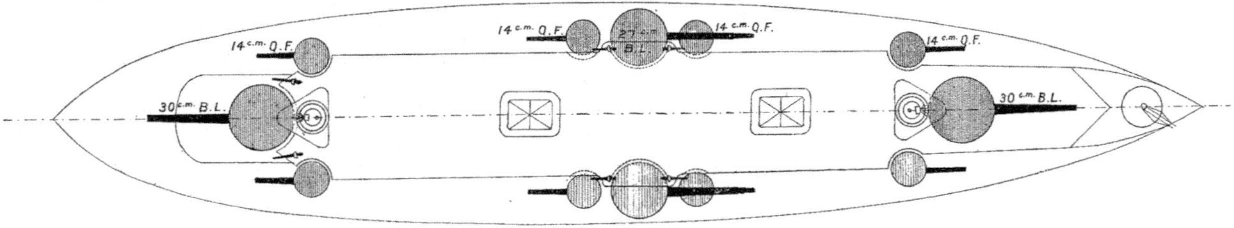

FRANCE

"Charlemagne"
"St Louis"
"Gaulois"

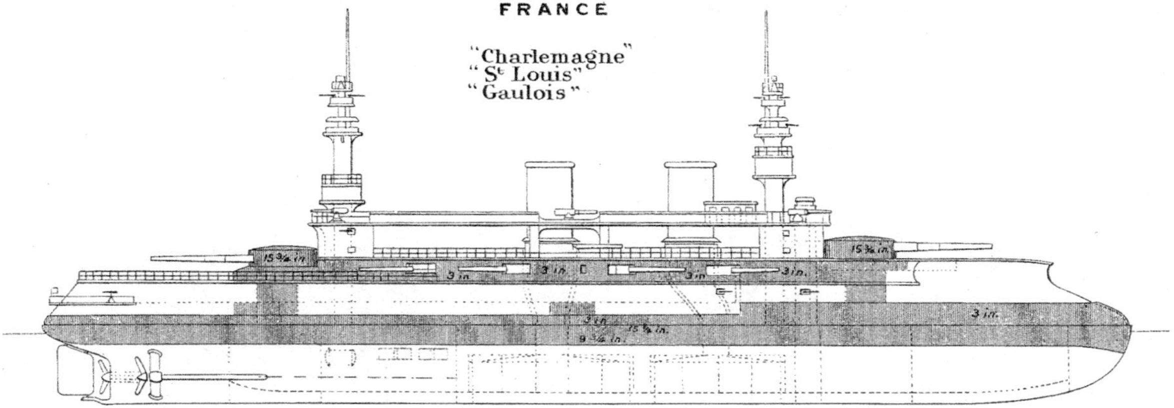

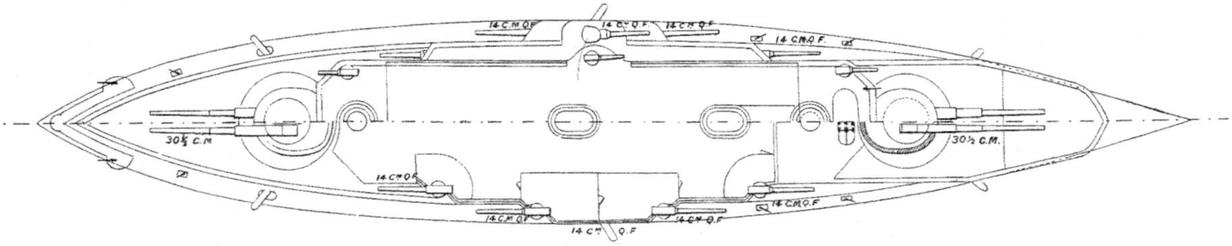

Plate 37.

FRANCE.

"Chasseloup Laubat."
"Bugeaud."
"Friant."

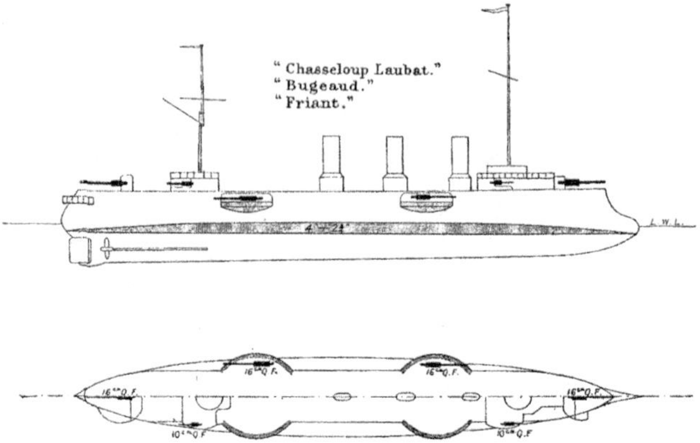

Devastation. Courbet. Redoutable.

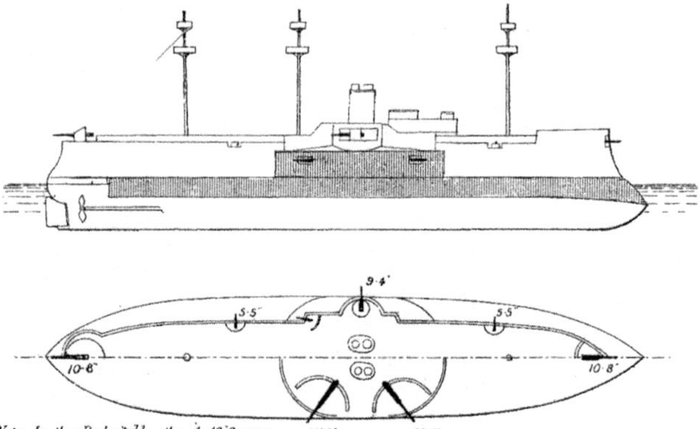

Note. In the Redoutable, the 4 10·6 guns in the central redoubt are replaced by 9·4 guns and the 6 5·5 guns by an equal number of 3·9" guns.

FRANCE.

Descartes.

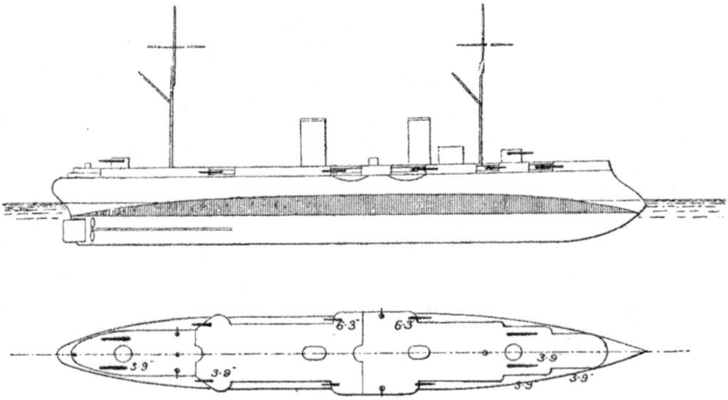

D'Estrées

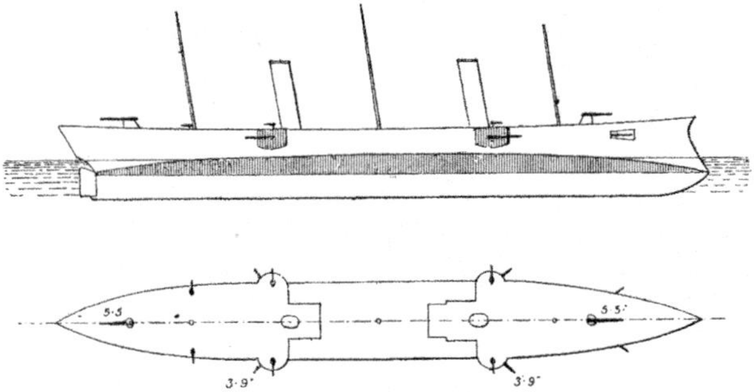

FRANCE.

"Dupuy de Lôme"

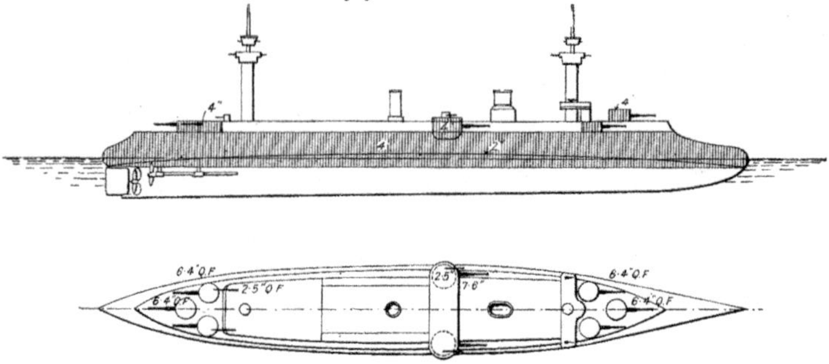

D'ENTRECASTEAUX.

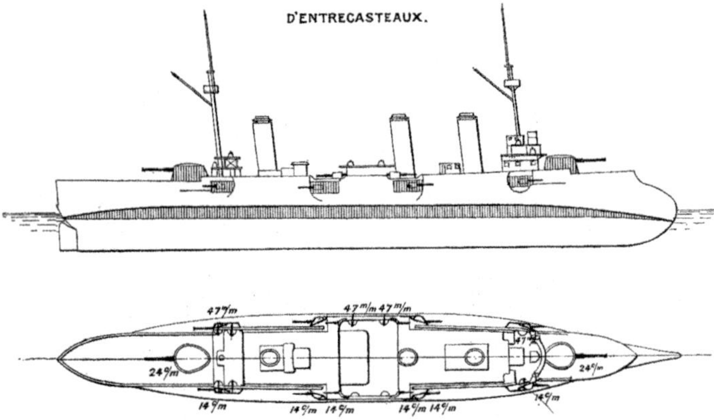

FRANCE.

"Châteaurenault."

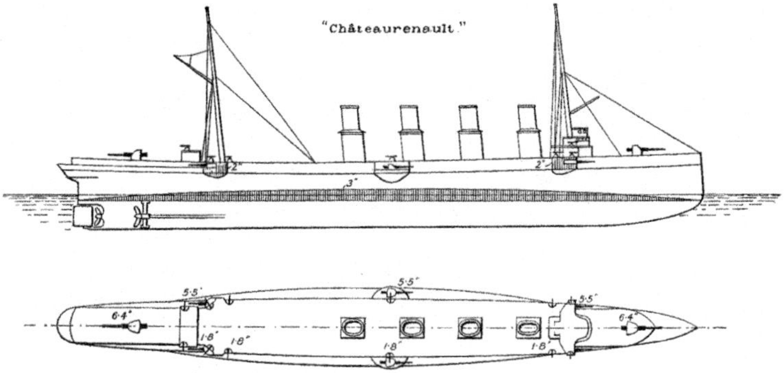

"Desaix"
"Kléber"
"Dupleix"

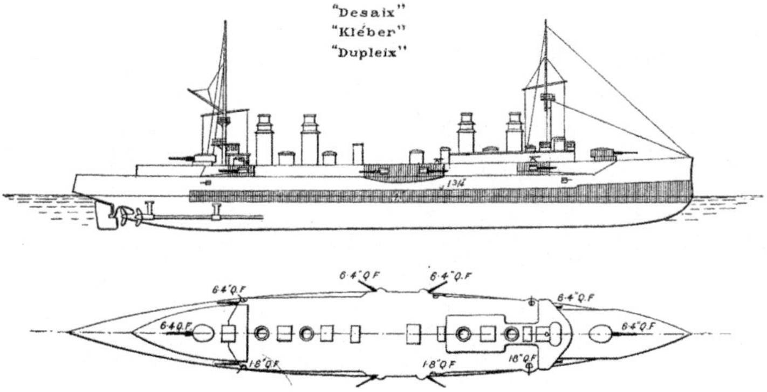

PLATE 41.

France

ARMOURED CRUISERS.

"Condé"
"Gloire"
"Sully"

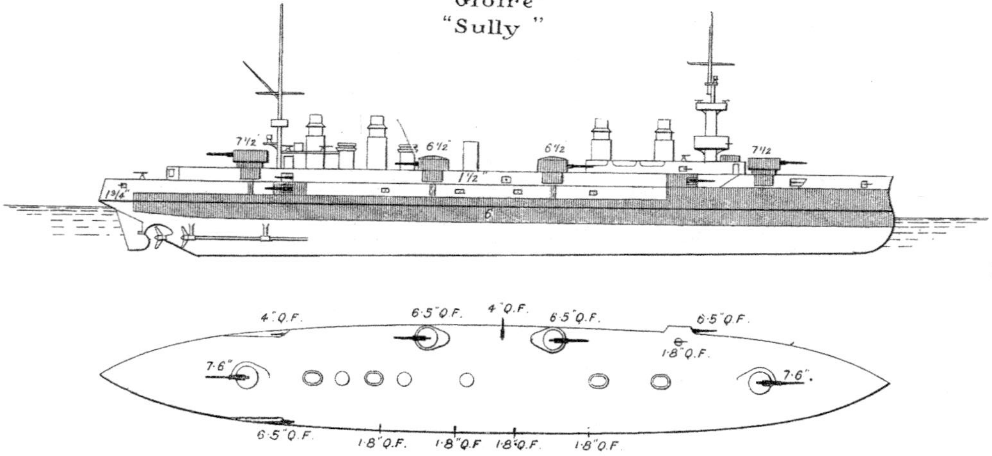

PLATE 42.

FRANCE.

"Gueydon"
"Du Petit-Thouars"
"Montcalm"

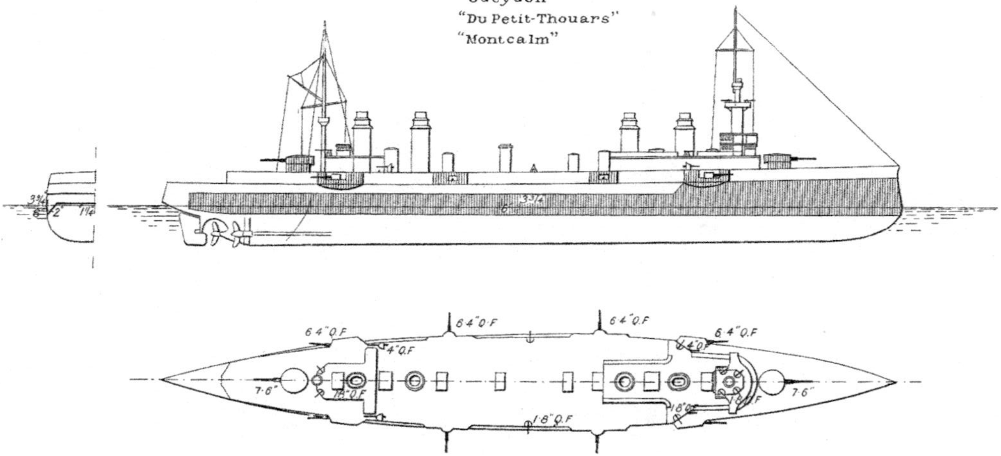

PLATE 43.

FRANCE

BATTLE-SHIP. SECOND CLASS.

Henri IV.

On Special Scale
75 Feet to the Inch.

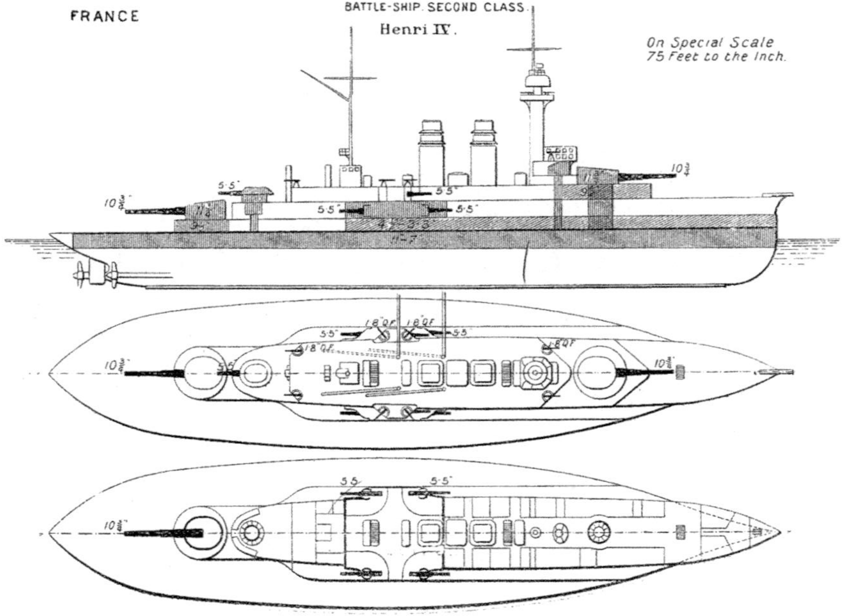

PLATE 44.

FRANCE

"Hoche"

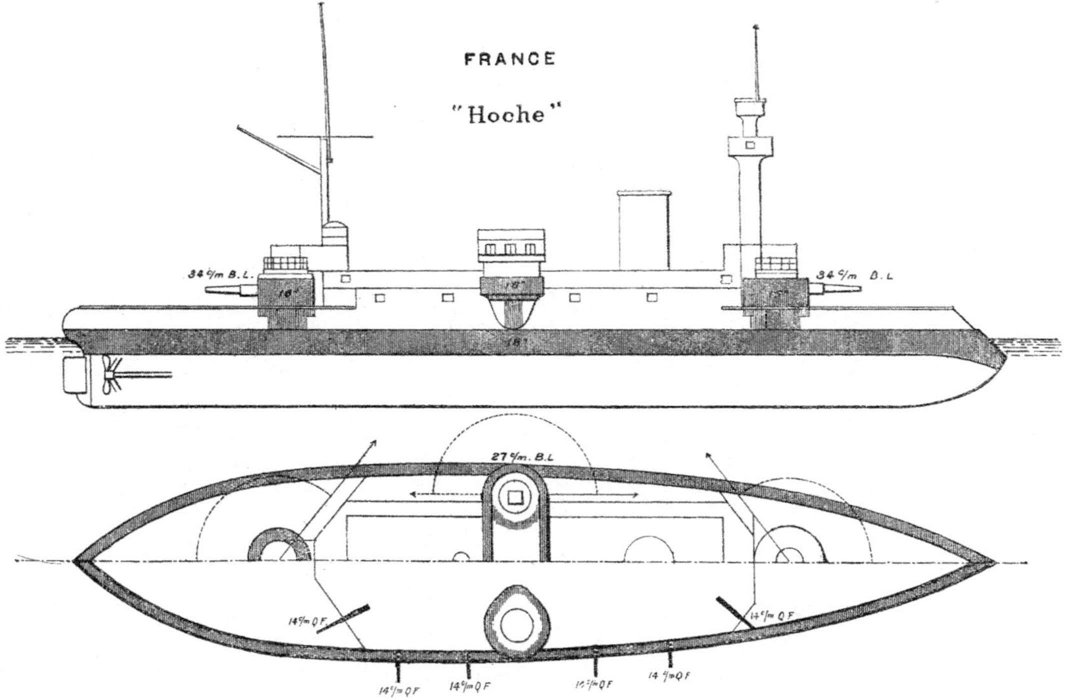

PLATE 45.

FRENCH SHIPS.

"Jauréguiberry."

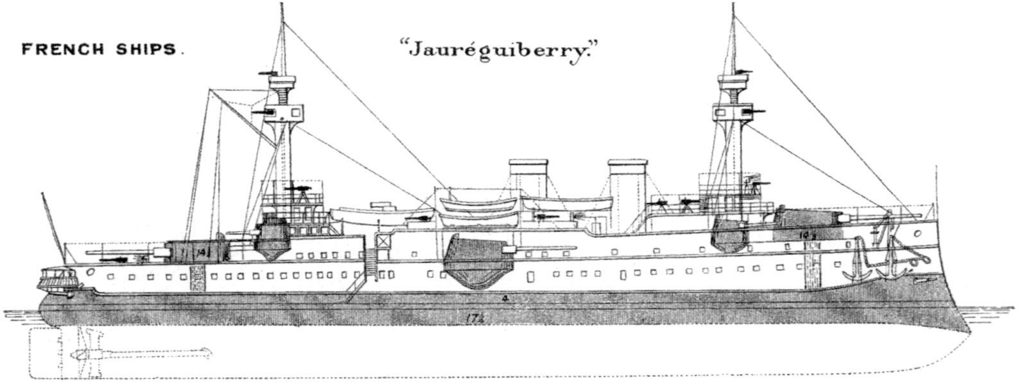

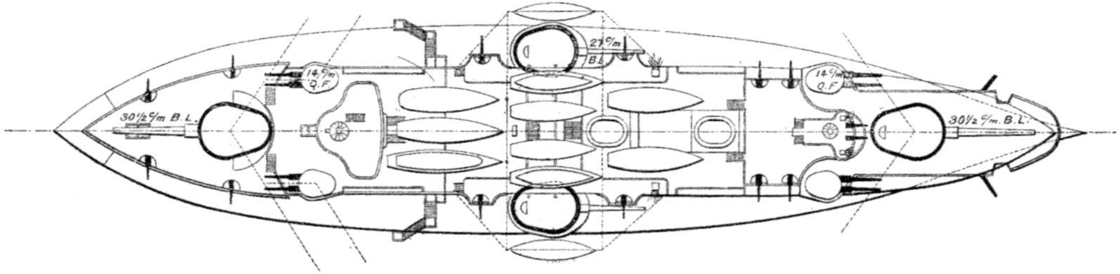

Plate 46.

FRANCE.

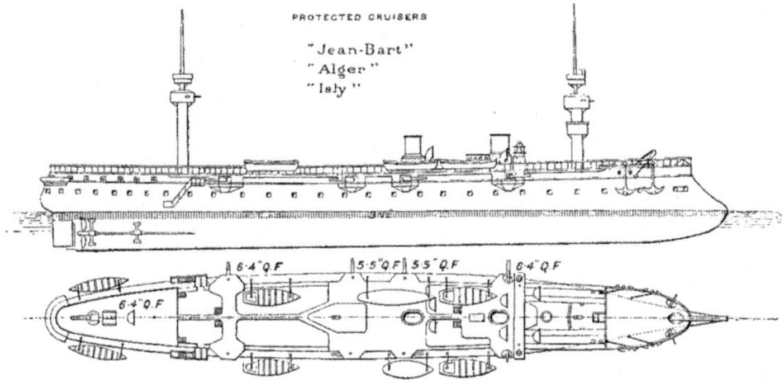
PROTECTED CRUISERS
"Jean-Bart"
"Alger"
"Isly"
6·4" Q.F
5·5" Q.F
5·5" Q.F
6·4" Q.F
6·4" Q.F

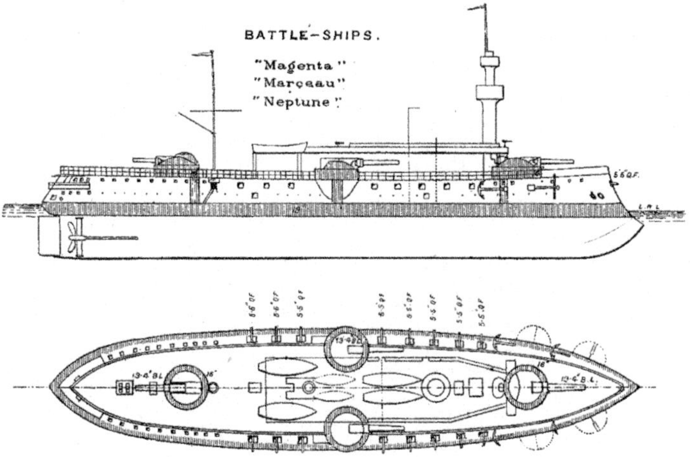
BATTLE-SHIPS.
"Magenta"
"Marçeau"
"Neptune"
5·5" Q.F.
13·4" B.L
13·4" B.L.
16"

France

ARMOURED CRUISER
"Jeanne d'Arc"

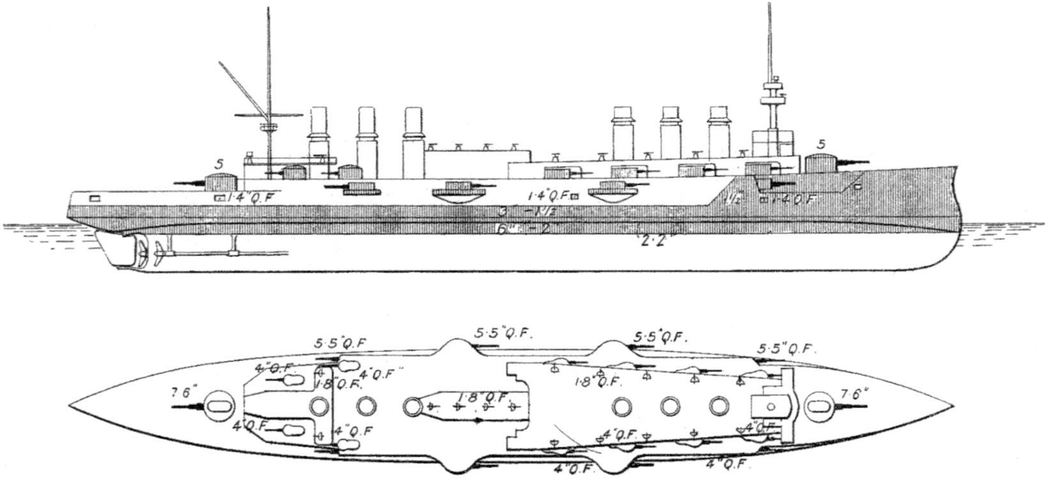

PLATE 48.

France

BATTLE-SHIP
"Suffren"

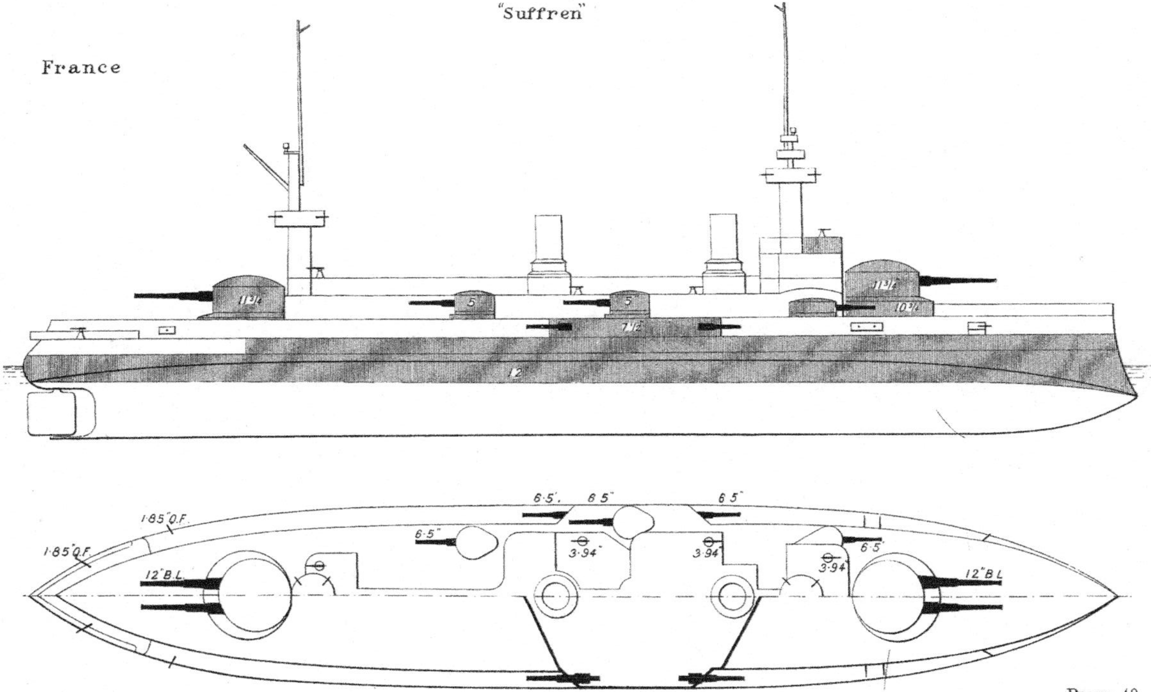

PLATE 49.

Germany.

"Freya"
"Hansa"
"Hertha"
"Victoria Louise"
"Vineta"

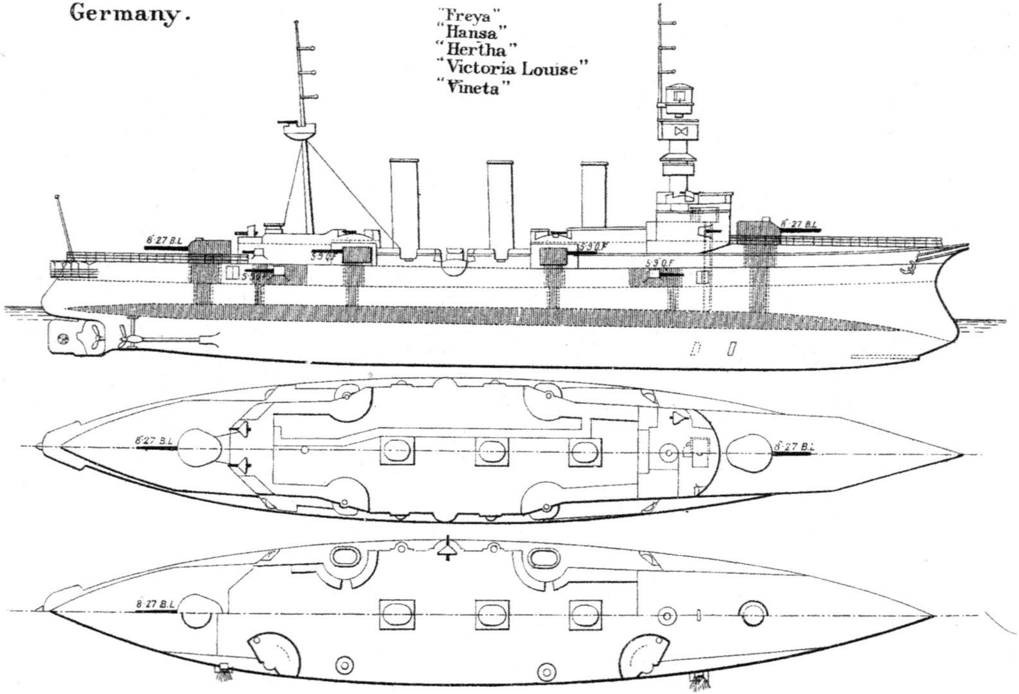

PLATE 50.

Germany

"Fürst Bismarck"

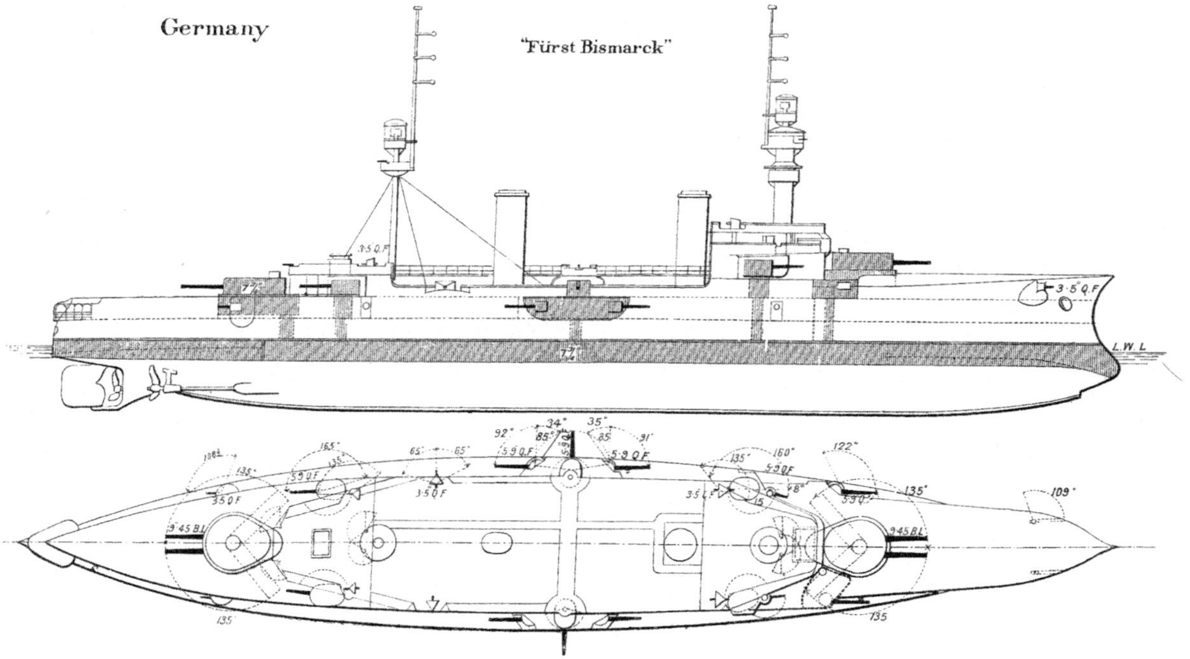

PLATE 51.

GERMANY.

Protected Cruiser "Gefion"

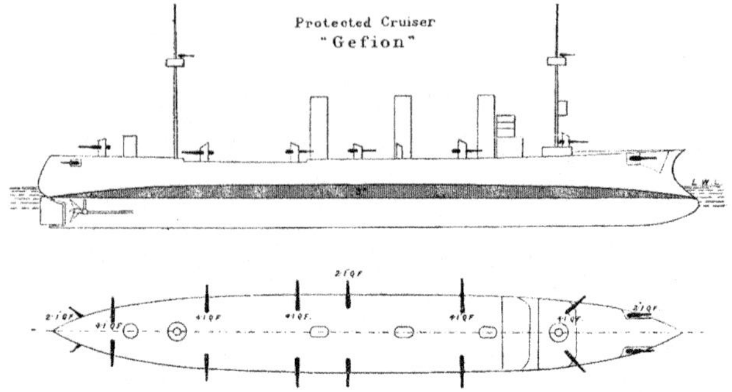

"Kaiserin Augusta"

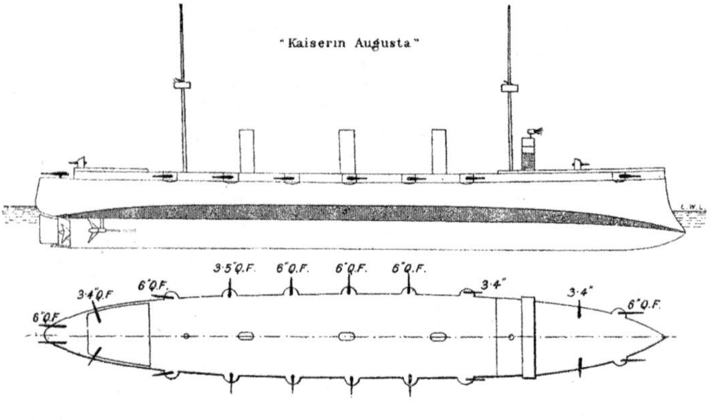

Germany.

"Kaiser Friedrich III."
"Kaiser Karl der Grosse."
"Kaiser Wilhelm II."
"Kaiser Wilhelm der Grosse."

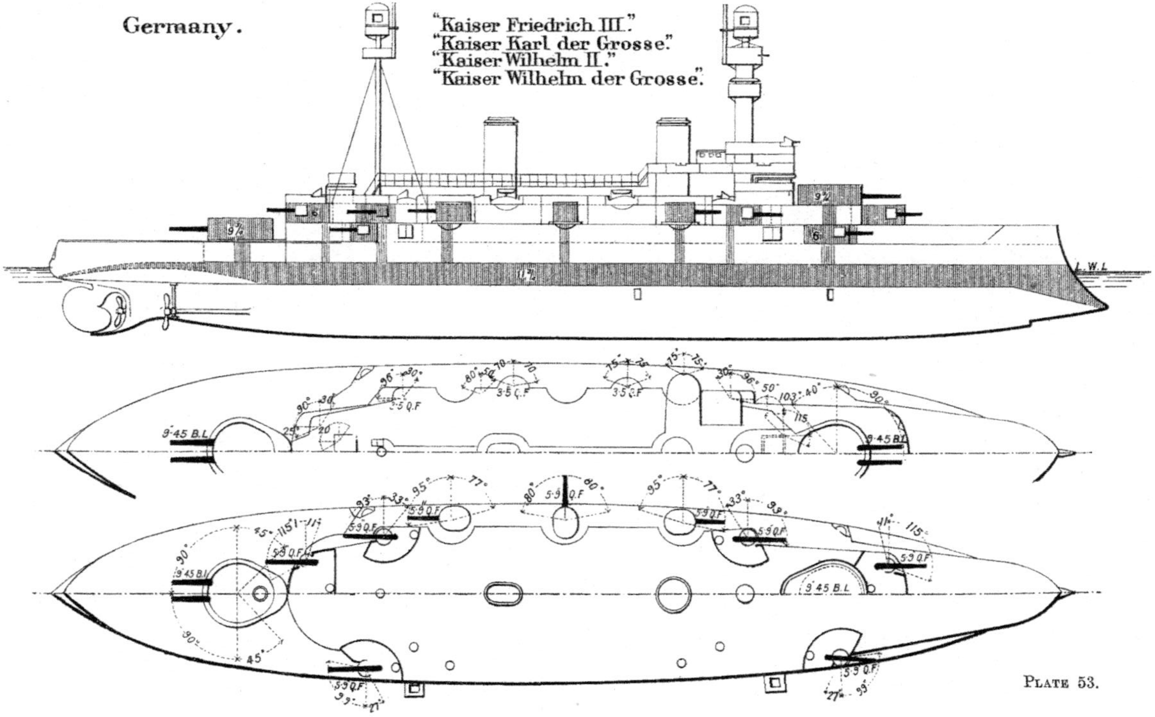

Plate 53.

GERMANY.

Kürfurst Friedrich Wilhelm.
Brandenburg.
Weissenburg.
Wörth.

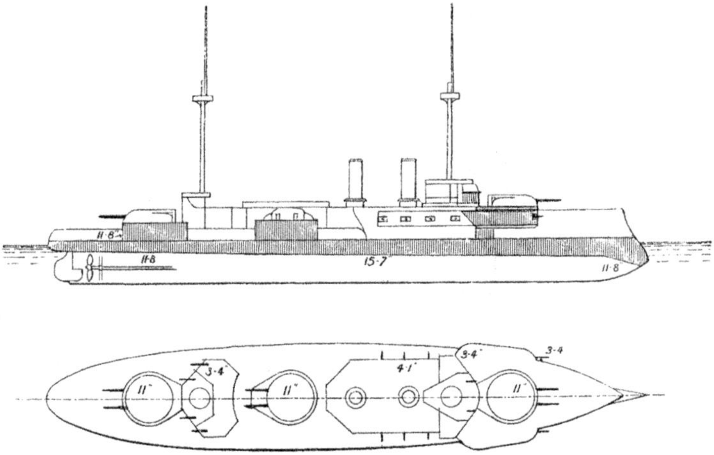

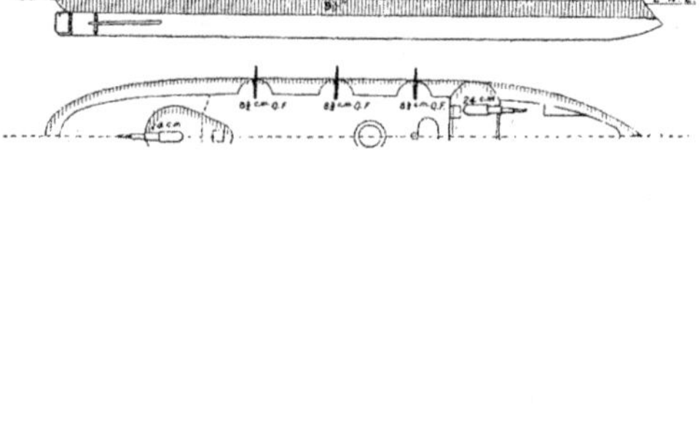

GREECE.

Hydra. Psara. Spetsai.

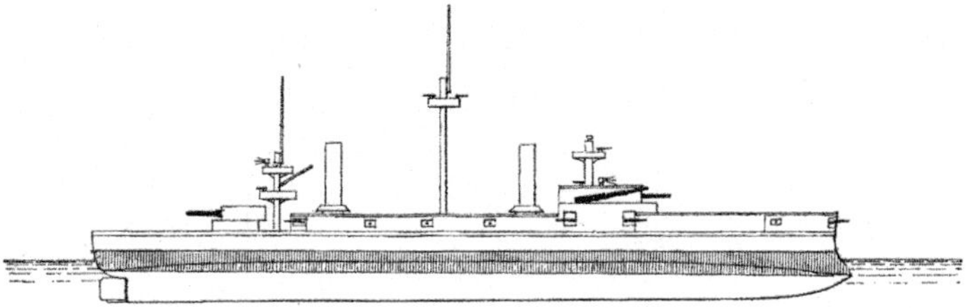

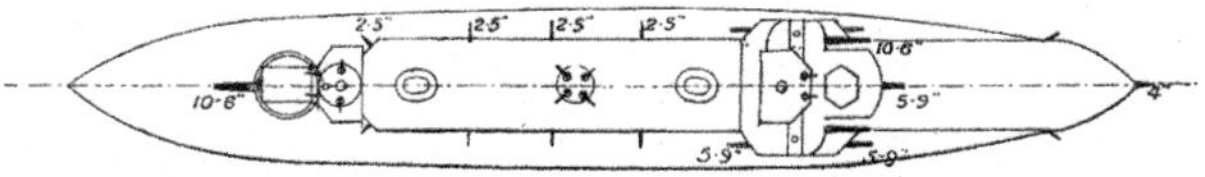

PLATE 55.

ITALY.

"Andrea Doria"
"Francesco Morosini"
"Ruggiero di Lauria"

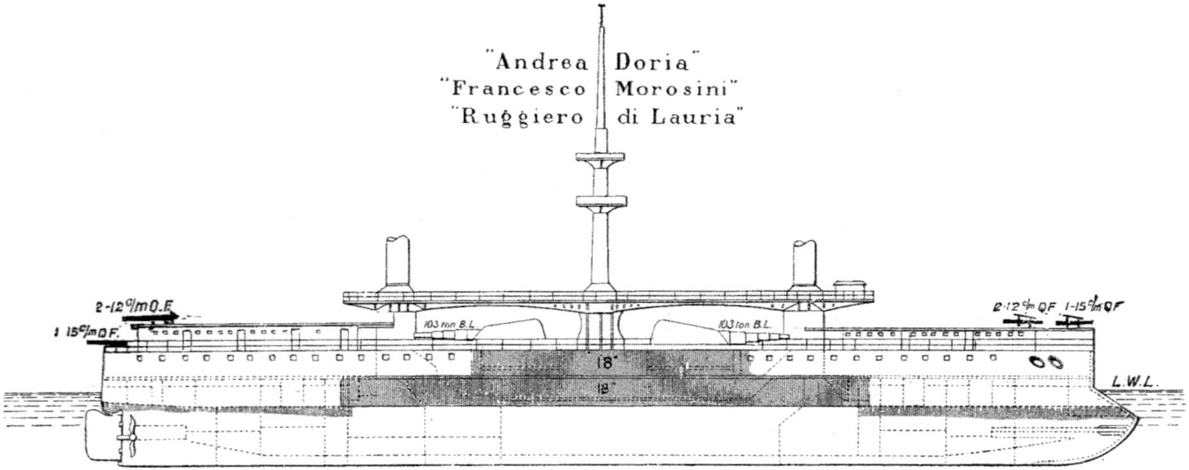

Plate 56.

ITALY,

BATTLESHIP NEW TYPE.

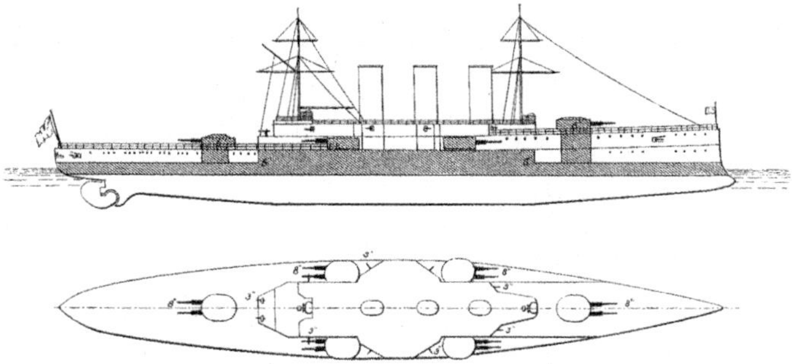

Benedetto Brin.
Regina Margherita.

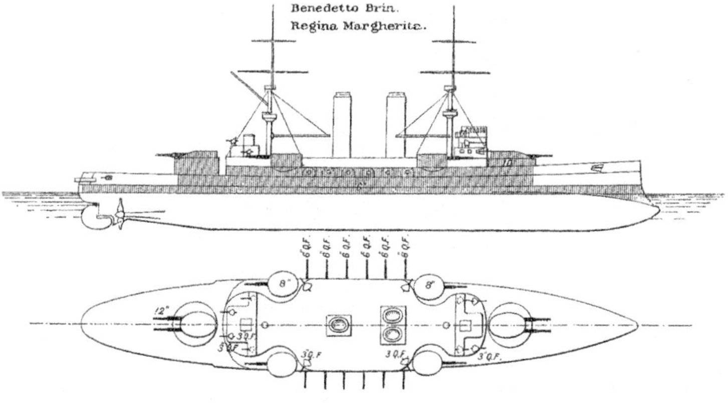

ITALY.

"Duilio"
"Dandolo"
(As refitted and rearmed)

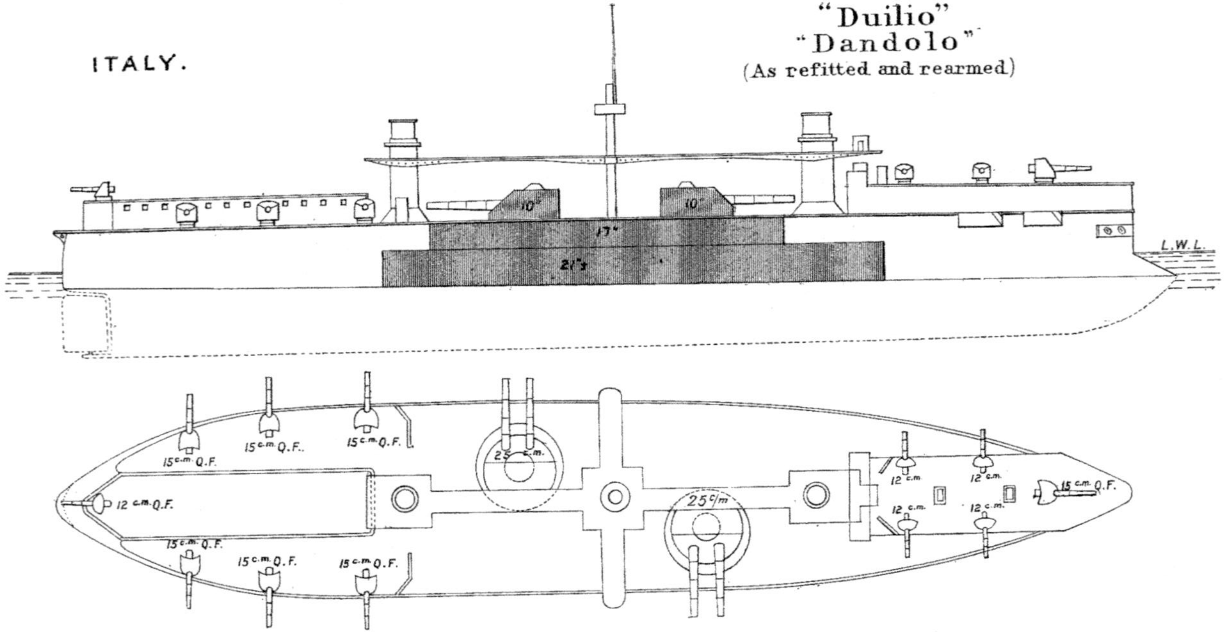

Plate 58.

"Guiseppe Garibaldi"
"Varese"
"Francesco Ferrucio."

ITALY

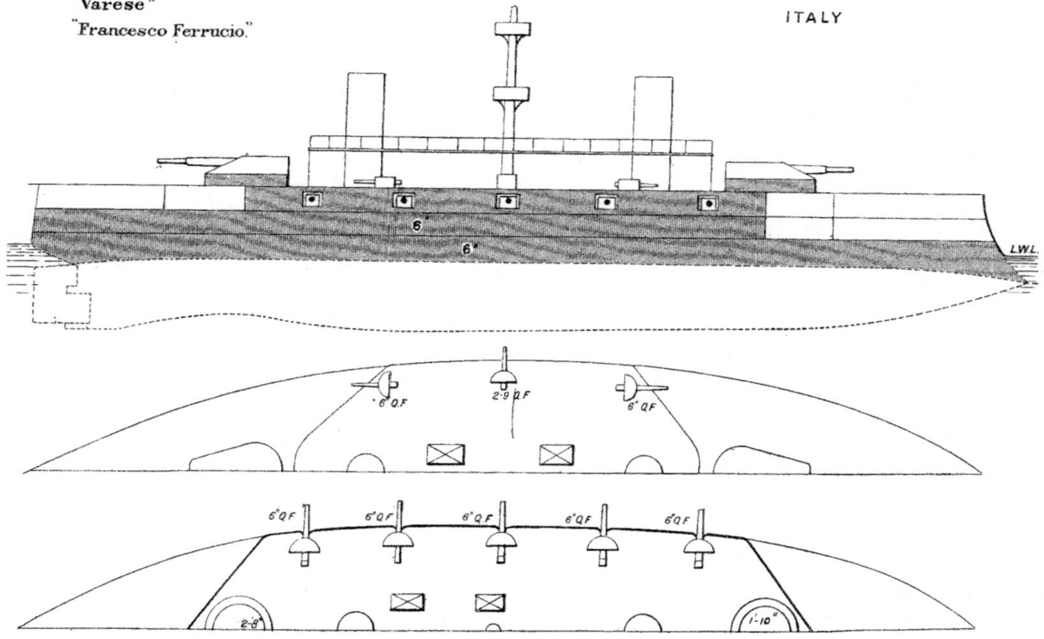

Plate 59.

ITALY.

Italia.
Lepanto.

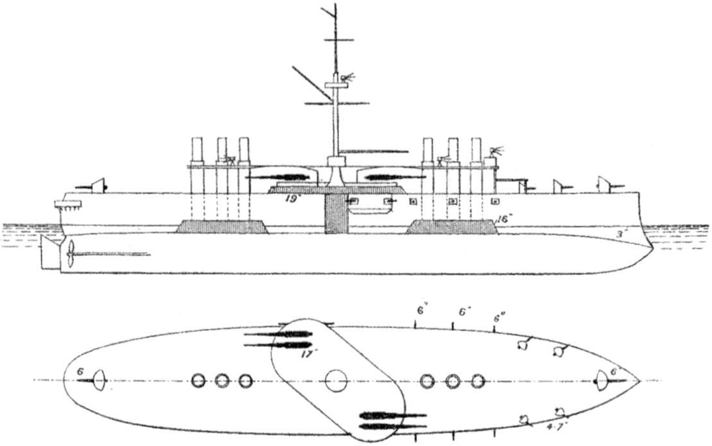

Marco Polo.

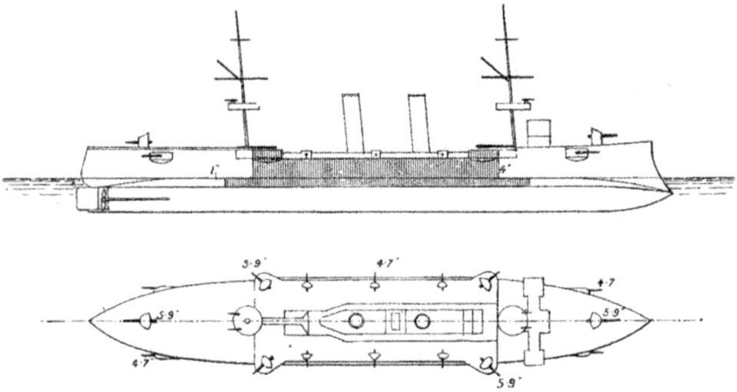

ITALY.

Ammiraglio Di St. Bon.
Emanuele Filiberto

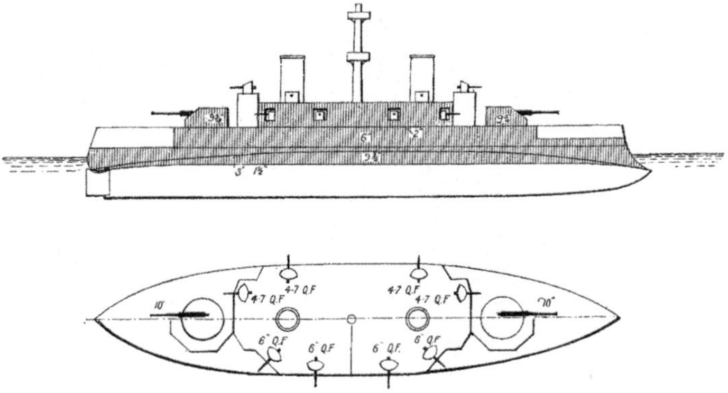

"Carlo Alberto"
"Vettor Pisani"

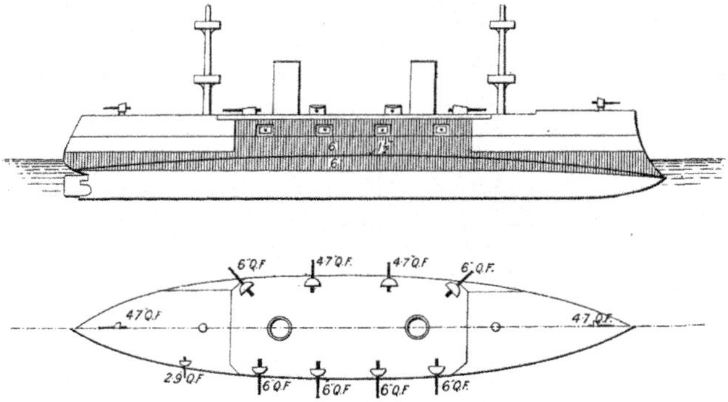

"Sardegna"

"Re Umberto"
"Sicilia"

ITALY

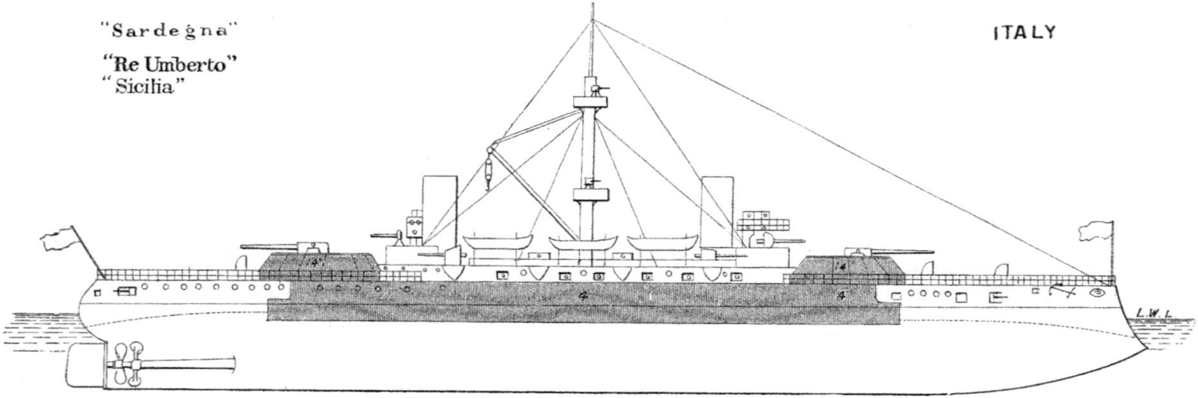

Note. *Sardegna is 9 ft. 10in. longer and 3 ft. 3in broader than the other two.*

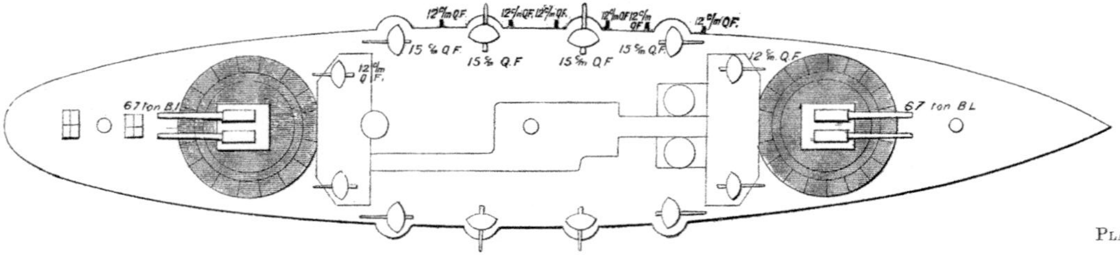

Plate 62.

JAPAN.

Asama and Tokiwa

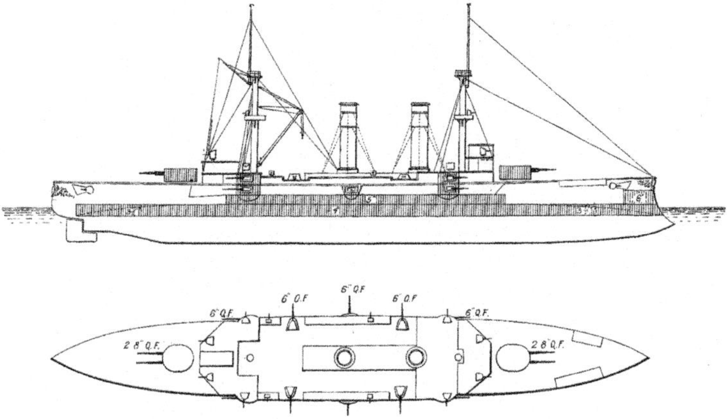

TAKASAGO.

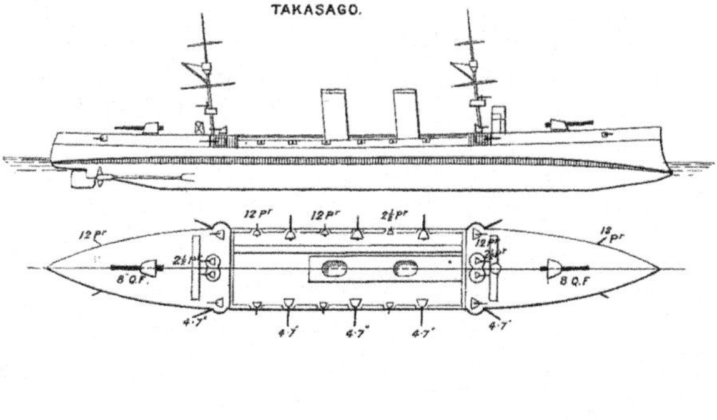

JAPAN.

"Chin Yuen"

Taken from the Chinese

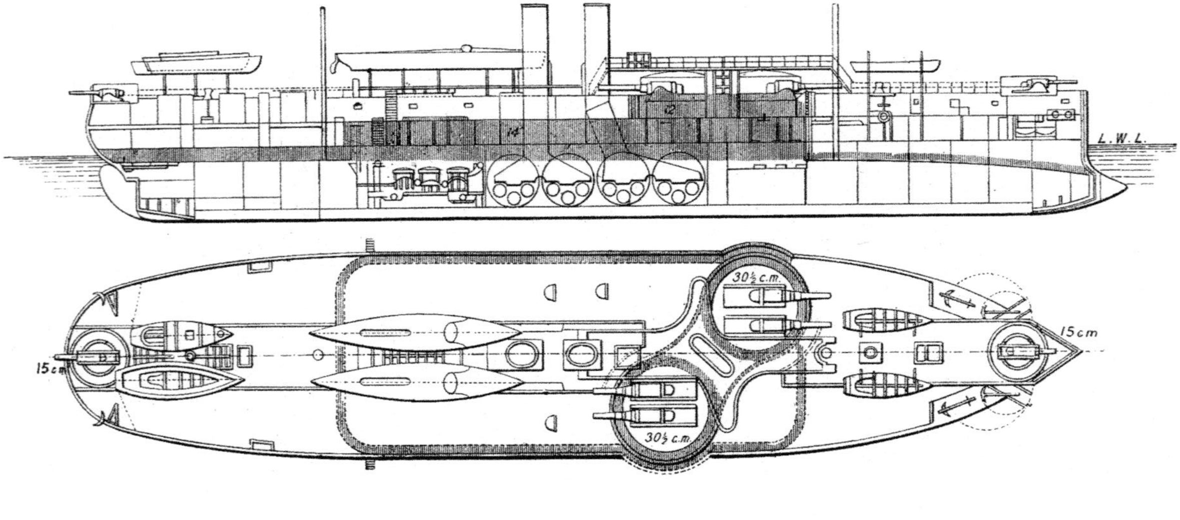

Plate 64.

JAPAN

First Class Battleship

"Fuji."

"Yashima".

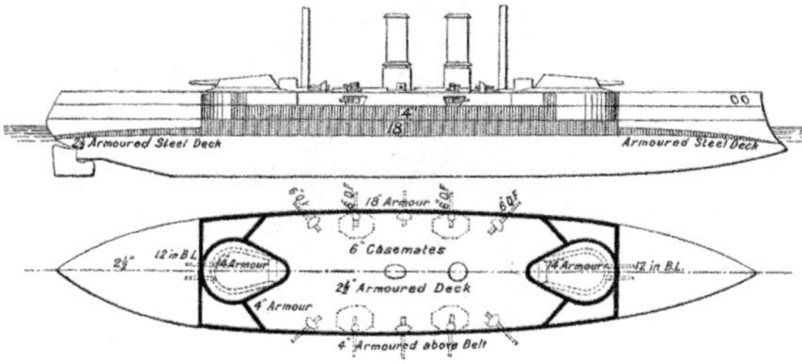

Battleship

"Mikasa"

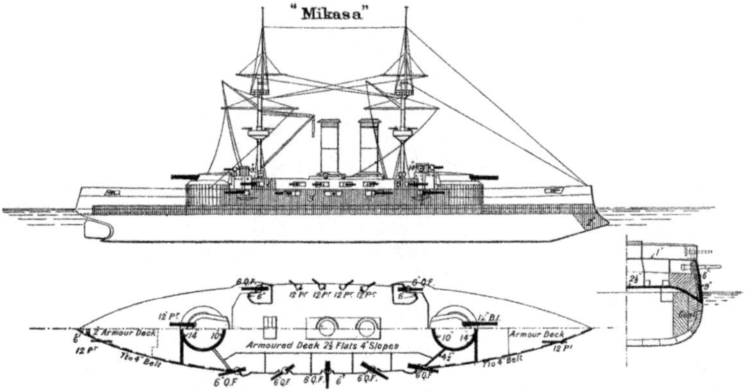

Plate 65.

Japan.

BATTLE-SHIP.
"Hatsuse."
15,000 Tons. 14,500 I.H.P.

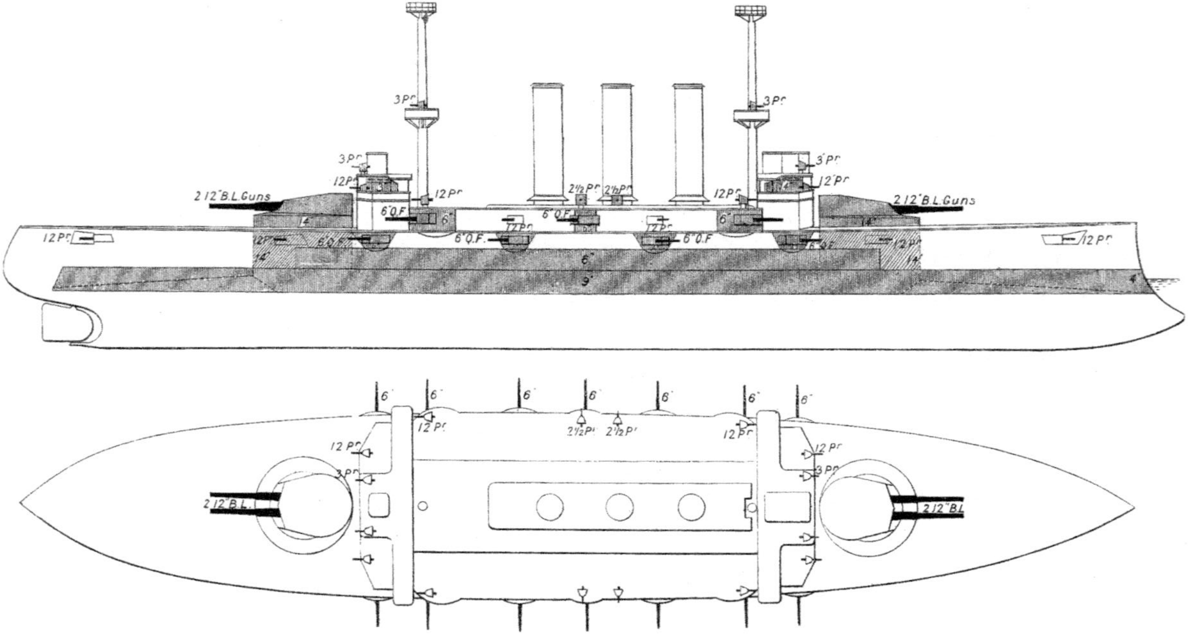

Plate 66.

JAPAN.

ARMOURED CRUISER.

"Idzumo"

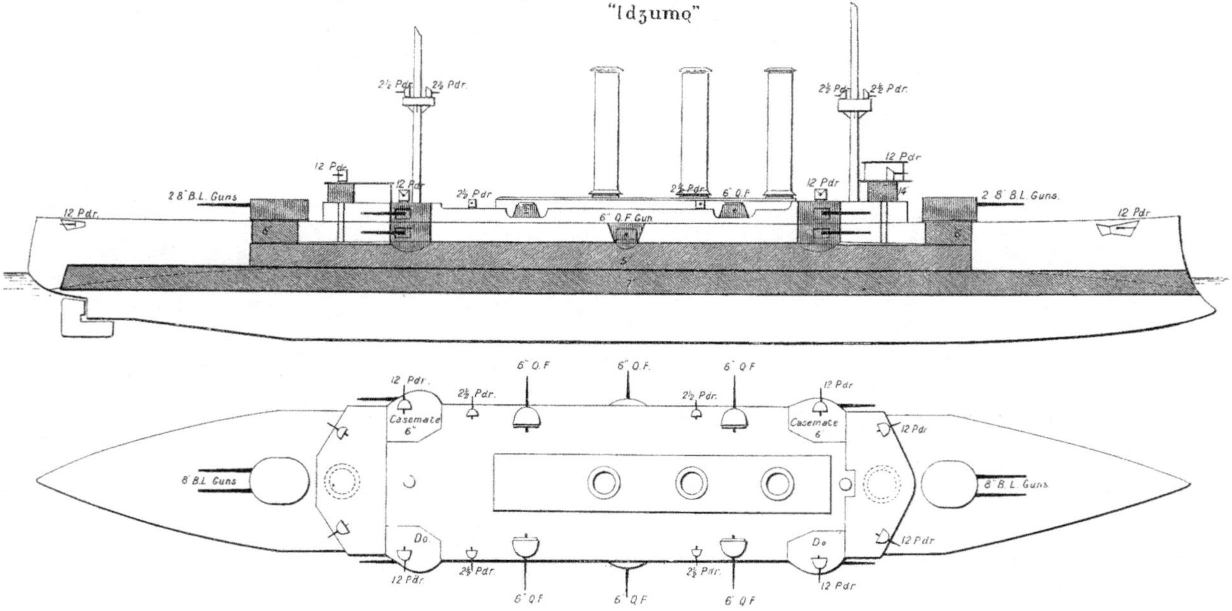

PLATE 67.

JAPAN.

"Itsukusima"
"Hashidate"

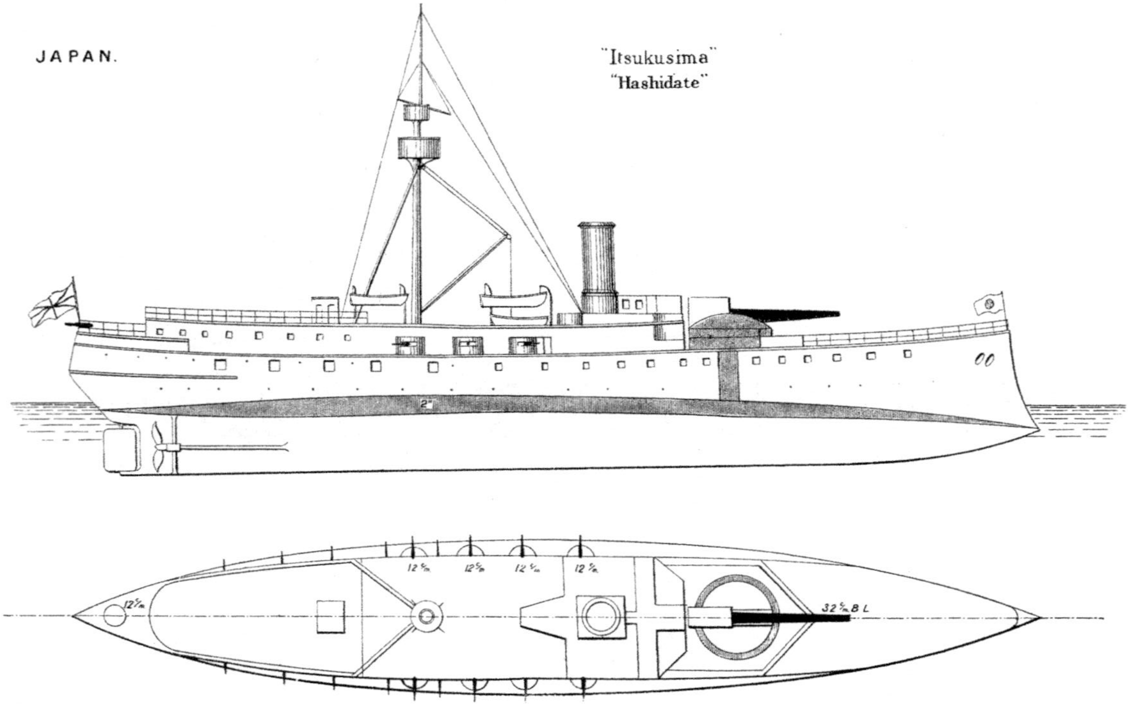

PLATE 68.

Japan.

"Shikishima"

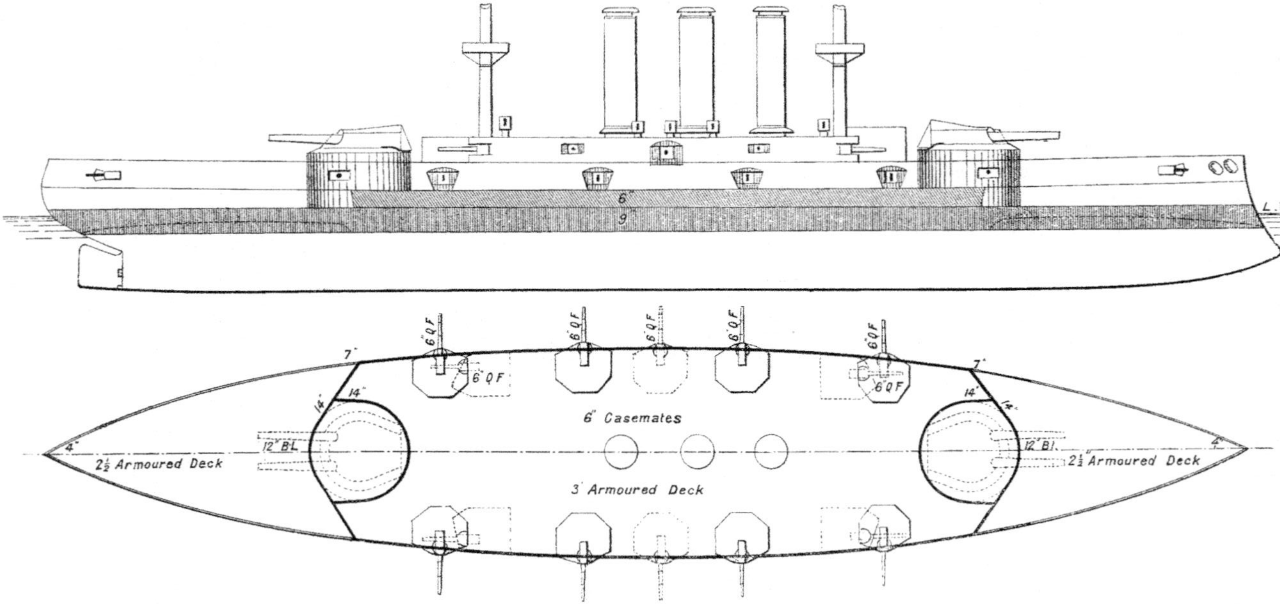

PLATE 69.

JAPAN

"Tatsuta."

On Special Scale

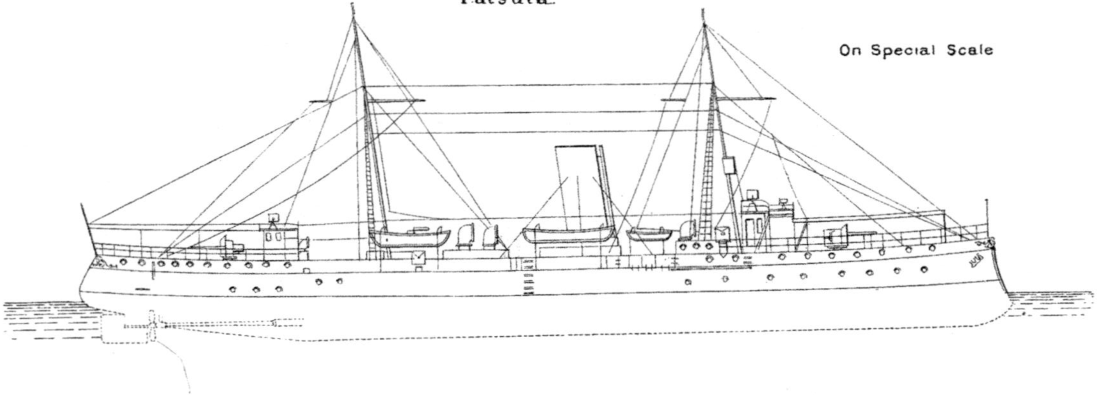

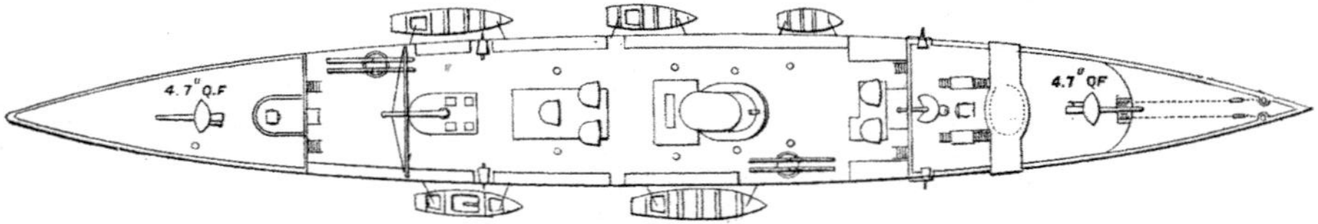

Plate 70.

JAPAN

"Yoshino"

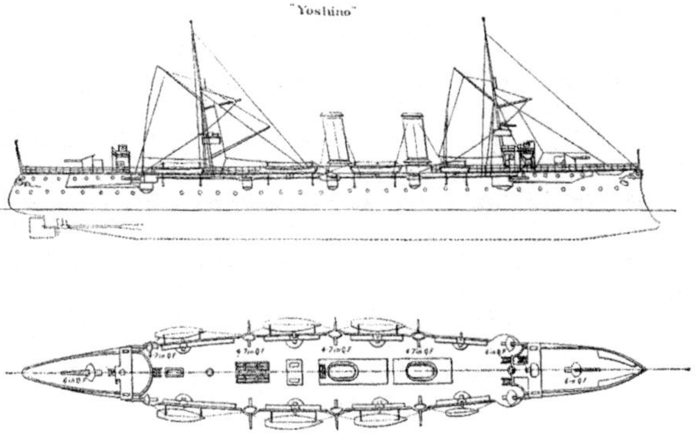

Kasagi.

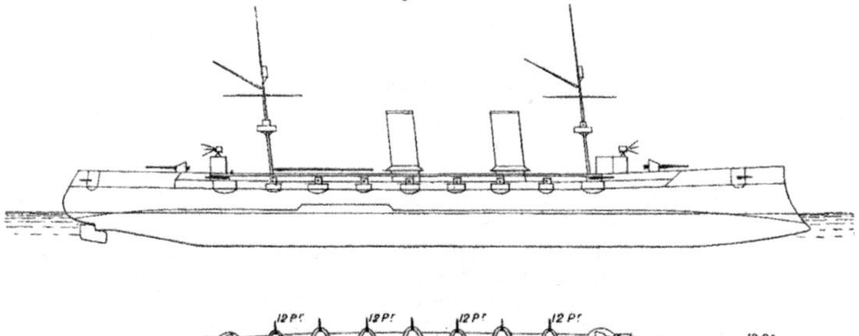

Plate 71.

NETHERLANDS.

"Evertsen"
"Piet Hein"
"Kortenaer"

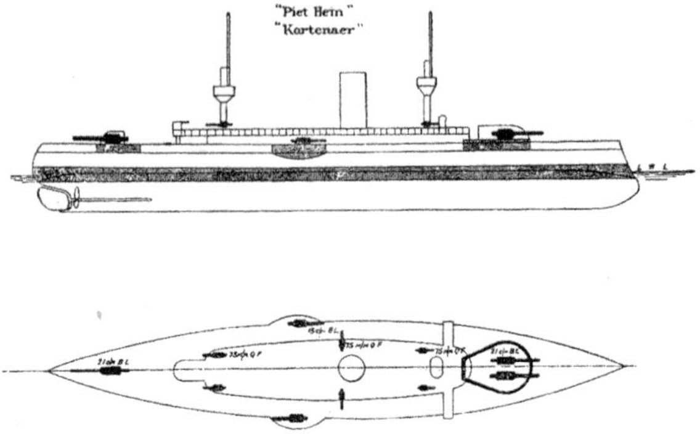

Koningin Wilhelmina de Nederlanden

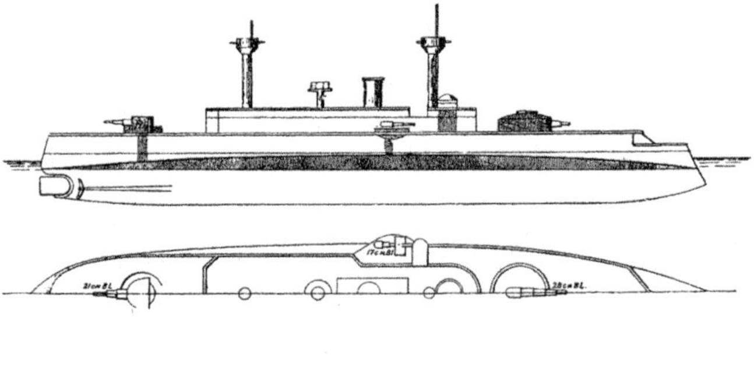

NETHERLANDS

"Reinier Claesen"

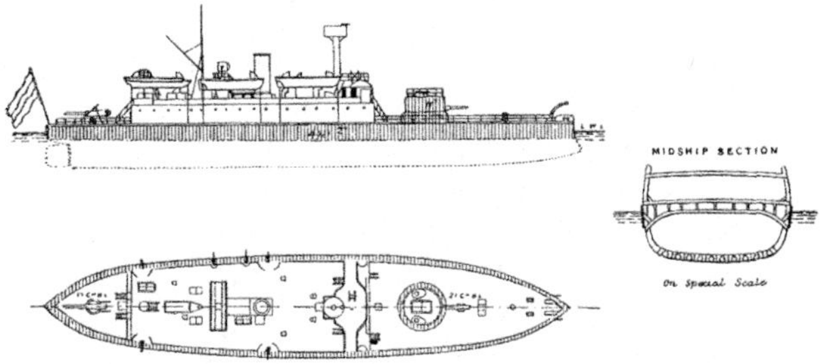

Holland.

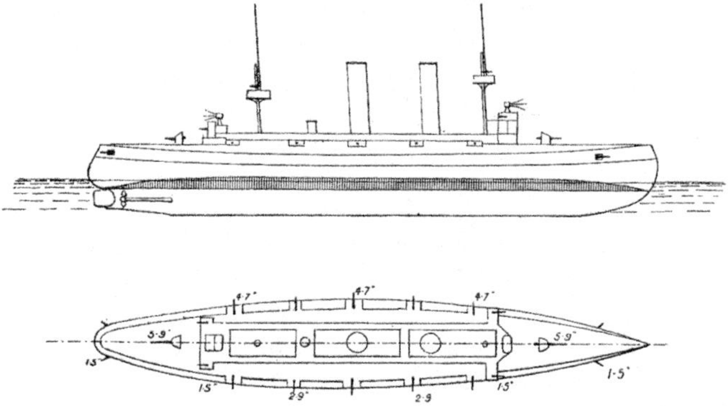

NORWAY.

ARMOUR-CLADS.

"Norge"

"Eidsvold"

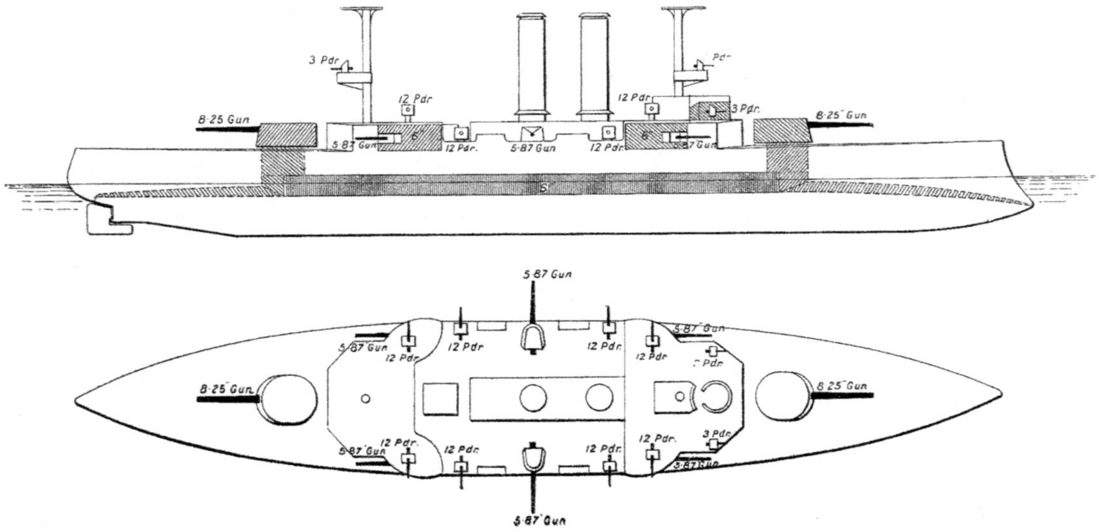

Plate 74.

RUSSIA.

ARMOURED CRUISER
"Admiral Nachimoff"

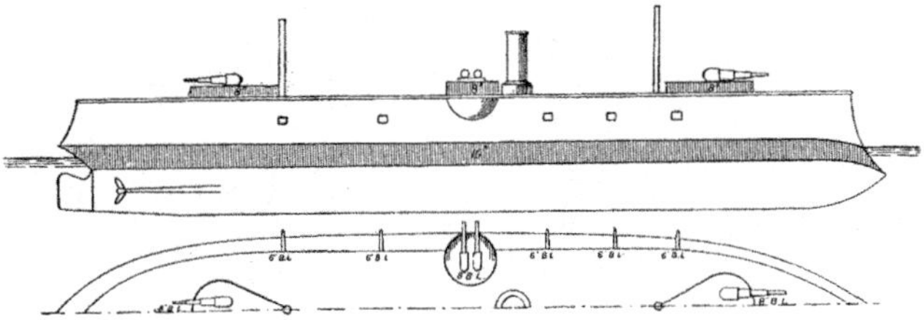

" Alexander II "

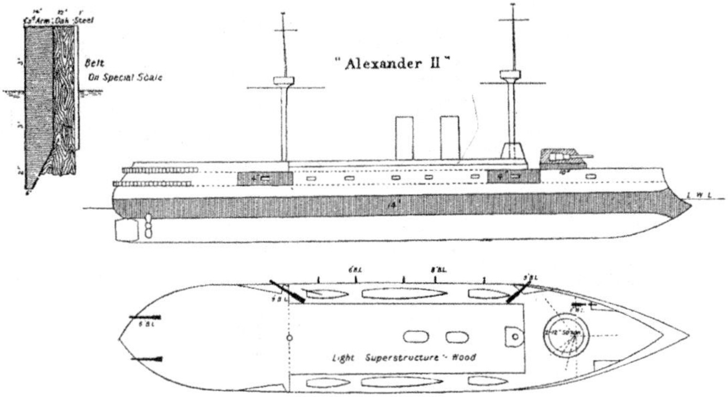

RUSSIA.

"Catherine II."
"Tchsmé."
"Sinope"

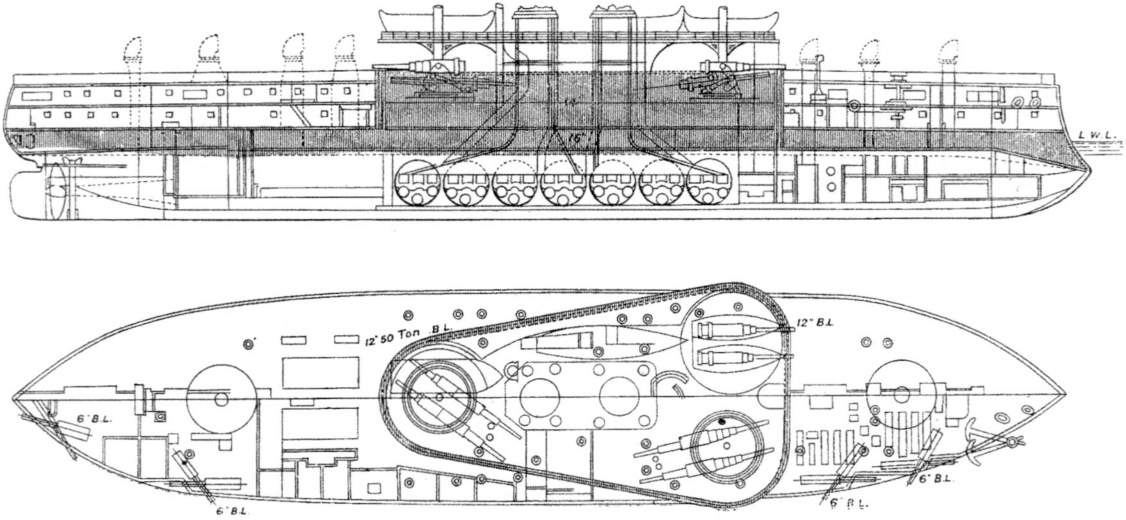

Plate 76.

Russia.

ARMOURED CRUISER.

"Gromoboi."

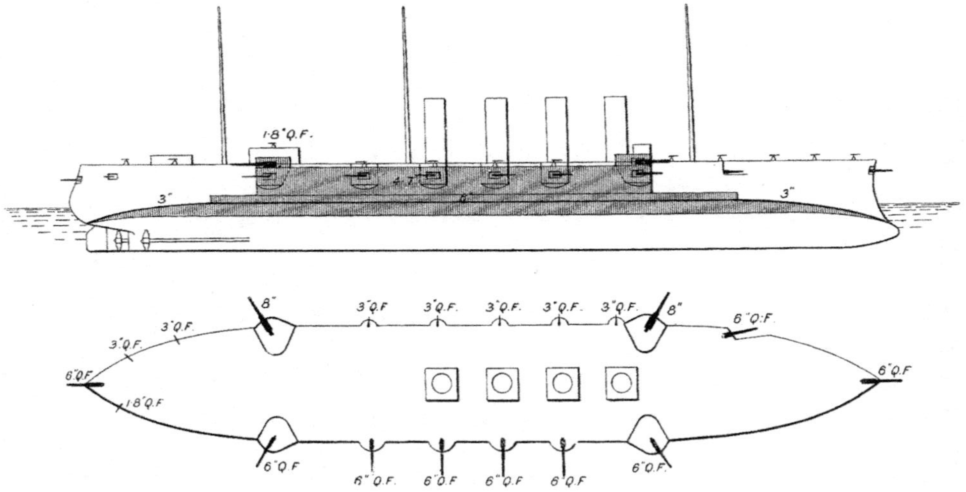

PLATE 77.

RUSSIA.

BATTLE SHIP

"Nicolai I"

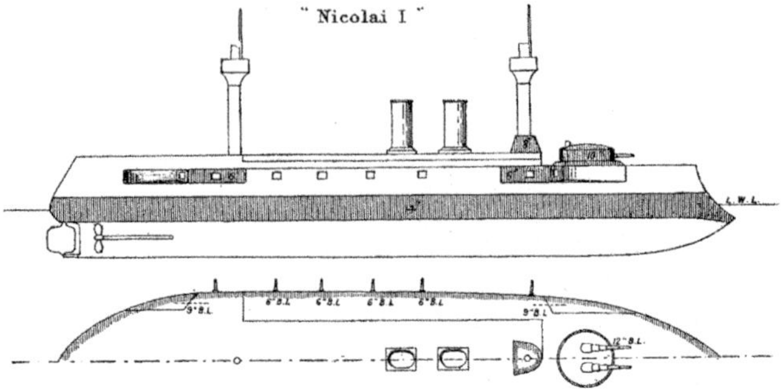

ARMOURED CRUISER

"Pamyat Azova."

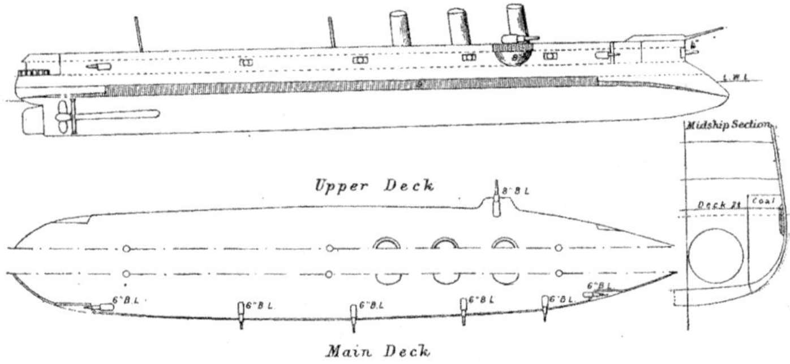

Russia.

FIRST CLASS BATTLE-SHIPS

"Oslabya"

"Peresviet"

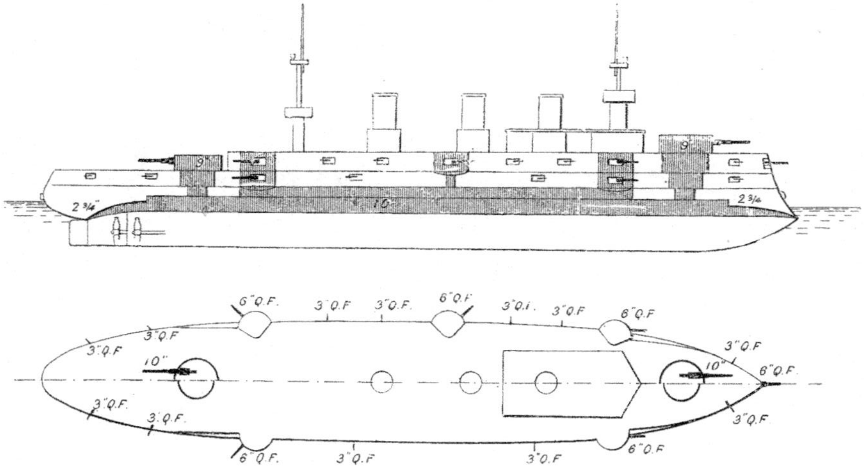

PLATE 79.

RUSSIA.

PETROPAVLOVSK
POLTAVA
SEVASTOPOL

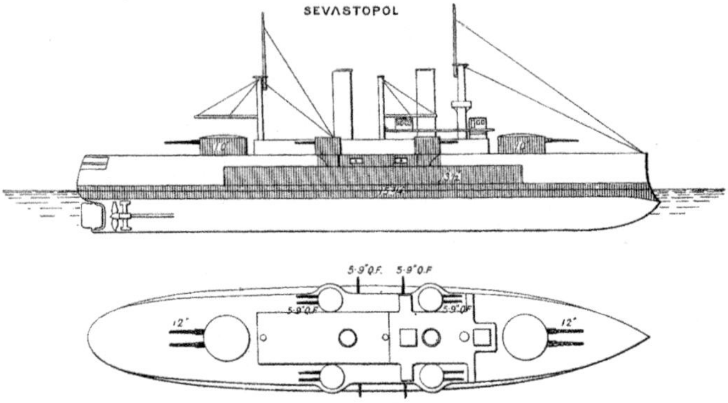

"Retwisan"

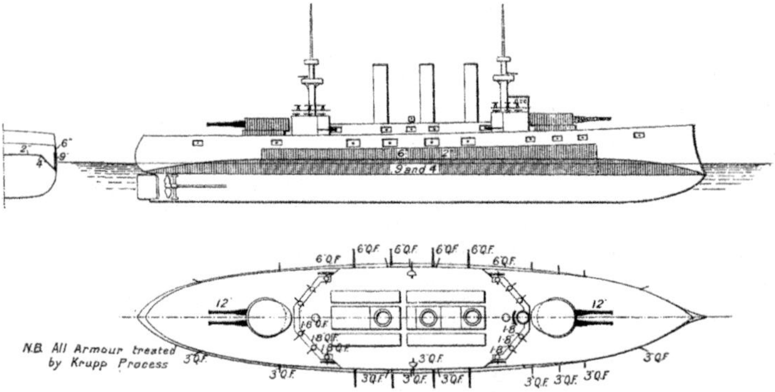

First Class Cruiser "Rurik."

PLATE 81.

RUSSIA.

"Sissoi Veliky"

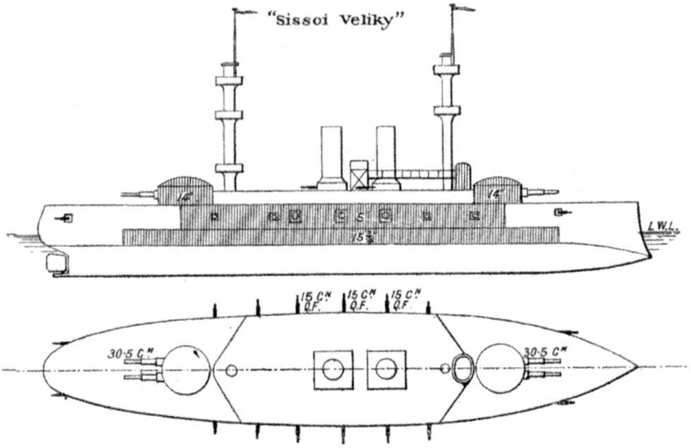

"Waryag"

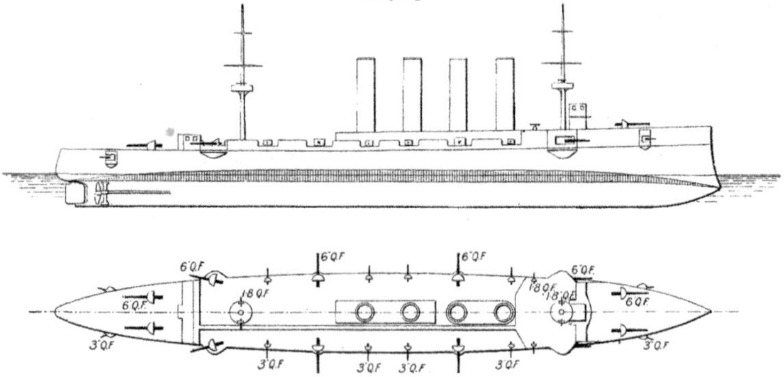

PLATE 82.

RUSSIA.

"Tria Sviatitelia"

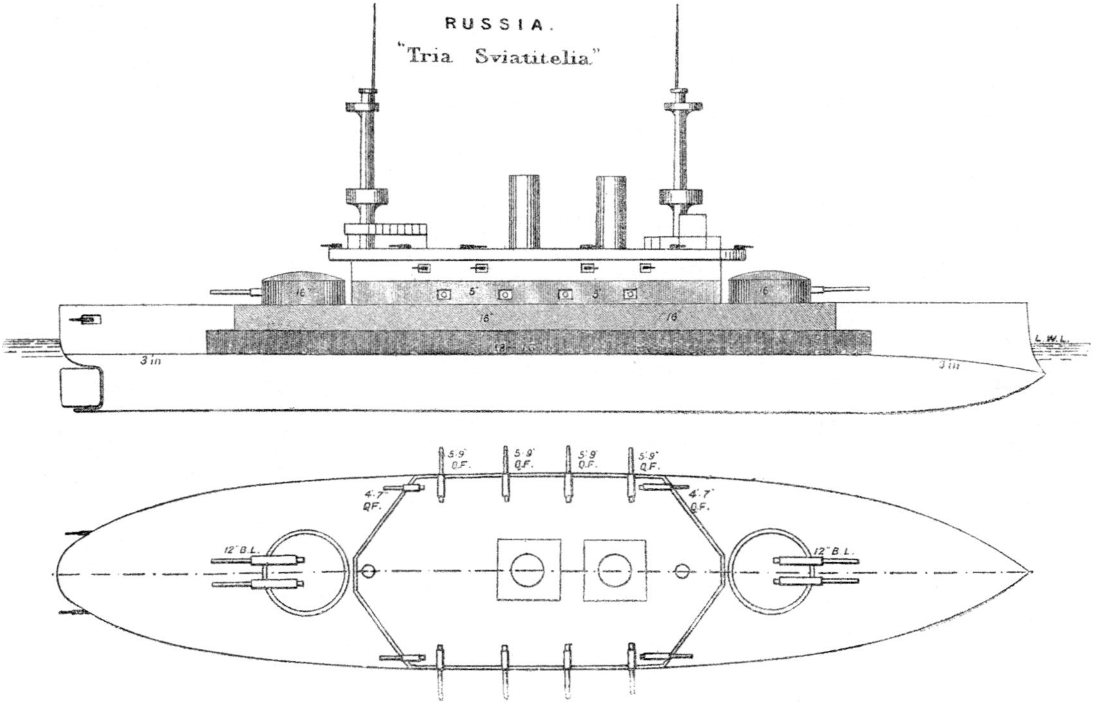

PLATE 83.

SPAIN

"Emperador Carlos V."

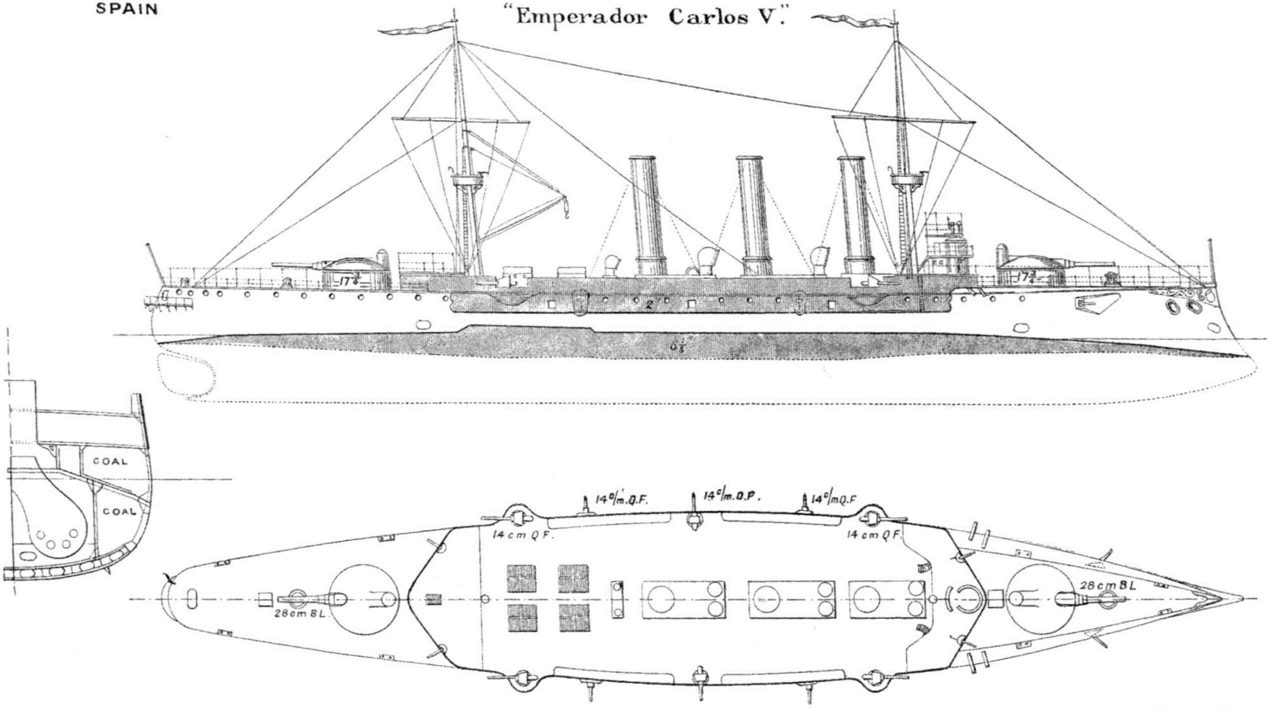

Plate 84.

SPAIN

"Cataluña"
"Cardenal Cisneros"
"Princesa de Asturias"

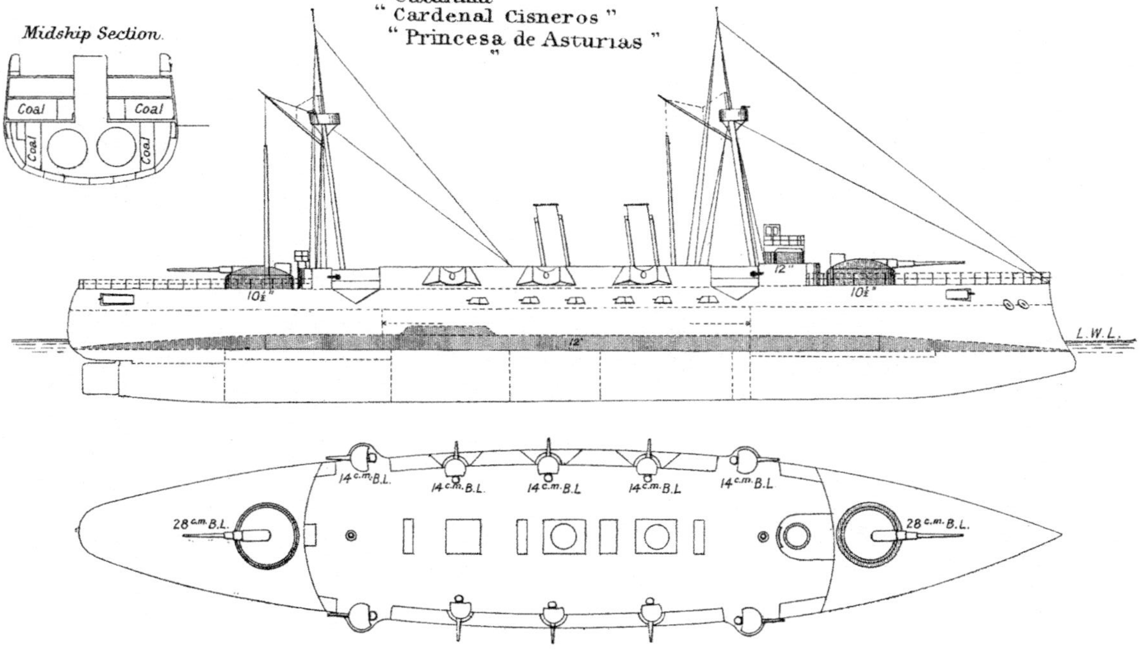

PLATE 85

SPAIN

"Pelayo"

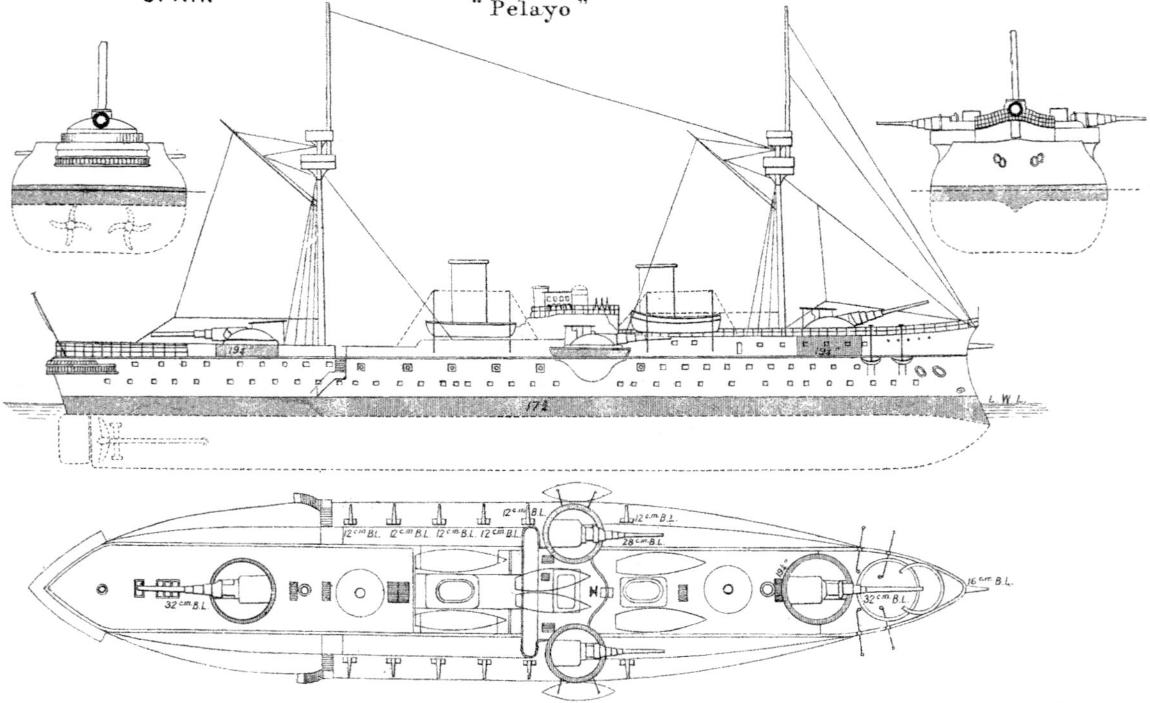

Plate 86.

SWEDEN

COAST DEFENCE SHIPS.
"A." "B" and "C."

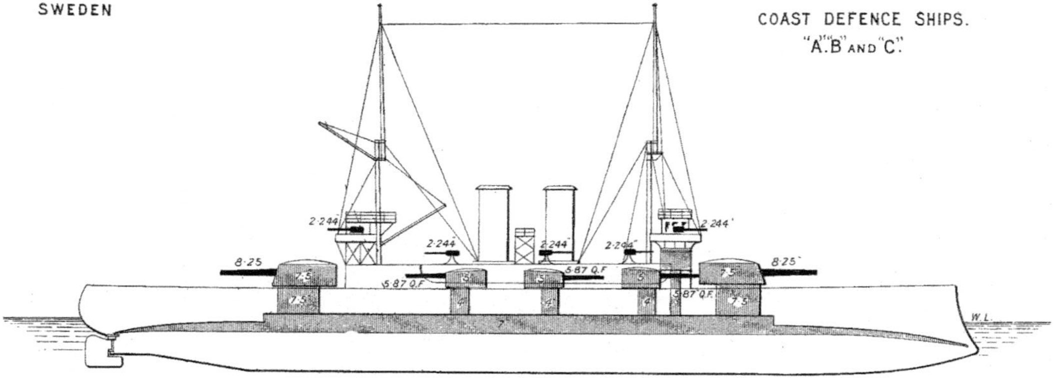

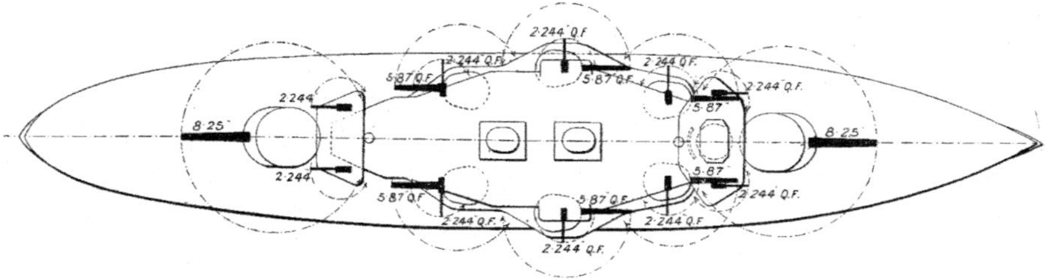

PLATE 87.

UNITED STATES.

Alabama

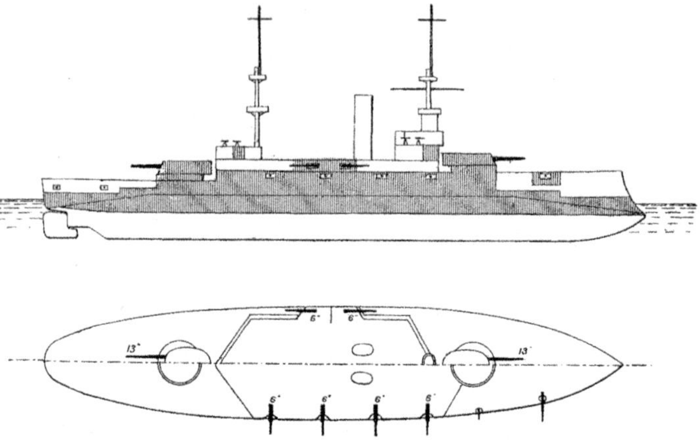

"Brooklyn"
(Armoured Cruiser)

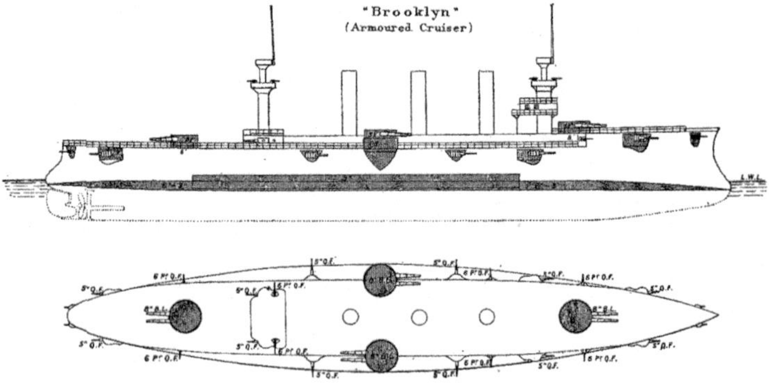

UNITED STATES.

ARKANSAS

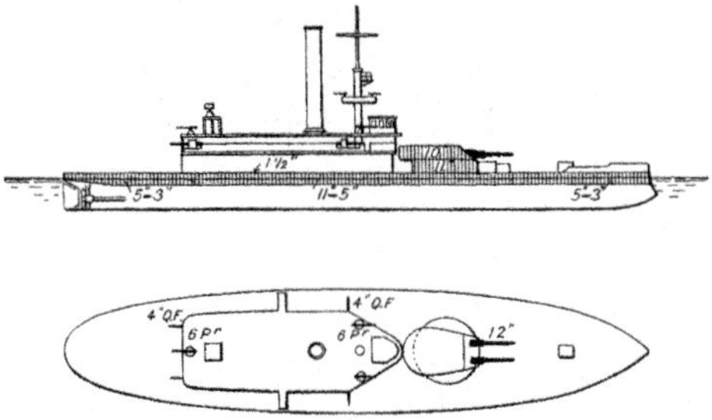

Maine
Missouri.
Ohio.

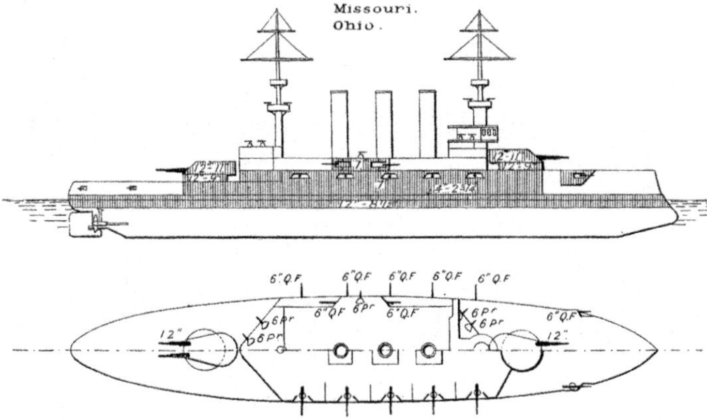

UNITED STATES

TRIPLE SCREW PROTECTED CRUISERS COLUMBIA & MINNEAPOLIS.

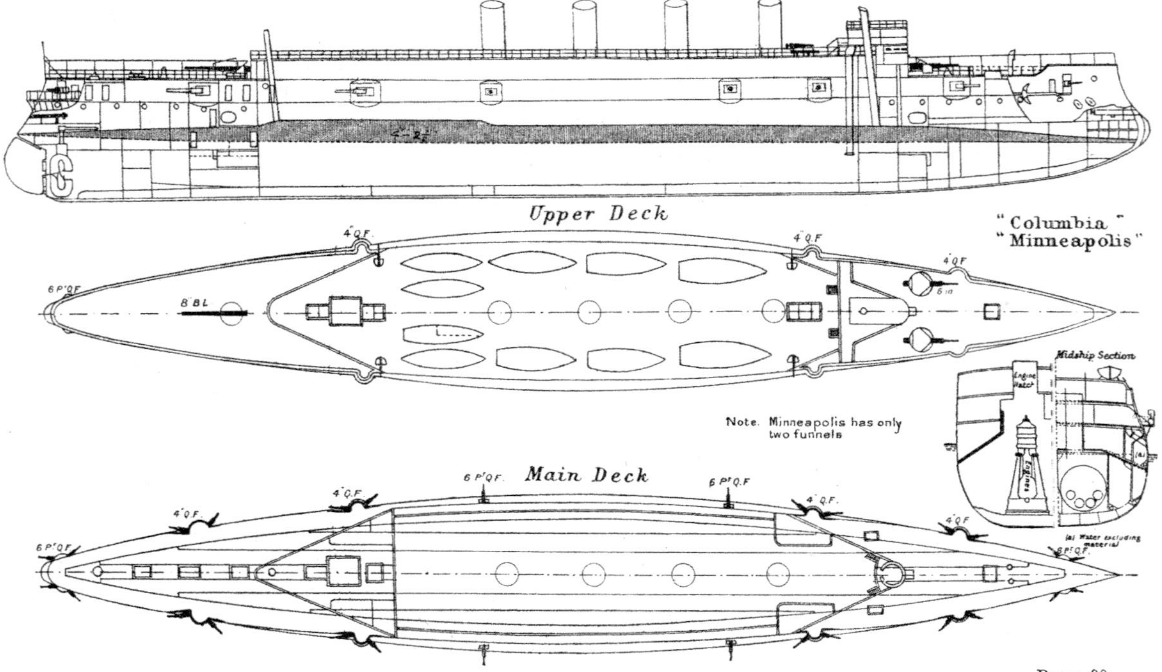

PLATE 90.

UNITED STATES.

"Iowa."
(Battle Ship)

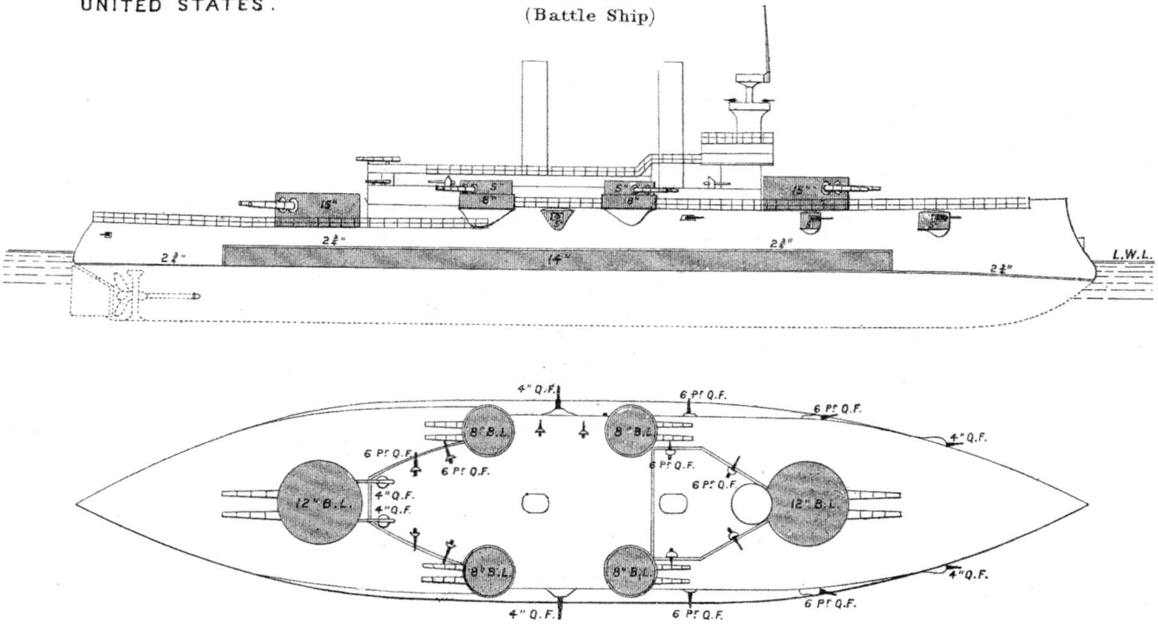

PLATE 91.

UNITED STATES.

Nº 5 (Kearsage)
" 6 (Kentucky)
(Battle ships)

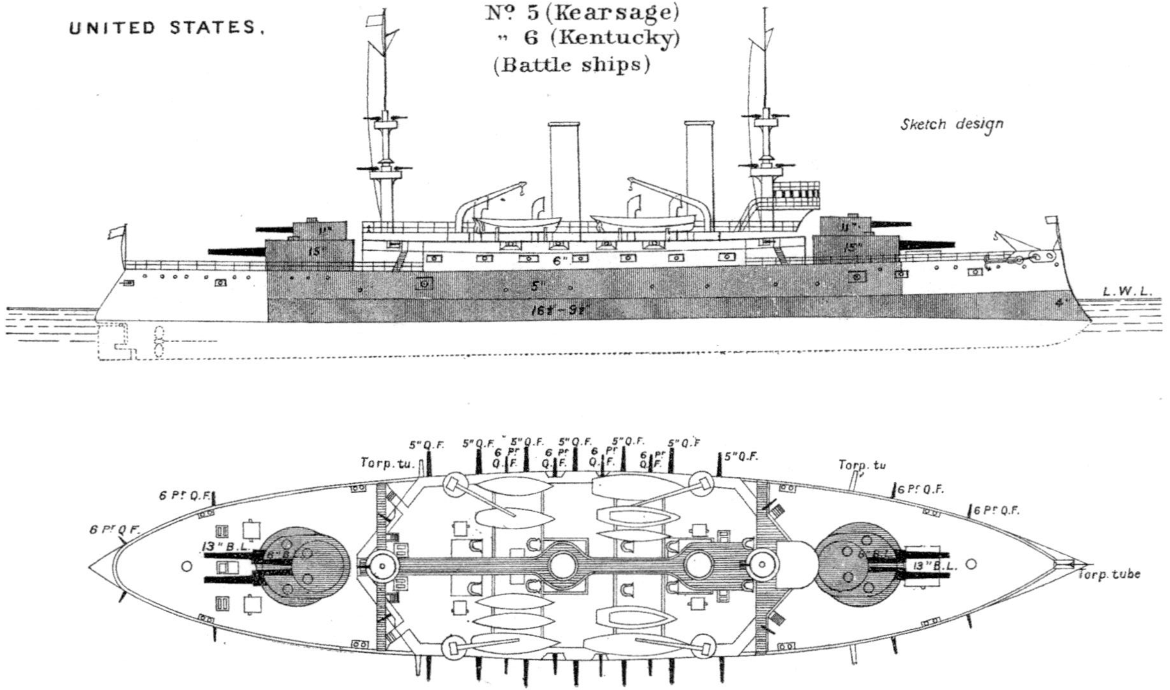

PLATE 92.

UNITED STATES

Coast Defence Ship
"Monterey"

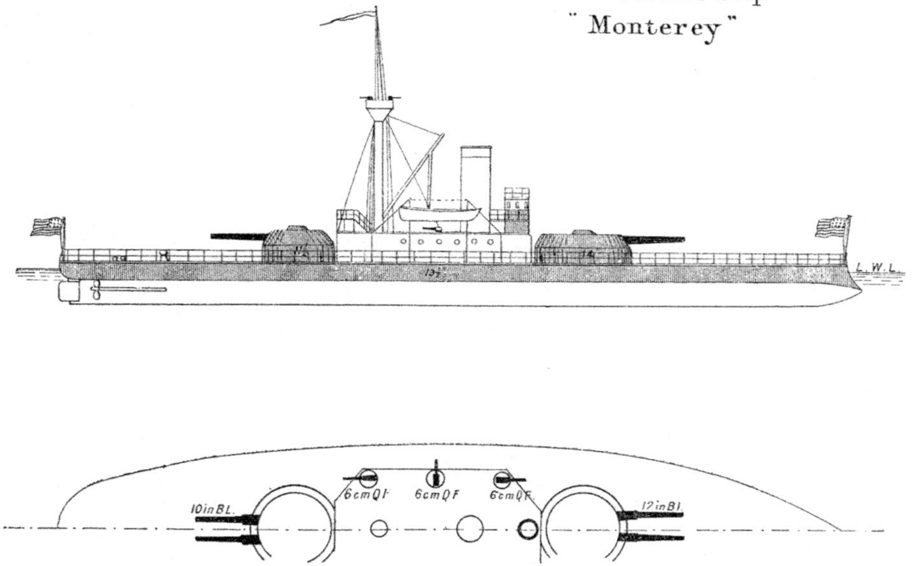

Midship Section.

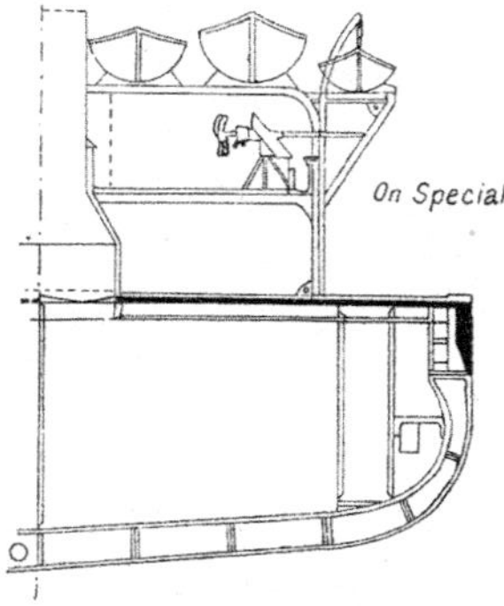

On Special Scale

PLATE 93.

UNITED STATES

ARMOURED CRUISER.

"New York"

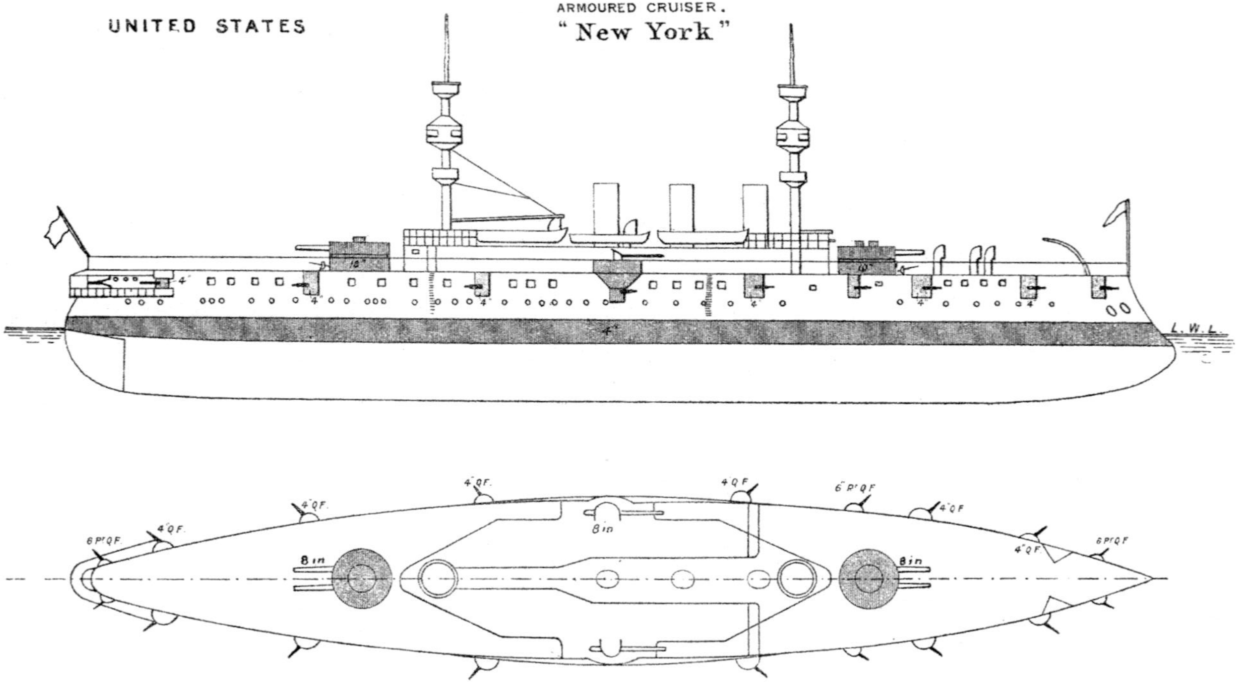

PLATE 94.

UNITED STATES

BATTLE SHIPS.

"Oregon"
"Massachusets"
"Indiana"

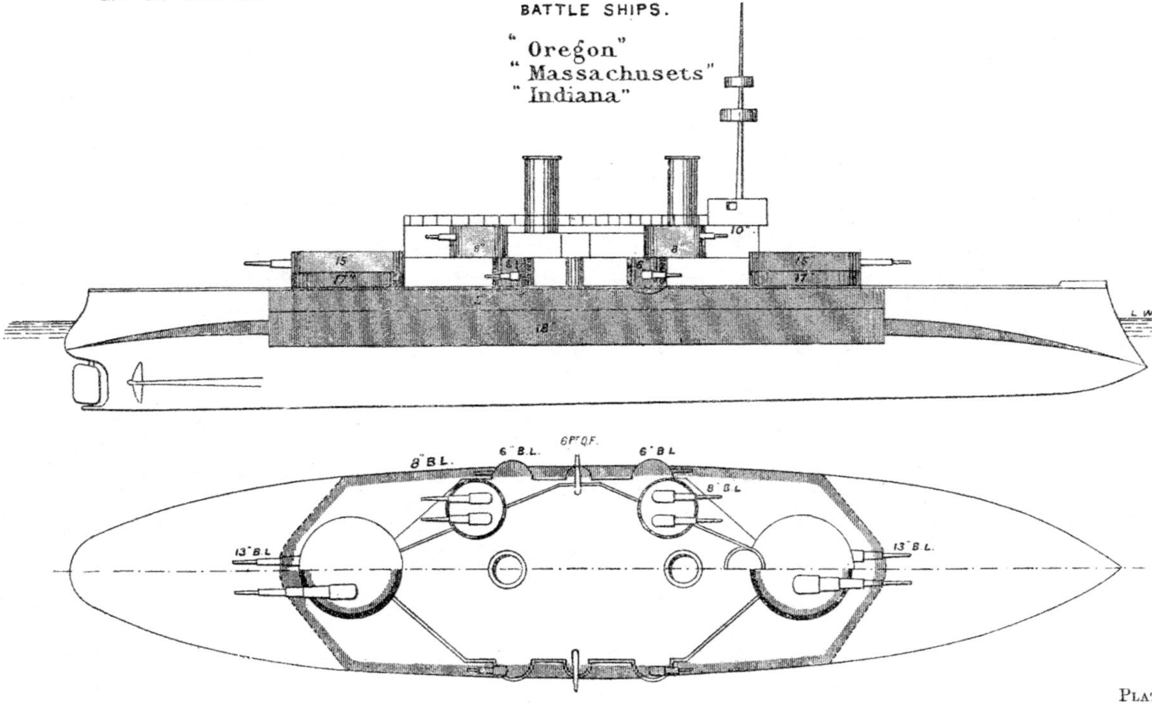

PLATE 95.

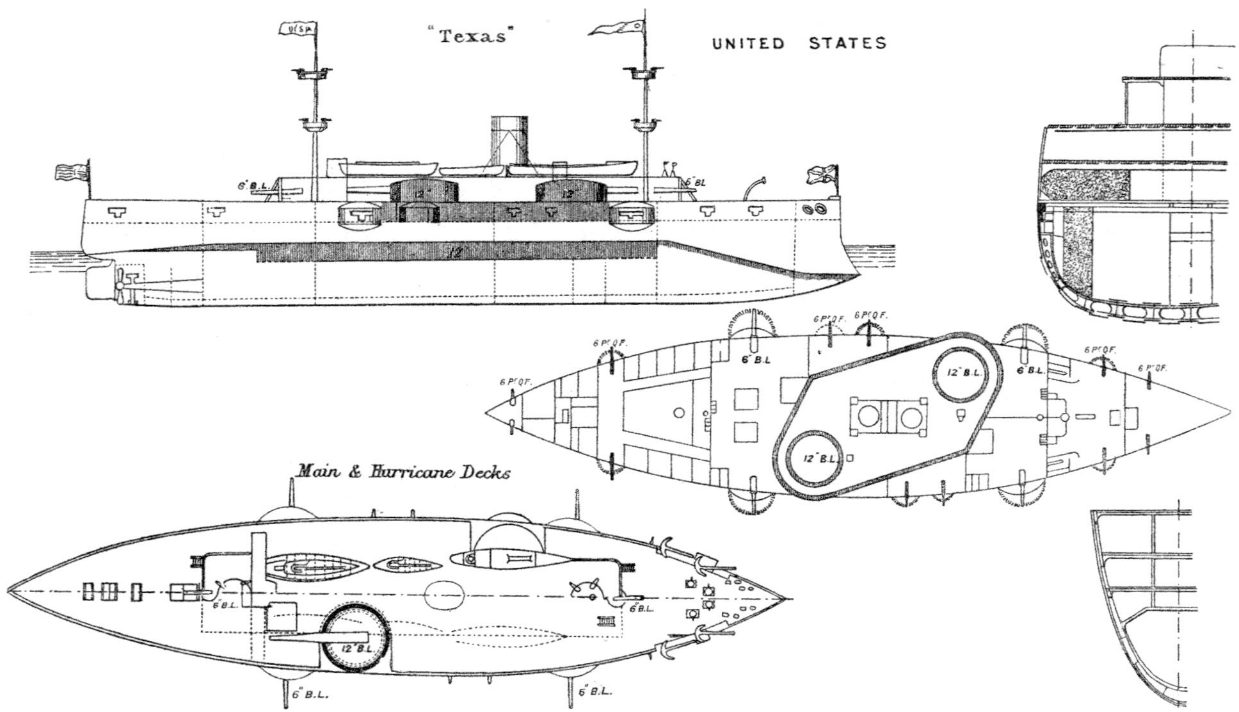
"Texas"
UNITED STATES
6" B.L.
12"
12
6 Pr Q.F.
12" B.L.
Main & Hurricane Decks
6" B.L.
12" B.L.

Plate 96.

PART III.

ARMOUR AND ORDNANCE.

PART III.

Armour and Ordnance.

CHAPTER I.

ARMOUR.

THE chief features in the construction of ships' armour and armour-piercing projectiles that seemed to be foreshadowed last year, were first, the extension of medium armour over the hulls of war ships, made possible by the adoption of Krupp's process plates, by which means a comparatively thin plate could be made of such resisting power that the attack of common shells might be altogether defeated; secondly, the final adoption of caps for the point of armour-piercing projectiles; and, thirdly, the introduction of a broadside gun heavier and more powerful than our 6-in. Q.-F. piece.

It cannot be doubted that the extension of armour over the hull above referred to is a mere matter of time, though the shape in which it comes may not be always the same. The belt has also increased in many cases; our own London has a kind of continuous belt and has no weak places under casemates. The Mikasa, building for the Japanese at Vickers' Yard at Barrow, however, is apparently slightly the stronger and the best protected battleship existing, as well as the largest except the Asahi, which has equal displacement, namely, 15,200 tons. The Mikasa has 12 ins. on her belt, 14 ins. on her gun position, 9 ins. over her lower Q.-F. 6-in. broadside battery, and 6 ins. over the four upper 6-in. guns. Besides this the 9-in. casemates unite and form an unbroken armoured wall, so that the 6-in. gun positions have no weak spaces between them. This vessel may be said to exhibit the system brought in by Sir W. White, developed and strengthened at the cost of 200 tons additional displacement.

Authorities.—The *Engineer* and *Navy and Army Illustrated* for plates and matter. Information furnished by the courtesy of Captain Colwell, U.S. Naval Attaché; the *Belgique Militaire*, Messrs. Vickers, Messrs. Schneider, Messrs. Carnegie, Captain Tresidder, of Messrs. Brown; Sir Alexander Wilson, of Messrs. Cammell, and Herr Krupp.

The casemates, the solid support to the water line, the armour carried down at the stem below the spur, and the general plan, so closely resemble the arrangements of the Formidable class that the Mikasa might be supposed to be designed by Sir W. White. The Russian war ship, Retwisan, has nearly the same disposition of armour as the Sevastopol class, but the little French double turrets for 5-in. guns are replaced by 6-in. pieces distributed like our own along the broadside, but less well protected. The United States Alabama and Maine classes furnish examples of broadside guns behind medium armour running continuously up from the belt, while the favourite design for the New Jersey class has embodied in it the British casemates containing 6-in. Q.-F. guns, the double wall casemate armour having apparently been quite given up. With its complete belt and contiguous casemates, this new design closely resembles that of the Mikasa on a smaller scale, but it has four 8-in. guns placed in pairs in small turrets amidships.

Six-inch hard-faced armour.

Six inch hard-faced armour has played a very important part latterly, and has occupied a position that might have told decisively in Naval actions and in more than one way. In English battleships 6-in. plates came in to protect our heavy Q.-F. armaments at a time when those of foreign powers were covered by 3, 4, rarely 5, and, in one instance, 6-in. plates. Superiority in defence tells specially when the armour is nearly matched to the power of guns employed. This was obviously the case here, for the 6-in. guns were habitually used to test the 6-in. plates supplied, and under favourable circumstances they perforated ordinary steel easily. When, however, the Harvey process, and still more the Krupp process, came in, perforation of 6-in. plates, even with the best 6-in. shot, was practically out of the question. Now it happened that very great difficulty was experienced in treating the faces of plates thinner than 6 in., because the plates became contorted, so that British battleships for some time had 6-in. hard-faced armour, as compared with much thinner plates of ordinary steel on the secondary armaments of almost all foreign ships. The effect in action would have been startling, for our 6-in. guns could have penetrated our adversaries with armour-piercing shells, while they, on the other hand, could not have penetrated our armour even with shot from their Q.-F. batteries. The advantage thus indicated will tell for some time to come, but there can be now no harm in pointing it out, for in new constructions thicker plates are being introduced, and the difficulty in hardening the face of thinner plates has been now got over. As said above, it has been shown to be possible to construct ships nearly completely covered with 6-in. armour, and it can only be a matter of time before numbers of these exist, now that the hard-face has given such great power and importance to 6-in. armour. Consequently our authorities have, as noticed last year, recognised the necessity of introducing a Q.-F. gun capable of perforating it, namely, the 7·5-in. gun, noticed in the chapter on ordnance. It is high time to possess such a gun, for it may be seen by consulting the tables of ships that 6-in. hard-faced armour is coming in both for belts and protection of Q.-F. guns; thus, the French cruiser Sully has a 6-in. Harveyed belt and the same for her 16·4 c.m. (6·3 m.) Q.-F. guns in casemates and turrets. The U.S. battleship Maine has continuous side armour above her belt and covering her 6-in. Q.-F. guns. The Yakumo, Mikasa, and other Japanese vessels have 6-in Krupp plates for their 6-in. Q.-F. gun casemates, and the Russian cruiser Gromoboi has casemates of 6-in. Harveyed armour for 6-in. Q.-F. guns.

Standard for equation of armour.

Now that armour exists varying enormously in resisting power, it is more desirable than ever to form some ready system of comparison, and the relation to the equivalent thickness of wrought iron advocated by Captain Tresidder appears to be the safest. Doubtless mild steel can be made of uniform quality by any given firm, but practically different firms would have different views, and steel is capable of very great variation. Wrought iron is rough and in theory very imperfect; the limit of its variation, however, if fairly made, is narrow, and there can be little uncertainty in taking it as a standard. The following relations may be roughly laid down for steel of various classes.

Ordinary mild steel is equal to 1¼ times its thickness of wrought iron; Harvey steel is equal to double its thickness in thick plates, and perhaps 2½ in thin ones. This, however, is by no means accurate. Harvey steel certainly has considerably higher relation or figure of merit than 2, but hardly reaches 2½, even when the plate is thin, and 2 is fairly correct when it is thick. Krupp-process armour has been very seldom tested so severely as to produce perforation. Last year's Annual gave an example of an uncapped shot just getting into the backing of a Carnegie 6-in. plate with a figure of merit of 2·95. In this chapter will be found a shot getting its point through the backing of a Terni hard-faced plate with a figure of merit of 3·33. This seems to have been very like a Krupp-process plate, and for a rough approximation 2·80 to 3·00 seems to be nearly correct for the very best class of 6-in. hard-faced plate. The results shown on the Table herewith suggest that thicker plates may probably yield at a slightly lower figure, perhaps 2·5. However, the rule given above is as low as can be taken for a figure of merit which is to ensure perforation.

Capped shot.

Less has been heard of capped shot this year than might have been expected. The difficulty has always been with oblique impact. The value of a cap in a direct blow cannot be questioned.

With regard to the actual quality of armour, improvements continue to be made by the various makers, and there is a tendency to drop the name Krupp-process armour; but so long as the characteristic treatment of chromium steel by chilling processes forms a leading feature and produces the valuable qualities which have ensured the success of this kind of plate, it appears misleading to classify the armour otherwise than as improved or modified Krupp-process plate.

It may be well now to deal with some of the successful results that have been exhibited under firing trial during the past year.

The following table shows figures of merit attained by various Krupp-process plates—that is, the relation of the thickness of wrought iron that would have been perforated, to the thickness of the actual Krupp-process plate attacked.

Plate Makers.	Thickness in Inches.	Figure of Merit of		Date of Trial.
		Blow Defeated by Plate	Plate Perforated.	
Brown	6	2 72	...	Nov. 1899
Carnegie	6	2·73	2·95	July 1899
Krupp	6	2·41	2·86 (just)	Dec. 1899
Krupp	6	2·42	2·85 (just)	" "
Carnegie	8	...	2·57	June 1898
Bethlehem	8	2·46	2·56	Oct. 1898
Krupp	8	...	2·41 (just)	...
Krupp	10	2 35	...	...
Krupp	11·8	2·33	...	June 1896*

* An early "record" plate. There were no signs of yielding, and it would have borne a very much heavier blow.

Vickers armour for Japan.

On September 1, 1899, a steel plate, manufactured by Messrs. Vickers, for the Japanese armour-clad Shikishima, building at the Thames Iron Works, was tested at Whale Island. The plate was 8-ft. by 8-ft. by 8¾-in. It was termed a special nickel Harveyed plate, probably differing only in detail from the admirable Krupp-process plates made by Vickers for our own Government. The attack consisted of three rounds with 9·2-in. armour-piercing steel Holtzer projectiles. Two rounds were delivered with a velocity of 1700 foot-seconds, and the third with 1800 foot-seconds velocity. Each projectile weighed about 380 lb. By Tresidder's formula the penetration through wrought iron is 17·1-in. for the first two rounds, and 18·6-in. for the third, the tests implying a figure of merit of 1·95 and 2·13 respectively. As the projectiles, far from perforating, only penetrated to a depth of 3-in. and 3¾-in., the plate had a much higher figure of merit, but it is impossible to say what it was. The photo-process prints herewith, Figs. 1 and 2, show the front and back of the plate after the third round. The shots had evidently been thoroughly broken up. In the large official photograph a trace of white radiating lines is visible, and these always imply extreme

FIG. 1.—ARMOUR PLATES FOR JAPAN.

disintegration, being the mark of langridge skimming over the plate face. There are no cracks.

About the beginning of the year, Terni hard-faced steel plates were tested at Muggiano, Spezia. Each was 8-ft. by 5-ft. by 6-in. Five rounds were fired at each plate with 6-in. 100 lb. projectiles, Terni hard-faced plate.

with a velocity of about 1600 foot-seconds. Two entered to depths of 4-in. and 4·4-in., all were broken up. Subsequently a Krupp 112 lb. shot was fired, with a velocity of 2500 foot-seconds, which also broke up, but the point got through both plate and backing.

FIG. 2.—ARMOUR PLATES FOR JAPAN.

By Tresidder's formula, this projectile should perforate twenty inches of iron, so that to resist this round, a plate would have a figure of merit of 3·3, that is to say, it must be equal to $3\frac{1}{3}$ times its thickness in wrought iron. This plate was something less than that; perhaps like Krupp armour, it may be put down as equal to about three times its thickness of wrought iron. This, so far as it goes, is as good as any result published for any kind of armour.

Brown's plates for Norway.

On November 10, 1899, a firing trial of one of John Brown & Co.'s armour plates was made at the proving grounds of Armstrong, Whitworth & Co., Limited, selected by Captain Mörch from a lot manufactured for the armour of two Norwegian coast-defence ships building at Elswick. The sample was subsequently reduced to dimensions for the prescribed test, of which the conditions were as follows:—Plate 8-ft. by 6-ft. by 5·9-in.; backing, 24-in. oak and 1¼-in. skin plate; bolts, eight of 2-in. diameter; number of rounds, four; projectiles, steel armour-piercing 6-in. diameter and 100 lb. weight; striking velocity, 1960-ft. per second; striking energy, 2664 foot-tons; calculated perforation, 13·4-in. of iron. The plate was expected to resist this attack without being perforated or seriously cracked; see Figs. 3 and 4.

The projectiles provided by the Norwegian authorities were made at the Elswick Works on the Wheeler-Sterling process, and as they weighed 102½ lb. each, the specified velocity was reduced to 1936 foot-seconds, to compensate for the extra weight. The striking energy thus remained unaltered at 2664 foot-tons, and the calculated perforation became 14-in. of iron.

The four prescribed rounds having been easily defeated, and no cracks being developed in the plate, the trial was pronounced highly satisfactory, and the whole of the armour represented was approved. To obtain some idea, however, of the ultimate defensive power of the armour, the attack was supplemented by two extra rounds at increased velocities, which also failed either to perforate or to produce any cracks in the plate except a few superficial air lines on the face. The following are the particulars of the results of each round:—

Number of round.	Projectile.		Striking velocity.	Striking energy.	Penetration.	Height of bulge at back.
	Calibre.	Weight.				
	in.	lb.	Ft.-sec.	Ft.-tons.	in.	in.
1	6	102½	1902	2571·4	3⅝ (measured)	1¼
2	6	„	1958	2725·1	3¾ (estimated)	1⅜
3	6	„	1948	2697·3	3⅝ „	1¼
4	6	„	1940	2675·1	4¼ „	1⅝*

* Shot's point apparently not destroyed.
The above constituted the acceptance test.

5	6	102½	1997	2834·6	4⅛ (estimated)	1½
6	6	102¾	2182	3392·5	5 „	2⅜

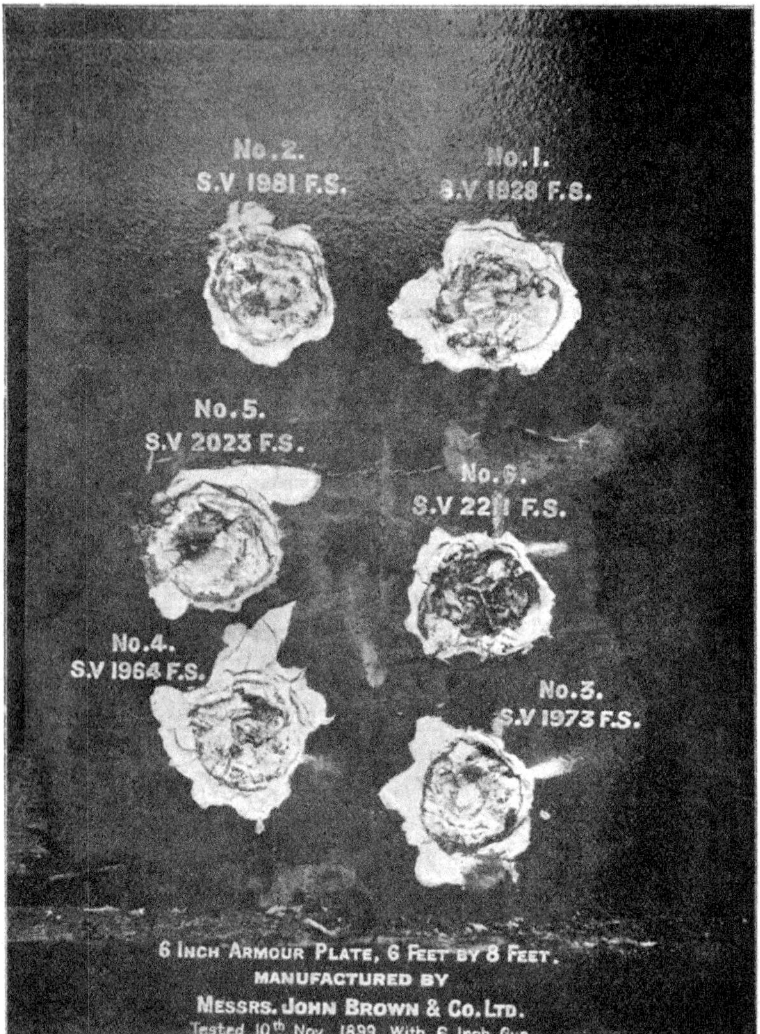

Fig. 3.—6-in. Armour Plate for the Norwegian Navy.

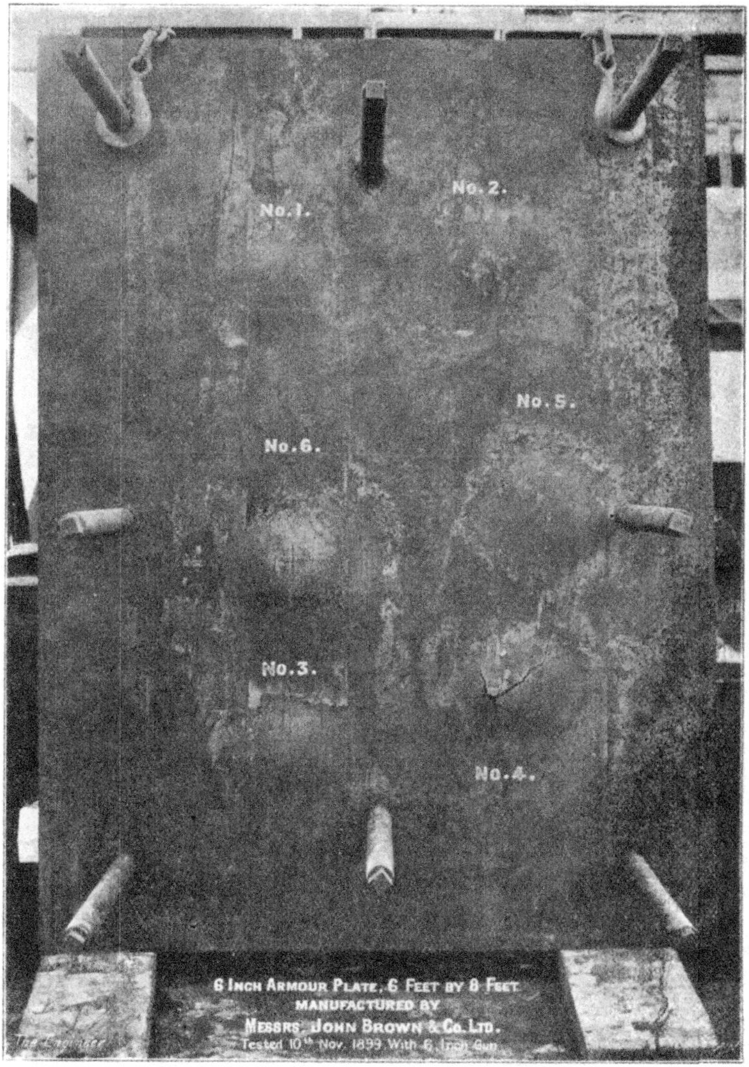

Fig. 4.—6-in. Armour Plate for the Norwegian Navy.

All projectiles were smashed to pieces. Except in the case of the first round, exact measurement of penetration was prevented by fragments of shot lodged in the indent. The last round was capable of piercing 17-in. of wrought iron by Tresidder's formula, that is, 2·7 times the thickness of the plate fired at.

There is one interesting feature in this trial to be noticed, namely, that the bulge made at the back by No. 4 round has in its centre a star-shaped tear, showing that the shot is in a measure boring a hole, which would be completed by sector-shaped pieces being torn and displaced outwards from the shot point. In the other five rounds apparently boring has been wholly defeated, and the shot point so far flattened as to be compelled to punch out a disc to get through the plate, entailing much more energy. It may be seen that No. 4 round happened to have less energy than all except No. 1, yet it may be questioned if its perforation would not be completed with less work than any of the other rounds. It is, in fact, an example of the case which occurs in certain rare instances, where the blow is so truly normal to the plate surface, and the projectile so good, that the point instead of being broken, retains much of its sharpness. Probably a capped shot would behave very nearly like this No. 4 projectile. If plates were really struck normally on service, as may be seen here, caps should certainly be adopted.

Carnegie Krupp-process armour for Russia.

Fig. 5 shows the result of a test at Indian Head of a Carnegie Krupp-process plate, measuring 11 ft. 2 in. by 7 ft. 9 in. by 8 in., for the Russian battleship Retwisan, carried out on December 28 last with an 8-in. gun firing projectiles under conditions stated below:—

No. of round.	Projectile Weight in pounds, and Makers.	Striking velocity.	Striking energy.	Perforation through wrought iron by Tresidder.	Figure of merit called for by each round.	Estimated Penetration.
		ft.-sec.	ft.-tons.			in.
1	252 Carpenter	1837	5896	16·8	2·10	2·33
2	253¾ Midvale	1791	5626	16·3	2·04	1·75
3	251 Carpenter	1815	5732	16·4	2·05	2·62
4	257 Wheeler	1834	5994	16·9	2·12	1·25

Seeing that the plate defeated the attack with insignificant penetration, the test did not bring out its powers; nor could it be expected to do so, for it only called for a plate with about double the resistance of wrought iron, whereas, as above stated, Krupp-process plates have shown themselves able to bear an attack which would perforate nearly three times the thickness of wrought iron. The fact

Fig. 5.—Armour Plate for Russia.

power of the plate. The chief interest of this trial lies in the fact that it furnishes an indication of the kind of armour carried by this new Russian ship on her barbettes, and we are driven to conclude that her barbette, though not invulnerable, is very strong. Our 12-in. IX. gun, it is true, perforates 36·8-in. of wrought iron at the muzzle when striking direct, but this comes down to 32·7-in. at 1000 yards. At

Fig. 6.—Krupp-Process Armour Plate for Russia.

30° with the normal this would still be over 24-in., but this means that the attack of our best gun would only get through the actual barbette plates, with little to spare, and any second defence would probably defeat it. We should prefer to attack such a ship elsewhere, using common shells. Russia generally keeps her trials dark, so that it is very unlikely that we shall get any other sample of her new battleships. At the same time, it is highly probable that Russia is very well advised in exhibiting this one, because the sample that we are thus able to study is a very excellent one. Russian makers are not likely to beat this; it is possible that they may fall considerably below it in excellence. The plate was slightly dished round the points of impact on each occasion, from $\frac{1}{4}$-in. to $\frac{1}{16}$-in., which seems to imply that the body of the steel was softer than would be expected.

Carnegie Krupp armour for Retwisan.

On September 19, 1899, was tested at Indian Head, a 5-in. Krupp-process plate made by the Carnegie Steel Company also for the Retwisan, which is building at Cramp's yard. The attack was made with five 5-in. steel projectiles of Wheeler-Sterling, Carpenter and Midvale make, the weight of each being 50 lbs., and the striking velocities successively 2060, 2086, 2057, 2099 and 2082 feet per second. The highest of these, 2099, had a striking energy of 1528 foot-tons and a calculated perforation of 11·5-in. of iron or 2·8 times the actual thickness of the plate. Fig. 6 shows the result. The plate certainly had a much larger figure of merit than this, as the projectile was only judged to have penetrated about two inches, and like the others, it failed to crack or try the plate seriously. This is the first Krupp-process 5-in. plate trial recorded in America. The importance of the introduction of this class of armour has been already noticed.

Schneider hard-faced armour.

Messrs. Schneider give the following (see fig. 7) as a sample of their hard-faced armour. The trial took place as long ago as May, 1898, but the result was very good. The plate was 258mm. (10·16 in.) thick, and was termed a special cemented plate. It was attacked first by one round from a 24 cm. (9·45-in.) gun, the projectile weighing 144 kilos. (317·5 lb.), striking with a velocity of 643 metres (2110 foot-seconds), and subsequently by two rounds from a 19·4 cm. (7·64-in.) gun, firing projectiles of the weight of 75 kilos. (165·3lb.), with striking velocities of 785 and 812 metres (2576 and 2664 foot-seconds), which produced effects shown in the figure herewith. The first round had a calculated perforation of 21·3 in. of wrought iron, or 2·09 times the actual thickness of the plate; the last two rounds, 23·1 in. and 24·4 in. of wrought iron, the last being 2·40 times the thickness of the plate. This was a tremendously severe attack. As shown in fig. 7, the plate entirely defeated it, but was cracked in thin cracks. The plate appears to have been rather harder and less tough than those made on the Krupp process. If the scale shows centimetres, the projectiles must have spread and set up considerably.

Russian deck armour.

The fig. 8 shows the result of a trial of deck armour for the Russian battleship Retwisan and cruiser Waryag. The plate is nickel steel 2 inches thick. It was attacked by three rounds of 6-in. Carpenter armour-piercing projectiles, weighing 100 lbs.; the third

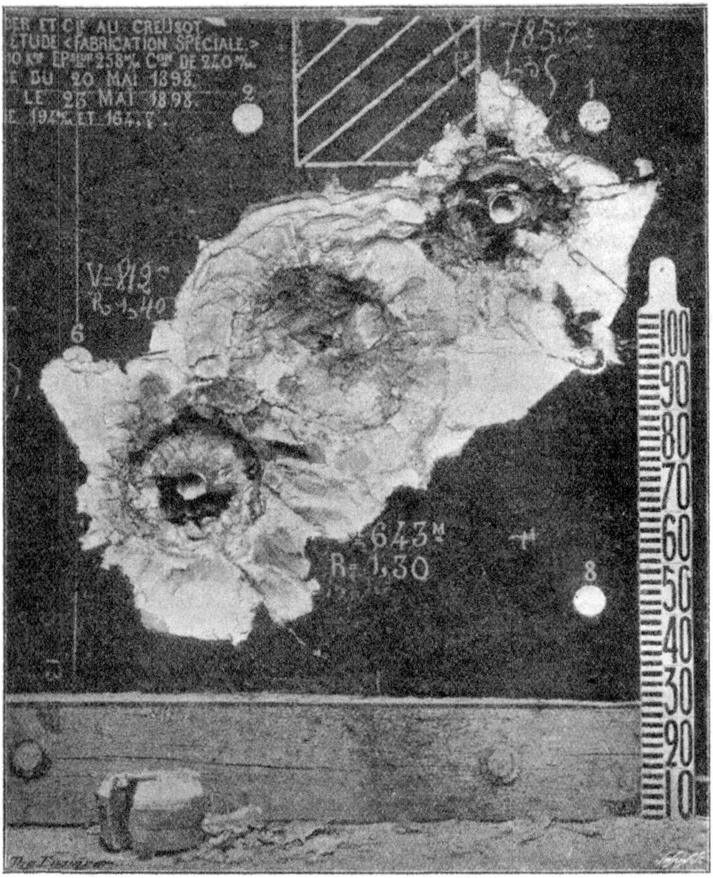

FIG. 7.—250mm. SCHNEIDER ARMOUR PLATE.

round, which was the most severe, was with a striking velocity of 1639 foot-seconds, impinging on the plate at an angle of 15° with the face of the plate, or 75° with the normal. An indentation was made 3 inches wide by 26 inches long and about 1·3 in. deep (see fig. 8). There was a through crack along the bottom of the indentation $19\frac{3}{4}$ inches long, but no other damage was done.

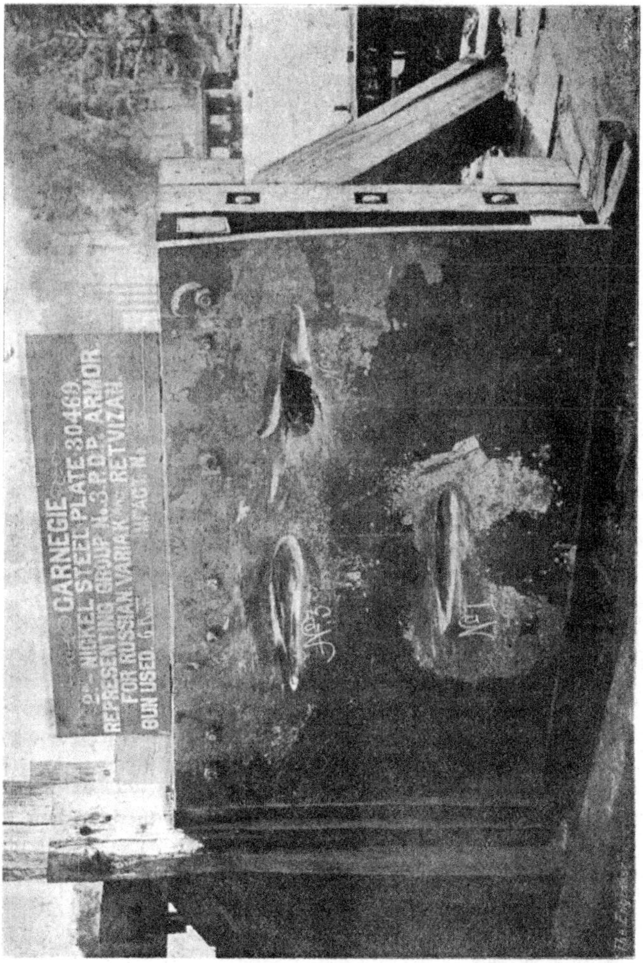

Fig. 8.—Carnegie Plate for Russia.

The Isham Shell.

In August, 1899, an experiment was made in America on the effect of shells charged with high explosives at armour. A projectile termed the Isham shell, weighing 1900 lbs., was employed. It contained over 129 lbs. of dynamite, by which the armour was broken in pieces and shattered. As no details as to dimensions are furnished, conjecture must be resorted to. The shell must have been of the largest calibre. Even the 13-in. gun fires an ordinary shell of 1100 lbs., so that a 1900-lb. projectile must have been a double or torpedo shell. It would be interesting to know the thickness of the plate. That an enormous shell would shatter this plate goes without saying, and a torpedo shell is generally inaccurate in flight and only good for short ranges.

Thorite.

About the same time, a new high explosive called "Thorite" was tried at Sandy Hook, and shown to be safe in discharge and capable of bursting a shell with great violence into very small fragments.

The *Belgique Militaire* states that since 1894 the Belgian Artillery have experimented with a view to discovering the composition of an explosive with a picric acid base closely resembling lyddite. A French engineer has submitted a method of charging shell with fused picric acid, with which trials are proceeding at Braschaet. This simply amounts to the adoption of melinite or lyddite, which are practically identical.

Bullet Proof Shields.

The South African War has called attention to the terrible effect produced by aimed rifle fire with smokeless powder. This may cause increased use of bullet-proof shields. Messrs. Cammell claim special success with plates $\frac{3}{16}$ in. thick, which are proof against the service Lee-Metford bullets at ten yards, while $\frac{1}{8}$ in. thick is proof at 400 yards range. These were experimented on for the Imperial Japanese Navy.

Hadfield Cast Steel A. P. Projectiles.

Messrs. Hadfield, of Sheffield, have succeeded by special methods in making their improved cast steel armour-piercing projectiles so well as to rival those of forged steel. A 4·7 in. cast steel projectile of this type, capped by their methods, perforated a 6 in. Krupp cemented plate at an angle of 20° in April, 1899, the striking energy being only 2290 foot-tons.

Projectiles.

Caps appear to be adopted in America for armour-piercing projectiles, and are said to give increased perforation to the extent of from 15 to 20 per cent. This calculation is, of course, based on results obtained against Harveyed armour. The value of a cap depends on the character of armour attacked, and should be tested with each kind.

PL. I

BRITISH M. L. GUNS

PERFORATIONS – STRIKING DIRECT.

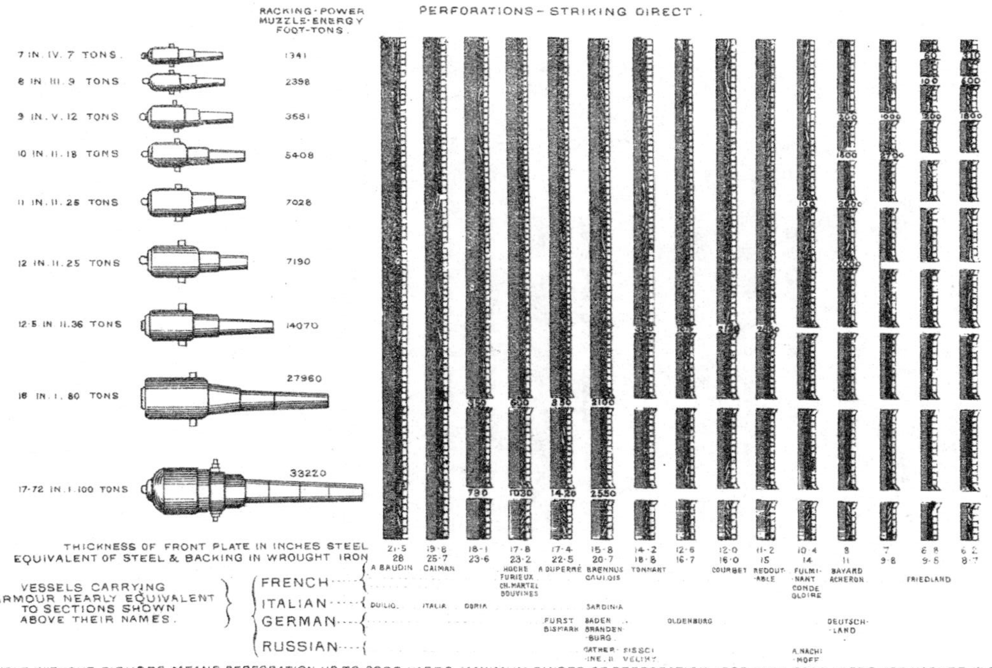

A HOLE WITHOUT FIGURES MEANS PERFORATION UP TO 3000 YARDS. MAXIMUM RANGES OF PERFORATION, LESS THAN 3000 YARDS, ARE MARKED IN FIGURES.

SCALE: 20 FEET TO AN INCH. SHADED GUNS ARE CORRECT IN GENERAL OUTLINE.

ALL BACKING IS 18 IN. OF WOOD WITH 1½ IN SKIN.

PL. II

BRITISH B.L. GUNS.

PERFORATIONS-STRIKING DIRECT.

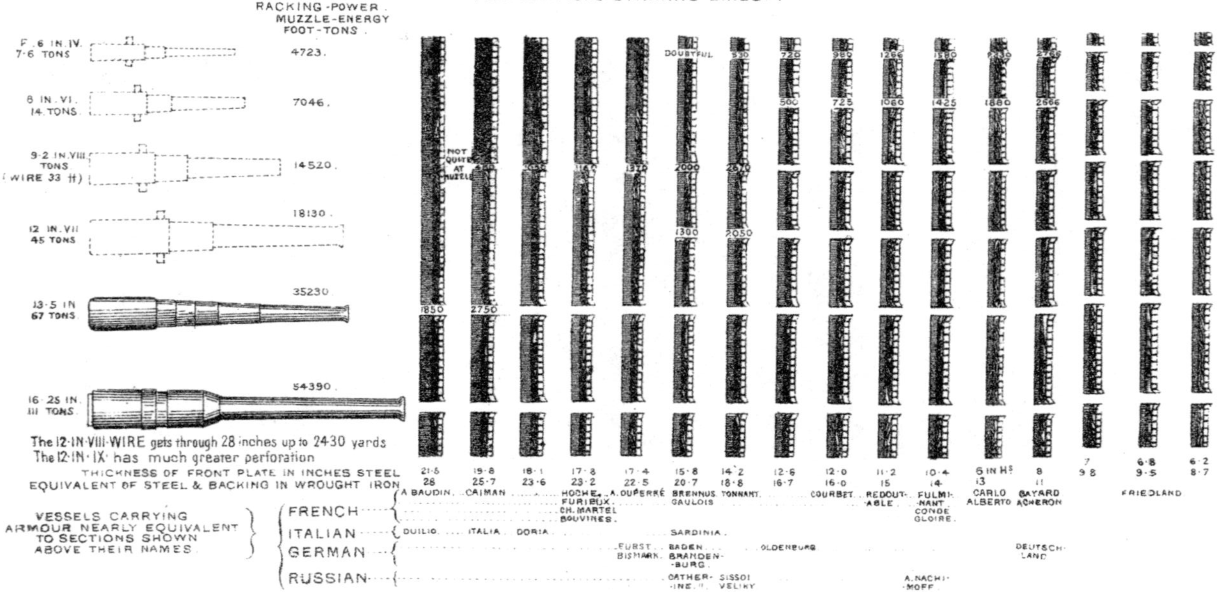

A HOLE WITHOUT FIGURES MEANS PERFORATION UP TO 3000 YARDS. MAXIMUM RANGES OF PERFORATION, LESS THAN 3000 YARDS, ARE MARKED IN FIGURES.

SCALE-·20 FEET TO AN INCH. *SHADED GUNS ARE CORRECT IN GENERAL OUTLINE.*

ALL BACKING IS 18 IN. OF WOOD WITH 1 IN SKIN.*

NOTE THE 29 TON 10 IN. GUN PERFORATES THE 18·8 IRON (TONNANT) SHIELD AT 1800 YARDS AND ALL THINNER SHIELDS AT ALL RANGES

THIS GUN IS TAKEN FROM ELSWICK Q.F. GUN TABLE. *†† PERFORATIONS CALCULATED BY TRESIDDERS-FORMULA*

FRENCH GUNS

PERFORATIONS–STRIKING DIRECT.

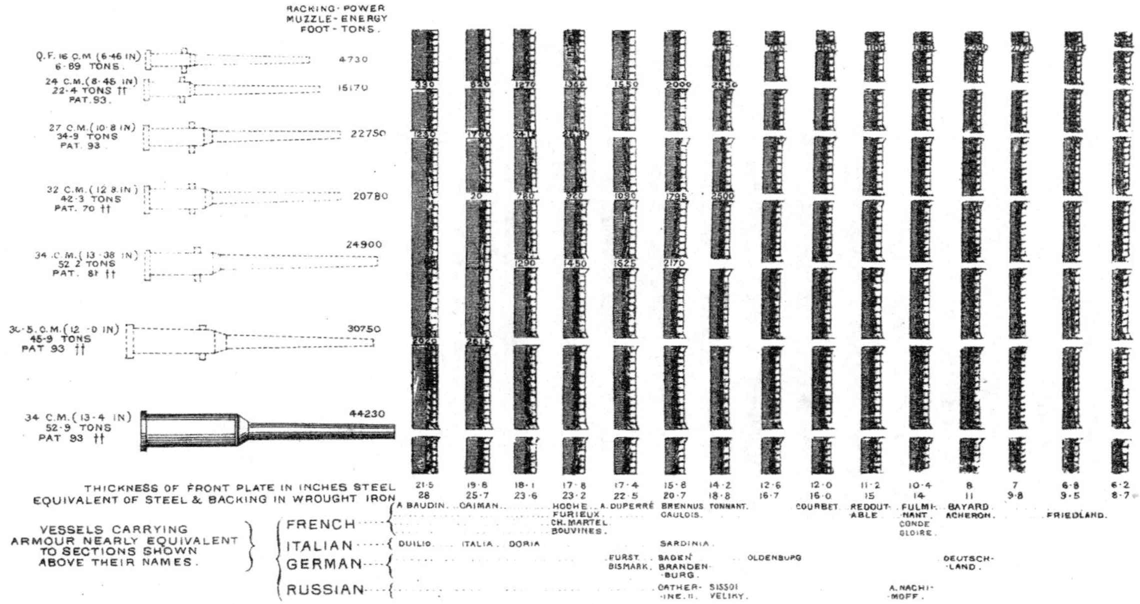

A HOLE WITHOUT FIGURES MEANS PERFORATION UP TO 3000 YARDS. MAXIMUM RANGES OF PERFORATION, LESS THAN 3000 YARDS, ARE MARKED IN FIGURES.

SCALE - 20 FEET TO AN INCH. *SHADED GUNS ARE CORRECT IN GENERAL OUTLINE.*

ALL BACKING IS 18 IN. OF WOOD WITH 1½ IN SKIN.

†† PERFORATIONS CALCULATED BY TRESIDDERS FOMULA.

PL. IV.

GERMAN GUNS.

PERFORATIONS – STRIKING DIRECT.

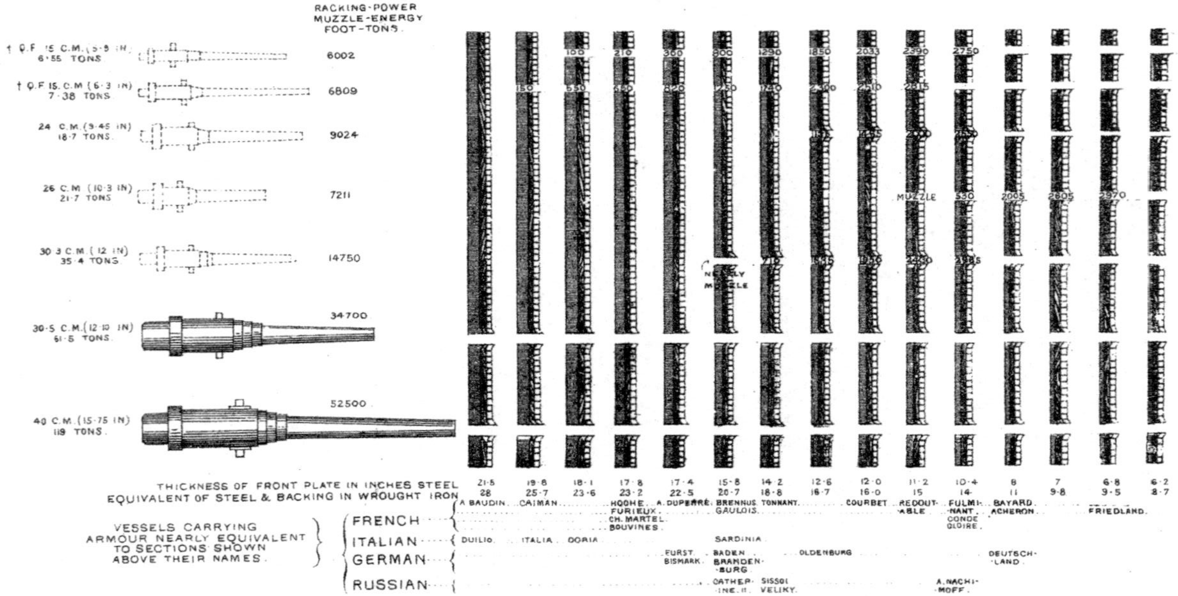

A HOLE WITHOUT FIGURES MEANS PERFORATION UP TO 3000 YARDS. MAXIMUM RANGES OF PERFORATION, LESS THAN 3000 YARDS, ARE MARKED IN FIGURES.

SCALE - 20 FEET TO AN INCH. *SHADED GUNS ARE CORRECT IN GENERAL OUTLINE.*

ALL BACKING IS 18 IN. OF WOOD WITH 1½ IN SKIN.

† TAKEN FROM KRUPP Q.F. GUN TABLE.

KRUPPS HEAVIEST 21 C.M. (8·27 IN.) 20¼ TON GUN PERFORATES 23·6 IN. OF IRON AT ALL RANGES.

CHAPTER II.

ORDNANCE.

Adverse criticism of our Ordnance.

DURING the last few months fierce attacks have been made on our guns in the press. Our field-pieces have been said to be "out-ranged," and to be slow in fire, and our heavy guns have been attacked in the House by Sir C. Dilke, as inferior to those of France; while statements to the same end have been made in the French Chamber of Deputies. The powder question has also been raised. An article written on Ordnance, at the present time, can hardly ignore any of these complaints. The field-guns do not strictly concern the Navy. As, however, guns were landed and used in the campaign in South Africa to meet our want in guns of position, the question will be dealt with briefly, by-and-by. In the first instance, however, attention must be given to the questions which primarily concern the Navy, that is, to the heavy guns and their powder. To begin then with the actual guns themselves. The proportions of guns have been calculated for so many years, that little new can be put forward. The French certainly keep the details of their new guns as secret as possible; nevertheless no uneasiness need be felt where no surprise is conceivable. Guns achieve higher velocity as years go by, because they are made longer and heavier in proportion to their calibre. The chamber may vary in size to suit special charges, but in actual proportions the possibilities are mainly limited to obtaining an increase of velocity if we like to pay for it in the shape of increased length and weight.

The result, however, is sometimes proved to be undesirable. A few years since, very long guns for coast service were advocated by M. Canet. In the table in the *Naval Annual* for 1899, may be seen pieces 80 calibres in length. It will be seen this year that these have disappeared from the tables of Schneider-Canet guns, so that this clearly illustrates the point, namely, that there is no magic in the

Authorities.—The ***Engineer*** for plates and matter. The ***Times***. The Bureau of Ordnance, United States. Sir Andrew Noble's paper, read before the Naval Architects at Newcastle, and another at the Royal Institution last March. Correspondence with Elswick. Lieut. Meigs, late United States Navy, of Bethlehem. Le Moniteur de la Flotte; and the ordinary official publications.

metal structure of a gun; that any good maker can produce a gun of increased power by giving it increased length and weight; but like our own 111-ton gun, it may be a white elephant, whose room is preferable to its company. To say that all metal is the same, is not to speak absolutely correctly; there is a little, although very little, difference between one and another gun steel; and much more, there is a substantial advantage which we claim in the wire or riband construction. This might be sufficient to take seriously into account, were it not that, owing to reasons to be noticed presently, we are at present unable to get the full benefit of it. At all events we may be satisfied that as regards the actual metal, the guns themselves are good, and that no surprises can await us and show us that we need new pieces. This is, at all events, satisfactory, as it means that our main provision in the way of ordnance is sound. The actual performance of the guns, however, depends not alone on the metal of the piece, but on the ammunition employed in it. Let us examine how our guns' performances compare with those of other Powers.

New Naval Guns for the United States.

The United States authorities have recently adopted guns of 40, 45, and 50 calibre of great power. These guns have the very high muzzle velocities of from 2800 to 3000 f.s.,* with projectiles of equal or greater weight than our own. Where the gun is longer than our own, a higher velocity naturally is attainable with an equal charge of powder. With equal lengths of bore the higher velocity requires explanation. For example, our 12-in. wire gun, like the American 12-in. piece, is 40 calibres long, and the projectiles of each weighing 850 lbs. Our British table now shows a muzzle velocity of 2481, and that of the United States 2800, and the energies naturally exhibit a corresponding difference. This might be attributed by a reader to superiority of the United States powder, or to their using a larger charge. At the present time no charge has been furnished, so that the weights cannot be compared. In England we have attained very high velocities with new guns, and the difference here seen is fully to be accounted for by the fact that hitherto the bore has worn so fast that it was thought misleading to give a velocity in our Service Tables which could only be realised with a new gun. For practical purposes a deduction should in some cases be made even on the velocity shown in our own table, if the perforations laid down are to be reckoned on as probable on service with worn guns. Tables supplied for the guns of other Powers must be given here as sent, but it is necessary to keep the conditions under which they are drawn up in mind, as has been pointed out before when velocities have been obviously estimated, and not actually obtained. This is

* *See* Table of United States Guns.

apparent from the fact that they are furnished in the same round number of mètres for a whole group of guns. When converted into British feet, the round number disappears, and is replaced by a precise or odd number of feet-seconds, suggesting the result of careful measurement, but where the same exact figures are repeated for several guns, it may be seen to be an estimate in round numbers converted into British units. What may be definitely said of these American guns of the 1899 type is that they agree in some cases closely with our own, though it is difficult to see how they can be better, but in other cases are longer and more powerful. As yet they have not all been tested, but up to the present time the velocities laid down have been easily attained within the prescribed limits of pressure. How far the velocity can be kept up has yet to be proved. As to the maximum attainable for a single round, our authorities would no doubt expect our *wire* guns to bear safely a higher strain than any piece otherwise constructed, but the difficulty is not *maximum pressure*, but *wear of bore*, so that a greater margin of safety remains the only advantage of the wire construction. If the United States guns keep up their high velocities without exposing the bore to intolerable wear, it would naturally appear as if ballistic qualities were much better realised by their powder than by ours. Nevertheless, it seems that the United States officers are not by any means satisfied, and it may be questioned if any powder at present is satisfactory. This seems curious in the face of the great results that have been achieved by powder during recent years. The fact is that while great powers have been exhibited, they have been accompanied by great evils. Enormous energy has been developed with reasonable pressure, but such terrible erosion has been produced on the bore that it may well be questioned if the heavier guns could discharge the ammunition supplied to them for service before their shooting had become seriously affected.

General Condition of the Powder Question.

The powder question stands generally in the following way: All powers have adopted powder in a great measure free from smoke, and the foundation of all such powder is gun-cotton. This substance, however, requires to be combined with some oxydising agent to act satisfactorily, and the most effectual one is nitro-glycerine. Unfortunately, nitro-glycerine is said specially to injure the bore, and America and Russia and France have hitherto endeavoured to use nitrate of barita or some other substitute, with the result that the powder is only semi-smokeless, and the products of combustion not being all gas, demand energy to get them out of the bore. Germany has a powder with perhaps about twenty-five per cent of nitro-glycerine, and some other continental nations have, more or less, about

Cordite. the same proportion. England adopted cordite with the enormous proportion of about 58·3 per cent. of nitro-glycerine. There was much to recommend cordite. It offers mechanical advantages; it bore long series of tests as to climatic influence very well, and this is of special importance to England. In smaller calibres the erosion is not very serious, and it was said that even in large bores it was at all events very symmetrical and even, and could be remedied by the use of augmenting strips of copper or gas checks.

The statement is made, and it appears to be true, that no nation, except perhaps the United States, fires its guns much in peace time; consequently it is argued that guns are not forcing this sad fact of wear on their owners, simply because wear has no opportunity of taking place. Nevertheless, it is surely mad to assume that because we experience an evil, our neighbours, whose conditions differ substantially from ours, must necessarily suffer the same evil and to the same extent. All we know is that they do not own it. Let us consider how we stand in the matter of powder and wear of bore. The wear has been classified under two heads—"erosion" and "wash." Erosion is the eating out of the surface of the bore by the charge pure and simple. Wash is the rush of gas between two surfaces, that is, between projectile and bore whenever a space is found. Erosion is due to gas moving rapidly at a very high temperature, and, of course, under pressure, although the latter need not be high. Thus, erosion is not found to be serious in the powder chamber, because, although the temperature is high, the gas is not in rapid motion. It is not serious in the forward part of the bore, because, although in rapid motion, the temperature has become considerably reduced. About the seat of the shot and a few calibres beyond it the action is—in truth—frightful. Wash can be remedied to a great extent by means of tight-fitting gas checks, and, as the bore gets enlarged, by augmenting strips of copper. The evil is serious enough, but erosion is the parent of wash, and erosion is the evil to be specially grappled with, and since we use powder of a very different character from other nations—that is cordite—surely we ought to put cordite on its trial again, not necessarily as to its behaviour in small pieces, but certainly it should be brought up on the charge of wearing out the bores of our heavy guns in an unprecedented manner. If the same thing is going on with our neighbours' powder, we may consider it eventually a necessary evil besetting the use of smokeless powder to obtain very high velocities; but we ought to make sure that this is so before we "sit down" under such a conclusion.

Doubtless it is quite possible that cordite is no worse than its rivals. Sir Andrew Noble, whose opinion should carry more weight than that of any one in this country, has always thought well of cordite, and manufacturers seem to like it. Sir Andrew in a paper read at the Royal Institution on March 23rd, said that brown powder would produce as much erosion as cordite if made to develop the same energy. Sir Andrew, however, did not speak of foreign new smokeless powders, and the following conclusion still appears to hold good. We know that our heavy guns wear so intolerably fast that we cannot reckon on the velocities we assign to them for any considerable number of rounds. We know that their shooting falls off rapidly. We use charges that bear no relation to the strength of our wire guns, at all events in many cases. We know also that other nations will not contemplate cordite, but use powders of widely different character, and do not complain of wear of bore to the same extent as ourselves. Surely therefore we ought to test our cordite carefully in comparison with other powders, especially having our heavy guns in view—guns in which, it may be observed, the absence of smoke is of less and less importance as the gun increases in size and its rate of firing becomes less rapid.

Spontaneous Combustion of Cordite.

An article evidently written by an expert, appeared in the *Times* of March 28th last, referring to the spontaneous combustion of cordite having occurred on board some man of war. In answer to a question by Sir C. Cameron, the First Lord of the Admiralty subsequently stated that this could not be attributed to hurried manufacture arising from the present demand, for the cordite had been made in 1894. So serious an accident makes it the more necessary to test cordite in comparison with its rivals to make sure that we have the best smokeless powder.

Lieut. Meigs on New Type Guns.

With regard to the best type of gun, Lieut. Meigs, late United States Navy, now of Bethlehem, has raised the question whether we have not gone too far in the development of small bore and length. On paper, the results are very good, even taking into account the great weight entailed by the increased length. But high velocity is a costly form of energy, especially in wear of bore, and it is urged that the increased size of the projectile, with a powerful bursting charge, which could be obtained from a piece of larger bore of the same weight, would offer advantages. Lieut. Meigs has been recently engaged in making a gun to order, of a type strongly contrasting with that now in fashion. The weight of the gun is, however, considerably in excess of that of the long 12-in. piece, the calibre is 18 in., and the projectile weighs 2000 lb. A muzzle velocity of 2000 f.s. was hoped for, though as yet one of 1800 f.s. only has been realised. This implies an energy of 44,920 foot-tons, which is second only to the 54,390 of our 111 ton gun. This being obtained from a gun weighing 59 tons, means 761 tons per ton of gun. The muzzle penetration would be lessened, namely 30·5-in. of iron, but belt attack is now so precarious with the increased resistance of armour that it seems quite possible that this gun might deliver a more terrible attack on the weaker parts of ships with its enormous bursting charge. The disadvantages would be the stowing and handling of such ponderous projectiles. It remains to be seen how far the gun realises further expectations, but it illustrates the view suggested by Lieut. Meigs. Some time since, a large-bore gun was proposed to our naval authorities and did not find acceptance. The fact is that the value of a gun now wholly depends on the work which it is called on to perform. As noticed in the chapter on armour with 6-in. plate coming in for broadside batteries, it is necessary to have a gun capable of perforating 6 in. of the best steel armour; anything less is almost useless. For our broadside Q.-F. armaments then a gun of great perforating power and moderate weight of projectile is absolutely necessary. This is, probably, the most important gun afloat. Other cases may arise when different conditions obtain. Perhaps the heaviest guns carried by cruisers might be of another type, they seldom are called upon to attack very thick armour, so that it is conceivable that pieces discharging large shells might be good for them.

Function of a Broadside Gun.

M. Claudinon on French, British and other Foreign Guns.

In March last a speech on French and foreign guns and ships, especially British, was made in the French Chamber of Deputies by M. Claudinon "Forge Master" of the Loire factories. M. Claudinon, has, as a maker of guns and war stores, the knowledge of a specialist. On the other hand, he is interested in defending material to which he has largely contributed. M. Claudinon stated that he quoted his figures from the *Naval Annual;* but this must be understood with reserve. He makes a comparison between certain British and American guns, taken from the *Annual* for 1899, with some Russian and French guns of newer design. He shows thus that of the 12-in. guns, the French stand first, with 12,200 metre tons (39,395 foot-tons); next comes the Russian, with 10,700 metre-tons (34,551 foot-tons); the British, with 10,600 metre-tons (or 34,228 foot-tons); and, lastly, the American, with 8100 metre-tons (26,155 foot-tons). He states that this is not the whole case, for the velocity has had to be reduced in the British gun. Indeed, according to M. Claudinon, our guns must be in a bad way, for, out of sixteen wire guns, three burst, and ten others were unable to continue firing. He says that it has been asserted that, owing to the lightness of the French projectile, the superiority at the muzzle would soon be lost; but this, he says, is disproved by two things; first, that the speed of a shell from a 305 mm. gun, at a distance of 6000 metres, was 563 metres, while that of a corresponding English shell was 505 metres*; and, again, that a French shell, of the 1893–1896 model, discharged at an angle of 20 deg., would, at 17,350 metres, still have a speed of 310 metres, which would enable it to pierce the deck of any English cruiser. Before passing on to M. Claudinon's application of his conclusions to the case of ships, it is well to deal with the above, which is too serious to let pass. M. Claudinon's statements as to the French and Russian guns may be accepted, and these guns will be found mentioned in the French and Russian Tables. It happens, however, also that the United States have now in hand a new 12-in. gun, giving, it is claimed, 2800 foot-seconds muzzle-velocity, with a projectile of the same weight as the British, namely, 850 lb., implying a muzzle energy of 46,246 foot-tons. Before this the French 39,395 foot-tons pales even at the muzzle, and every 100 yards range tells against the lighter French projectile. This speaks for itself so far as America is concerned. Both guns are new, and their figures may be equally authentic; at all events, there is nothing more to add except that such statements will have to be made good by results. We are, however, rather concerned with England's position, and, unfortunately, it will be seen that the velocity on the British Table for this year has been reduced to 2481 foot-seconds at 60 deg., or 2526 at 80 deg. temperature. Probably, however, 2600 foot-seconds, originally estimated, might be assigned as truly and fairly as the velocities which are generally given in tables which, in fact, are only estimated, or at best apply to new pieces; but the energy quoted by M. Claudinon, given in the *Annual*, is not that due to 2600, but to the figure of the earlier Mark VIII. gun, namely, 2367, and our newest gun should be compared with that of France. With 2526† foot-seconds, our Mark IX. gun has 37,600 foot-ton energy, that is to say, less than this new French gun at the muzzle, but our relative position improves with every 100 yards of range, owing to the lighter weight of the French projectile. M. Claudinon does not give the weight of this French projectile, but, as he speaks of a type of shot of a few years back, *i.e.*, 1893 to 1896, it is probably right to take it as 643·8 lb., as given in the Brassey and the Pola annuals. This means that he takes the muzzle velocity at 2959 foot-seconds, which, with this light shot, is conceivable, thus giving his 39,395 foot-tons muzzle energy. The velocity, however, comes down to

* It happens that with these velocities the British shot has the greater energy, viz., 16,180 foot-tons, as compared with 15,230 foot-tons, taking the weights as explained in the text.

† At 80° Fahrt. See note on Table.

2683 foot-seconds at 1000 yards, with a striking energy of 32,130 foot-tons, while the British shot at this range has 2353 foot-seconds velocity and 32,640 foot-tons energy. In other words, the inferiority at the muzzle has changed into superiority to the French blow even at 1000 yards.

But to pass on to the startling facts which M. Claudinon reports. We have accepted his figures in the case of France and Russia, but we must be excused for demurring to his statements as to British guns. Three wire guns could not burst without their possessors being aware of the fact, and it is not possible to detect any sign of bursting in any of our wire guns; neither have we been unable to continue firing. The beginning and ending of our causes of complaint with our guns is that the bores wear out much too fast. The new American gun is almost identical with our own; its proportions might, indeed, be taken from ours. It is rather heavier, but ours being of wire construction, it is probable that our gun would do all that the American would do with the same powder, and this is, as above said, much more than the French gun. The fact is that these high velocities, which are to be attained only with new guns, seem to be deceptive, and tables are not required merely for purposes of "bluff," or even of fair comparison of the possibilities of new pieces, but for service purposes, and very commonly the heavy guns on board foreign men-of-war remain year after year without firing full charges and shot, while ours have their annual sea practice. There remains the possibility that cordite may be putting our heavy guns at a disadvantage, and that it is desirable to make comparative trials with other powder. Sir Andrew Noble's paper read at the Royal Institution, as said above, did not deal with comparisons between our own and new foreign powders. Till that is done it is difficult to pronounce as to powder. As to the guns themselves we may be quite satisfied.

M. Claudinon follows our guns on board ship, and tells us that in our gunnery trials with the Mars, Resolution, Hannibal, and Jupiter we attempted to get a muzzle velocity of 731 metres (2400 foot-seconds), but had to abandon it for his alleged very sufficient reason, above referred to, namely, that three guns burst and ten had to discontinue firing. These lamentable accidents did not, therefore, occur in the dark places of Woolwich or Shoeburyness, where evils of this dye might, as he doubtless thinks, be carefully concealed, but in H.M. ships in broad daylight on the high seas. Is it possible for M. Claudinon to persuade himself and his listeners that three 12-in. wire guns could thus burst on board our new ships and the matter be kept dark until he told the fell secret to the Chamber? He adds also that only some of the results were made public. Does he mean that still worse remains to be told? We made fuss enough about the Thunderer accident, but that seems nothing to what now takes place without our even hearing of it.

M. Claudinon's Errors.

To pass on to ships. "The English," he said, "did not deem it necessary to protect battleships against high explosive shells, from which it could only be inferred that they were not satisfied with the results of their trials, and did not think that the shells would be dangerous except to those who handled them." This is an extraordinary statement. Any one who looks at drawings of British and French battleships will see that for the sake of maintaining a thick complete belt the French ships have had to leave many portions of their hulls unprotected. The British, by stopping the belt a little short of bow and stern, are enabled to be covered pretty completely upwards from belt to battery above. We made trials of shell fire at the Resistance, and adopted casemates and defences specially to keep it out. It has been said in this country that the French courted destruction by common shell, and this opposite statement seems inexplicable. Then how can we be supposed to be afraid to handle and use shells with high explosives? Does M. Claudinon never read the papers? Has he not seen that we are using lyddite shells in the South African war, and does he not know that lyddite is almost identical with melinite? It is hardly necessary to follow the deputy further. He makes an extraordinary supposition of the Powerful cruiser engaging the Charlemagne, and shows that the battleship ought to destroy this cruiser, which, considering that she is not even an armoured, but only a protected vessel, is not wonderful. The Powerful is constructed for entirely different work, and it is hardly necessary to examine this most improbable engagement, or to take M. Claudinon so seriously as to be unhappy about our ships and guns. His ability as a speaker may have tempted him to make statements which it would have been wise to question, for his ability as a specialist might well have suggested doubt as to their reliability.

Guns in Ladysmith.

Although not a question of naval warfare, the service rendered by our Navy in gunnery in the defence of Ladysmith deserves notice. We were taken greatly at a disadvantage in the fact that the forces in Ladysmith were called upon to sustain a regular siege in which fairly heavy siege guns were brought up against us, while we had for our defence only field pieces. Ordinarily, the besieged have, if not heavier, at least as heavy metal as the besiegers. In this instance, however, we were painfully deficient as compared with the enemy, from the fact that the Boers had long been preparing, and we had done nothing except despatch some troops hastily with the ordinary accompaniment of field guns. It is, however, a noteworthy fact that our naval officers, looking beyond their immediate and ordinary duties, devised a way of bringing up two comparatively heavy quick-fire guns and some 12-pounders, and further that the idea was put into action so rapidly that the guns were got into Ladysmith before it was invested. Special credit is due to Captain Scott for his ready adaptation of the quick-firing mounting to such a base as enabled the 4·7-in. gun to be utilised in defence of earthworks, it is stated, by simply bolting it on to four lengths of pile, as also for devising a wheel and trail mounting both for the 4·7-in. and 6-in. guns. At the same time, the 12-pounder 12 cwt. Q.-F. gun was also converted into a gun of position on a travelling carriage, by using a balk of timber for a trail attached to two wagon wheels. The value of the 4·7-in. piece was obvious; it discharges a 45 lb. projectile with a muzzle velocity of 2188 foot-seconds, and is a rather more powerful and accurate gun than the Boers possessed, their siege pieces being 40-pounders. This gun was therefore extremely valuable in dismounting and silencing the Boer guns at all ranges where accuracy was possible, though it is obvious that the besieger may elect to fire with safety at the fortress or town from a distance at which accuracy is out of the question, seeing that a fortress or town is a larger mark than a gun emplacement.

Naval Guns in South Africa.

The particular value of the 12-pounder gun is less immediately apparent, seeing that in Ladysmith we already had several field batteries with the 15-pounder gun. It was no doubt a good and useful thing to add to their number, but the 12-pounder naval gun possesses certain distinct advantages over our field guns for siege work. The field battery and horse artillery guns are pieces of the same calibre as the naval 12-pounder, that is 3-in., but the field battery gun weighs only 7 cwt. and the horse artillery gun 6 cwt., while the naval gun weighs 12 cwt., their muzzle velocities being respectively 1574ft., 1553ft., and 2210ft. per second. Thus the naval piece, while too heavy for the work for which the field artillery exists, has the advantage of greater energy and range and of a flatter trajectory, added to which, it is supplied with common shell instead of shrapnel only. In short, it is more of a siege gun than the others. The 15lb. shell, of course, possessing increased length and weight, gains on the naval 12lb. projectile as the range increases, but certainly does not overtake it at any range where accuracy is attainable.

Shrapnel *versus* Common Shells.

The question of the relative powers of shrapnel and common shell, to which attention has been called by the South African War,

From the "Navy and Army Illustrated."

THE NAVAL 4·7-IN. GUN MOUNTED FOR LAND SERVICE.

The carriage is the later form designed by Capt Percy Scott, R.N., and was manufactured in the railway works at Durban. The upper picture shows the extreme angle of elevation.

concerns land operations chiefly, but is not without interest to Naval artillery, both because Naval gunners and guns have taken a distinguished part in the war, and also because ships and boats may fire shrapnel against troops on land, while common shell are the safest projectile to fire against ships in case of doubt in any Naval engagement. In England, as in Germany, shrapnel with time fuses has been found the projectile giving the best results when used with skill on open ground. On hard flat ground, so destructive is the bullet-shower when skilfully delivered, that troops would indeed be ill-advised to remain exposed in close formation within the range of shrapnel. On the other hand it is impossible to effect much against men in trenches. Shrapnel is usually thus employed by bursting it about fifty yards short, so as to deliver showers of bullets along the crest of the parapet. The chief effect of this is to make men keep their heads down under cover, because if they have not a large supply of sandbags to make loopholes, they continue their fire under difficulty. By this means, then, troops may be enabled to advance for some distance, but as soon as the attacking party get near the works shrapnel fire must cease, for one shrapnel shell burst by accident on the exposed backs of the attacking force might kill more men than many hours of firing against the enemy under cover. The cessation of the shrapnel is the signal for the defenders to pour in the fire of their magazine rifles, and this with smokeless powder can be continued with fatal precision and speed until the attack reaches the actual parapet, a feat which is consequently often impossible of attainment. Common shells are more likely to strike men under cover than shrapnel. If burst just over the parapet, the fragments fly in all directions, and some, though not many, come back at the unprotected backs of the defenders. When burst by lyddite or melinite, the number of fragments is much greater than with powder, and such shells are proportionately more formidable. Nevertheless, the results against men distributed along a trench, must at best be poor, and the crest is not swept so well with common shell as with shrapnel; indeed, the men are hardly more exposed firing over the parapet than lying still, so that the enemy's fire may be less hindered, and the attacking party would be less efficiently protected by common than by shrapnel shell.

Russian Siege of Plevna.

How difficult it is to injure men in earthworks by artillery fire may be seen from the fact that at Plevna the Turks successfully defied all Russian attacks for many months, although for the last three months Todleben had employed against them 508 field and 45 siege guns, and 30,000 Russians had been killed or wounded in the early assaults before he took command. Plevna eventually was reduced by having been surrounded and starved into surrender. At this time, it is to be observed, smokeless powder had not come in, and the Turks used ordinary breechloading rifles, except in repelling assaults, when each man took up a magazine arm lying ready to hand, and discharged its fire at random through the smoke, as a rule, killing more men at a distance than near. The actual quantity of bullets, however, pouring through the air rendered success impossible; yet this fire cannot be compared with that of the Boers, which is well directed throughout.

It will be seen, then, that both common and shrapnel shell-fire produce but little effect against troops in earthworks, who may finally only be overcome by counter trenching, surrounding, and starving out. The difference in effect between the two kinds of projectiles is that common shells, especially from the curved fire of howitzers, search behind cover, and may kill many men if crowded in a work, while shrapnel keep down an enemy's fire better in the earlier part of an attack. Against batteries common shells, however, possess the great advantage of being suited to dismount or destroy the actual gun and carriage attacked.

Smoke Shells.

It has been suggested that some shrapnel might have their bullets taken out and replaced by pellets of smoke-giving composition, and that when shrapnel fire ceases in an attack, these smoke shells might be fired. The risk to the attacking party would be practically nothing, seeing that shrapnel do not break into fragments, but project their contents forward, and it is urged that defenders, just when they expected to pour in their fire, would be discouraged by smoke pellets discharged in their faces, hiding everything and masking troops advancing with the bayonet. The success of this would depend on the quantity of thick smoke thus capable of being generated, and also in some measure on wind and atmospheric conditions, but it is suggested that it is worth testing at home. Something of this kind seems likely to come in sooner or later for the special task of carrying earthworks by assault. It may be seen that an ordinary smoke ball would not answer the purpose, for, fired horizontally, it would ricochet, and vertically it would be difficult to drop it where desired. Moreover, the smoke would be too much concentrated on one spot, whereas pellets, liberated in succession to shrapnel bullets, would fall soon after liberation about the required place, and at once form smoke from many points on the ground, which would be but little affected by wind. They would only burn for a short time, but the guns supporting the attack could probably fire a considerable number of such shells after shrapnel became unsafe.

A paper read by Sir Andrew Noble, last summer, before the

Naval Architects at Newcastle, gives valuable examples of modern Naval Mountings, from which the following are taken:— SirAndrew Noble on Gun Mounting.

Figs. 1 and 2 show an 8-in. central-pivot mounting for swift cruisers. The man laying the gun can look over the top of the shield, thus commanding a good field of view, his head being protected by a hood I in Fig. 2. The hand rammer D and the sights and various parts are seen in Figs 1 and 2. Electric and auxiliary hand training gear is provided, either of which can be applied at once, should the other be disabled. The elevating gear is worked entirely by hand, the trunnions being mounted on Mr. Brankston's anti-friction arrangement, with knife edges supported on springs to relieve the shock when the gun is fired. It may be readily seen, that if a trunnion rests on a knife edge which is kept slightly projecting above a broad plate by a spring, under any violent pressure produced by firing or other force, the knife edge gives back and leaves the broad bearing to take the shock. So easily does this gear work that one man can elevate or depress the gun at the rate of 2 deg. per second. With the hand training gear, one man can train the mounting through 60 deg. in 25 seconds, and with the electric gear through 180 deg. in 30 seconds. The shield is 4½ in. thick, and is supported on elastic stays in the usual manner. The powder supply is brought up the centre, and is delivered at the side under cover of the shield. The axial hoist for this purpose is shown in Fig. 3, and is so arranged that when one charge is going up, the empty case is going down, thus effecting a great saving of time and labour, as the weight of the two cases balance each other, and there is thus only the actual weight of the charge to lift. Fig 3 shows the position for the cases to be recharged at I. This would come into the vertical position as the door J, on which it rests is closed, and passes up, eventually taking the position shown at A at the top. Four rounds have been fired in a minute from this gun.

In 1889, Mr. Vavasseur with Sir A Noble, submitted to the Admiralty the design of a mounting so arranged that the gun could be fired at all elevations up to 35 deg. or 40 deg., the firm having been requested by a foreign Government to consider whether or not such a gun was feasible. Cases may arise, such as the passage of the Dardanelles, once carried out, and a few years since seriously contemplated, when guns on an ordinary mounting could not be brought to bear on the coast batteries of high command. The Naval authorities were much pleased with this design, but as the arrangement was altogether novel, it was not unreasonably stipulated that, before it could be introduced into the service, its success must be proved by an experimental mounting being made, and by passing a satis-

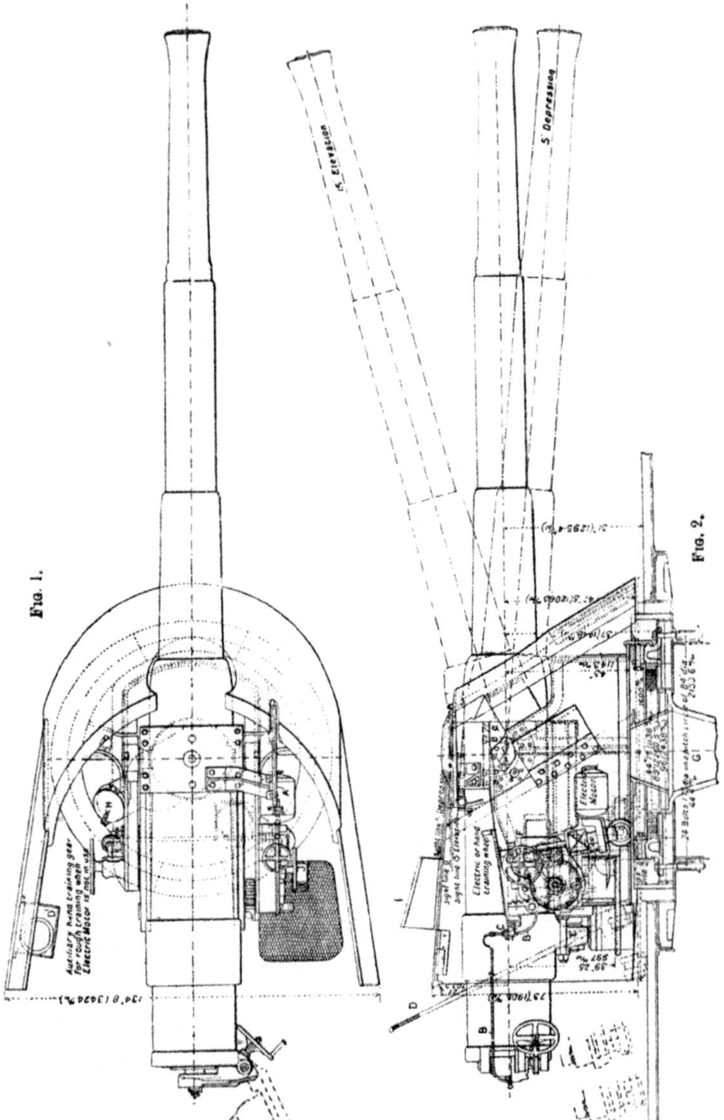
15° Elevation
5° Depression
Auxiliary hand training gear for rough training when Electric Motor is not in use
Electric Motor
Electric or hand training wheel

FIG. 1.

FIG. 2.

FIG. 3.

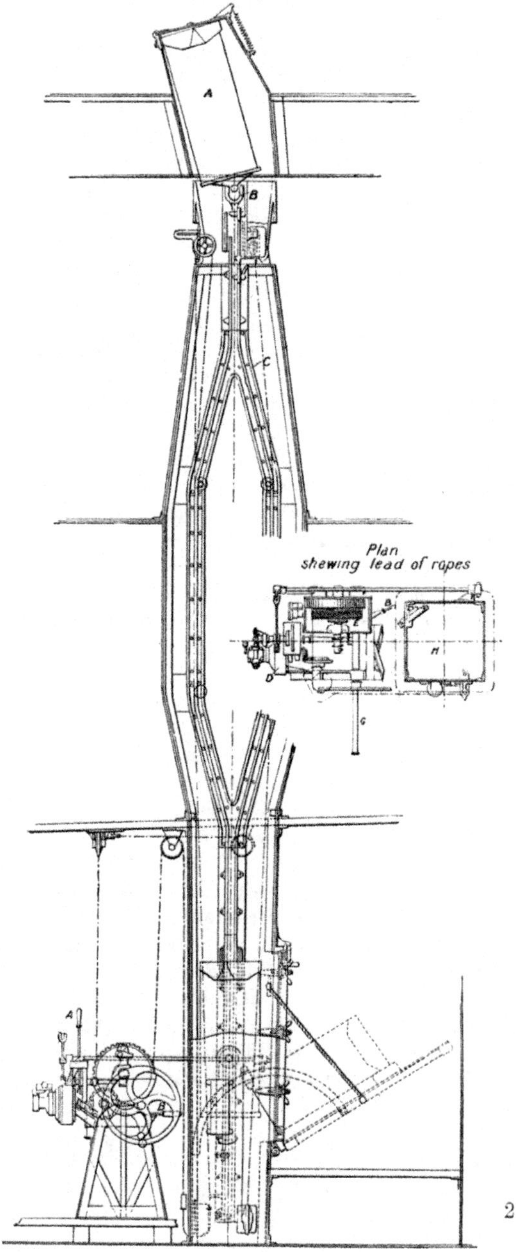

factory trial. Acting on this stipulation, Elswick made a high angle-mounting for a 9·2-in. gun, and, fitted up in the Handy, it passed a most satisfactory firing trial in April, 1890.

The total weight of this gun and mounting was 54 tons, and it could be trained easily by hand power. The range of elevation extended from 5 deg. depression to 40 deg. elevation, and an arc of 45 deg. could be traversed in thirty seconds by one man. At the trial the gun was fired at angles varying from 5 deg. depression to 39 deg. elevation, and the result was remarkable. The range of three of the rounds at 39 deg. was estimated to be about ten miles, but the shot could not be seen to strike the water.

In this mounting, the slide was horizontal, and the carriage was of the Vavasseur type, the recoil press and carriage being in one piece of forged steel; the gun consequently did not recoil in the line of fire, that is, parallel to the axis of the bore, but horizontally, and was returned to the firing position by means of springs, the force of the springs being regulated by means of a controlling ram in the recoil press. An illustration of this type of mounting is shown in Fig. 4, the position of the gun at 35 deg. elevation being indicated. The cordite charge case I, with its ball joint J, is shown at the top of the lift. A considerable number of vessels, chiefly in foreign navies, are fitted with this form of mounting, for example, the Italian ships, Dandolo, St. Bon, and Emanuele Filiberto. The sighting arrangements being on a line between the two guns, are not interfered with by the elevation of the guns' muzzles.

Gun Mountings for Italian Battle-ships.

Passing over the 6-in. Q.-F. mountings, and others very well known, the following are examples of mountings for heavier pieces:

The last five ships of the Majestic class might have been fitted with all-round main loading positions for their 12-in. guns, had it not been that the oval barbette frames were well advanced, and it was feared that any alteration of design might cause delay. The Canopus mounting designed at Openshaw was therefore the first for the English Admiralty to be carried out with all-round loading. In this is a roomy working chamber below the gun turn-table, into which the powder and projectiles are brought by suitable central hoists. These hoists are fixed in relation to the ship. Hydraulic cranes are fitted to transfer the shot to the gun hoists. In the working chamber there are shell bins holding twenty-four rounds per gun, which are commanded by the hydraulic cranes, and these shell bins could all be exhausted first, and be replenished by the central hoists at leisure.

Canopus and Alternative Mountings

There is no doubt some disadvantage in having two sets of hoists, the central ones and those behind the guns, as it involves an

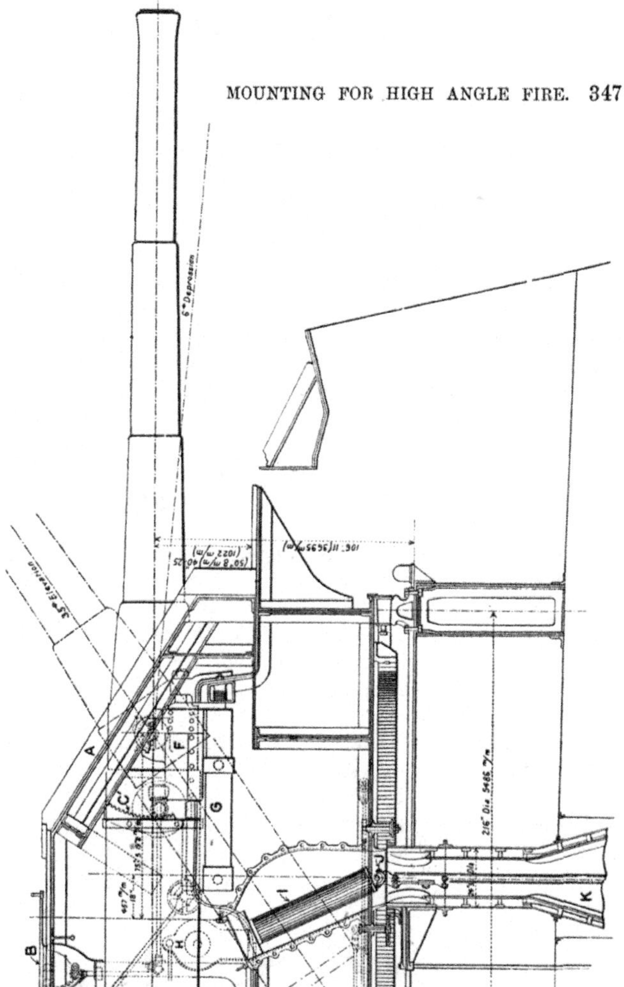

FIG. 4.
SECTIONAL ELEVATION

additional set of operations to transfer the charges from one to the other, but against this must be set the fact that, having a large store of projectiles immediately under the gun turntable, an ordinary action might be fought before this store would be used up, so that the central hoists might never be required for use during the action, and this store might be so increased as to render the central hoists unnecessary. Fig. 5 shows a design which is advocated by Mr. Watts, the chief constructor at Elswick, to effect the bringing-up of projectiles directly from the bottom of the ship to the gun.

This mounting is fitted with a pair of main hoists, each carrying a projectile and powder charge from the bottom of the ship to the rear of the guns. It has also as an alternative a pair of shot hoists—shown in Fig. 5—reaching from the bottom of the ship to the rear of the gun shield, and arranged to deliver or pick up shot from the working chamber.

The alternative shot hoists work either by hydraulic power or hand. In Fig. 5, L is the hydraulic cylinder, and M the hand gear. The projectile is brought up end on, as seen at K. Also there are a pair of powder hoists G, reaching from the bottom of the ship to a platform placed between the two guns; these work by hand only. The powder is brought up in bronze cases, H. In addition it is provided that either shot or powder can be hoisted from below to the working chamber by hand tackle as a last resource. Besides all this three rounds per gun are stowed in the gun house and eight in the working chamber.

The weight of gear for transporting shell in the shell room and also of the shell hoists and their gear, is about 54 tons per ship, or half the weight of the projectiles stowed in the shell rooms. In order to charge the main hoists, a revolving platform is provided in the shell room, having on each side trays for carrying a couple of projectiles. This revolving platform is first locked in one particular position to the ship, and shell are placed in the trays by overhead tackle in the shell room. The platform is then unlocked and moved to whatever position is necessary to bring the shot trays opposite the hoist doors. It is then locked to the hoist trunk until the shot are required to be passed into the hoist cages.

To manage this heavy platform in a seaway it has been thought necessary to revolve it by hydraulic engines. These platforms, and the gear for working them, have a total weight per ship of nine tons.

Mounting for 12-in. gun in the Mikasa.

In the 12-in. mounting shown in Fig. 6 for the Japanese ship Mikasa, the outer casing of the hoists is built water-tight at the middle, lower, and platform decks, which are therefore strengthened and bound together. The interior and bottom portions

Fig. 5.

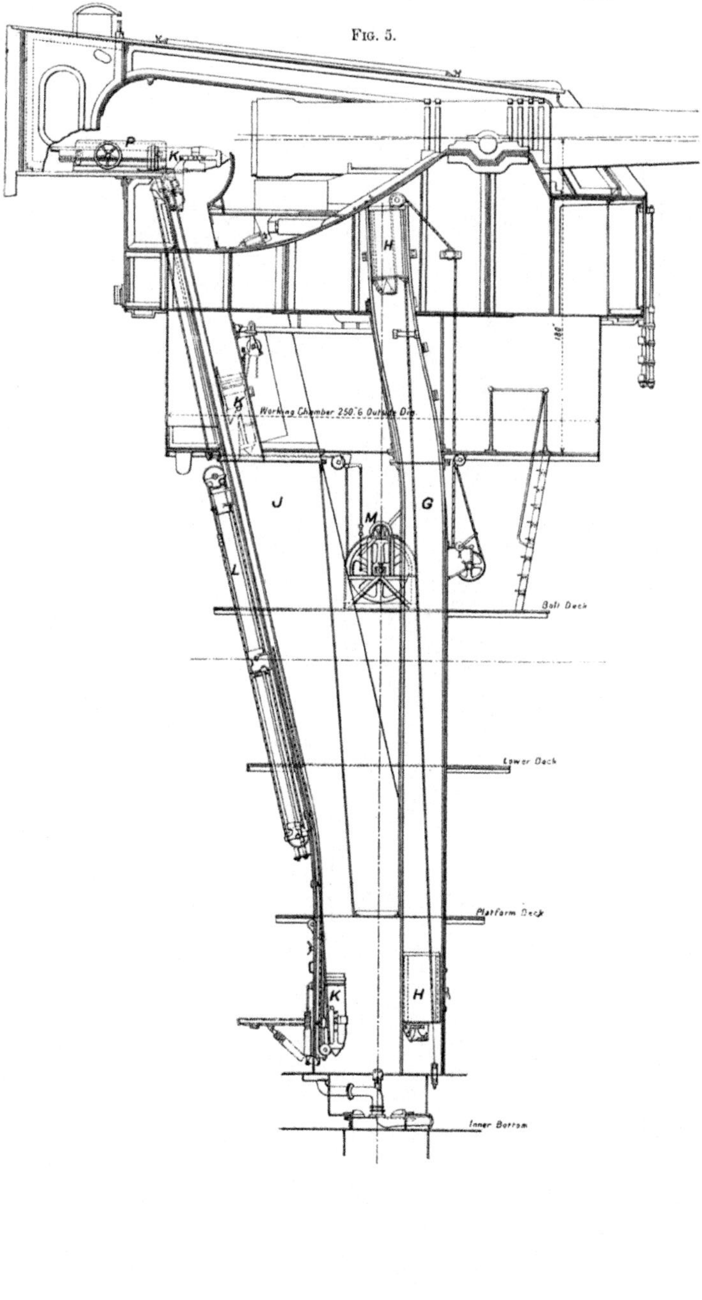
P
K
H
K
Working Chamber 250"6 Outside Dia
18'6"
J
G
M
L
Bolt Deck
Lower Deck
Platform Deck
K
H
Inner Bottom

Fig. 6.

SECTIONAL ELEVATION

40 Calibre Guns

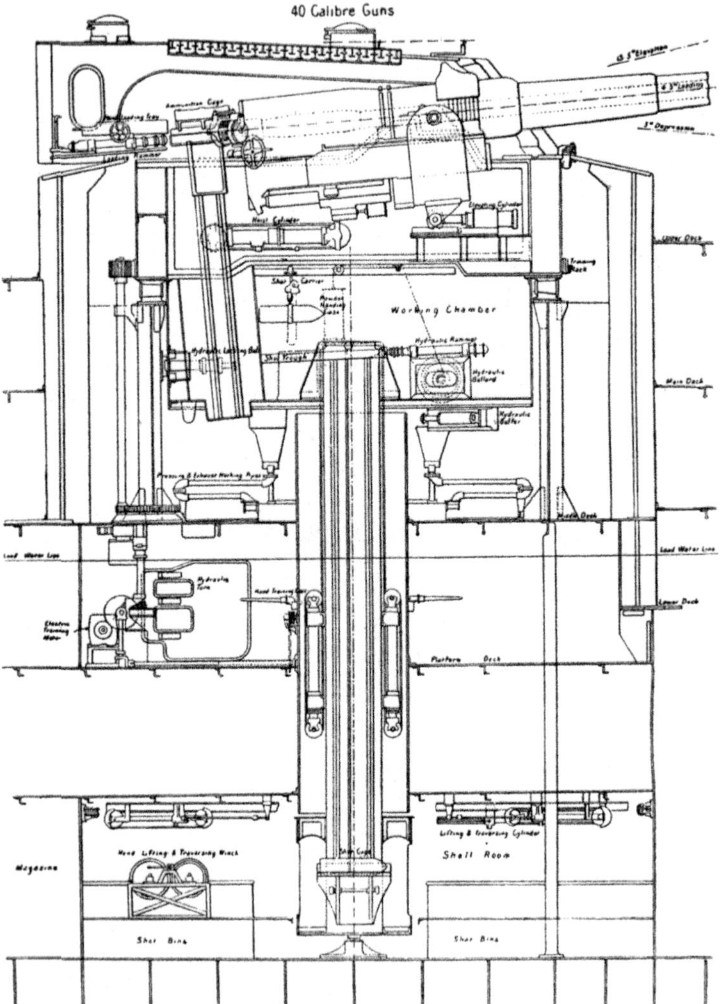

"The Engineer" Swain Eng

of the hoist are practically the same, and revolve within the fixed casing. This design of hoist is also adopted for the Italian ships Regina Margherita and B. Brin. A chain rammer saves room in a fixed loading position by reducing the length of the oval barbette; but, as now applied to all-round loading, its advantage is questionable. In the hydraulic rammer the machine is in the line with the work, and delivers a direct stroke in a straight line; there can be no advantage in converting rectilinear motion into circular and then back into rectilinear. Moreover, the hydraulic rammer, Fig. 6, is made up of far fewer pieces, and weighs only one-fourth as much as the chain rammer of similar power.

In the Re Umberto the difficulty of supplying a system of all-round loading in every position of the turntable, and in spite of the latter moving, or being liable to move at any moment, is met by using a turntable top to the central hoist, and sliding the projectiles radially outwards into the gun hoists; in the Canopus, by overhead travelling cranes placed above the central hoists; in the Albion, by surrounding the bottom of the ammunition trunk by a revolving platform running on rails on the ship's bottom—and capable of being locked either to the ship or the hoist trunk; in the Shikishima, by using a double overhead rail, half of which moves with the hoist and the other half a fixture to the ship; and in the Formidable, by having two shot carriages running on rails carried at the bottom of the trunk of the hoist. All these systems are open to objections. If the hoist cage carrying the projectile can be made to vary its position according to the training of the gun during its ascent from the shell room, all difficulty will be overcome.

Figs. 7 and 8 show a design in which the central hoist does not stop at the working chamber, but runs direct to the breech of the gun. It has been objected that this opens a direct path from gun to magazine, but Sir Andrew Noble sees no ground for fear, and if shot are to be kept at the bottom of the ship, the most perfect arrangement is to convey any shot to either gun in any position by mechanical means. The line of guide rails is indicated by dotted lines.

However necessary it has become to bring in complicated designs of machinery, every effort is made to provide such as secure simplicity in use, combined with efficiency, as far as possible. It may be readily understood that gear which is simple in design may be anything but simple to work. Simplicity in working on service, and freedom from liability to get out of order, are the first objects to secure; occasionally, what may appear rather to add to complication in design, and no doubt entails additional work in construction, serves to simplify

Fig. 7.

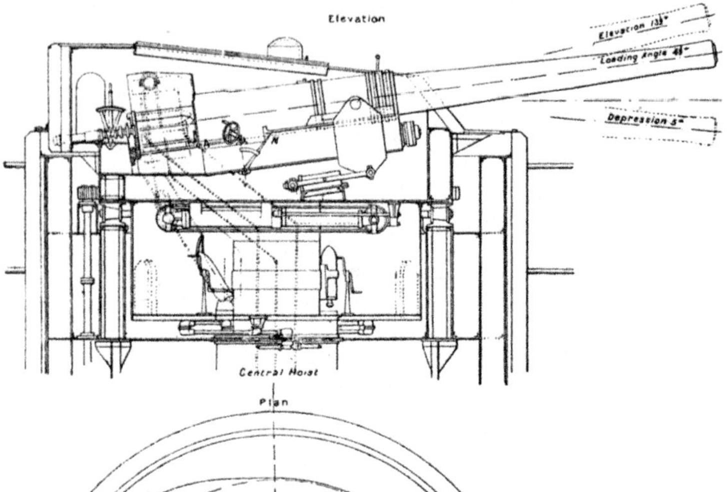
Elevation
Elevation
Loading Angle
Depression 5°
Central Hoist
Plan

the service of the gun in action and to prevent mistakes and casualties.

At the conclusion of his paper Sir Andrew gave some conclusions on the use of explosives and projectiles, which are particularly valuable as coming from him. Sir Andrew, in dealing with explosives for bursting charges of shells, confines himself to three kinds—gunpowder, gun-cotton, and melinite or lyddite. Gun-cotton and lyddite are capable of detonation, and also possess a very much greater potential energy than gunpowder. Distinguishing between the action of shells on unarmoured and armoured structures, he points out that in the former, that is in the attack of unarmoured structures, shells charged with gunpowder do not generally explode until they are some short distance within the side of the vessel, but "with gun-cotton and lyddite two alternatives have to be considered." "The shell may either be fired with a fuse and detonator, so arranged that the shell will burst immediately on impact, or it may be so arranged as to give rise to a slight delay or hang fire. In the first alternative the shell will burst instantaneously on impact—a result impossible to obtain with gunpowder—and in such cases a hole of very large dimensions, and impossible to plug, will be made in the side of the ship, while innumerable small fragments to which the shell is reduced sweep the deck in the wake of the shell.

SirAndrew Noble on Shells.

"In the second alternative, the shell will probably burst inside, making only a small hole in the side of the vessel, but the full effect of the explosion and the destruction to the crew from the fragments of the shell would undoubtedly be serious, and the case of dispersion of the fragments much larger from the explosion taking place inside the vessel.

"Shells charged with gunpowder against unarmoured structures possess, however, one great advantage. The shell will probably burst from 2 ft. to 4 ft. inside the vessel, and although the dispersion of fragments is not nearly so great as with high explosives, the large fragments into which the shell parts are capable of doing much more serious damage to any portion of the ship's structure with which they may come in contact.

"If fired at armoured structures, the results will greatly depend upon the thickness and resistance of the plates, and on the size and energy of the attacking projectile." Generally, it may be stated, that armour is a most effective protection against high explosives, the shell in the large majority of circumstances bursting comparatively harmlessly against the armour. Even if non-fused but with detonators, and possessing "sufficient energy to penetrate the plate, the shell will burst in passing through, but the dispersion of fragments is not very great. If fired without fuse or detonator, wet gun-cotton will not explode, but melinite or lyddite probably will, the result to a great extent depending on the thickness of the armour."

Sir Andrew then draws the following conclusions, based on the results of a number of experiments:—" (1) To attack unarmoured structures, a shell charged with high explosive is a most formidable weapon. The large quantity of explosive that can be carried and the power of immediate detonation permit the vessel to be attacked either by making large holes at or near the water-line, or if the shell should burst on board the effect of the explosion and the destruction to everything in the wake of the shell would be very serious. (2) But with high explosives the shells are reduced to very small fragments, and even very thin steel plates resist penetration. Hence the importance of traverses; and supposing a first-class cruiser to engage two smaller cruisers firing high explosives, one on each broadside, a longitudinal traverse of very moderate thickness would be a protection, the importance of which could hardly be overrated. (3) Having regard to the size of the holes made by high explosives in unarmoured structures, I regard it of great importance that, wherever possible, the water-line should be protected from stem to stern with such armour as can be conveniently carried, and that the same protection should be afforded where the guns are carried on the main deck. On the upper deck effective shields, and as thick as can be conveniently carried, should be attached to the mountings. (4) Where an attack is made against thin armour, shell charged with gunpowder are more effective than high explosive shell, as, dependent on circumstances, the former can be got to pass through thin armour and burst inside. I doubt if shell charged with any explosive can be got to pass through thick armour without bursting. (5) There is one serious objection to certain high explosives as bursting charges which is not shared by wet gun-cotton, and that is the liability to detonate if struck by another projectile, or even by a large fragment. Wet gun-cotton is quite safe in this respect, and yet if fired, for example, by a fulminate, it detonates even more rapidly than in the dry state. This property has led certain Governments to adopt it as the high explosive for use on board ship."

In concluding this paper the author defends the Elswick practice which is " sometimes heard attacked, of mounting as many guns on the broadside as can be conveniently carried." Personally he shared strongly the opinion which a distinguished Admiral once expressed to him—that, supposing a fight between two cruisers equally ably commanded, the victory would remain with the ship that got in first her second broadside, and the victory would be more assured if the broadside were the more powerful. It must also be remembered that with our modern weapons allowance must be made for a gun or two being disabled without altogether crippling the broadside. For these reasons he preferred to carry as many guns as possible, even if the number of rounds carried per gun were reduced.

Trial of Mounting for 12-in. gun, Mark IX.

On April 5th last the Elswick mounting for the 12-in. IX. gun was tested at Portsmouth on the experimental gunboat Drudge. After six rounds the plates at the bows of the vessel cracked, but the mounting stood its test admirably. This design is for the Formidable class. The loading position is fixed at 4½° elevation instead of 13½°; this change will save, it is hoped, from 7 to 8 seconds in loading. A chain rammer is furnished which rams home in less than 6 seconds, as compared with 8 formerly occupied. Consequently as each round is twice rammed, between 11 and 12 seconds in time should be saved. The loading in the Cæsar and Illustrious occupied about 1 min. 42 seconds.

Kearsarge Firing Trials.

The Kearsarge went through her firing trials this spring satisfactorily, the four guns in each double turret being fired simultaneously in a rolling sea.

Report of U.S. Chief of Bureau of Ordnance.

The Report of the Chief of the Bureau of Naval Ordnance of the United States, dated October 1st, 1899, shows the form which progress has made since the war.

New Guns.

There are new models of guns introduced which have enlarged powder-chambers and increased length, the 12-in. and 10-in. being 40 calibres long, like our own 12-in., Mark IX., and 9·2, Mark VIII. The 8-in. of 45 calibres and lighter guns of 50 calibres exceed ours in length. The weight of the projectile in the cases of the 12-in., 10-in., and 6-in. guns agrees with ours; in other cases it is heavier. This being so, the muzzle velocities of 3000 in lightest and 2800 in heaviest guns is very high indeed, and if achieved in new pieces cannot be long maintained by any means at present known. Conversion from ordinary to quick-fire action is in progress with 6-in. and 4-in. guns, and the calibres of machine guns and small arms for land and sea service are being assimilated, the Navy adopting the 0·30-in. small arm.

Smokeless Powder.

A smokeless powder, known as pyro-cellulose, has been tried and has given excellent ballistic and keeping qualities. The Board recommend that for a year or two rather small charges should be used, which, although giving higher velocities than brown powder, will be a safeguard against the erosion produced by cordite and other nitro-glycerine powders. This course will probably appear to most of us very questionable, because if objectionable erosion is produced, the sooner the better. The effect of deferring it surely will only be to discover it at a much more serious stage of progress when it has come into the service. What is the object of testing powder at all before adopting it except to discover its faults and properties as far as possible? Smokeless powder, as noticed last year, generally gives flattering results at first and afterwards shows its ugly side. Surely the sooner the worst is known the better. On the other hand, is it not also desirable to know as soon as possible any happy immunity from this vice which a powder may possess?

The Dum Dum bullets. The question of bullets in war has attracted great attention this year, so that a short account of the matter may be desirable. As noticed in the *Annual* for 1898, the small bore 0·3 in. nickel-sheathed bullets pass through non-vital parts of the body with the infliction of less pain and injury than would be readily believed. It has been shown that such a bullet may pass through a man without his being aware of it. The Japanese and ourselves have found the inconvenience of this on service, and men have nicked or removed the nickel from the bullet point with their knives, to cause it to spread on impact and stop a charging enemy. This was afterwards done in the manufacture of ammunition at Dum Dum. This bullet, however, is apt to break and fly, and has been termed a cruel bullet, and the Peace Conference at the Hague condemned all incisions or removal of the nickel sheath. This would apply to our own Service Mark IV., if not also to Mark III. bullet, each of which has a cylindrical hole in the apex, though in the latter lined with nickel. For the Cape War, our troops received the Mark II. ammunition with solid nickel sheathed head. The Mauser bullet used by the Boers appears from experiment to be even more harmless than our own, so that the temptation to Dum Dum their bullets to stop our men from charging them home, must be strong, but a slash given by an individual man to the sheath of his bullet must be distinguished from the deliberate manufacture and issue of bullets intended to burst and fly, much more from the case of bullets with explosive composition in their heads.

Notes on Heavy Guns in South Africa.

Reference has been made in the preceding chapter to the service rendered by the Naval guns, and in particular to the platform mounting of the 4·7-in. gun used in the defence of Ladysmith. A similar mounting was devised by Capt. Percy Scott, C.B., of the Terrible, for the 4·7-in., which has been used in a railway truck, the ordinary ship's mounting being bolted to the truck. The gun could be fired in any direction, the strain when it was at right angles to the line being met by placing props to meet the recoil. This mounting was practically like those used for the same gun in Ladysmith. One form of platform mounting used by the relief column was made of beams 1 ft. square and 16 ft. long, bolted together, the gun pedestal being in the centre, with two beams bolted on to secure it, the whole mounting being 16 ft. square and weighing about 3 tons.

Wheel carriages designed by Capt. Scott gave the same gun mobility, and a discussion upon the question has taken place in the *Times*. The following is from the letter of a "Naval officer" in Natal, published on May 2nd, in reply to another correspondent:—

> I had the honour of accompanying the 4·7 Naval field guns with Sir Redvers Buller's column, and I can assure him that there was no occasion on which the 4·7 guns could not "keep up with the infantry." On the contrary, we were on several occasions ordered to halt in order to permit the advance guard of our escort to keep ahead of us, and I am sure that if the general officer commanding himself were asked he would certainly say that if, at dusk for example, he directed our guns to be at a certain spot by daybreak the next morning, whether the spot involved hill-climbing or no, he could always rely on the guns being there and ready before the time named. As regards "travelling fast," our speed was an average one of three miles an hour.

Illustrations are given of the wheel carriages for the 4·7-in. and 6-in. guns.* The latter is perhaps the heaviest gun ever put into the field on wheels, and it is reported to have been a complete success in Natal, and "that 200 men with drag-ropes could take it anywhere." The mounting shown of the 4·7-in. gun is of the later type. The following is from the *Times*, May 10, 1900, supplied by a correspondent in Natal:—

> In this carriage lightness and mobility were the points aimed at, and Royal Artillery officers pronounce it to be far better than anything that Woolwich has produced. The wheels are made of steel plates strengthened by angle irons, with a very broad tyre. The axle tree is bent down to allow of a great range of elevation. The elevating gear consists simply of an arc, a manipulating wheel, and a lever. A novel feature in the mounting is a single wheel in the rear part of the trail for travelling the gun; this wheel has a tyre one foot broad. It is stated that Captain Scott's idea was to unscrew the nut of the recoil press, let the gun come back, and so distribute the weight over three wheels instead of

* From the *Navy and Army Illustrated.*

two. An elevation of 32° is allowed for, which means a range of about 15,000 yards. The rear wheel is mounted on a claw-shaped bracket and comes up through a transom at the rear end of the trail, on the same principle as in the mounting of the naval 6-in. gun which Captain Scott had previously designed. For extreme elevation or to reduce the recoil this rear wheel is unshipped, and to do so it is only necessary to lift the trail, and the wheel disengages. Every part is well calculated to stand the strain, and the workmanship reflects the greatest credit upon the Natal Government Railways, who have now constructed this and various other mountings on Captain Scott's designs. The absence of the necessity of a limber for transport is a tremendous advantage. With twenty-five men we run the gun about easily. General Buller has wired for six more since Captain Scott left, but I understand that drawings have been left by him with the railway company. . . . It was predicted that these extempore mountings for the 4·7-in. and 6-in. guns would not stand the strain of up-country work, but they have now been up hill, down hill, into dongas, up kopjes, across rivers, and are none the worse, though one has fired over 2000 rounds.

Analogous mountings from designs by Captain Scott have been made in the Cape Colony for the 4·7-in. and other naval guns which went up to the Modder River and have accompanied Lord Roberts's advance.

The following interesting account of land service 6-in. guns on railway mountings, which have proceeded to Warrenton on the Vaal, is from the notes of a correspondent published in the *Times*, May 10, 1900:—

It has hitherto been considered advisable that, as far as possible, the fact that two 6-in. quick-firing guns had been mounted at Cape Town and sent to the front should be kept from the Boers. No authoritative information respecting them has, therefore, before been published. They are land service guns, mounted by Major H. de T. Phillips, R.G.A., to whom belongs the credit of having mounted the biggest guns in the Cape Colony. These guns were lying in an unfinished fort at Seapoint, Cape Town, waiting to be mounted, and early in January they were entrusted to Major Phillips for mounting and equipment. The original requirement demanded was that they should be able to fire two degrees right and left of a railway line. The first gun was made to fire five degrees each way, and subsequently, by special additions, Major Phillips has made both guns to fire practically in any direction from a railway line. The range of these guns is 11,000 yards at 20° elevation, and their velocity is 2200 ft. per second. The shells are lyddite, common shell, shrapnel, and armour-piercing, and they weigh 100 lb. The guns are fired electrically, or by percussion if the electric gear gets broken down. They are fitted with improvised telescopic sights. The recoil of the gun is largely taken up by a hydraulic buffer, and the actual recoil of the mounting can be reduced to about 3 ft. on the level. Night firing from these guns is very simple, as the gun, being run up to one spot on rails, and placed to a previously marked training and elevation on the truck, can be laid very rapidly, and as often as possible in any spot selected and laid on in daylight. The guns are always either ready for action or for attaching to any train for transport. In the event of a big siege a railway would probably be run round the besieged town, and the guns on railway mountings could work all round the fortress, constantly changing their positions as the enemy began to find their range with his artillery. The guns were first mounted at Salt River Locomotive Works, near Cape Town, on railway trucks, and a trial of the first gun took place in January at Simons Town, and proved so successful that no alterations were required, and it was at once ordered to the front in charge of Captain Austin, and arrived at Magersfontein the day before it was evacuated by the Boers. Lyddite shell is probably more efficient from this gun owing to the greater force of impact, besides which the cartridge for a 6-in. howitzer has only a *maximum* charge of 1lb. 12oz. cordite, whereas a 6-in. Q.-F. gun has a charge of 13lb. 4oz. cordite. Major Phillips is now on the central line of advance with the two guns, while two more are being completed at Cape Town, which will be sent to him, and will, it is to be hoped, do good work on the advance to Pretoria.

From the "Navy and Army Illustrated."

THE NAVAL 6-IN. GUN MOUNTED FOR LAND SERVICE.

The carriage was designed by Capt. Percy Scott, R.N., and was made in the locomotive works of the Natal Government Railways. The second picture shows the gun in action near Chieveley.

NOTES ON TABLES OF ORDNANCE.

THE authorities on which the data in the Ordnance Tables are based are as follows:—

Speaking generally, the British and United States Tables contain figures from official sources. Through the courtesy of the Chief of the Bureau of Naval Ordnance the United States Table has been examined and brought up to date by the Intelligence Department. The Tables for the Continental powers are mainly taken from the *Austrian Marine Almanack.* The energies and perforations, however, are worked out independently, as explained below. The Q.-F. Ordnance Tables of Elswick, Vickers, Schneider Canet, and Krupp guns are obtained directly from the manufacturers, and the data in them are given on their authority. In justice to British manufacturers, the compiler would call special attention to the fact that the very high velocities, 2740 ft.-secs. and over, which occur often in the foreign tables, are very rarely found in the guns marked with an asterisk, that is, existing guns, and when actually achieved with new guns cannot be long maintained. In the course of time lower velocities are generally substituted. Elswick explains the possibility of obtaining such velocities in a footnote, but limits the columns to existing guns, except one 12-in. gun under manufacture.

There are very few alterations this year in the British and foreign service Tables, except in the case of the United States Table, in which a number of new pieces of greatly increased power appear. The Russian Table is the best obtainable, but is certainly not up to date, and readers are cautioned not to be misled by it. Much more powerful guns than those shown must now exist. Elswick, Krupp, and Canet (Schneider) furnish new Tables.

Tresidder's formula being now recognised and used in official papers, and this even for velocities below 2000 ft.-secs., it has been concluded that it is best to follow the course indicated in the *Annual* for 1896, p. 363, that is to say, to employ Fairbairn's or Maitland's formula only for velocities up to 1580 ft.-secs. For these low velocities they have been thoroughly tested and found good, and for these it would be a mistake to alter the existing tables based on them. About 1580 ft.-secs. the formulæ of De Marre, Krupp, Tresidder, Maitland and Fairbairn all agree fairly well. At this point, then, it is convenient to "shunt," as it were, from the Fairbairn, on to the Tresidder curve, for British Tables. Krupp's formula gives nearly the same results as Tresidder's, and it makes little difference which of the two is employed, and in some foreign Tables where Krupp's formula has been used it is left undisturbed, Tresidder's being often added to enable a comparison to be made. In Krupp's formula weight tells more in comparison to velocity than in Tresidder's. The actual formula used is nearly always stated on the face of the Table.

BRITISH RIFLED ORDNANCE.

(Compiled from the official "List of Service Ordnance, 1898," and supplemented by subsequent information.)

B.L. GUNS.

Ordnance. Nature. Calibre or Pr.	Nature. Weight.	Nature. Mark and Service.*	Total length in inches.	Length of Bore, including Chamber.	Chamber. Diameter (at largest).	Chamber. Length to base of projectile.	Rifling. Twist one turn in. Least at breech.	Rifling. Twist one turn in. Greatest at muzzle.	System.†	Charge (full). Weight.‡	Charge (cordite). Weight.	Charge (cordite). Size.	Projectile. Diameter.	Projectile. Weight.§	Projectile. Bursting Charge of Common Shell.	Projectile. Value of $\frac{d^2}{w}$.	Projectile. Value of $\frac{w}{d^3}$.	Muzzle velocity.	Ballistics (with full charges). Total muzzle energy.	Muzzle energy per ton of gun.	Perforation of wrought iron. At muzzle.	At 1000 yards range.	At 2000 yards range.
				calbs.	ins.	ins.	cals.	cals.		lbs.	lbs. oz.		ins.	lbs.	lbs.			f. s.	ft. tons.	ft. tons	ins.	ins.	ins.
16·25-in.	110½ tons.	I. II. & III.	524·0	30·0	21·125	84·5	Various in the newer patterns.	30	P. Elswick, Hook, or Modified Pl. Section, the last in the new guns.	960 S.B.C.	..	..	16·25	1800	‡‡193 / **179¼	0·147	0·420	2087	54,390	492	38·0	34 6	31·7
13·5-in.	69 & 67 tons.	I. II. III.& IV.	433·0	30·0	18·0	66·5		30		630 S.B.C.	..	..	13·5	1250	**85	0·146	0·508	2016	35,230	511 / 526	33·0	30·2	27·6
12-in.	45 & 46 tons.	III. IV. V. & V_w	328·5	25·25	16·0	48·0		35		295 P.Br.	88 8	30	12·0	714	31$\frac{13}{16}$ / ‡‡95 / **79	0·202	0·413	1914	18,130	403 / 394	24·4	21·5	18·9
12-in.	46 tons.	VIII. Wire	445·5	35·43	16·0	70·0		30		..	167 8	50	12·0	850	80$\frac{9}{11}$	0·169	0·492	2367	33,020	718	35·8	32·7	29·4
12-in.	50 tons.	IX. Wire	496·5	40·0	17.5	87·2		..		..	201 8 / 9 8	50 / 3¾						2481 a	36,290	726	39·7	35·4	31·6
10-in.	29 tons.	II. III. III.A & IV.	342·4	32·0	14·0	54·0		30		252 P.Br.	76 0	30	10·0	500	37¾	0·200	0·500	2040	14,430	498	24·8	21·8	19·3
9·2-in.	21 & 22 tons.	I. & II.	255·8	25·56	11·0	44·0		35		140 P Br.	42 0	30	9·2	380	18 / ‡‡33 / **30$\frac{6}{16}$	0·223	0·488	1781	8,356	411 / 392	18·3	15·9	14·4
9·2-in.	24 & 22 tons.	III. V. VI. VI.A&C & VII.	310·0	31·5	12·0	43·0		30		164 P.Br.	53 8	30	9·2	380		0·223	0·488	2065	10,910	455 / 496	22·9	19·8	17·2
9·2-in.	25 tons.	Wire VIII.	384·0	40 08	10·5	53·15		..		..	63 0	40	9·2	380	..	0·223	0·488	2347	14,520	581	27·6	23·9	20·7
9·2-in	27 tons.	Wire IX.	445·25	46·74	13·0	71·215		..		..	..	..	9·2	380	..	0·223	0·488	2600	17,810	660	32·4	28·1	24·2
8-in.	14 tons.	III.	222·5	25·1	10·5	34·5		35		104 P.Br.	28 12	20	8·0	210	13$\frac{10}{16}$ / ‡‡29 / ‡‡18½	0·305	0·410	1953	5,554	427 / 397	16·8	13·8	9·5
8-in.	15 tons.	IV.	254·5	29·61	10·5	38		35		118 P.Br.	32 10	20	8·0	210		0·305	0·410	2150	6,730	449	19·3	16·0	13·2
8-in.	14 tons.	VI.						30										2200	7,046	503	20·0	16·6	13·7
6-in.	82 cwt.	80 pr.) I.	162·6	25·53	7·5	28·05		40		34 S.P.	..	..	6·0	80	..	0·450	0·370	1880	1,961	484	11·2	8·4	6·8
6-in.	5 tons.	III.	170·7	25·53	8·0	26·75		35		36 E.X.E. / 48 E.X.E.	14 12	20	6·0	100	7$\frac{4}{16}$ / **9$\frac{13}{16}$	0·360	0·463	1960	2,665	533	13·4	10·7	8·9
6 in.	5 tons.	IV. / VI.	173·5	26·0	8·0	26·75		35 / 30		48 E.X.E.			6·0										
5-in.	38 cwt. / 40 cwt.	II. / III. IV. & V.	139·15	25·07 / 25·0	5·75	19·05		25		15·5 S.P.	4 7¼	7·5	5·0	50	7$\frac{12}{16}$ / 3$\frac{5}{16}$	0·500	0·400	1750	1,062	559 / 531	8·8	6·6	5·3
4-in.	23 cwt. / 26 cwt.	II. II.A III. III.A IV. V. & VI.	120·0	27·0	5·3	18·5	120	30		12 S.P.	3 1	5	4·0	25	1$\frac{12}{16}$ / 3$\frac{3}{16}$	0·640	0·391	1900	625	544 / 481	7·7	5·4	4·0
12-pr.(3·0)	6 cwt.	Wire I. (L.)	66·75	19·66	3·2	8·35	105	28		..	0 12$\frac{7}{16}$	5	3·0	‖12·5	..	0·720	0·463	1553	209	497	..	..	..

M.L. GUNS.	17·72-in.	100 tons.	I. L.	391·85	20·48	19·7	59·72	150	50	P.Pl.	450 Pr[1] Bl. or Pr[2].	..	..	17·72	2000	0	106	0·157	0·359	1548	33,220	332	25·2	23·5	21·7
	16-in.	80 tons.	I.	321·0	18·0	18·0	59·6	0	50	P.Pl.	450 Pr.[1] Br	..	..	16·0	1700	0	$75\frac{6}{16}$ / c$112\frac{8}{16}$	0·151	0·415	1540	27,960	350	24·3	22·5	20·9
	12·5-in.	38 tons.	I.	230·0	15·84	Unchambered		0	35	W.	130 Pr.[2] or 165 Pr.[1] Bl	..	..	12·5	818·0		$37\frac{8}{16}$	0·191	0·419	1442	11,820	311	17·4	16·0	14·7
	12·5-in.	38 tons.	II.	222·8	15·84	14·0	41·125	438	35	W.	200 E.X.E 210 Pr.[2] or 190 Pr.[1] Bl									1575	14,070	371	19·5	17·8	16·3
	12-in.	35 tons.	I.	195·0	13·54	Unchambered		0	35	W.	110 Pb.[2]	..	..	12·0	714·0		$34\frac{10}{16}$	0·202	0·413	1340	8,890	273	15·5	14·4	13·2
	12-in.	25 tons.	II.	182·5	12·09	,,		100	50	W.	85 Pb.[2]	..	..	12·0	614·0		$28\frac{2}{16}$ / c$61\frac{10}{12}$	0·235	0·355	1292	7,109	288	13·4	12·1	11·0
	11-in.	25 tons.	II.	180·0	13·18	,,		0	35	W.	85 Pb.	..	..	11·0	548·0		27	0·221	0·411	1360	7,028	281	14·2	12·9	11·7
	10-in.	18 tons.	II.	180·0	14·55	,,		100	40	W.	70 Pb.	..	..	10·0	410·0		$23\frac{1}{2}$	0·244	0·410	1379	5,408	300	13·1	11·7	10·5
	9-in.	12 tons.	V.	156·0	13·89	,,		0	45	W.	50 Pb.	13 4	$7\frac{1}{2}$	9·0	256·0		17	0·316	0·351	1440	3,681	307	11·4	9·8	8·5
	8-in.	9 tons.	III.	144·0	14·75	,,		0	40	W.	35 Pb.	..	..	7·92	179·0		→ $17\frac{12}{16}$	0·358	0·350	1391	2,398	266	9·7	8·2	7·1
	7-in.	90 cwt.	I.	131·0	15·86	,,		35	35	W.	22 Pb.	..	..	6·92	114·6		→ $11\frac{1}{16}$	0·428	0·334	1325	1,395	310	7·9	6·5	5·7
	7-in.	6½&7 tons.	III. & IV.	133·0	15·86	,,		35	35	W.	30 Pb.	..	..	6·92						1525	1,848	284	9·2	7·2	5·9
	64-pr.	64 cwt.	III.	118·0	15·47	,,		40	40	Pl.	10 R.L.G.[4]	..	..	6·28	64·5		→ $8\frac{5}{16}$	0·588	0·271	1390	865	270	..	..	..
	64-pr.	71 cwt.	I.	122·72	16·42	,,		40	40	Pl.	$8\frac{1}{4}$ R.L.G.[4]	..	..	6·29	64·5		→ $8\frac{8}{16}$	0·588	0·271	1260	715	201	..	..	..
	40-pr.	35 cwt.	II. L.	120·0	22·0	,,		35	35	W.	$6\frac{3}{4}$ R.L.G.[4]	..	..	4·75	38·8		→ $3\frac{6}{16}$	0·582	0·362	1425	546	312	..	..	..
	25-pr.	18 cwt.	I. L.	98·0	22·0	,,		35	35	W.	$4\frac{1}{2}$ R.L.G.[4]	..	..	3·94	24·96		→ $1\frac{13}{16}$	0·622	0·408	1350	315	350	..	..	..
	16-pr.	12 cwt.	I. L.	78·0	19·0	,,		30	30	F.M.	$3\frac{1}{16}$ R.L.G.[4]	..	..	3·54	16·1		→ $1\frac{2}{16}$	0·771	0·366	1355	205	342	..	..	..
	13-pr.	8 cwt.	I. L.	92·0	28·0	3·15	14·13	100	30	P.	$3\frac{1}{4}$ R.L.G.[4]	..	..	3·0	13·0		$\frac{11}{16}$	0·689	0·484	1595	232	580	..	..	..
	9-pr.	8 cwt.	I. & II.	72·0	21·17	Unchambered		30	30	F.M.	$1\frac{3}{4}$ R.L.G.[2]	..	..	3·0	9·1		→ $8\frac{12}{16}$	0·996	0·335	1330	119	298	..	..	..
	9-pr.	6 cwt.	I.	61·0	17·67	,,		30	30	F.M.	$1\frac{1}{2}$ R.L.G.[2]	..	..	3·0						1250	97	323	..	..	..
	9-pr.	6 cwt.	II. III. IV.	74·5 / 74·875	22·0	,,		30	30	F.M.	$1\frac{3}{4}$ R.L.G.[2] / $1\frac{1}{2}$ R.L.G.[2]	..	..	2·94	9·1			0·956	0·356	1330	119	397	..	..	..
	2·5-in.	400 lbs.	I. & II. L.¶	70·45	26·6	2·56	11·07	80	30	P.	$1\frac{1}{2}$ R.L.G.[2] or $1\frac{10}{16}$ R.L.G.	..	..	2·5	7·25		$\frac{4·5}{16}$	0·819	0·488	1440	101	377	..	..	..
	7-pr.	200 lbs.	IV.	41·0	12·0	Unchambered		20	20	F.	$\frac{3}{4}$ R.F.G.	..	..	2·94	7·29		→ $\frac{7·25}{16}$ / $\frac{14·5}{16}$	1·185	0·287	950	46	515	..	..	..
R.B.L. GUNS (screw).	7-in.	82 cwt.	..	120·0	14·21	7·2	16·0	37	37	P.	11 R.L.G.[2]	..	..	7·0	91·25 / 108·3		$8\frac{4}{16}$ / $10\frac{5}{16}$	0·537	0·266	1100	909	222	..	..	..
	40-pr.	35&32cwt.	..	121·0 & 120	22·39	4·96	13·5	36½	36½	P.	5 R.L.G.[2]	..	..	4·75	40·7		$2\frac{8}{16}$	0·554	0·380	1160	380	217	..	..	..
	20-pr.	15&13cwt.	..	66·125	14·43	3·94	11·0	38	38	P.	$2\frac{1}{2}$ R.L.G.[2]			3·75	21·8		$1\frac{5}{16}$	0·645	0·414	1000	153	203	..	..	..
	12-pr.	8 cwt.	..	72·0	20·458	3·2	8·5	38	38	P.	$1\frac{1}{2}$			3·0	11·25		$\frac{8}{16}$	0·800	0·417	1239	200	299	..	..	..
	9-pr.	6 cwt.	..	62·0	17·5	3·2	7·0	38	38	P.	$1\frac{1}{8}$	..	..	3·0	8.56		$\frac{6}{16}$	1·052	0·317	1055	66	220	..	..	..

L., Land service only. * The Roman numeral is the number of the pattern given. Further differences in pattern are indicated by letters a, b, and c.
† P. means Polygroove; Pl., Plain; W., Woolwich; F., French; F.M., French modified; H., Henry; E.O.C., Elswick Ordnance Co.
‡ S.B.C. (in column for charge) means Slow-burning Cocoa; P.Bl. stands for Prismatic Black; P.Br. for Prismatic Brown; Pb., Pebble; R.L.G., Rifle Large Grain; L.G., Large Grain; E.X.E., Experimental letter E.
◊ For the higher natures the weight of projectile given is for Palliser shot; for the lower natures it is for filled common shell. ** Cast steel. ‡‡ Forged steel. †† Double shell. ‖ Shrapnel.
¶ Mountain-service jointed gun. → Studded projectiles. c For steel common shell. d Estimated.
Note.—For 6-in. (Vickers) see Q.F. table. There is also under construction a 7·5-in. B.L. gun, firing a 200-lb. projectile, with about 2700 f.s. muzzle velocity. Length, 45 calibres, weight about 17 tons.
a At 60° Farht. At 80°, V = 2526 f.s. E = 37,600 ft. tons. Perftn. = 40·6 in.

BRITISH RIFLED ORDNANCE—*continued.*

(Compiled from the official "List of Service Ordnance, 1898." Supplemented by subsequent information.)

Ordnance. Nature. Calibre or Pr.	Weight.	Mark and Service.*	Total length in inches.	Length of Bore, including Chamber.	Chamber. Diameter.	Chamber. Length to base of projectile.	Rifling. Twist one turn in Least at breech.	Rifling. Twist one turn in Greatest at muzzle.	Rifling. System.†	Charge (full). Weight.‡	Charge (cordite). Weight.	Charge (cordite). Size.	Projectile. Diameter.	Projectile. Weight.§	Projectile. Bursting Charge of Common Shell.	Projectile. Value of $\frac{d^2}{w}$.	Projectile. Value of $\frac{w}{d^3}$.	Muzzle velocity.	Ballistics (with full charges). Total muzzle energy.	Muzzle energy per ton of gun.	Perforation of wrought iron. At muzzle.	At 1000 yards range.	At 2000 yards range.
				cals.	ins.	ins.	cals.	cals.		lbs. lbs. oz.	lbs. ozs.		ins.	lbs.	lbs.			f. s.	ft. tons.	ft. tons.	ins.	ins.	ins.
UICK-FIRING GUNS		I. & III.	249·25	40	..	..	60	30										1882	2457	351	12·6	10·1	8·0
·0 in.	7 tons	II. (Wire)							P.	29¾ E.X.E.	13 4	30	..	100 0	..	0·360	0·463	2200	3356	479	15·9	12·7	10·2
·0 in. converted ..	5 "	I. to VI.	169·1 166·6	26·2 26·6	.. ..	.. ..	.. ..	.. ..										1913	2537	362	13·0	10·3	8·2
·0 in. (Vickers) ‡‡ ..	$7\frac{8}{20}$ tons	"BL. VII."	..	45	..	..	..	..	..	..	..	..	..	100·0	..	0 360	0·463	2642	††4840	654	21·0	16·7	13·4
·7 in.	41 cwt. 42 "	I. II. III. & IV. Wire	194·1	40	..	...	100 0	34·4 30	E.O.C.	12 0 S.P.	5 7	20	4·72	45·0	..	0·495	0·428	1786 2188	995·4 1494	485·7 711	9·2 12·4*a*	6·7 9 2	4·9 6·6
in.	26 cwt.	I. II. III. Wire converted guns	165·25 120	40 28	..	..	..	30	M.Pl.	..	3 9	15	..	25 0	..	0·640	0·390	2456 2177	1046 961	805 739	10·5*a*	6·9	4·9
2-pr.	12 cwt.	I.	123·6	40	..	..	120 0	28 30	E.O.C. M.Pl.	..	1 15	10	3·0	12·5	..	0 667	0·500	2210	423	677	8·1	5·3	3·5
2-pr.	8 cwt.	I.	87·6	28	..	..	60 0	28 30	E.O.C. M.Pl.	..	13¾	10	3·0	12 5	..	0·667	0·500	1667	223·8	544	4·9	3·2	2·4
Iotchkiss . 6-pr...	8 cwt	I. & II.	97·63	40·0	..	..	180	29·9	M.Pl.	1 15 Q.F.	*b*7¾	5	2·24	6·0	..	0·836	0·534	1818	137·5	344·8	4·8	2·8	..
Iordenfelt . 6-pr...	6 cwt.	I. II. & III.	104·4	42·3	..	..				1 15 Q.F.													
Iotchkiss . 3-pr ..	5 cwt.	I. & II.	80·63	40	..	..	25	25	M.Pl.	1 8 Q.F.	*b*6⅜	5	1·85	3·3	..	1·037	0·521	1873	80·3	321·2	4·1	2·1	.
Iordenfelt . 3-pr...	4 cwt.	I. L.	91·5	45·4									..	..	..	..	..	1920	84·3	337·2	4.3	2·2	..
MACHINE GUNS.										grains.				oz. grs.									
Iordenfelt, 2 bar 1-in.	180 lbs.	I.	52		..	..	35	35	H.	625 M.G.	..	..	1·0	7¼ 0	..	2·207	0·153	..	..	..	¼-in. at 200 yards.		
" 4 bar 1-in.	447 lbs.	III.**	57·0		..	..	35	35	H.														
" 5 bar 0·45-in.	160 lbs.	I. G. G.	46·0		..	..	22	22	H.														
	143 lbs.	II. G. G.	42·25		..	..	22	22	H.														
Iardner, 1 bar 0·45-in.	76 lbs.	I. G. G.	47·0		..	..	22	22	H.	85 R.F.G₂.	..	..	0·450	480	..	2·952	0·751	..	..	..	Same as MH.Rifle, which perforates ⅛ in. wrought iron plate at 600 yds., $\frac{3}{16}$ in. at 400 yds., ¼ in. at 100 yds.		
" 2 bar 0·45-in.	120 lbs.	I. G. G.	47 0		..	..	22	22	H.														
" 5 bar 0·45 in.	268 lbs.	I. G. G.	53·5		..	..	22	22	H.														
Iatling, 10 bar 0·65 in.	787 lbs.	..	66·5		..	..	30	30	H.	270 R.F.G₂.	..	..	0·65	1422	..	2·109	0·730	..	..	..	Not known.		
" 10 bar 0·45-in.	402 lbs.	I. G. G.	59·0		..	..	22	22	H.	85 R.F.G₂	..	..											
" (Acclesfeed)	266 lbs.	I. G. G.	51·0		..	..	22	22	H.														
			43·75				22	22	H.														
Iaxim, 1 bar 0·45 in. Iaxim, converted ·45	63 lbs.	I. G. G.	45·0 42·38		..	..	10 10	27 25·6	Enfi'ld Metf'd	31 Cordite	..	..	0·450	480	..	2·952	0·751	..	..	..	Same as MH.Rifle.		

* L., Land-service only, but might concern navy when serving on land. The Roman numeral is the number of the pattern given. † P. means Polygroove; M.Pl., Modified plain; W., Woolwich; F., French; F.M., French modified; H., Henry; E.X.E., Experimental letter E. § For the higher natures the weight of projectile given is for Palliser shot; for the lower natures it is for filled common shell.

** I. and II. differ chiefly in being 7 lbs. lighter; I. has a pitch of rifling of 1 in 60 in.; G.G., Gardner Gatling; H., Henry. *a* The 4·7 and 4-in. shot must perforate 5 inch, and the 12-pr. shot 4 inch water-hardened steel plate.

b With 4 drs. R.F.G. †† With a new gun at Portsmouth and must be regarded as the maximum effect. ‡‡ This gun has no metal cartridge. §§ There is a Gardner and also a Nordenfeldt ·303 bore gun for L.S.

Note.—An armour-piercing shell has now come in for the 6-in. guns.

AUSTRIAN NAVAL ORDNANCE.

	Krupp Steel B.L. Guns.											Uchatius. St., Br.					Cast Iron BL.	Q.F. Krupp.				Q.F. Skoda.	
Designation by Calibre, in centimètres	30·5 L. 35 C. 80	26 L. 22	24 L. 35 C. 86	24 L. 22	21 L. 20	15 L. 35 C. 86	15 L. 35 C. 80	15 L. 26 K.Z.	15 L. 26 P.Z.	12 L. 35 C. 80	12 L. 35 C. 87	15 L. 25	15 L. 37	12 L. 35	9 L. 24	7 L. 15	15 L. 21	24 L. 40	15 L. 40	15 L. 35	12 L. 35	15 L. 40	12 L. 40
Calibre, in inches	12·01	10·24	9·45	9·27	8·24	5·87	5·87	5·87	5·87	4·72	4·72	5·87	4·72	4·72	3·43	2·60	5·87	9·45	5·87	5·87	4·72	5·87	4·72
Length: Total, in Feet	35·11	18·77	27·60	17·16	13·73	17·13	17·13	12·63	12·63	13·8	13·8	12·19	14·38	13·97	6·76	3·28	10·13	31·5	19·55	17·13	13·85	19·55	15·78
Length: Rifled Portion, in ins.	314·8	148·4	237·7	135·9	105·0	151·4	153·6	106·8	112·4	128·5	126·3	111·4	123·2	123·8	57·5	23·8	91·1	287·8	182·8	153·6	128·2	182·8	147·9
Length: Powder Chamber ,,	69·9	46·1	65·2	41·7	37·0	37·3	35·4	28·3	22·6	24·0	26·3	22·8	36·6	31·65	15·3	11·0	15·7	63·8	35 7	35·7	28·5	35·7	28·5
Length: Of bore in calibres	35	19·0	35	22·0	20 0	35	35·0	25·8	25·8	35·0	35	24·9	37	35·0	23·7	15·0	18·4	40	40	35	35	40	40
No. of Grooves	68	32	25	32	30	36	36	24	36	32	32	36	32	32	24	18	30	72	44	44	36	44	36
Twist in calibres	45·25	70	26·9	70	[illegible]	45—25	25	68·4	45	25	25	45	45—25	25	45	30	63·2	25	45·25	45·25	45·25	45·25	45·25
Weight: Gun, tons	47·2	21·7	26·6	14·5	8·68	5·7	4·69	3·94	3·94	2·25	2·31	3·35	3·15	2·85	0·479	0·09	2·81	27·8	4·43	3·75	1·97	4·33	19·7
Weight: Breech Block, in lbs.	3306·9	1951	1776·9	1422	1080	445·3	463·0	321·9	321·9	253·5	211·6	209·4	211·6	211·6	55·1	18·7	176·4	1450·6	328·5	328·5	211·6	291·0	189·6
Weight: Steel Shell ,,	1003·1	395·7	474·0	292·1	206·6	112·5	86·0	72·8	84·9	57·3	57·3	84·9	57·3	57·3	..	..	..	474·0	100·3	100·3	52·4	100·3	52·4
Weight: Common Shell ,,	1003·1	354·2	474·0	263·5	172·0	112·5	69·9	67·2 / 65·5	69·4	57·3	57·3	69·45	57·3	57·3	14·02 / 14·22	6·42	6·11	474·0	100·3	100·3	52·4	100·3	52·4
Weight: Shrapnel Shell ,,	..	..	..	..	..	112·4	71·9	67·2 / 65·5	69·4	57·3	57·3	69·45	57·3	57·3	15·76	6·88	6·79	474 0	100 3	100·3	52·4	100·3	52·4
Weight: Case Shot ,,	..	..	..	..	..	..	..	..	..	..	..	..	..	..	16·53	6·94	38·4	..	..	..	..	..	..
Weight of Bursting Charge: Steel Shell ,,	10·6	8·8	5·1	6·6	4·4	1·3	1·76	1·65	2·09	0·55	0·55	2·09	0·53	0·55	..	..	..	5·07	2·0	2·0	1·1	2·0	1·1
Weight of Bursting Charge: Common Shell ,,	97·7	20·3	17·9	15·0	15·0	5·29	3·86	3·86	5·07	2 2	2·2	5·07	2·20	2·21	0·46	0·18	1·87	4·72	8·82	8·82	4·6	8·82	4·6
Weight of Bursting Charge: Shrapnel Shell ,,	..	..	..	..	..	1·26	1·10	1·08	1·08	0·57	0·57	1·08	0·57	0·57	0·20	0·09	0·90	..	1.0	1·0	0·53	1·0	0·53
Weight of Firing Charge: Steel Projectile, in lbs.	156·5 24 cm.N	89·3 B	99·2N	76·1 B	50·7 C	22·5 15 cm.N	38·8*	21·6 C	20·9 C	19·8 B	12·13N	20·9 C	30·0 B	19·8 B	..	..	..	91·5 (for Steel Projectile and Common Shell)	18·3	17·9	9·7	18·3	17·9
Weight of Firing Charge: Common Shell, in lbs.	156·5 N 24 cm.N	59·5	99·2N	44·1	30·9	22·5 15 cm.N	38·8*	14·3 / 21·6 C	20·9 C	19·8 B	12·13N	20·9 C	30·0 B	19·8 B	3·31 / 0·97	0·77 O / 0·35	4·74O						
Weight of Firing Charge: Shrapnel, in lbs.	..	..	..	..	..	..	..	..	..	..	..	..	..	..	..	..	..	..	11·9	11·9	4·6	11·9	4·4
Weight of Firing Charge: Exercising, in lbs.	154·3B	59·5	..	44·1	30·9	28·7	19·6	21·6 C	20·9 C	11·0	6·6 N	20·9 C	19·8 B	12·13 B	3·31	0·77 O / 0·35	4·74O	63 9	11·9	11·9	4·6	11·9	4·4
Weight of Firing Charge: Saluting ,,	19·8 O	19·8 O	15·4O	15·4 O	8·82 O	4·74	4·74 O	4·74 O	4·74 O	2·4 O	2·4 O	4·74 O	2·4 O	2·4 O	0·88 O	0·35	4·74O		..	..	..	..	..
Muzzle Velocity, in feet	1969	1575	2100	1587	1519	2133	1969	1641	1562	1755	2133	1562	1755	1755	1470	978	1017	2264	2264	2133	2133	2264	2264
Muzzle Energy: Total, foot-tons	26,970	6808	14,500	5104	3306	3549	2312	1358	1435	1215	1808	1435	1224	1224	..	..	..	11,850	3565	3163	1647	3565	1856
Muzzle Energy: Per inch circumference, foot-tons	714·8	211·6	488·3	175·8	127·7	192·5	125·4	73·7	77·9	82·5	122·2	77·9	82·45	82·45	..	..	..	567·6	193·3	171·6	111·1	193·3	125·1
Thickness of Iron, perforated inches at Muzzle	28·0§	15·0§	27·4‡	13·7§	11·6§	14·5§	11·7§	8·8§	9·1§	9·4§	11 6§	9·1§	9·5§	9·5§	..	..	..	25·0	14·6	13·7	11·1	14·6	11·7
Ditto by Tresidder's formula	30·1	14 7	25·8	13·4	..	16·1	12·6	8·9	8 9	9·7	12·9	8·9	9·7	9·7	..	..	..	29·0	17·0	15·5	12·5	17·0	13·7

NOTE.—C for cube powder; * prismatic powder; O, ordinary powder; B, brown prismatic. N, nitro-glycerine smokeless powder.
‡ By Krupp's formula. § By Fairbairn's formula.
There are also Q.F. Skoda 7 cm., Skoda and Hotchkiss 47 mm., another 47 mm. and Hotchkiss 37 mm.

DANISH NAVAL ORDNANCE.

	Krupp B.L. Guns designated.											Armstrong M.L.					Finspong.
Designation by Calibre	35·5 cm.	30·5 cm.	26 cm. long.	26 cm. short.	21 cm.	15 cm. long.	15 cm. medium.	15 cm. short.	12 cm. long.	12 cm. short.	8·6 cm.	10 in.	10 in.	10 in.	9 in.	8 in.†	6 in.
Calibre, in inches	13·98	12·01	10·24	10·24	8·24	5·91	5·91	5·91	4·72	4·72	3·43	10·0	10 in.	10 in.	9·0	8·0	6·04
Total length, in feet	29·1	22·0	32·8	18·77	24·04	17·1	12·63	10·7	11·8	9·6	6·9	17·0	14·5	14·0	13·0	10·8	9·5
Length of Bore, including Powder Chamber, in inches	304·7	227·2	327·6	194·5	264·5	190·3	135·0	112·9	128·8	102·4	73·6	175·5	145·5	140·0	125·0	104·2	100·8
Length of Bore, including Powder Chamber, in calibres	21·8	18 9	32·0	19·0	35	32·2	22·8	19·1	27·3	21·7	21·3	17·5	14·55	14·0	13·3	13·1	16·7
Number of Grooves	80	68	60	60	48	36	36	36	32	32	24	7	7	7	6	6	6
Twist of Rifling, in calibres	45	45	70–25	45	70–25	70–25	45	45	25	40	45	40	40	40	40	50	40
Total weight, including Breech-gear, tons	51·3	35·4	27·6	21·6	13·3	4·7	4·4	3·5	2·13	1·39	0·49	20·0	18·5	18·0	12·5	8·65	2·46
Weight of Breech Block, lbs.	4695·8	2910	2006	1940	903·9	390·2	330·7	324·1	229·2	176·4	101·4	..	..	..	..	..	..
Weight of Steel Shell, "	1157·4	725·3	451·9	451·9	238·1	112 4	..	86·0	..	44·1	..	400	400	400	..	..	..
Weight of Chilled Shell, "	1157·4	725·3	..	451·9	..	..	86·0	86·0	..	44·1	..	400	400	400	250·2	165·3	..
Weight of Common Shell, "	1157·4	725·3	451·9	451·9	238·1	112·4	69·4	69·4	57·3	36·2	15·2	400	400	400	250·2	131·2	55·1
Weight of Shrapnel Shell, "	1157·4	725·3	451·9	451·9	238·1	112·4	86·0	86·0	57·3	44·1	..	..	..	..	..	..	15·4
Weight of Case Shot, "	..	..	..	..	..	..	..	..	..	..	..	191·8	..	..	154·3	127·9	58·4
Weight of Bursting Charge, Common Shell, "	57·3	39·7	25·4	25·4	12·8	6·2	3·0	3·0	1·7	1·4	0·44	26·5	26·5	26·5	18·5	7·5	5·0
Weight of Firing Charge, Steel or Chilled Shell, lbs.	330·7	180·2	191·8	101·4	105·8	41·9	19·3	21·8	17·4	8·8	..	71·7	71·7	71·7	44·1	29·8	..
Weight of Firing Charge, Common Shell, "	330·7	180·2	191·8	112·4	105·8	41 9	19·3	21·8	17·4	8·8	3·3	71·7	71·7	71·7	44·1	19·8	60·6
Muzzle Velocity, Armour-piercing Projectile, feet	1762	1675	2018	1640	2021	1800	1565	1542	..	1416	..	1457	1368	1368	1368	1378	..
Muzzle Velocity, Common Shell, "	1762	1675	2018	1640	2021	1890	1683	1690	1720	1549	1457	1457	1368	1368	1368	1320	1076
Muzzle Energy, Total foot-tons	24910	14110	12770	8428	6745	2784	1461	1418	..	6130	..	5889	5192	5192	3246	2177	..
Muzzle Energy, Per inch circumference, foot-tons	568·3	374·1	396·8	262·0	260·6	150·0	78·7	73·0	..	32·8	..	189·0	165·3	165·3	115·8	86·9	..
Perforation at Muzzle, in ins. by Fairbairn's forma.	24·8	20·0	20·1	16·7	16·9	12·8	9·1	8·8	..	5·8	..	14·1	13·1	13·1	10·9	9·5	..
ditto ditto Tresidder's forma.	25·6	20·1	22·9	16·8	18·5	15·6	12·6	..	..	..	..	..	..	..	..	..	..

NOTE.—Chilled projectiles will gradually be replaced by steel. † There is another Armstrong gun differing very little from this one.
Krupp has supplied 12-cm. and 8·7-cm. Q.-F. guns, and Bofor's Q.-F. piece have been adopted and manufactured.

DUTCH NAVAL ORDNANCE.

	Krupp Breech Loading.							Armstrong Muzzle Loading.			Dutch Breech Loading.		
Designation by Calibre, in centimètres	28	21	17	15 No. 1.	No. 2.	12 No. 1.	12	28	23	18	12 No. 2.	12*l*	7·5
Calibre, in inches	11·02	7·91	6·80	5·87	5·87	4·72	4·72	11·00	9·00	7·00	4·72	4·72	2·95
Total Length, in feet	20·01	24·04	13·94	12·63	17·13	6·89	13·78	14·42	13·00	11·00	6·89	13·78	7·87
Length of Rifled Portion of bore, in inches	170·8	222·2	112·7	111·8	151·4	61·4	128·5	119·0	104·0	95·5	61·4	..	43·2
Length of Powder Chamber "	36·4	42·4	36·0	23·2	37·7	13·0	24·0	26·0	21·9	15·5	13·0	..	6·7
Length of bore, in Calibres	18·8	35	21·9	23·0	35	15·8	35	12·1	14·0	15·9	15·8	35	17·5
Number of Grooves	64	48/64	42	36	44	12	32	9	6	3	12	32	20
Depth of Grooves, inches	0·069	0·059	0·118	0·118	..	0·049	..	0·20	0·18	0·18	0·118	0·06	0·049
Twist of Rifling in Calibres	45	∝ 25	45	40	25	40	25	∝ 45	∝ 45	35	40	∝ 45	∝ 30
Total Weight, in tons	27·21	12·79/13·98	5·51	3·94	4·72	0·79	2·26	24·46	12·50	7·17	0·93	2·31	0·21
Firing Charge { Armour-piercing Projectile, in lbs.	121·3	99·2	27·6	20·9	49·6	..	19·8	86·0	50·7	30·0	..	19·5	..
Firing Charge { Common Shell "	121·3	99·2	27·6	20·9	49·6	2·43	19·8	86·0	50·7	13·9	2·43	19·8	0·82
Weight { Armour-piercing Projectile "	560·0	308·6	132·3	86·0	112·2	41·0	57·3	533·5	249·1	114·6	..	57·3	..
Weight { Common Shell "	476·2	308·6	112·4	69·4	112·2	29·5	57·3	535·7	262·4	116·8	29·5	57·3	9·5
Weight { Case Shot "	273·4	..	63·9	41·9	..	26·5	57·3	185·2	149·9	68·3	26·5	..	9·3
Bursting Charge { Armour-piercing Projectile "	6·6	4·6	2·2	1·1	..	0·44	..	4·4	2·2	2·2	..	..	..
Bursting Charge { Common Shell "	26·5	12·3	6·6	6·6	..	2·0	..	28·7	17·6	8·8	1.8	..	0·44
Muzzle Velocity, feet	1558	1739	1558	1558	2001	971	1755	1332	1476	1558	951	1804	958
Muzzle Energy { Total, in foot-tons	9423	6471	2226	1447	3115	..	1224	6563	3763	1929	..	1264	..
Muzzle Energy { Per inch Circumference, foot-tons	272	260·7	104	84	169·0	..	82·5	191	134	89	..	85·2	..
Perforation at Muzzle, in inches	17·0	{16·8 / 17·1†}	10·5	9·1	{13·6 / 14·8†}	..	{9·4 / 10·1†}	14·0	11·9	9·7	..	9·6	..
Metal employed or system of construction	Steel Jacket and Hoops.		Steel-hooped.		Steel Jacket and Hoops.	Steel-hooped.	Steel Jacket and Hoops.	Steel Tube and Wrought Iron.			Bronze.		

NOTE.—The 23-cm. ML. guns also discharge 113-Kg. (249·1 lbs.) steel shells and 113-Kg. solid shot. The 18-cm. ML. guns discharge steel shells of 51-Kg. (112·4 lbs.) and segment shells of 53-Kg. (116·8 lbs.). The 7·5-cm. BL. guns discharge ring-shells of 4·3 Kg. (9·5 lbs.). Of the older guns there are yet extant three sorts—rifled [illegible]-cm. muzzle-loader (mostly bronze), and rifled bronze 7-cm. and 5-cm. † By Tresidder's formula.

FRENCH NAVAL ORDNANCE.

Date and Pattern of Gun.	Model 1893.					Model 1887.				1870–81.	70–81. 70–84.	1884.					1881.									
Desig. by Calibre, in cms.	34·0	30·5	27·44	24·0	19·4	34	30·5	27	19	27	32	34	27	24	16	14	34 long.	34 short.	27	24	16 heavy.	16 light.	14	10	mm. 90	mm. 65
Calibre, in inches	13·39	12·0	10·8	9·45	7·64	13·39	12·0	10·80	7·64	10·80	12·6	13·39	10·80	9·45	6·49	5·45	13·39	13·39	10·8	9·45	6·49	6·49	5·46	3·94	3·54	2·57
Total length, in feet	..	..	..	..	..	..	..	..	..	23·97	27·93	..	28·47	24·89	17·04	..	33·69	25·32	27·12	23·70	15·14	15·14	14·3	8·6	7·1	3·58
Length of Bore, in inches	..	..	..	..	..	..	..	..	..	269·0	313·8	..	..	..	..	..	380·6	280·2	306·9	269·3	180·9	180·9	162·6	102·6	77·9	41·2
Length of Bore, in calibres	35	40	45	40	40	42	45	45	45	25	25	30	30	30	30	30	28·5	21·0	28·5	28·5	28	28	28	26	22	16
Number of Grooves	..	..	..	..	..	..	..	..	..	..	..	..	..	..	..	..	..	..	..	..	50	50	42	30	28	20
Depth of Grooves, inches	..	..	..	..	..	..	..	..	..	0·059	0·059	..	..	..	..	..	0·067	0·067	0·059	0·055	0·039	0·039	0·035	0·028	0·024	0·020
Rifling Twist	..	..	..	..	..	..	..	..	..	7°	7°	..	..	..	..	..	7°	7°	7°	7°	7°	7°	7°	7°	7°	8°
Total weight, in tons	52·9	45·9	34·9	22·4	10·6	60·0	49·2	37·1	10·6	24·6	42·3	50·8	27·7	17·9	5·4	3·15	52·2	47·2	27·4	17·7	4·9	3·9	3·2	1·18	0·54	0·09
Weight of Firing Charge: Armour-piercing Projectile lbs.	220·5	198·4	114·6	110·2	44·1	220·5	198·4	114·6	44·1	154·3	282·2 249·1	388·0	200·6	..	42·5	..	388·0	337·3	203·9	149·9	42·5	32·6	..	..	..	..
Weight of Firing Charge: Common Shell "	..	..	..	..	..	..	..	..	..	154·3	282·2 249·1	..	200·6	..	42·5	27·1	337·3	368·2	203·9	149·9	42·5	32·6	27·1	9·9	3·6	0·79
Weight: Armour-piercing Projectile * lbs.	925·9	643·8	476·2	317·5	165·3	925·9	643·8	476·2	165·3	476·2	760·6	925·9	476·2	317·5	99·2	..	925·9	925·9	476·2	317·5	99·2	99·2	..	..	..	..
Weight: Common Shell "	..	..	..	..	..	..	..	..	..	396·8	630·5	771·6	396·8	264·6	99·2	66·1	771·6	771·6	396·8	264·6	99·2	99·2	66·1	30·9	17·6	5·95
Weight: Case Shot "	..	..	..	..	..	..	..	..	..	..	..	..	..	..	..	..	..	..	..	..	130·7	130·7	61·7	39·0	19·5	7·7
Muzzle Velocity, in ft.-sec.	2625	2625	2625	2625	2625	2625	2625	2625	2625	1887	1985 1804	1969	1969	1969	1969	1969	1969	1804	1969	1969	1969	1821	1936	1673	1493	1135
Muzzle Energy: Total, in foot-tons	44230	30750	22750	15170	7898	44230	30750	22750	7898	11760	20780 17160	24900	12800	8539	2668	1777	24900	20880	12800	8539	2668	2080	..	..	..	..
Muzzle Energy: Per in. circ., foot-tons	1052	815·8	670·7	511·1	329·1	1052	815·8	670·7	329·1	346·6	525·0 433·5	591·9	377·5	287·7	130·8	103·9	591·9	496·6	377·5	287·7	130·9	121·3	..	..	..	..
Perforation at Muzzle, inches	42·5†	37·3†	33·7†	29·4†	13·4†	42·5†	37·3†	33·7†	23·7†	20·6†	26·1† 22·6†	27·6†	22·0†	19·2†	13·0†	10·7	27·6†	24·2†	22·0†	19·2†	13·0†	11·6†	..	..	..	..
,, ,, ,,	..	..	..	..	..	..	..	..	..	..	..	26·6‡	21·1‡	18·4‡	12·4‡	..	26·6‡	..	21·1‡	18·4‡	..	..	..	..	..	..

* Steel or chilled iron. † By Tresidder's formula ‡ By Krupp's formula.

NOTE.—M. Claudinon stated in March, in the French Chamber of Deputies, that France has a 12-in. (30·5 c.m.) gun with a muzzle energy of 12,200 metre tons or 39,395 ft. tons. This for a projectile of 643·8 lbs. implies a velocity of 2971 f.s.

FRENCH NAVAL ORDNANCE—*continued.*

Date and Pattern of Gun.	75–78.	Jacketed. 1870.		Jacketed. 75	75	1875.				1870.					Q.-F. Guns. 16§	16‡	14§	14	Mod. 92. 10	Mod. 91. 10	Mod. 81. 10‡
esig. by Calibre, in cms.	37	27	14	27 No. 1	10	42†	34	27 No. 1	10	27	24	19 long.	16	14	16·47		13·86		10 00		
alibre, in inches	14·57	10·79	5·46	10·8	3 94	16·54	13·39	10·8	3·91	10·8	9·45	7·64	6·49	5·46	6·46		5·44		3·94		
otal length, in feet	36 7	17·7	10·3	19·3	9·3	32·5	22	19·3	9·3	17·7	16·21	13·6	12·2	10·3							
ength of Bore, in inches	414 0	194·3	115·0	213·4	104·3	366·0	241·5	213·4	104·3	194·3	179·1	151·0	137·3	115 0							
ength of Bore, in calibres	28·5	18	21	19·7	26	22	18	19 8	26	18·0	19	19·7	19	21	45	30	45	30	60	50	26
umber of Grooves	..	54	28	54	20	84	68	54	20	54	48	28	50	28							
epth of Grooves, inches	0·079	0·059	0·047	0·059	0·032	0·079	0·059	0·059	0·032	0·059	0·059	0·059	0·039	0·047							
fling Twist	7°	4°	4°	4°	7°	7°	4°	4°	7°	4°	4°	4°	7°	4°							
otal weight, in tons	**75·1	22 8	2·6	27·9	1·18	74·8	47·6	27·6	1·18	22·8	15·4	7·9	4·92	2·66	6·89	4·92	4·13	3·84	2·19	1·62	1·18
eight of Firing harge: Armour-piercing Projectile * . Pro- lbs.	463	136·7	..	165·3	..	604·1	304·2	136·7	..	92·6	62·8	33·1	39·7	..	30·2	19·0	16·1	12.8	8·16	8·16	5·07
Common Shell „	463	126·8	11·2	145·5	10·1	..	231·5	121·3	7·1	92·6	62·8	33·1	39 7	9·0							
eight: Armour-piercing Projectile . Pro- lbs.	1235	476·2	..	476·2	..	1719·6	925·9	476·2	..	476·2	317·5	165·3	99·2	..	99·21		66·14		30·87		
Common Shell „	1014	396·8	61.7	396·8	30·9	1433·0	771·6	396·8	26·[illegible]	396·8	264·6	137·8	99·2	46·3							
Case Shot „	..	321·9	42·8	321·9	18·7	..	..	321·9	18·7	321·9	211·6	..	68·3	39·7							
uzzle Velocity, in ft.-sec.	1969	1608	1523	1640	1673	1663	1722	1641	1591	1424	1444	1470	1782	1332	2625	2100	2625	2100	2625	2625	2034
uzzle nergy: Total, in foot-tons	33210	8515	..	8880	..	17750	19160	8865	..	6695	4592	2477	2183	..	4730	3061	3160	2022	1475	1475	885·5
Per in. circ., foot-tons	725·4	251	..	261·7	..	422	456	261	..	197·3	154·7	103·2	107	..	233·5	150·9	184·9	118 7	119·2	71·6	71 6
rforation at Muzzle, inches	30·5†	16·2†	..	16 7†	..	26·3†	22·6†	16·7	..	20·53	20·3	10·4	10·8	..	20·0†	14·4†	17 7†	12·7†	14·3†	14·3†	9·8

* Steel or chilled iron.
** Made at St. Chamond. The Creusôt gun weighs 71·4 tons.
† By Tresidder's formula.
‡ Models 1881 and 1884 converted guns.
§ There are three models of the years 1887 1891 and 1893, of slightly different weight from the above.

GERMAN NAVAL ORDNANCE.

	Krupp Steel Breech-loading Guns, designated by calibre.																					Bronze B.L.
Designation in centimètres	30·5 jack'd.	28	28	26 long.	26 jack'd.	26 short.	24 lon".	24 long.	24 long.	24 short.	21 long.	21 long.	17 long.	15 long.	15 short jack'd.	15 short.	15 short jack'd.	12·5 hoop'd.	10·5 long.	8·7	6¶	8
Calibre, in inches	12·01	11·02	11·02	10·33	10·33	10·33	9·45	9·45	9·45	9·37	8·24	8·24	6·80	5·87	5·87	5·87	5·87	4·92	3 96	3·43	2·36	3·19
Length: Total, in feet	21·98	36·75	32 15	18·77	18·77	17·06	31·50	27.56	23·63	15·45	24·0	20·61	13·94	14·67	10·73	10·73	10·68	9·60	12·08	6·89	4·1	5·15
Length: Rifled portion, in ins.	181·9	§407·9	§352·8	149·8	150·0	129·3	349·6	302·4	201·6	116·2	218·2	176·5	117·1	128·5	93·3	87·1	87·1	85·7	113·6	62·7	44·3	45·9
Length: Powder Chamber†, „	45·3			44·7	44·4	44·7	..	..	53·5	40·9	75·3	46·7	31·5	31·1	19·0	25·1	25·1	16·7	19·5	10·7	..	9·73
Length: Bore, in calibres	18·9	40	35	18·8	18·8	16·8	..	..	26·1	16·8	35·6	27·1	21·9	27·2	19·1	19·1	19·1	20·8	33·6	21·4	..	17·4
Number of Grooves	72	..	..	36	48	36	..	..	56	48	48	48	30	36	36	36	36	32	32	24	24	12
Depth of Grooves, in inches	0·079	..	..	0·077	0·079	0·077	..	..	0·059	0·061	0·059	0·059	0·063	0·059	0·061	0·061	0·061	0·059	0·049	0·049	..	0·051
Twist, in calibres	45	..	..	50	50	50	..	..	25*	45	25*	25*	45	25*	45	45	50	40*	25*	40*	..	46
Weight: Gun, including Breech Gear, tons	35·4	43·4	43·2	21·7	18·7	17·7	25·4	21·7	18·7	14·6	13·03	12·3	5·51	4·04	3·44	3·44	3·15	1·38	1·15	0·44	0·10	0·23
Weight: Breech Block, in lbs.	2954	..	..	2050	1973	1973	..	..	..	1378	831·1	908·3	496·0	390·2	324·1	324·1	324·1	163·1	149·9	86·0	..	55·1
Weight: Armour-piercing projectile,‡ in lbs.	725·3	562·2	562·2	412·3	412·3	412·3	474·0	474·0	474·0	306·4	308·6	308·6	117·9	112·4	76·1	76·1	76·1	..	..	..	..	..
Weight: Common Shell, in lbs.	725·3	474·0	474 0	357·1	357·1	357·1	474·0	474·0	474·0	261·5	308·6	308·6	112·9	112·4	65·0	65·0	65·0	40·1	39·7	14·9	6·61	8·3
Weight of Bursting Charge: Armour-piercing Shell, in lbs.	7·7	..	..	5·3	5·3	5·3	7·05	7·05	6·6	3·2	5·5	5·5	1·3	1·5	0·8	0·8	0·8	..	..	..	..	..
Weight of Bursting Charge: Common Shell, in lbs.	19·8	25·4	25·4	14·3	14·3	22·0	16·5	16·5	15·4	15·4	12·1	12·1	5·1	4·3	4·2	4·2	4·2	2·4	0·9	0·4	..	0·6
Weight of Firing Charge: Armour-piercing Shell, in lbs.	202·8	352·7	297·6	105·8	105·8	125·7	..	..	152·1	67·2	103·6	103·6	30·9	33·1	14·3	17·1	17·1	..	..	..	..	..
Weight of Firing Charge: Common Shell, in lbs.	202·8	352·7	297·6	105·8	105·8	125·7	..	..	152.1	50·7	103·6	103·6	30·9	33·1	14·3	17·1	17·1	8·8	8·8	3·3	0·88	0·9
Initial Velocity: Armour-piercing projectile, ft.-sec.	1713	2362	2133	1588	1588	1578	2067	1903	1657	1493	1739	1657	1608	1624	1463	1463	1463	..	..	..	..	..
Initial Velocity: Common shell, ft.-sec.	1713	..	..	1641	1641	1654	2067	1903	1657	1391	1739	1657	1654	1624	1555	1555	1555	1545	1526	1545	1545	1053
Muzzle Energy: Total, foot-tons	14,750	21,750	17,740	7211	7211	7119	14050	11910	9024	4736	6471	5876	2112	2055	1131	1131	1131	..	..	..	..	..
Muzzle Energy: Per in. circ., ft.-tons	391	628·4	512·4	223	223	220	473·3	401·2	304	161	250·0	227	98·9	111·5	61·3	61·3	61·3	..	..	..	..	..
Perforation at Muzzle, in ins.	20·5	‖30·7	‖26·3	15·4	15·4	15·3	‖26·8	21·0	18·1	13·0	16·4	15·6	10·3	11·0	8·0	8·0	8·0	..	..	..	..	..
Ditto by Tresidder's formula	20·8	30·6	26·7	15·1	15·1	15 0	25·3	22·2	18·0	..	16·8	15·6	10·2	..	..	..	..	..	..	..	..	..

Note.—There are also quick-firing guns; see Table of Krupp Q.-F. guns. * Maximum twist. † Including taper entrance into bore.
‡ In most cases steel shells. § Length including powder chamber. ‖ Iron by Krupp's formula. ¶ Boat gun for landing and working ashore.

ITALIAN NAVAL ORDNANCE.

	Armstrong Breech Loading.				B.L.	Armstrong Muzzle Loading.							Muzzle Loading. Old Pattern.		Breech Loading.		Armstrong Quick Firing.			
Designation by Calibre, in centimètres	43·1‡ New Pattern.	43·1‡ Early Pattern. 1882.	34·3	12·0	12·0	45·0	27·9 New Pattern.	25·4 No. 1 Long.	25·4 No. 2 Short	25·4 No. 2.	22·8	20·3	16	16	7·5 No. 1.	7·5 No. 2.	15·2†	14·9	12·0*	12·0§
Calibre, in inches	17	17	13·5	4·72	4·72	17·72	11	10	10	10	9	8	6·5	6·5	3	3	6·0	5·87	4·7	4·7
Length: Total, in feet	40·75	39	36·09	8·5	9·25	32·7	14·4	14·4	14	13·8	13·8	10·8	11·8	10·6	5·8	3·3	13·8	13·87	16·2	13·0
Length: Rifled Bore, in inches	346·8	315·7	..	75	88	302	121	120	114	112	106	89	96	87	52	27	126	..	189 (Rifled Bore and Powder Chamber)	..
Length: Powder Chamber, in inches	84·5	98	..	10·8	22	56 5	24·5	26·0	26·0	14 0	19·5	15·7	21·3	21 3	10·2	7·9	28	..		..
Length: Bore, in Calibres	27	26	..	20·2	23·5	20·5	13·2	14·6	14·0	12·6	13·9	13·1	16·8	15·5	20·7	11·7	26	..	40	35
No. of Grooves	82	82	56	37	36	28	9	7	7	8	6	6	6	6	12	12	28	28	22	22
Twist of Rifling, in Calibres	50	50	..	40	42	50	35	40	40	55	45	45	42·5	27·3	48	48	40	40	34·4	..
Total Weight, in tons	104·3	101·5	67·9	1·20	1·38	100	25·0	18·0	18·1	12·1	12·6	6·99	5·12	3·54	0·29	0·095	4	4·2	2·05	1·69
Firing Charge: Armour-piercing projectile, lbs.	900·0	725	630·5	5·5	9·9	551	95·2	77·6 (No. 1 and No. 2 Short)		63·9	59·7	37·7	19·8	..	..	..	39 7	26·5	12·0	..
Firing Charge: Common Shell, ,,	600	480	..	5·5	9·9	63·0	66·6	52·9		41·9	37·7	26·7	7·3	7·1	1·9	0·7	26·5	40	12·0	..
Weight: Armour-piercing projectile, ,,	2000	2000	1250	52·0	52·2	2000	540·1	451·9		331·8	315·3	191·8	103 6	..	..	..	80		45·0	36·0
Weight: Common Shell, ,,	2000	2000	1250	31·7	36·8	2000	526·9	399·0		284·4	250·0	180·0	64·6	65·7	9·4	9·4	80	about 80·0 (Common Shell and Shrapnel)	..	36·2
Weight: Shrapnel ,, ,,	2017	2017	1250	37·3	37·37	2180	533·5	399·0		284·4	250·0	180·0	68·3	..	9·4	9·4	80		..	29·8
Weight: Case Shot ,,	..	..	..	32·4	35·9	..	200·1	188·1		135·6	99·6	79·4	33·1	33·1	9·0	9·0	70	..	..	..
Bursting Charge: Armour-piercing projectile, ,,	32	32	17 4	2 31	2·31	32 ?	15·0	12·3		8·4	6·5	3·8	..	..	..	..	1 5	..	..	1·83
Bursting Charge: Common Shell, ,,	60	60	87·1	2·2	2·2	78 ?	26 0	23·8		18·2	18·8	9·7	2·87	2·87	0·31	0·31	5	..	..	3·02
Bursting Charge: Shrapnel ,, ,,	5	5	4·25	0·35	0·35	5 ?	2·2	2·20		1·96	1·80	1·17	0·55	..	0·03	0·03	0·16	..	..	0·35
Muzzle Velocity, in feet	1992	1935	2016	1345	1591	1700	1353	1388		1373	1284	1311	1290	1024	1335	..	1946	..	1786	..
Muzzle Energy: Total, foot-tons	55,030	51,930	35,230	650·4	916·4	40,060	6857	6035		4369	3604	2286	1195	..	..	..	2100	..	995·4	..
Muzzle Energy: Per inch circumference, foot-tons	1035	976·3	830·8	43·9	61·8	753·4	198·5	192·2		139 1	127·6	91·0	58·5	..	..	..	114·1	..	67·1	..
Perforation at Muzzle, inches of iron	33·7	32·8	30·2	6·7	8·1	27 8	14·3	14·1		12 0	11·4	9·6	7·7	..	..	..	**11·4	..	**8·8	..
,, by Tresidder's formula	36·7	35·0	33·0	..	8·3	28·2	..	..		..	..	..	..	..	..	..	11·8	..	9·2	..
Metal employed in structure	St.	I. & St.		St.	St.	Steel tube in Wrought Iron jacket.							I. & St.	Cast I.	Br.	Br.			St.	

St. stands for steel, I. for iron, Br. for Bronze.

* For Piemonte. ** By Krupp's formula. † For Piemonte, Fieramosca, Re Umberto, Ancona, Doria.

‡ There are four types of these bores, viz.—types Lauria, Lepanto, Italia, Valente. § For Duilio, Dandolo, Formidabile. The Piemonte has a 40-calibre gun.

RUSSIAN NAVAL ORDNANCE.

	Obukhoff Steel Breech Loading Hooped Guns.																Steel B.L. Guns.		
Designation by Calibre, in inches	12	12 Long.	12 M. 77.	11 M. 67.	11 Pattern 77.	9	9 (a) M. 67.	9 (b)	8	8	8 M. 67.	6 Long.	6	6·03	6	Long 9-pdr.	4·2 (9-pdr.)	3·43 Long (4-pdr.)	3·43 (4-pdr.)
Calibre in centimètres	30·48	30·48	30·48	27·94	27·94	22·86	22·86	22·86	20·32	20·32	20·32	15·24	15·24	15·32	15·24	10·67	10·67	8·70	8·70
Total Length, in feet	**35	30	20	18·3	20·0	**26·25	15·0	13	**23·33	**20	14·6	**17·5	14	12·2	11·7	6·9	7·0	6·9	5·8
Length of Rifled Portion of Bore, in inches	..	..	165·0	152·0	158·0	..	124·0	..	..	..	128·0	..	118·7	106·0	98·0	61·5	65·0	62·6	53·0
Length of Powder Chamber, in inches	..	..	38·5	35·0	50·4	..	28·5	..	..	..	23·0	..	30·5	22·4	22·2	10·5	8·0	10·7	..
Length of Bore in calibres, including Powder Chamber	**35	..	17	17	18·9	**35	16·9	..	**35	**30	18·9	**35	24·9	21·3	20	17·1	17·4	21·4	..
Number of Grooves, in inches	..	..	36	36	64	..	32	32	..	..	30	..	..	24	24	24	16	24	12
Depth of Grooves "	..	0·070	0·135	0·135	0·135	..	0·110	0·110	..	..	0·090	..	0·060	0·085	0·070	0·055	0·055	0·050	0·050
Twist of Rifling in calibres	..	..	73·5	70	..	..	60	60	..	..	70	..	*24	60	68	*40	50	40	41
Total Weight, in tons	55·7	50·45	39·9	28·2	28·2	19·44	15·0	12·5	13·64	12·74	9·65	6·26	4·08	4·35	4·03	0·60	0·87	0·45	0·35
Weight of: Steel Shell, in lbs.	..	..	665·8	515·9	..	..	249·1	275·6	..	..	172·0	..	90·9	97·6	86·0	..	..	..	..
Weight of: Chilled Shell, "	..	731·9	665·8	515·9	562·2	..	275·6	264·7	..	193·1	169·8	..	119·0	86·0	86·0	..	..	..	..
Weight of: Common Shell, "	626·4	..	639·3	496·0	520·3	268·2	266·8	266·8	192·3	172·4	172·0	73·35	..	81·6	81·6	27·6	24·2	15·2	12·6
Weight of: Case Shot, "	..	..	293·2	216·1	..	..	176·4	176·4	..	..	134·5	..	..	57·3	57·3	27·6	22·3	15·2	11·0
Weight of Firing Charge: Steel Shell, "	..	..	144·4	115·3	..	..	64·2	47·0	..	..	31·5	89·38	..	14·3	18·1	..	..	..	..
Weight of Firing Charge: Chilled Shell, "	..	246·9	144·6	90·6	132·2	..	47·0	47·0	..	72·0	29·3	39·6	37·8	14·3	18·1	..	..	..	..
Weight of Firing Charge: Common Shell, "	..	..	117·3	81·6	132·2	180	42·1	42·1	88·2	72·0	28·4	39·6	..	10·8	14·3	4·5	2·6	3·1	1·3
Muzzle Velocity, in feet	..	1942	1470	1486	1516	2376	1463	1260	1925	1796	1352	2080	‡1739	1206	1463	1225	..	1444	..
Muzzle Energy: Total, foot-tons	..	19140	9974	7903	8960	10500	4095	3035	..	4321	2180	2682	1905	982	1276	..	..	..	..
Muzzle Energy: Per Inch Circumference, foot-tons	..	508·4	264·6	228·8	259·3	371·4	144·7	107·4	..	172·0	86·7	142·3	101·1	51·8	67·74	..	..	..	..
Perforation at Muzzle, in inches	..	23·6	16·7	15·5	16·5	20·2	12·3	10·5	..	13·5	9·5	12·50	10·5	7·2	8·4	..	..	..	..
Ditto by Tresidder's formula	..	25·3	..	..	..	†24·0	..	..	†15·7	†14·2	..	..	†12·2	..	..	..	..	..	..

** It is doubtful if this refers to the total length of gun or of bore. * Maximum of increasing twist. † The weight of the projectile is uncertain. ‡ With pyroxiline. *a* New. *b* Converted. There exist also 15 and 10·7-cm. Krupp guns.

NOTE.—The Russians certainly possess some more powerful pieces than are here shown.

NOTE.—M. Claudinon in the French Chamber recently stated that Russia has a 12-in gun (30·5 c.m.) with a muzzle energy of 10,700 metre tons or 34,551 ft. tons.

SPANISH NAVAL ORDNANCE.

	Hontoria, Pattern 79. B.L.				Hontoria, Pattern 83. Breech Loading.							
Designation by Calibre	18-cm	16-cm No. 1.	16-cm No. 2.	16-cm No. 3.	32-cm.	28-cm.	24-cm.	20-cm.	18-cm	16-cm.	14-cm	12 cm.
Calibre, in inches	7·09	6·30	6·30	6·30	12·60	11·02	9·45	7·87	7·09	6·34	5·51	4·72
Length: Total length, in feet	15·57	13·8	9·50	9·65	38·7	33·8	29·0	..	21·75	19·3	16·91	14 5
Length: Rifled Portion, in inches	141·2	125·6	83·1	..	352·4	309·1	..	..	..	170·6	149·1	126·0
Length: Powder Chamber, in inches	..	31·9	17·3	..	86·8	77·1	..	..	..	49·8	53·9	39·4
Length: Bore, in calibres	..	25	17	..	50	50	30	..	30	35	35	35
No. of Grooves	42	38	3·8	..	80	70	60	50	45	40	35	30
Depth of Grooves, in ins.	0·06	0·06	0·06	..	0·06	0·06	0·05	0·06	0·04	0·04	0·04	0·04
Twist of Rifling, in cals.		Increasing from 16 to 40.	56	..	From 9 to 90.							
Total Weight, in tons	7·87	5·6	48·0	..	47·3	32·5	20·7	11·5	8 71	6·1	4·1	2·6
Weight: Armour piercing projectile, in lbs.	135·6	93·7	93·7	..	1041	694·3	438·7	253·5	187·4	130·1	86·0	53·1
Weight: Common Shell, in lbs.	113·8	83·6	83·6	..	879·6	586·4	370 4	213·8	..	112·4	75·0	47·2
Weight: Ring Segment, in lbs.	..	83·8	83·6 / ..	63·9 / 63·9	886·3	590·8	370·4	211·6	..	112·4	75·0	47·6
Firing Charge: Armour-piercing projectile, in lbs.	..	26·5	..	15·4	485·0	352·7	220·5	112·4	94·8	66·1	44·1	28·7
Firing Charge: Other projectiles	..	24·8	..	15·4	463·0	319·7	220·5	..	..	61·7	..	28·7
Muzzle Velocity, in feet	..	1631	1493	..	2034	2034	2034	2034	2034	2051	2001	1988
Muzzle Energy: Total, in ft.-tons	..	1550	1448	..	29850	24030	12580	7271	5374	3806	2386	1511
Muzzle Energy: Per inch circumference, ft.-tons	..	87·4	3·2	..	754·3	694·0	423 9	294·1	241·4	191·1	137·8	101·9
Perforation at Muzzle, in inches	..	†9·3	8·8	..	†32·9	†28 7	†24·6	†20·5	†18·6	†16·6	†13 9	†11·6
Metal and Construction	St. & Cast Iron.				St. Jacket and Hoops.							

	Armstrong, Pattern 83.				Armstrong. Muzzle Loading.		Armstrong. Pattrn. 81 B.L.	Krupp. Breech Loading.				Converted. Q.F. guns.			Garcia de Loma. Q.F.
Designation by Calibre	15-cm.	12-cm.	8·7-cm	7·5-cm long.	22·86-cm	20·3-cm	6-in.	15-cm.	12-cm.	8·7-cm	7·5-cm	14-cm. M. 83	12-cm. M. 83	7-cm. M. 79	10c m
Calibre, in inches	6·00	4·72	3·4	2·95	9·00	8·00	6·00	5·87	4·72	3·43	2·95	5·51	4·7	2·76	3·9
Length: Total length, in feet	16·97	13·75	7·9	7·50	13·0	11·0	14·5	17·13	11·81	6·9	6·6	17·4	..	3·71	18·2
Length: Rifled Portion, in inches	158·3	135·8	75·0	70·7	104·0	102·0	126·9	..	..	..	57·6	..	..	..	..
Length: Powder Chamber, in inches	31·4	19	13	13	..	..	29·7	..	..	..	..	..	..	..	..
Length: Bore, in calibres	32	33	27	28·7	14	14·75	26·1	35*	30*	24*	25·8*	..	..	..	..
No. of Grooves	28	22	20	18	6	4	28	36	32	24	24	..	..	..	..
Depth of Grooves, in ins.	0·037	0·03	0·03	0·03	0·18	0·18	..	0·06	0·06	0·05	0·05	..	..	..	..
Twist of Rifling, in cals.	30	40	30	35	45	40	100 to 40	25	25	40	36	..	..	..	..
Total Weight, in tons	5·0	2·2	0·45	0·35	12·0	9·0	4·0	4·7	2·1	0·44	0·30	4·23	..	0·98	1·67
Weight: Armour piercing projectile, in lbs.	97·0	39·2	..	..	250·0	180·0	78·3	84·9	43·65	..	..	‡154·3	‡112·1	..	32·4
Weight: Common Shell, in lbs.	92·6	36·4	14·1	11·5	250·0	180·0	73·6	65·5	34·61	14·6	9·48	‡145·5	‡108	7·8	28·2
Weight: Ring Segment, in lbs.	..	38·6	15·4	11·7	..	..	83·6	..	34·61	14·6	9·04	‡145·5	‡108	8 2	28·7 / 17·2
Firing Charge: Armour-piercing projectile, in lbs.	48·5	16·0	..	..	50·0	35·0	34·0	37·48	19·29	..	..	..	..	..	17·2
Firing Charge: Other projectiles	30·0	11·9	4·0	4·0	33·0	21·0	24·9	25·4	..	10·3	10·4	..	..	..	..
Muzzle Velocity, in feet	2070	2000	1625	1709	1339	1339	1929	2001	1887	1539	1552	..	..	1136	..
Muzzle Energy: Total, in ft.-tons	2882	1087	258	233	3105	2239	2018	2357	1076	..	..	..	..	..	..
Muzzle Energy: Per inch circumference, ft.-tons	153·3	73·33	..	..	110·0	89·1	1071	127·8	72·6	..	..	..	..	..	..
Perforation at Muzzle, in inches	†14·4	†9·3	..	..	10·6	9·6	†11·0	†12·7	†9·7	..	..	..	..	..	..
Metal and Construction	St.				St. and Wt. I.		St.	St.							

18 and 16-cm. Palliser guns and 16 and 13-cm. Parrot guns also exist, and some bronze muzzle loaders. St. stands for Steel; I. for Iron.

There is also a 20-cm. (7·87-in.) B.L. Hontoria, Pattern 79, weighing 0·8 tons, firing an armour-piercing projectile weighing 180·8 lbs. with a charge of 61·73 lbs.

* Total length, the length of bore not being supplied. † By Krupp's formula. ‡ These weights include the charge and case.

NAVAL ORDNANCE OF SWEDEN AND NORWAY.

Sweden.

	Breech Loaders.			Model 78.		Model 81.		Model 83.		M. 85.	M. 86.	M. 89.	M. L.
Designation by Calibre, in cms.	27	24	17	27	24	27	12	15	8	25	6·5	15	12
Calibre, inches	10·80	9·45	6·58	10·80	9·45	10·80	4·72	6·00	3·31	10·00	2·60	6·0	4·80
Total Length, feet	17·46	14·96	11·27	17·65	16·24	23·10	10·29	13·87	7·37	28·33	3·79	16·98	8·87
Length { Rifled Portion of Bore, ins.	160·8	137·0	107·8	159·2	150·5	191·6	94·5	124·1	71·3	2609	35·0	155·2	83·3
Length { Chamber, "	29·9	25·9	16·5	32·3	28·1	66·2	20·6	31·1	9·7	58·1	4·2	35·2	13·6
Length { Bore in calibres, "	17·2	17·1	18·7	17·8	18·9	23·9	24·0	25·7	24·3	32·9	15·4	32	20·2
Number of Grooves	5	5	5	42	36	45	30	28	24	42	26	28	8
Twist of Rifling	30*	30*	30*	45*	..	40*	30*	30	33*	40*	22*	30	40
Total Weight, tons	23·6	14·4	5·2	23·6	16·4	27·1	1·9	4·2	4·2	29·8	9·4	5·2	1·9
Weight of { Armour-piercing Shell in lbs.	476·2†	317·5†	107·1†	476·2†	317·5†	476·2†	..	100·0	..	449·7	..	100	..
Weight of { Common Shell, in lbs.	396·8	224·9	97·7	396·8	273·4	396·8	48·5	100·0	14·8	401·2	6·2	100	34·6
Weight of Firing Charge { Armour-piercing Shell, in lbs.	83·8	59·5	22·0	90·4	56·2	206·4	..	35·3	..	242·5	..	54·0	..
Weight of Firing Charge { Common Shell, lbs.	83·8	59·5	22·0	90·4	56·2	145·5	16·0	35·3	3·3	242·5	0·9	..	6·6
Muzzle Velocity, feet.	1322	1312	1365	1378	1365	1788	1640	1663	1542	2100	1148	2067	..
Total foot-tons	5771	3789	1384	6272	4102	10550	..	1918	..	13750	..	2964	..
Muzzle Energy { Per inch Circumference.	170·1	127·6	66·9	184·9	138·2	311·3	..	101·7	..	437·7	..	157·2	..
Muzzle Energy { Perforation through Iron**	13·19	11·4	8·3	13·8	11·9	18·4	..	10·4	..	21·9	..	13·1	..
Muzzle Energy { Ditto by Tresidder's formula	..	..	..	..	..	18·9	..	10·5	..	24·5	..	13·9	..

Norway.

	Krupp, B.L.					Armstrong, M.L.				Palliser, M.L.	
Designation by Calibre, in cms.	26 No. 2.	26 No. 1.	15	12 No. 2.	12 No. 1.	26·7 No. 3.	26·7 No. 2.	26·7 No. 1.	20·2	16·7	15·5
Calibre, inches	10·24	10·24	5·91	4·72	4·72	10·51	10·51	10·51	7·94	6·58	6·11
Total Length, feet	25·59	18·77	12·63	13·78	9·60	16·87	14·65	13·45	10·82	11·58	10·30
Length { Rifled Portion of Bore, ins.	218·9	160·4	112·4	128·6	85·9	138·7	121·0	110·6	85·7	92·4	91·7
Length { Chamber, "	55·4	34·1	22·6	36·8	16·5	36·8	24·0	20·6	18·5	19·3	10·8
Length { Bore in calibres, "	30	19·0	22·8	35	25	16·7	13·8	12·5	13·2	17·0	16·8
Number of Grooves	60	60	36	32	32	8	8	8	6	3	3
Twist of Rifling	∝25	45	45	∝25	40	55	55	55	50	34	34
Total Weight, tons	24·8	21·7	3·9	2·31	1·38	21·7	19·7	18·2	7·4	4·9	3·4
Weight of { Armour-piercing Shell in lbs.	606·3	463·0	86·0	57·3	44·1	448·6	393·5	384·9	157·4	109·8	..
Weight of { Common Shell, in lbs.	606·3	381·4	69·4	57·3	36·1	316·4	316·4	316·4	153·9	82·7	59·1
Weight of Firing Charge { Armour-piercing Shell, in lbs.	191·8	99·2	22·0	19·8	9·9	110·2	82·7	66·1	29·8	22·0	..
Weight of Firing Charge { Common Shell, lbs.	191·8	81·6	20·9	19·8	9·9	77·2	77·2	48·5	29·8	16·5	7·72
Muzzle Velocity, feet.	1722	1575	1624	1804	1493	1549	1444	1296	1247	1329	1116
Total foot-tons	12460	7966	1573	1290	680	7463	5692	4484	1696	1345	..
Muzzle Energy { Per inch Circumference.	387·4	247·7	84·7	87·1	45·9	226·0	172·4	135·8	68·0	65·1	..
Muzzle Energy { Perforation through Iron**	20·4	16·2	9·5	9·7	7·0	15·5	13·4	11·8	8·3	8·2	..
Muzzle Energy { Ditto by Tresidder's formula	20·9	15·9	9·4	10·5	..	15·1	..	..	..	..	..

Sweden.—The breech-loaders have breech screw-stoppers. The whole of the guns which do not fire shrapnel discharge case-shot.
Norway.—Besides the chilled shell, there are chilled solid shot for the 26·7-cm. and the 20·2-cm. guns, and for all muzzle-loaders case-shot also, besides steel shrapnel for some Krupp guns. * Maximum rate of increasing twist. ** By Fairbairn's formula. † The 16·7 muzzle-loading gun fires steel solid shot.

UNITED STATES NAVAL ORDNANCE.

NATURE OF GUN.	Calibre.	Weight.	Total Length.	Total Length of Bore.	Length of Rifling.	Twist of Rifling.	Length of Chamber.	Weight of Service-charge (not Smokeless Powder.)	Weight of Projectile.	Muzzle Velocity (Service). Brown Powder.	Muzzle Energy. Brown Powder.	Perforation of Wrought Iron at Muzzle.‡
	inch.	tons.	feet.	inch.	inch.		inch.	lbs.	lbs.	ft.-seconds.	ft.-tons.	inch.
3-in. (11 pr.)	3	0·37	12·5	149·7	125·5	..	21·3	..	14	3000‡	·874	13·5
4-in. Q.-F., Mark I.	4	1·5	13·7	157·3	130·3	zero to 1 in 25	24·7	12 to 14	33	2000	915	9·8
4-in. Q.-F. Gun	4	1·5	13·7	157·5	128·1	..	25·4	..	33	2000*	..	9·8
4-in. Q.-F., Mark VII., of 50 Cals.	4	2·56	17·0	200·0	168·4	..	31·6	..	32	3000†	1,909	17·6
5-in. Q.-F., Mark I.	5	2·8	13·5	150·3	120·8	1 in 180 to 1 in 30	27·1	26 to 29	60	2000	1,660	11·8
5-in. Q.-F. Gun	5	3·1	17·4	191·5	164·4	zero to 1 in 25	32·0	28 to 30	50	2300	1,834	13·2
5-in. Q.-F., Mark V.	5	4·46	21·3	250	212·9	..	37·2	..	60	2900†	3,503	20·5
6-in. B.L.R., Mark I.	6	4·8	15·8	176·0	136·7	1 in 180 to 1 in 30	36·9	50	100	2000	2,773	..
6-in. B.L.R., Mark II.	6	4·9	16·1	180·1	144·9	..	32·7	45 to 48	100	2000	..	13·8
6-in. B.L.R., Mark III., of 30 Cals.	6	4·8	16·3	183·8	147·3	zero to 1 in 25	34·0	44 to 47	100	2000	..	..
6-in. B.L.R., Mark III., of 35 Cals.	6	5·2	18·8	213·8	177·3	..	34·0	..	100	2080	2,990	14·7
6-in. B.L.R., Mark III., of 40 Cals.	6	6·0	21·3	243·8	207·3	..	34·0	..	100	2150	3,204	15·4
6-in. Q.-F. Gun	6	6·0	21·3	243·8	204·3	..	37·0	44 to 47	100	2150	3,200	15·4
6-in. Q.-F., Mark VI.	6	8·17	25·0	293·7	245·3	..	48·4	..	100	2900†	5,838	24·2
8-in. B.L.R., Mark I.	8	12·3 12·9	21·5	239·9	195·2	1 in 180 to 1 in 30	42·1	105 to 115	250 250	2000 2000	6,932	19·0
8-in. B.L.R., Mark II.	8	13·0	21·5	239·9	195·2	..	42·1	..	..	..	..	19·0
8-in. B.L.R., Mark III., of 35 Cals.	8	13·1	25·4	290·5	242·8	zero to 1 in 25	45·1	..	250	2080	7,498	20·1
8-in. B.L.R., Mark III., of 40 Cals.	8	15·2	28·7	330·5	282·8	..	45·1	..	250	2150	8,011	21·1
8-in. B.L.R., Mark V., of 45 Cals.	8	18·0	28·6	335·0	271·0	..	64·0	..	250	2800†	13,602	31·4
10-in. B.L.R., Mark I., of 30 Cals.	10	25·7	27·4	306·3	247·3	1 in 180 to 1 in 35	57·2	225 to 240	500	2000	13,864	24·0
10-in. B.L.R., Mark I., of 35 Cals.	10	27·1 28·2	30·5	343·8	283·7	zero to 1 in 25	57·2	..	500	2060	14,709	25·0
10-in. B.L.R., Mark II., of 30 Cals.	10	25·1	27·4	307·3	247·3	zero to 1 in 26·8	57·2	..	500	2000	13,864	24·0
10-in. B.L.R., Mark II., of 35 Cals.	10	27·6	31·2	354·9	294·9	zero to 1 in 25	57·2	..	500	2100	15,285	25·8
10-in. B.L.R., Mark III., of 40 Cals.	10	33·4	33·3	389·0	313·4	..	75·6	..	500	2800†	27,204	42·0
12-in. B.L.R., Mark I.	12	45·2	36·8	419·2	343·1	..	74·1	425	850	2100	25,985	30·8
12-in. B.L.R., Mark III., of 40 Cals.	12	52	41·8	480·1	388·1	..	91·9	..	850	2800†	46,246	47·2
13-in. B.L.R., Mark I.	13	60·5	40·0	454·5	370·5	..	80·9	550	1100	2100	33,627	33·5

NOTE.—The weight of fixed ammunition for Q.-F. 4-in. and 5-in. guns is 58 and 95 lbs. respectively. * With smokeless powder 4″ = 2200, 5″ = 2650, 6″ = 2550, 8″ = 2300, 10″ = 2200, 12″ = 2300, 13″ = 2300. The charges are kept down to suit the sight bars. † It should be pointed out that these high velocities cannot be fairly compared with the velocities on the British table, because the former are obtained with new guns, while the latter are purposely taken as what might be expected on service after the gun has fired a considerable number of rounds. ‡ By Tresidder's formula.

ELSWICK QUICK-FIRING GUNS.

This Table is supplied by the Manufacturers.

	*	*	*	*		*		*	* Field and Naval	* Field and Naval	*
Diameter of Bore, ins.	1·46	1·46	1·46	1·85		2·244		2·244	3·0	3·0	3·0
do. do. m.m.	37	37	37	47		57		57	76·2	76·2	76·2
Length of Bore, cals.	20	25	45	40		40		50	23	28	40
do. Gun, cals.	22·7	27·8	46·7	43·6		43·6		53·6	24	29·2	41·2
Weight of Gun	lbs. 73	lbs. 79	lbs. 268	lbs. 506		lbs. 800		cwt. 10	cwt. 5	cwt. 7·5	cwt. 12·0
do. Projectile, lbs.	1·1	1.1	1·5	3·3				6	12·5	12·5	12·5
do. Battering Charge ...	ozs. 1·125	ozs. 1·125	ozs. 4·5	ozs. 6·375	ozs. 7·94	ozs. 7·75	ozs. 9·2	...	ozs. 13·62	ozs. 13·5	lbs. 1·62
Muzzle Velocity, f.-s.	1319	1540	2300	1873	2132	1820	1968	2592	1550	1585	2210
Velocity at 2,500 yards, f.-s. ...	629	564	819	867	923	932	971	1168	945	955	1082
Muzzle Energy, f.-t.	13·3	18·1	56·0	80·3	104	137·8	161	279·5	208·2	217·8	423·3
Energy at 2,500 Yards, f.-t. ...	...	...	...	17·2	19·5	36·1	39·2	56·6	77·4	79·1	101·5
Perforation by Tresidder's formula, ins.	1·5	1·9	...	4·3	5·2	6·[illegible]	5·6	8·6	4·8	4·9	8·1
Rounds per Minute, Guns should be capable of	...	...	25	30	30	25	25	25	20	20	20

	*		*	*	*	*	*	*	*	*	*	*
Diameter of Bore, ins.	3·5	4	4	4	4·7	4·7	4·7	6	6	6	6	8
do. do. m.m.	88·9	100	100	100	120	120	120	152	152	152	152	203
Length of Bore, cals.	40	40	40	48·7	40	43·9	48·9	40	40	45	50	40
do. Gun, cals.	41·3	41·3	41·3	50	41·1	45	50	41·54	41·54	46·54	51·54	41·63
Weight of Gun	cwt. 24	cwt. 32	cwt. 32	cwt. 36	cwt. 42	cwt. 53	cwt. 55	tons. 6·6	tons. 6·6	tons. 7	tons. 7·6	tons. 15·5
do. Projectile, lbs.	20	25	30	27·5	45	45	45	100	100	100	100	210
do. Battering Charge ...	lbs. 3·75	lbs. 4·8	lbs. 4·8	lbs. 4·8	lbs. 5·5	lbs. 8·15	lbs. 8,15	lbs. 15	lbs. 18·8	lbs. 18·3	lbs. 18·8	lbs. 32
Muzzle Velocity, f.-s.	2420	2540	2325	2500	2150	2570	2630	2300	2500	2570	2610	2250
Velocity at 2,500 yards, f.-s. ...	1296	1313	1332	1366	1282	1535	1574	1594	1712	1763	1792	1647
Muzzle Energy, f.-t.	812·2	1118	1124	1192	1442	2061	2158	3668	4334	4580	4723	7372
Energy at 2,500 Yards, f.-t. ...	232·9	299	369	356	513	735	773	1762	2032	2155	2227	3950
Perforation by Tresidder's formula, ins.	10·9	11·8	11·8	12·3	11·6	15·3	15·6	16·1	19·5	20·4	20·7	20·3
Rounds per Minute, Guns should be capable of	15	15	15	15	10	10	10	7	7	7	7	4

	*		*		*	*	*	*		*	‡
Diameter of Bore, ins.	8		8		8·24	9·2	9·449	10		12	12
do. do. m.m.	203		203		203	234	240	254		305	305
Length of Bore, cals.	40		45		43.8	40	40	40		40	40
do. Gun, cals.	41·63		46·7		45·5	41·85	41·8	41·64		41·7	41·7
Weight of Gun	tons. 16·5		tons. 18·5		tons. 18.1	tons. 27·65	tons. 26·3	tons. 30		tons. 48·9	tons. 50·8
do. Projectile, lbs.	210	250	210	250	308·6	380	352	450	500	850	850
do. Battering Charge ...	lbs. 47		lbs. 47		lbs. 47	lbs. 67	lbs. 67·25	lbs. 81·56		lbs. 141	lbs. 220
Muzzle Velocity, f.-s.	2575	2440	2650	2480	2300	2360	2500	2400	2300	2400	2580
Velocity at 2,500 yards, f.-s. ...	1883	1893	1935	1902	1848	1878	1927	1892	1874	2022	2179
Muzzle Energy, f.-t.	9655	10321	10226	10662	11320	14676	15256	17973	18341	33949	39233
Energy at 2,500 Yards, f.-t. ...	5163	6212	5452	6271	7308	9293	9063	11170	12176	24098	27985
Perforation by Tresidder's formula, ins.	24·9	26·0	26·0	26·8	27.0	28·6	28·4	29·9	30·4	38·4	42·8
Rounds per Minute, Guns should be capable of	4	3	4	3	3	2	2	2	2	...	...

Guns from 3 to 6 inches can be fitted with either a metallic cartridge case or modified De Bange pad.
* Existing or service guns.
Velocities of 2813 and 2600 f.-s. are obtained with the 210 and 250 lb. projectiles, respectively, with Battering charges. With special charges and suitable cordite a velocity of 2940 f.-s. has been obtained with 100 lb. projectiles. This high velocity, however, is not desirable, except on very rare occasions, on account of the excessive wear of the gun.

SOME RESULTS ACTUALLY OBTAINED.

4·7-in. 42 cwt. gun, with single motion breech mechanism, 5 rounds in 22 seconds, at Silloth, at a target, 2 hits, range 1,000 yards; 7 rounds in 25 seconds at drill.

6-in. Admiralty gun, with three-motion breech mechanism and E.X.E. powder, 10 rounds in 85 seconds, at sea, on board gunboat Kite; 18 rounds in 3 minutes, H.M.S. Royal Arthur, 14 hits on target, ship steaming 8 knots, range from 1,600 to 2,200 yards; 18 rounds in 3 minutes, H.M.S. Blake, 15 hits on target, ship steaming 8 knots, range from 1,600 to 2,200 yards. Total number of rounds fired from 10 guns in same time 148, of which 110 hit the target.

6-in. 6·6-ton gun, with single motion breech mechanism, 7 rounds in 61 seconds, at Silloth, cordite charge; 4 rounds in 20 seconds, at drill.

8-in. 15·5-ton gun, with single motion breech mechanism, 3 rounds in 28 seconds at drill; 4 rounds in 62 seconds, on board cruiser Blanco Encalada, ammunition supplied from magazine.

13·5-in. 68-ton B.L. gun, with hydraulic breech mechanism, 7 rounds in 12 minutes, H.M.S. Royal Sovereign, 6 hits on target, ship steaming 8 knots, range from 1,600 to 2,200 yards; 4 rounds in 6 minutes, H.M.S. Empress of India, with an interval between rounds of only 1 minute 27 seconds.

12-in. 46-ton B.L. gun, interval between 2 rounds, 1 minute 19 seconds, H.M.S. Majestic; 1 minute 4 seconds, H.M.S. Cæsar.

12-in. gun, interval between 2 rounds, 49 seconds, H.M.S. Illustrious; 6 rounds were fired from one turret in 1 minute 47 seconds.

VICKERS, SONS AND MAXIM'S Q.F. GUNS AND MOUNTINGS.

This Table is supplied by the Manufacturers.

	—	* 37 m.m. 30 cal.	* 37 m.m. 42·5 cal.	* 47 m.m. 40 cal.	* 47 m.m. 47·2 cal.	* 57 m.m. 42·3 cal.	* 57 m.m. 50 cal.	* Naval. 76·2 m.m. 45 cal.	* Naval. 76·2 m.m. 50 cal.	* Field. 76·2 m.m. 23·5 cal.	* Mountain. 75 m.m. 10·7 cal.	* 101·6 m.m. 45 cal.	* 101·6 m.m. 50 cal.	* 12 c.m. 40 cal.	* 12 c.m. 45 cal.	* 15·24 c.m. 40 cal.	* 15·24 c.m. 45 cal.	‡ 20·3 c.m. 45 cal.	‡ 23·36 c.m. 45 cal.	‡ 25·4 c.m. 42 cal.	* 30·48 c.m. 40 cal.
Gun.	Diameter of Bore (in ins.)	1·457	1·457	1·85	1·85	2·244	2·244	3	3	3	2·953	4	4	4·724	4·724	6	6	8	9·2	10	12
	Length of Bore (in ins.)	43·5	62·0	73·72	87 34	95	112·2	135	150	70·5	31·6	180	200	188·98	212·58	240	270	360	414	405·15	480
	Total length of Gun (in ins.)	73·75	94·0	77·95	91·5	104·4	116·4	140	155	75·55	35·85	186·1	206	193·28	217	249·2	279·2	372·1	426·8	120	496·5
	Diameter of Chamber	1·44	1·6	2·04	2·04	2·45	2·8	3·6	3·6	3·4	3·0	5	5	5·1	5·5	6·8	8·5	11	13·5	11·5	17·5
	Length of Chamber	2·64	3·73	12·93	12·93	10·2	14·2	15·4	15·4	9·6	4·575	21·2	21·2	25·5	25·75	32·5	33	43	67	63·35	872
	Maximum pressure in Chamber	13 tons	14 tons	13 tons	13 tons	15 tons	15 tons	16 tons	16 tons	14 tons	8 tons	17 tons	17 tons	16 tons	17 tons	16 tons	17 tons	17 tons	17 tons	17 tons	17 tons
	Nature of Charge	Cordite	Cordite	Cordite	Cordite	Cordite	Cordite	Cordite	Cordite	Cordite	Cordite	Cordite	Cordite	Cordite	Cordite	Cordite	Cordite	Cordite	Cordite	Cordite	Cordite
	Weight of Charge	oz. grs. 1 110	oz grs 3 0	oz. 9	oz. 11	oz. 15	lb. oz. 1 4	lb. oz. 2 9	lb. oz. 2 9	lb. 1	ozs. 6·25	lbs. 6	lbs. 6	lbs. 8·5	lbs. 9	lbs. 19	lbs. 25	lbs. 50	lbs. 94·5	lbs. 100	lbs. 207
	Weight of Projectile	1 lb.	1·25 lbs.	3·3 lbs.	3·3 lbs.	6 lbs.	6 lbs.	12·5 lbs.	12·5 lbs.	12·5 lbs.	12·5 lbs.	25 lbs.	25 lbs.	45 lbs.	45 lbs.	100 lbs.	100 lbs.	250 lbs.	380 lbs.	450 lbs.	850 lbs.
	Total weight of Gun, including Breech Mechanism	c. q. l. 4 0 22	c. q. l. 5 1 26	c. q. l. 4 0 0	c. q. l. 4 2 6	c. q. l. 6 2 0	c. q. l. 8 0 0	c. q. l. 14 2 14	c. q. l. 15 3 0	c. q. l. 5 2 23	c. q. l. 2 0 13	t. c. q. l. 1 13 0 0	t. c. q. l. 1 16 0 0	t. c. q. l. 2 10 0 0	t. c. q. l. 2 14 0 0	t. c. q. l. 6 15 0 0	t. c. q. l. 7 8 0 0	t. c. q. l. 18 16 2 0	t. c. q. l. 26 16 0 0	t. c. q. l. 28 4 0 0	t. c. q. l. 50 7 0 0
	Muzzle Velocity in feet per second	1800	2350	2125	2400	2300	2500	2600	2700	1700	918	2700	2800	2494	2600	2530	2775	2625	2750	2580	2600
	Muzzle Energy in foot tons	22·5	48	103	132	220	260	586	632	250·4	73	1263	1359	1940	2109	4437	5340	11945	19927	20811	19843
	Penetration of Wrought Iron Plate at Muzzle by Gavre's formula	1·9	3·3	4·5	5·3	6·2	7	9·2	9·7	..	..	11·6	12·3	13·3	14·1	18·5	21·1	27·6	34·3	32·3	42·3
	Penetration of Steel Plate at Muzzle by Gavre's formula	1·5	2·6	3·5	4·1	4·8	5·4	7·1	7·5	..	..	9	9·5	10·3	10·9	14·4	16·4	21·4	26·6	25·0	32·8
	†Perforation of Wrought Iron at Muzzle (Tresidder's formula)	..	..	..	..	7·1	7·9	10·25	10·8	5·4	..	13·3	14·0	14·6	15·5	19·8	22·6	28·5	32·7	33·4	42·4
	Rounds per minute	300	300	30	30	28	28	20	20	20	14	15	15	12	12	8	8	5	..	..	..
Mounting.	Weight of Mounting complete with Shield	c. q. l. 3 0 22	c. q. l. 3 2 5	c. q. l. 9 3 0	c. q. l. 10 0 18	c. q. l. 12 2 0	c. q. l. 12 0 0	t. c. q. l. 1 6 0 0	t. c. q. l. 1 7 0 0	Carriage and limber with accessories and 40 rounds of ammunition.	c. q. l. 5 3 5	t. c. q. l. 4 4 3 0	t. c. q. l. 4 10 0 0	t. c. q. l. 5 17 0 0	t. c. q. l. 6 5 0 0	t. c. q. l. 8 15 0 0	t. c. q. l. 9 5 0 0	t. c. q. l. 16 8 0 0			
	Thickness of Shield (in ins.)	¼	no shield	¼	¼	3/16	¼	2	2		no shield	4	4	4	4	4	4	4		Depending on type of mounting used.	
	Weight of Shield	c. q. l. 0 3 11	..	c. q. l. 1 3 0	c. q. l. 1 3 0	c. q. l. 1 0 0	c. q. l. 1 3 9	c. q. l. 9 2 0	c. q. l. 10 0 0	t. c. q. l. 1 4 0 0	..	t. c. q. l. 2 6 0 0	t. c. q. l. 2 6 0 0	t. c. q. l. 3 18 0 0	t. c. q. l. 3 18 0 0	t. c. q. l. 4 0 0 0	t. c. q. l. 4 0 0 0	t. c. q. l. 7 0 0 0			
	Angle of Elevation	16°	13°	18°	18°	20°	20°	20°	20°	17°	26°	20°	20°	20°	20°	16°	16°	15°			
	Angle of Depression	25°	25°	15°	12°	20°	15°	10°	10°	5°	10°	7°	7°	7°	7°	7°	7°	5°			

Older types of 9·2-in., 10-in., 12-in. and 13·5-in. guns, as manufactured by Vickers, Sons & Maxim, Limited, are not enumerated in the table, but only modern guns making use of smokeless powders.

* Existing guns. † The perforations through wrought iron (Tresidder's) are added by the compiler for purposes of comparison. ‡ Under manufacture.

SCHNEIDER - CANET QUICK - FIRE GUNS.

This Table is supplied by the Manufacturers.

* Guns in actual use.

Calibre, in centimètres	24		20			152·4		15			14	12			
Calibre, in inches	9·45		7·87			6		5·91			5·5	4·72			
Length, in calibres	40	* 45	40	* 45	50	* 45	* 50	* 40	* 45	* 50	* 45	* 30	* 40	* 45	* 50
Length, in feet	31·5	35·4	26·2	29·5	32·7	25	27	19·6	22·2	24·6	20·8	11·8	15·7	17·7	19·7
Weight of Gun, in tons	21·45	22·73	11·61	13·18	14·96	5·65	5·08	4·72	5·61	6·20	4·72	1·61	2·36	2·75	3
Weight of Projectile, in lbs.	331	331	198	198	198	94 9	94·9	83·2	88·2	88·2	88·2	39·7	46·3	46·3	46·3
Muzzle Velocity, in ft.-secs.	2690	2790	2630	2760	2860	2670	2750	2560	2660	2740	2370	1970	2530	2630	2710
Muzzle Energy, in ft.-tons	16612	17838	9484	10450	11258	4559	4959	4000	4330	4590	31432	1114	2048	2216	2348
Velocity at 2000 mètres, in ft.-secs.	2075	2147	1944	2039	2111	1860	1900	1744	1813	1858	1690	1170	1581	1645	1691
Energy at 2000 mètres, in ft.-tons	9900	10600	5250	5750	6120	2254	2390	1860	2000	2150	1739	387	800	870	955
Perforation at muzzle, Gavre formula, in ins.	34·1	36·04	27·3	29·5	31·1	22 2	23·3	19·6	20·8	21·8	19·3	9·7	15·1	16·2	16·8
,, ,, Tresidder formula, through wrought-iron, in ins.	31·3	33·1	25·6	27·5	29·0	21·0	22·0	19 1	20·2	21·0	17·6	10·4	15·2	16·1	16·8
Perforation at 2000 mètres, in ins.	22·9	24	17·2	18·5	19·5	12·7	13·3	10 9	11·5	12·1	12·4	4·3	6·3	6·7	6·9

Calibre, in centimètres	10			9	7·5			6·5			5·7		47	37
Calibre, in inches	3·94			3·5	2·95			2·57			2·24		1·85	1·46
Length, in calibres	* 45	* 50	* 60	* 30	* 45	* 50	* 60	* 45	* 50	* 60	* 50	* 60	* 60	* 60
Length, in feet	14·8	16·4	19·7	8·8	11·04	12·3	14·8	9·59	10·65	12·8	9·34	11·2	9·24	7·27
Weight of Gun, in tons	1·69	1·83	2·18	0·71	0·8	0·88	1·08	0·49	0·54	0·68	0·45	0·59	0·26	0·14
Weight of Projectile, in lbs.	28·6	28·6	28·6	16·7	13·2	13·2	13·2	9·8	9·8	9·8	5·9	5·9	3·3	1·7
Muzzle Velocity, in ft -secs.	2490	2630	2820	1970	2430	2530	2760	2360	2459	2721	2426	2721	2688	2622
Muzzle Energy, in ft.-tons	1235	1367	1587	451	540	587	696	342	372	455	243	307	167	34·5
Velocity at 2000 mètres, in ft.-secs.	1475	1557	1672	1020	1291	1337	1462	1167	1203	1327	1101	1209	1045	826
Energy at 2000 mètres, in ft.-tons	439	485	560	121	155	165	197	83·5	90	108	51	50	25	8·3
Perforation at muzzle, Gavre formula, in ins.	12·6	13·7	15·3	7·4	9·3	9 99	11·3	7·75	8·3	9·4	6·95	8·30	6·4	5·08
,, ,, Tresidder formula, through wrought-iron, in ins.	12·7	13·8	15·3	7·2	9·4	10·1	11·5	8·5	9·0	10·5	7·3	8·7		
Perforation at 2000 mètres, in ins.	5·06	6·16	6·8	3·10	3·55	3·75	4·3	2·62	2·78	3·9	2·06	2·3	1·5	0·85

KRUPP QUICK-FIRE GUNS, Model 1897.

Table supplied by Manufacturers.

LIGHT GUNS.

Calibre, in centimetres	7·5			10·5			12			15			21			24			28			30·5		
Calibre, in inches	2·95			4·13			4·72			5·91			8·27			9·45			11·02			12·01		
Total Length of Gun, in cals.	40	45	50	40	45	50	40	45	50	40	45	50	40	45	50	40	45	50	40	45	50	40	45	50
Total Length, in feet	9·84	11·07	12·30	13·78	15·5	17·22	15·75	17·7	19·69	19·55	22·00	24·44	27·56	30·9	34·45	31·50	35·4	39·37	36·75	41·3	45·93	40·03	45·0	50·03
Length of Bore, in inches	108·66	123·43	138·19	153·55	174·21	194·89	175·20	199·25	222·45	218·12	247·49	276·78	305·91	347·49	388·59	350·80	398·28	445·28	409·46	464·62	519·70	445·67	505·95	565·76
Weight of Piece, in lbs.	1488·2	1710·8	1935·7	4078	4740	5401	6107	7055	7937	11767	13448	15212	32294	37254	41890	48722	56222	63273	77103	89062	100753	100312	115081	130070
Weight of Piece, in tons	0·66	0·76	0·86	1·82	2·12	2·41	2·73	3·15	3·54	5·23	6·00	6·79	14·42	16·63	18·70	21·75	25·10	28·25	34·65	39·76	44·98	44·78	51·14	58·07
Weight of Steel Projectile, in lbs.	11·5 14·6	11·5 14·6	11·5 14·6	30·86 39·68	30·86 39·68	30·86 39·68	46·30 59·52	46·30 59·52	46·30 59·52	90·39 112·4	90·39 112·4	90·39 112·4	249·1 308·6	249·1 308·6	249·1 308·6	374·8 474·0	374·8 474·0	374·8 474·0	595·2 760·6	595·2 760·6	595·2 760·6	771·6 981·0	771·6 981·0	771·6 981·0
Weight of Charge, in lbs.	2·23	2·58	3·04	5·95	7·10	8·38	11·22	12·74	14·88	21·61	24·27	28·66	59·20	66·27	78·70	85·98	101·45	114·42	142·64	161·05	189·6	184·1	209·44	244·7
Muzzle Velocity, in ft.-secs.	2249 2083	2539 2254	2703 2402	2362 2087	2556 2254	2723 2402	2569 2267	2772 2444	2956 2608	2556 2267	2726 2444	2907 2608	2566 2303	2766 2487	2969 2664	2592 2303	2795 2487	2995 2664	2602 2303	2808 2487	3012 2664	2598 2303	2802 2487	3005 2664
Muzzle Energy, in ft.-tons	418·6	512·1	581·5	1197	1420	1588	2121	2465	2807	4007	4663	5803	11253	12779	15192	17439	20298	22334	27985	31528	37443	36090	40660	48290
Perforation through Steel, in ins.	5·87	6·58	7·20	8·4	9·37	10·28	10·95	12·21	13·4	13·7	15·24	16·7	20·0	22·32	24·6	23·5	26·18	28·9	27·9	31·14	34·3	30·5	34·06	37·6
Perforation through Iron, Tresidder's formula	8·1	9·1	10·0	11·4	12·5	13·8	14·6	16·3	18·0	17·9	20·0	22·1	25·6	28·8	31·8	30·0	33·3	36·9	35·1	39·1	43·2	38·1	42·6	47·1

HEAVY GUNS.

Calibre, in centimetres	7·5			10·5			12			15			21			24			28			30·5		
Calibre, in inches	2·95			4·13			4·72			5·91			8·27			9·45			11·02			12·01		
Total Length of Gun, in cals.	40	45	50	40	45	50	40	45	50	40	45	50	40	45	50	40	45	50	40	45	50	40	45	50
Total Length, in feet	9·84	11·07	12·30	13·78	15·5	17·22	15·75	17·7	19·69	19·55	22·0	24·44	27·56	30·9	34·45	31·50	35·4	39·37	36·75	41·3	45·93	40·03	45·0	50·03
Length of Bore, in inches	108·66	123·43	138·19	153·55	174·21	194·89	175·20	199·25	222·45	218·12	247·49	276·78	305·91	347·29	388·59	350·80	398·28	445·28	409·46	464·62	519·70	445·67	505·95	565·76
Weight of Piece, in lbs.	1860·7	2091·4	2325·9	5115	5732	6393	7628	8598	9524	14639	16535	18298	40345	45633	50486	58642	68560	73193	97000	109132	121250	125224	140874	156580
Weight of Piece, in tons	0·83	0·94	1·04	2·28	2·56	2·85	3·41	3·84	4·25	6·54	7·38	8·17	18·01	20·37	22·54	26·18	30·61	32·68	43·31	48·72	54·13	55·9	62·44	69·9
Weight of Steel Projectile, in lbs.	11·5 14·6	11·5 14·6	11·5 14·6	30·86 39·68	30·86 39·68	30·86 39·68	46·40 59·52	46·30 59·52	46·30 59·52	90·39 112·4	90·39 112·4	90·39 112·4	249·1 308·6	249·1 308·6	249·1 308·6	374·8 474·0	374·8 474·0	374·8 474·0	595·2 760·6	595·2 760·6	595·2 760·6	771·6 981·0	771·6 981·0	771·6 981·0
Weight of Charge, in lbs.	2·40	2·73	3·10	7·91	9·00	10·41	11·86	13·27	15·54	22·71	24·45	29·76	62·83	70·73	82·23	94·36	108·14	119·27	151·02	174·16	197·3	194·7	224·9	255·7
Muzzle Velocity, in ft.-secs.	2448 2172	2635 2326	2805 2490	2658 2349	2845 2510	3035 2684	2710 2392	2904 2566	3104 2740	2667 2392	2864 2566	3051 2740	2664 2392	2854 2566	3051 2740	2687 2392	2887 2566	3084 2740	2707 2392	2904 2566	3100 2740	2697 2392	2894 2566	3087 2740
Muzzle Energy, in foot-tons	476·6	551·5	626	1519	1736	1984	2361	2720	3099	4461	5131	5853	12246	14102	16070	18812	21653	24685	30191	34690	39611	38935	44758	51090
Perforation through Steel, in ins.	6·22	6·93	7·56	9·96	10·94	12·0	11·8	13·07	14·3	14·8	16·30	17·9	21·1	23·35	25·6	24·8	27·40	30·1	29·4	32·56	35·7	32·2	35·59	39·1
Perforation through Iron, Tresidder's formula	8·5	9·5	10·5	13·3	14·7	16·4	15·7	17·6	19·3	19·5	21·6	23·8	27·0	30·1	33·0	31·4	35·0	38·4	36·9	41·0	45·1	40·1	44·6	49·1

NOTE.—Every one of the Guns included in the Tables has been actually constructed and can be supplied on order. Fresh orders are all executed according to model 1899.

KRUPP QUICK-FIRE GUNS, Model 1899.

Table supplied by Manufacturers.

LIGHT GUNS.

Calibre, in centimètres	7·5			10·5			12			15			21			24			28			30·5		
Calibre, in inches	2·95			4·13			4·72			5·91			8·27			9·45			11·02			12·01		
Total Length of Gun, in cal.	40	45	50	40	45	50	40	45	50	40	45	50	40	45	50	40	45	50	40	45	50	40	45	50
Total Length, in feet	9·84	11·07	12·30	13·78	15·5	17·22	15·75	17·7	19·69	19·55	22·0	24·46	27·56	30·9	34·45	31·50	35·4	39·37	36·75	41·3	45·93	40·03	45·0	50·03
Length of Bore, in inches	108·66	123·43	138·19	153·55	174·21	194·89	175·20	199·25	222·45	218·12	247·49	276·78	305·91	347·29	388·59	350·80	398·28	445·28	409·46	464·62	519·70	445·67	505·95	565·76
Weight of Piece, in lbs.	1488·2	1710·8	1935·7	4078	4740	5313	6107	7055	7937	11707	13448	15212	32294	37254	41890	48722	56222	63273	77603	89062	100974	100312	115081	130290
Weight of Piece, in tons	0·66	0·76	0·86	1·82	2·12	2·37	2·78	3·15	3·54	5·23	6·00	6·79	14·42	16·63	18·70	21·75	25·10	28·25	34·65	39·76	45·08	44·78	51·14	58·17
Weight of Steel Projectile,	11·5	11·5	11·5	30·86	30·86	30·86	46·30	46·30	46·30	90·39	90·39	90·39	249·1	249·1	249·1	374·8	374·8	374·8	595·2	595·2	595·2	771·6	771·6	771·6
in lbs.	14·6	14·6	14·6	39·68	39·68	39·68	59·52	59·52	59·52	112·4	112·4	112·4	308·6	308·6	308·6	474·0	474·0	474·0	760·6	760·6	760·6	981·0	981·0	981·0
Weight of Charge, in lbs.	2·69	3·11	3·53	7·41	8·51	9·66	12·21	14·31	16·56	23·42	27·44	31·76	64·55	75·64	87·55	97·35	114·09	132·0	152·2	181·88	210·5	200·4	234·79	271·7
Muzzle Velocity, in ft.-secs.	2530	2723	2890	25·46	2740	2910	2766	2982	3199	2720	2936	3143	2776	2995	32·15	2805	3025	3248	2818	3038	3264	2812	3035	3255
	2247	2418	2566	2247	2418	2566	2438	2631	2822	2438	2631	2822	2494	2690	2887	2494	2690	2887	2494	2690	2887	2494	2690	2887
Muzzle Energy, in foot-tons	509·2	589 6	664·8	1388	1607	1810	2453	2856	3285	4631	5389	6196	13296	15490	17845	20427	23783	27397	32786	38143	43952	42274	49312	50666
Perforation through Steel, in ins.	6·54	7·2	7·91	9·3	10·4	11·3	12·1	13·4	15·0	15·2	16·9	18·7	22·8	24·9	27·6	26·5	29·2	32·4	31·5	34·7	38·5	34·5	38·0	42·0
Perforation through Iron, Tresidder's formula	9·0	10·0	11·0	12·5	13·9	15·2	16·2	18·2	20·1	20·1	22·5	25·0	28·9	32·1	36·0	33·4	37·4	41·9	39·1	44·0	49·0	42·8	47·8	53·5

HEAVY GUNS.

Calibre, in centimètres	7·5			10·5			12			15			21			24			28			30·5		
Calibre, in inches	2·95			4·13			4·72			5·91			8·27			9·45			11·02			12·01		
Total Length of Gun, in cals.	40	45	50	40	45	50	40	45	50	40	45	50	40	45	50	40	45	50	40	45	50	40	45	50
Total Length, in feet	9·84	11·07	12·30	13·78	15·5	17·22	15·75	17·7	19·6[illegible]	19·55	22·0	24·46	27·56	30·9	34·45	31·50	35·4	39·37	36·75	41·3	45·93	40·03	45·0	50·03
Length of Bore, in inches	108·66	123·43	138·19	153·55	174·21	194·89	175·20	199·25	222·45	218·12	247·49	276·78	305·91	347·29	388·59	350·80	398·28	445·28	409·46	464·62	519·70	445·67	505·95	565·76
Weight of Piece, in lbs.	1860·7	2094·3	2325·9	5115	5732	6393	6680	7628	8598	12787	14661	16535	35270	40341	45413	53241	60843	68340	84882	97000	109132	109573	125224	140874
Weight of Piece, in tons	0·83	0·93	1·04	2·28	2·56	2·85	2·98	3·41	3·84	5·74	6·55	7·38	15·75	18·01	20·27	23·32	27·16	30·51	37·89	43·30	48·72	48·92	55·90	62·44
Weight of Steel Projectile,	11·5	11·5	11.5	30·86	30·86	30·86	46·30	46·30	46·30	90·39	90·39	90·39	249·1	249·1	249·1	374·8	374·8	374·8	595·2	595·2	595·2	771·6	771·6	771·6
in lbs.	14·6	14·6	14·6	39·68	39·68	39·68	59.52	59·52	59·52	112·4	112.4	112·4	308·6	308·6	308·6	474·0	474·0	474·0	760·6	760·6	760·6	981·0	981·0	981·0
Weight of Charge, in lbs.	2·91	3·26	3·66	9·22	10·41	12·02	14·11	15·72	18·50	27·07	30·16	35·49	74·65	83·11	97·84	112·6	125·36	147·6	179·45	199·82	235·2	231·66	257·96	303·7
Muzzle Velocity, in ft.-secs.	2638	2822	2999	2907	3123	3324	2940	3143	3350	2891	3091	3294	2884	3087	3288	2913	3117	3320	2930	3133	3337	2920	3127	3330
	2343	2507	2661	2562	2753	2930	2592	2772	2953	2592	2772	2953	2592	2772	2953	2592	2772	2953	2592	2772	2953	2592	2772	2953
Muzzle Energy, in foot-tons	552·8	633·1	715·1	1807	2085	2362	2772	3172	3598	5228	6002	6809	14360	16458	18652	22073	25171	28656	35400	40531	45953	45662	52277	58205
Perforation through Steel, in ins.	6·93	7·6	8·31	11·3	12·5	13·7	13·2	14·6	16·0	16·5	18·2	19·6	23·6	26·3	28·4	27·7	30·6	33·7	32·9	36·5	39·9	35·9	39·8	43·6
Perforation through Iron, Tresidder's formula	9·6	10·6	11·6	15·3	17·0	19·0	17.8	19·7	21·7	22·0	24·3	26·9	30·5	33·5	37·0	35·5	39·1	43·2	41·1	45·9	50·5	45·4	50·0	55·1

NOTE.—Every one of the guns included in the Tables has been actually constructed and can be supplied on order.

Table Relating to Conversion of Measures.

Length.

Metric to English. English to Metric.

I. Mètres.	II. Yards.	III. Feet.	IV. Inches.	V. Yards.	VI. Mètres.	VII. Feet.	VIII. Mètres.	IX. Inches.	X. Centimètres.
1	1·0936	3·2809	39·37	1	0·91438	1	0·30479	1	2·5400
2	2·1873	6·5618	78·74	2	1·82877	2	0·60959	2	5·0799
3	3·2809	9·8427	118·11	3	2·74315	3	0·91438	3	7·6199
4	4·3745	13·1236	157·48	4	3·65753	4	1·21918	4	10·1598
5	5·4682	16·4045	196·85	5	4·57192	5	1·52397	5	12·6998
6	6·5618	19·6854	236·22	6	5·48630	6	1·82877	6	15·2397
7	7·6554	22·9663	275·60	7	6·40068	7	2·13356	7	17·7797
8	8·7491	26·2472	314·97	8	7·31507	8	2·43836	8	20·3196
9	9·8427	29·5281	354·34	9	8·22945	9	2·74315	9	22·8596

Explanation.—To convert any number from one measure to the other, take the values of the different multiples of 10 by shifting the position of the decimal point, and add together. Thus, find the number

of yards in 2354 mètres (see cols. I. & II.).	of feet in 12·4 mètres (see cols. I. & III.).	of inches in 30·5 centimètres (see cols. I. & IV.). Note, 1 m.=100 cm.	of mètres in 1026 yards (see cols. V. & VI.).	of mètres in 1742 feet (see cols. VII. & VIII.).	of centimètres in 17·72 ins. (see cols. IX. & X.)
mètres. yards.	mètres. feet.	cms. inches.	yards. mètres.	feet. mètres.	inches. cms.
2000=2187·3	10 =32·809	30·0=11·811	1000=914·38	1000=304·79	10·0 =25·400
300= 328·09	2 = 6·562	·5= ·197	20= 18·29	700=213·36	7·0 =17·780
50= 54·68	0·4= 1·312		6= 5·49	40= 12·19	0·7 = 1·778
4= 4·37				2= 0·61	·02= ·051
∴ 2354=2574·44	∴ 12·4=40·083	∴ 30·5=12·008	∴ 1026=938·16	∴ 1742=530·95	∴ 17·72=45·009

Note.—A ready way of approximately converting all French measures into English inches is to multiply by 4 and apply the decimal point by common sense—Thus for a 15-cm. gun; 15 × 4 = 60. Now this Calibre cannot be 60 inches, nor can it be 0·6 inch; therefore it must be 6 inches. (The exact value is 5·906 in.)

Weight.

Metric to English. English to Metric.

I. Kilogrammes.	II. Tons.	III. Pounds Avoirdupois.	IV. Grains Troy.	V. Tons.	VI. Milliers.	VII. Pounds Avoirdupois.	VIII. Kilogrammes.	IX. Grains. Troy.	X. Gramme.
1	·000984	2·2046	15432·3	1	1·016	1	0·4536	1	·0648
2	·001968	4·4092	30864·7	2	2·032	2	0·9072	2	·1296
3	·002953	6·6139	46297·0	3	3·048	3	1·3608	3	·1944
4	·003937	8·8185	61729·4	4	4·064	4	1·8144	4	·2592
5	·004921	11·0231	77161·7	5	5·080	5	2·2680	5	·3240
6	·005905	13·2277	92594·1	6	6·096	6	2·7216	6	·3888
7	·006889	15·4323	108026·4	7	7·112	7	3·1751	7	·4536
8	·007874	17·6370	123458·8	8	8·128	8	3·6287	8	·5184
9	·008858	19·8416	138891·1	9	9·144	9	4·0823	9	·5832

Explanation.—To convert any number from one measure to the other, take the values of the different multiples of 10 by shifting the position of the decimal point, and add together. Thus, find the number

of tons in 35 milliers (see cols. I. & II. Note, 1000 kg. =1 millier).	of pounds in 56·3 kilogrammes. (see cols. I. & III.).	of grains in 120 grammes (see cols. I. & IV. Note, 1000 grms. = 1 kg.)	of milliers in 38 tons (see cols. V. & VI.).	of kilogrammes in 68 pounds (see cols. VII. & VIII).	of grammes in 85 grains (see cols. IX. & X.).
milliers. tons	kgrms. lbs.	grammes. grains.	tons. milliers.	lbs. kgs.	grains. grammes.
30 = 29·53	50 =110·231	100=1543·23	30 = 30·48	60 = 27·216	80 = 5·184
5 = 4·92	6 = 13·228	20= 308·65	8 = 8·13	8 = 3·629	5 = 0·324
	0·3= ·661				
∴ 35 = 34·45	∴ 56·3=124·120	∴ 120=1851·88	∴ 38 = 38·61	∴ 68 = 30·845	∴ 85 = 5·508

Note.—7000 grains troy=1 pound avoirdupois.

PRESSURE.

Metric to English.			English to Metric.				Atmospheric to English.		English to Atmospheric.	
I.	II.	III.	IV.	V.	VI.	VII.	VIII.	IX.	X.	XI.
Kilo-grammes per square centi-mètre.	Pounds per square inch.	Tons per square inch.	Pounds per square inch.	Kilo-grammes per square centi-mètre.	Tons per square inch.	Kilo-grammes per square centi-mètre.	Atmo-spheres.	Tons per square inch.	Tons per square inch.	Atmo-spheres.
1	14·223	·00635	1	·07031	1	157·49	1	·00656	1	152·38
2	28·446	·01270	2	·14062	2	314·99	2	·01313	2	304·76
3	42·668	·01905	3	·21093	3	472·48	3	·01969	3	457·14
4	56·891	·02540	4	·28124	4	629·97	4	·02625	4	609·52
5	71·114	·03175	5	·35155	5	787·47	5	·03281	5	761·91
6	85·337	·03810	6	·42186	6	944·96	6	·03938	6	914·29
7	99·560	·04445	7	·49217	7	1102·45	7	·04594	7	1066·67
8	113·783	·05080	8	·56248	8	1259·95	8	·05250	8	1219·05
9	128·005	·05715	9	·63279	9	1417·44	9	·05906	9	1371·43

Note.—One atmosphere is taken to be 14·7 lbs. per square inch.

Explanation.—To convert any number from one measure to the other, take the values of the different multiples of 10 by shifting the position of the decimal point, and add together. Thus, find the number

of pounds per square inch in 32·1 kilogrammes per square centimètre (see cols. I. & II.).

kgs. per sq. cm.		lbs. per sq. in.
30	=	426·68
2	=	28·45
0·1	=	1·42
32·1	=	456·55

of tons per square inch in 3210 kilogrammes per square centimètre (see cols. I. & III.).

kgs. per sq. cm.		tons per sq. in.
3000	=	19·05
200	=	1·27
10	=	·06
∴ 3210	=	20·38

of kilogrammes per square centimètre in 15 lbs. per square inch (see cols. IV. & V.).

lbs. per sq. in.		kgs. per sq. cm.
10	=	·7031
5	=	·3516
∴ 15	=	1·0547

of kilogrammes per square centimètre in 18·3 tons per square inch (see cols. VI. & VII.).

tons per sq. in.		kgs. per sq. cm.
10	=	1574·9
8	=	1259·95
0·3	=	47·25
∴ 18·3	=	2882·10

of tons per square inch in 3254 atmospheres. (see cols. VIII. & IX.).

atmospheres.		tons per sq. inch.
3000	=	19·69
200	=	1·31
50	=	·33
4	=	·03
∴ 3254	=	21·36

of atmospheres in 14·6 tons per square inch (see cols. X. & XI.).

tons per sq. in.		atmospheres.
10	=	1523·8
4	=	609·5
0·6	=	91·4
∴ 14·6	=	222.7

ENERGY.

Metric to English.		English to Metric.	
I.	II.	III.	IV.
Mètre-tons.	Foot-tons.	Foot-tons.	Mètre-tons.
1	3·2291	1	0·3097
2	6·4581	2	0·6194
3	9·6872	3	0·9291
4	12·9162	4	1·2388
5	16·1453	5	1·5484
6	19·3743	6	1·8581
7	22·6034	7	2·1678
8	25·8324	8	2·4775
9	29·0615	9	2·7872

1 mètre-ton is termed a "dinamode" in Italy.

Explanation.—To convert any number from one measure to the other, take the values of the different multiples of 10 by shifting the position of the decimal point, and add together. Thus find the number

of foot-tons in 4367 mètre-tons (see cols. I. & II.).

mètre-tons.		foot-tons.
4000	=	12916·2
300	=	968·72
60	=	193·74
7	=	22·60
∴ 4367	=	14101·26

of mètre-tons in 3592 foot-tons (see cols. III. & IV.).

foot-tons.		mètre tons.
3000	=	929·1
500	=	154·84
90	=	27·87
2	=	·62
∴ 3592	=	1112·43

PERFORATION THROUGH IRON AND STEEL WITH THE FACE NOT HARDENED.

To obtain perforation through steel equivalent to a given perforation through iron, and *vice versâ*.

1 inch steel = $1\frac{1}{4}$ inches iron ;

that is, 4 inches steel = 5 inches iron.

Thus, given 9·4 inches perforation through iron,

$$9{\cdot}4 \times \frac{4}{5} = 7{\cdot}52 \text{ inches steel;}$$

or, given 5·2 inches steel,

$$5{\cdot}2 \times \frac{5}{4} = 6{\cdot}5 \text{ inches iron.}$$

PART IV.

STATISTICS, OFFICIAL STATEMENTS AND PAPERS.

Statement of the First Lord of the Admiralty explanatory of the Navy Estimates for 1900–1901.

THE Navy Estimates for 1900–1901 amount to a net total of £27,522,600, being an increase of £928,100 beyond the amount of £26,594,500 voted for the year 1899–1900.

Of this sum of £928,100 the various Votes connected with the *personnel* account for £447,600. The greater part of this amount is due to the increase in the numbers voted. Among other causes which have contributed to produce this increase are the grant of higher pay to the Marines sanctioned last year, and various improvements in pay and allowances to the Medical Service.

The miscellaneous Votes show an increase of £27,900.

The Works Vote shows an increase of £50,700.

The Vote for Ordnance is increased by £293,900, which, however, includes a sum of £117,000 for ammunition transferred to the Army in this financial year, and which, though paid for, could not be replaced until next year.

Vote 8 shows a net increase of £108,000. The three sub-heads of which it is composed show the following comparison:—

Section I.—Dockyard wages, &c.—Increase . . .	£95,000
Section II.—Naval stores, including steam vessel coal—Increase	285,000
Section III.—Contract work—Decrease	272,000

A considerable amount of unforeseen and exceptional expenditure, for which otherwise a Supplementary Estimate would have been necessary, has, with the consent of the Treasury, been defrayed out of the unspent balance on the vote for contract work. Among the principal items are upwards of £350,000 excess expenditure on coal, due partly to increased cost, partly to operations in connection with South Africa, partly to strengthening of stocks; about £300,000 on the victualling vote; £50,000 for extra dockyard labour; nearly £300,000 for increased prices and increased purchases of Naval Stores other than coals. Smaller additional sums have also been spent on sea conveyance, telegraphic communication, and miscellaneous items. The state of war with the South African Republics, and its attendant circumstances, naturally caused a considerable expenditure beyond that provided for in the original Estimate. The total additional expenditure thus incurred, or still to be incurred during the financial year, amounts to a little more than £1,000,000.

NUMBERS.

The total number of Officers, Seamen and Boys, Coastguard, and Royal Marines voted for the year 1900–1901 is 114,880, being an increase of 4,240 as compared with the number voted in the year 1899–1900.

The total number borne on the 1st February, 1900, was 110,273, leaving 367 to be added in the last two months of the financial year. The number similarly borne on 1st February, 1899, was 105,280. Recruiting generally has been good, and it is expected that the total number voted will be probably reached by the 1st April next, though there may still be a falling off in some of the artificer ratings.

The additions proposed are composed as follows:—

220	Officers.
3,050	Petty Officers and Seamen.
150	Engine-room Staff.
200	Miscellaneous.
300	Marines.
320	Apprentices (Artizan ratings).
4,240	

Consequent on the gradual increase in the list of Captains and Commanders, sanctioned by Order in Council of November, 1898, the number of Officers eligible for higher rates of full and half pay has been proportionately increased.

The promotion, *status*, and pay of Engineer Officers have been recently considered by a Committee of the Board, with the result that the following changes have been approved:—

The list of Chief Inspectors of Machinery has been increased from 5 to 8, and that of Inspectors of Machinery from 8 to 13.

The Engineer-in-Chief has been given the relative rank of a Rear-Admiral.

The rank of Staff Engineer has been abolished.

Chief Engineers will rank *with* Lieutenants of and above 8 years' seniority, while Engineers on promotion will rank *with* Lieutenants of less than 8 years' seniority, instead of as now, with but *after* Lieutenants.

In other respects the relative rank of Engineer officers remains unchanged.

Engineers will be given a new scale of pay, viz.:—

On promotion	10*s.* a day.
After 4 years	11*s.* ,,
After 8 ,, ...	12*s.* ,,

And the allowance of 1*s.* a day at present paid to Senior Engineers for all ships will be replaced by a scale varying, according to responsibility, from 1*s.* to 2*s.* 6*d.* a day.

In order to further encourage signalling in the Navy, an allowance of 8*d.* a day for instructional duties will be granted to 40 Chief or other Yeomen of Signals in Battleships, first-class Cruisers and Depôt Ships, and the "higher standard" allowance of 3*d.* a day will be granted to an increased number of petty officers and men of the signal class above the rating of signalmen.

The numbers of Inspectors-General and Deputy Inspectors-General of Hospitals and Fleets have been increased, and the conditions required for promotion to these ranks have been modified.

The period of the course of instruction at Haslar Hospital for surgeons on entry has been extended, and the award of prizes at the end of each session introduced.

The number of medical officers allowed to undergo periods of study at medical schools has been considerably increased, and the privilege extended to the senior ranks.

An additional professor has been appointed on the instructing staff at Haslar in connection with the study of diseases of foreign stations.

Medical officers newly entered will in future be only required to provide themselves with a pocket case of instruments, as all ships bearing medical officers, and naval and marine barracks, will be supplied with surgical instruments at the public cost within the next three years.

To meet the difficulty of recruiting suitable men for the rating of sick berth attendant, the pay and prospects of this class have been improved, and the rank of warrant officer granted to the position of ward master at the three principal hospitals. An extended course of training will be introduced for probationers.

The boys' training ship Ganges, which had been stationed at Falmouth, has been removed to Harwich, which it is hoped will encourage the recruiting of boys on the East Coast.

Consequent on the recommendations of a Committee appointed by the Board, the pay and position of domestics have been considerably improved with the object of attracting a more desirable class of men, and of relieving officers of many expenses and difficulties inherent in the former system.

The Royal Marines.

2585 recruits were raised for the corps during the year, of whom 676 joined the artillery branch. The average height of these men on entry ranged from 5 ft. 7½ in. to 5 ft. 8 in. for the Royal Marine Artillery, from 5 ft. 5½ in. to 5 ft. 6 in. for the Royal Marine Light Infantry.

The wastage of the corps for the twelve months amounted to 2078 men.

An additional annual gratuitous issue of a pair of canvas shoes has been approved for all non-commissioned officers and men serving on shore for the purpose of relieving the feet after marching, &c. This issue will commence on the 1st April next.

The barrack room accommodation at Walmer is still insufficient for the quartering of the whole of the recruits at the depôt, and the temporary removal of the Royal Marine Artillery recruits to Eastney therefore still continues. Additional accommodation, which is also required at the latter establishment, is being provided by the conversion of the married quarters into barrack rooms. This work will be completed shortly.

The building of the new hospital at Walmer is nearing completion. Steps will then be taken to provide additional accommodation by the conversion of the old hospital into single men's quarters.

A gymnasium has been built at Eastney and taken over for use.

The Eastney rifle range has been completed up to 500 yards, and good progress is being made with the remaining portion.

The negotiations for procuring a rifle range at Plymouth for the use of the Royal Marines have not yet been concluded. Consequently, for the greater part of the year the practice was carried out at Browndown. The Army ranges at Tregantle, however, are now temporarily available.

The armament of the gun-drill batteries at the several divisions has been improved by the addition of several quick-firing guns; several of the ordinary breech-loading guns have also been replaced by quick-firing guns.

At the Chatham division a new drill battery for light quick-firing and machine guns is building, and will be finished shortly.

A battalion of Marines took part in the army drills and manœuvres on Salisbury Plain during the month of July. The men were attached to the 3rd Brigade, consisting of Guards and Lincolnshire Regiments.

The detachment of Royal Marine Artillery lent to the Dominion Government for submarine and defence work at Vancouver Island were finally withdrawn in September last after six years' useful work at that station.

Eight field and six company officers have been lent to the Army for special service during the war in South Africa.

Naval Reserves.

The total number of executive officers now on the active list, who have served for twelve months or more in the Navy, or who are now undergoing twelve months' training, is 267, an increase of twenty-four since last year.

Vacancies for executive officers are filled up as soon as they occur. There are now 302 qualified candidates on the list of applicants for appointment to the Reserve, but only about fifty vacancies can be filled in the course of a year, twenty of which are given to cadets nominated from the mercantile training ships, Worcester and Conway.

The establishment of engineer officers, Royal Naval Reserve, is fixed at 400. The present number borne is 380, being an increase of 29 over last year, and it is anticipated that all vacancies will be filled within the next six months at the present rate of volunteering.

The Instructional Classes for Engineers in the Home Dockyard Reserves, commenced in 1898, have been continued. Three courses of six officers are held each year, and the number of applicants is far in excess of the number authorized. The officers who have been through the course have been well reported on for conduct and attention, and have appreciated the instruction and experience afforded.

During the year 1899, 1,292 seamen and 544 firemen were enrolled, as compared with 2,536 and 621 respectively in 1898.

On the 31st December, 1899, the numbers borne, as compared with the numbers voted, were :—

—	Voted.	Borne.
Qualified Seamen, and 1st Class (old system)	11,700	11,001
Seamen Class and 2nd Class (old system)	11,300	10,961
Firemen	3500	3494
Boys	250	256
Totals	26,750	25,712

The numbers voted for 1899 included an increase of 1,000 beyond the numbers voted for 1898, and the above figures show that they have not been obtained.

During the year ending 31st December, 1899, 980 men were embarked for six months' naval training. In 1898 the number was 1,711.

The decrease in the numbers entered and also in those embarked for naval training occurred chiefly in the earlier months of the financial year, when the fishing industry was unusually active. It is also stated that latterly unusually good wages have prevailed owing to the large number of transports employed, and that there have been many vacancies at the docks, owing to Reservists being called up for service with the colours.

The result of the six months' training in a man-of-war on the men is remarkably good, and it is quite easy to distinguish among the men on drill at a Royal Naval Reserve Battery those who have been through the six months' training. The number now serving afloat is 448.

500 Reserve men, including 53 firemen, served in ships during the manœuvres, and were well reported on.

All Royal Naval Reserve drill ships and batteries have now received 3-pr. Q.-F. guns, and will all have received 6-in. Q.-F. C. guns by the end of the present financial year. Ten Maxim guns have also been supplied. It is in contemplation to supply all Royal Naval Reserve drill ships and batteries with Maxim guns, and with 5-in. Q.-F. C. guns to replace the remaining obsolete M. L. R. guns. A commencement will be made during the forthcoming financial year.

The gunboat Gleaner having proved a success in training the Royal Naval Reserve officers and men of the President by taking them to the Nore for target practice with modern 4·7-in. and 3-pr. guns, the gunboat Antelope has been stationed at Portishead to train the Royal Naval Reserve officers and men of the Dædalus in the same manner.

Considerable correspondence has taken place with regard to a larger and more thorough development of a system of Naval Reserves in the Australian Colonies, but it has been considered most expedient to await the establishment of one central authority under the scheme of federation, so as to organize a central system of Colonial Naval Reserves if an agreement can be arrived at.

Negotiations are still in progress as to the formation of a Naval Reserve in Canada. It is sought to overcome the difficulty created by the requirement of a six months' training at sea in a man-of-war, which is one of the conditions of service in the United Kingdom.

Other plans for strengthening and developing our system of Reserves are under consideration.

MOBILISATION.

The number of vessels and torpedo boats taking part in the manœuvres in 1899 was 117, manned by 28,839 officers and men.

The total number of coastguard men embarked was 1,433, and of Royal Naval Reserve, officers and men, 571.

Detachments consisting approximately of 98 officers and 1,192 seamen and marines have been landed from the ships at the Cape for temporary service with the land forces in South Africa.

The number of ships in commission for home and foreign service on 1st January, 1899, was 339; on 1st January, 1900, the number was 348.

The effective strength of the number of ships in commission for active service has been increased by the transfer of the officers and men of the ships formerly belonging to the training squadron to four modern cruisers, viz., the St. George, Juno, Minerva, and Cambrian. These ships are still kept together as a separate squadron under a Commodore.

During the past year several ships of a new type have been commissioned. The Canopus has relieved the Anson on the Mediterranean Station, and the Ocean is about to relieve the Hood. The Hermes has joined the North America Squadron, and her sister ship, the Highflyer, is now in commission, and will proceed later to the East Indies as flagship.

As regards other reliefs, the Renown has been replaced on the North America Station by a first-class cruiser, and is now the flagship of the commander-in-chief on the Mediterranean Station. The Terrible, which was destined for the China Station, and the Powerful, which was on her way home, have been temporarily detained at the Cape. The squadron in South African waters has been further reinforced by five cruisers detached from the Channel, Mediterranean, and East Indies Squadrons.

The four river gunboats sent out to the China Station in sections in the autumn of 1898 were all commissioned during 1899, three for the Yangtsze River, and one for the West River. Another of these gunboats is now on passage out, and, on arrival, will be put together for service in the West River.

Two torpedo-boat destroyers have been attached to each of the gunnery schools and one to each torpedo school for training purposes. It is intended later to attach a sea-going cruiser to each gunnery school, in place of one of these destroyers, for the purpose of enabling firing to be carried out under modern conditions.

Steps are being taken to provide kits for a further number of 10,000 men of the Royal Naval Reserve and pensioners when called out for active service, making a total of 20,000 kits that will be ready for immediate issue from the victualling yards should the occasion for their use arise.

Coaling of the Fleet.

Arrangements connected with the coaling of the fleet have been under the special consideration of the board. Steps have been taken to increase the reserve stocks at certain of our coaling stations, and experiments are in progress with the object of selecting a patent fuel suitable as a special reserve on the more distant stations.

Arrangements are being made for the institution of a system of supply of coal to certain fleets and dockyards by colliers directly under Admiralty control, a successful experiment in this direction having been made in the course of the year. Efforts are being made to widen the area of supply as far as practicable, and to take advantage of the coal resources of the Colonies when local coal can be shown to be of suitable quality for Her Majesty's ships, and can be supplied at reasonable rates. Steps have been taken to provide for a certain quantity of New Zealand coal from the West Port Collieries for use on the China station, and local Australian coal is now used on the Australian station as far as circumstances permit.

Transport Service.

Since the beginning of July 181 transports and freight ships (including second voyages) have been engaged in the conveyance to South Africa of a force of 132,384 officers and men, 23,345 horses, and 26,364 mules, exclusive of the troops, horses, &c., conveyed from India and the Colonies under local arrangements, and of certain volunteer detachments and special corps.

New Construction.

The abnormal activity in shipbuilding and engineering, which was described in the statement of last year, has continued during 1899–1900, and has seriously affected progress and expenditure on ships, machinery, and armour. Delays in delivery of material, difficulties in securing adequate numbers of workmen, and other circumstances, have caused the aggregate earnings on contract work to fall short of the estimated amount by about £1,400,000, though the estimate was carefully calculated on the basis of actual earnings in past years on ships of similar character and on very close investigation of the possible output of armour.

While the most marked effect of these exceptional conditions is shown in the contract vote, progress in the construction of dockyard ships has also been somewhat affected, especially in connection with the supply of propelling machinery and armour. It is the rate of this supply which will practically determine the dates of completion of several important ships.

The armour-plate makers have considerably increased their output during the year, and have made large extensions of plant. Their total output will be nearly 50 per cent. greater than it was in 1898–99, but nevertheless the deliveries will fall short of the amount anticipated by £420,000 in value. All of the firms have been fully employed, and there are large orders yet to be executed. The situation is one of some difficulty; but there is every reason to hope that it will still further improve during the next financial year, as new extensions of plant come into use.

The fact that so large a number of ships now in construction are designed for exceptionally high speeds, and will, therefore, be equipped with propelling machinery of great power, also tends to affect the rate of progress. Machinery of this kind can only be produced by firms of the first rank, who are limited in number, and who, in many cases, have other important contracts in hand. Longer periods are required for the manufacture and erection of the machinery, with the natural result of more time being necessary for the completion of the ships.

Battleships.

The Canopus was commissioned in December last, and is on service in the Mediterranean. The Ocean and Goliath will be completed this financial year. The Glory has been delivered by the contractor, and is now making her steam trials as a preliminary to being taken in hand for completion at Portsmouth. The completion of the Albion has been delayed by financial difficulties of the contractors for the engines, but special arrangements have been made for finishing their work, and it is hoped that the vessel will be delivered shortly. The sixth ship of the Canopus class, the Vengeance, is well advanced, and her builders anticipate delivering her at the contract date in July next.

Good progress has been made with the six battleships of the Formidable class which are building in the dockyards, especially with the last three which were launched in the autumn—namely, the London, on the 21st September, in nine months after laying down, the Bulwark on 18th October, in seven months, and the Venerable on the 2nd November, in ten months. Their completion will depend on the delivery of certain portions of their armour. It is hoped that the Formidable, Irresistible, and Implacable will be completed by March, 1901, and the London, Bulwark, and Venerable during the autumn of that year.

Four vessels of the Duncan class, building by contract, have been advanced during the financial year, but not to the extent that would have been possible under ordinary conditions.

Two others, the Albemarle and Montagu, have been begun in the dockyards.

Armoured Cruisers.

Fourteen of these vessels are now in course of construction. Particulars of their designs have been furnished in previous statements.

The Cressy class were the first laid down, all being built by contract. Two of the six vessels of this class have been launched, and it is hoped that they will be delivered by the contractors during the year 1900–1901. Two others are well advanced.

The Drake class includes four vessels, the Drake being under construction at Pembroke, and the remaining three vessels in private yards. Considerable delay occurred in the commencement of the contract ships; good progress is now being made, but it is too soon to forecast the dates of completion with certainty.

The new class, described in my statement last year as vessels of a displacement of 9,800 tons, now called the Monmouth class, includes four vessels, two building by contract and two in the Dockyards, the two latter only lately commenced.

First-class Protected Cruisers.

The vessels of this class which were building at the commencement of 1899–1900 have all been completed, with the exception of the Spartiate, which, like the Albion, has been delayed by financial difficulties of the contractors for the engines. It is anticipated that she will be finished in the autumn of this year.

Second-class Protected Cruisers.

Four vessels of this class have been under construction during the financial year. Three—the Gladiator, Hermes, and Highflyer—are already completed, the last two being commissioned.

The Hyacinth has passed through her trials and will be completed at an early date.

Third-class Cruisers.

Seven of these vessels (Pelorus class) were under construction at the commencement of this financial year. All except the Pandora will be completed before the 31st March.

It was contemplated to lay down three new third-class cruisers of rather larger dimensions than the Pelorus class, and of high speed, and a small sum was inserted in the Estimates for their commencement.

A thorough investigation has since been made of alternative designs for the proposed vessels, and of the designs for vessels of a similar class projected or building for foreign Powers. The action taken by each Power, according to the latest information in respect of such designs, has also been under review. After full consideration of all the circumstances, and weighing the fact that the cost involved in building such vessels is out of proportion to their fighting value and sea-keeping qualities, it has been deemed expedient not to proceed with their construction. In their place it is proposed to build an improved second-class cruiser of the Hermes type, with about 21 knots speed, to be built in a dockyard, the cost of which would remain within the limits of liability for the three third-class cruisers sanctioned by Parliament. This design is not yet complete.

Sloops and Gunboats.

Eight sloops have been in progress. Two will be completed this year, and four next year. The remaining two have been recently commenced.

Of the four gunboats (Dwarf class) two have been completed, one being in commission. The other two, which were delayed at contractors' works, are now approaching completion.

Torpedo-boat Destroyers.

The total number of vessels in this class is 108. Forty-two have trial speeds of 26 to 27 knots; 62 have trial speeds of 30 knots; and 4 have contract-speeds ranging from 31 to 33 knots.

Forty of the first category and 44 of the 30-knot class have been completed and passed through their trials.

Of the two of the first category mentioned in the statement of last year as having new boilers fitted, one has passed the official speed trials, obtaining $26\frac{3}{4}$ knots, and should be delivered shortly. The sister vessel is being prepared for trial.

Of the eighteen incomplete vessels of the 30-knot class, three have been delivered, but have still to pass their speed trials. Twelve of the remainder were ordered last year under the Supplementary Programme and are being advanced. The first of these will probably be delivered immediately, and be ready to undergo her official trials.

Of the four remaining Destroyers, intended to develop speeds exceeding 30 knots, one obtained 32 knots on her preliminary trials, and has been delivered. She will soon undergo her official trials. Another vessel of equal contract speed is in an early stage of construction. A third, designed for 33 knots, has been undergoing a long series of preliminary trials, but has not yet attained the full speed.

The fourth experimental vessel is the Viper, which has been fitted with Parsons' Steam Turbine. The contract speed in this case is 31 knots, but it was anticipated that a considerably higher speed would be attained. On preliminary trials (for short periods) speeds of about 35 knots have been reached. It is hoped that the vessel will soon be ready for her official trials; and it is proposed, after these trials are completed, to make exhaustive experiments with her, as great importance attaches to this novel system of propulsion.

Torpedo Boats.

The two torpedo-boats included in the programme for 1899-1900 are under construction. The contract speed, at ordinary load displacement, is 25 knots.

Royal Yacht.

The new Royal Yacht was ready for her steam trials at the beginning of January, but an accident which occurred to her while undocking at Pembroke, besides damaging the vessel's bottom, revealed a serious miscalculation of weight which will make considerable alterations necessary before she can proceed with her trials.

NEW SHIPBUILDING PROGRAMME.

In the coming financial year it is proposed to lay down—

2 Battleships.
6 First-class armoured cruisers.
1 Second-class cruiser (improved Hermes).
2 Sloops (twin screw).
2 Light draught gunboats.
2 Torpedo boats.

Of these, 2 battleships, 2 armoured cruisers, 1 second-class cruiser, and 2 sloops will be built in the dockyards, and 4 armoured cruisers, 2 gunboats, and 2 torpedo boats will be built by contract.

Summary.

The following ships will thus be under construction in the course of the year:—

17 Battleships.
20 Armoured cruisers.
1 First-class protected cruiser.
2 Second-class protected cruisers.
1 Third-class cruiser.
8 Sloops.
2 Light draught gunboats.
4 Torpedo boats.
21 Torpedo-boat destroyers.
1 Royal yacht.

Financial Provision for New Construction.

The estimated expenditure on new construction for the coming year, exclusive of establishment charges, is less by £395,335 than the sum voted for the same purpose in the present year, the figures being £8,460,146, against £8,855,481 for 1899–1900, but it is larger by £1,131,179 than the anticipated *actual* expenditure for this year—viz., £7,328,967—and if this larger sum should be *spent* in 1900–1901, it will represent an expenditure larger by more than a million than has ever yet been reached. The experience gained in recent years that, after the most careful calculations as to the probable earnings of contractors for hulls of ships, machinery, and armour, the expenditure for new construction has continually failed to reach the sum voted, has been taken into account in framing the Estimate for 1900–1901.

If the contractors should earn more instalments than are estimated for in the proposed vote, a Supplementary Estimate would, of course, be necessary.

It should be observed that only a portion of the short earnings in the present year will affect the liabilities of 1900–1901. The financial effect of slower work by contractors is mainly to throw part of the cost on later years.

Machinery and Boilers.

The following vessels have completed their contract steam-trials during the present financial year:—

First-class battleships.—Canopus, Ocean, and Goliath.

First-class cruiser.—Amphitrite.

Second-class cruisers.—Hermes, Highflyer, and Hyacinth.

Third-class cruisers.—Perseus, Pyramus, Pioneer, and Prometheus.

Sloop.—Rosario.

Gunboats.—Dwarf, Thistle, Bramble, and Britomart.

Torpedo-boat destroyers.—Eleven, in addition to one which completed its trial between the date of the last Statement and 1st April, 1899.

The battleship Glory, the sloop Condor, and four torpedo-boat destroyers will probably complete their trials before the 1st April, 1900.

The third-class cruisers Barham and Bellona, and the torpedo gunboat Seagull, have been re-boilered with water-tube boilers, the two first with Thornycroft boilers, and the last with Niclausse boilers, and have satisfactorily completed their full-power steam trials during the present year.

The torpedo gunboats Skipjack and Speedwell have been fitted with new water-tube boilers of the Reed or small tube type, and with new engines of 6000 I.H.P. The adoption of water-tube boilers in conjunction with light and quick-running machinery of the torpedo-boat destroyer type has enabled the I.H.P. developed in these vessels to be increased from 3500 to 6000. They have both successfully completed their contract machinery trials.

The Sheldrake, which is fitted with water-tube boilers of the Babcock and Wilcox type, having tubes of a diameter intermediate between those of the large and small tube type, has completed an exhaustive series of sea-going trials, similar to those carried out by the Sharpshooter, with satisfactory results, and with a view to further experience with these boilers, orders have been given for a set for the new sloop Espiegle now building.

The Seagull, fitted with water-tube boilers of the Niclausse pattern, is now in commission and is undergoing a series of sea-going trials similar to those carried out by the Sharpshooter and Sheldrake, which are fitted with the Belleville and Babcock and Wilcox types of water-tube boiler respectively.

DOCKYARDS.

The work in the royal dockyards continues to be performed in a satisfactory manner.

The numbers of men employed in the six home dockyards have had to be increased considerably during the current year, to meet the requirements of the Service. The total number to be employed in 1900–1901 in dockyards at home and abroad is 32,340.

Special care is being taken to continue to introduce labour-saving machines and tools of the newest description and make.

During the present year changes have been made in the *status* and pay of the subordinate officers of the dockyards, and the office of "Inspector" has been introduced in all trades with a view to improving the supervision of the workmen and removing certain anomalies which existed. It is believed that this change will be generally beneficial.

The scales of piece-work rates for the several classes of workmen have been very much amplified in order to ensure greater uniformity in the rates of pay for different kinds of work. The results of the working of the new scheme have been very satisfactory.

The general tendency of modern cruisers to increase in length has made it necessary that longer building slips should be provided at the dockyards. The action taken is referred to in this statement under "New Works."

The scheme for adapting Haulbowline Dockyard for repairing Her Majesty's ships, and making the best use of the fine dry dock there, is now drawing near completion, and the work done at that dockyard is increasing. It has, therefore, been decided to make some additions to the professional staff, and to place a certain number of men on the establishment as at other yards.

The Naval yards abroad have been fully employed in carrying out the work of repairing ships for re-commission, and also on the ordinary necessary repairs, &c., to commissioned ships.

The increase of the fleets on the Mediterranean and China Stations has rendered the large extensions of the dockyards at Malta and Hong Kong, which are being carried out under the Works Loan, urgently necessary.

Large Repairs during 1899–1900.

The following ships have been or will be completed in the home yards:—

Dreadnought.	Archer.
Blake.	Scout.
Blenheim.	Cossack.
Gibraltar.	Barham.
Grafton.	Bellona.
Hawke.	Calypso.
Immortalité.	Basilisk.
Talbot.	Speedy.
Astræa.	Skipjack.
Fox.	Speedwell.
Intrepid.	Harrier.
Pique.	Hussar.
Scylla.	Seagull.
Forte.	Magpie.
Sybille.	Redbreast.
Pallas.	Partridge.
Phœbe.	Lizard.

The following ships are in hand:—

Anson.	Hecate.
Camperdown.	Blanche.
Imperieuse.	Spider.
Narcissus.	Sparrow.
Amphion.	

The details of the repairs and refits to be carried out in 1900–1901 appear in the Appendix to the Navy Estimates.

Naval Ordnance.

The manufacture of guns is proceeding, and the production has so far kept pace with the requirements of the Fleet.

The new 9·2-in. gun will shortly be ready for trial, and will be completed before it is required for ships.

A new gun of 7·5-in. diameter, throwing a projectile of 200 lbs. weight, is under manufacture for trial, and will shortly be delivered.

The conversion of the 6-in. B. L. guns to quick-firers has been completed for all ships which are considered to be worth the expense.

Conversion for the Royal Naval Reserve batteries and drill ships is proceeding, and will, it is hoped, be completed by June, 1900.

Trials of the Vickers-Maxim automatic 37 m.m. (1½-in.) gun in competition with the existing 3-pr. quick-firing gun were carried out during the past year, and it was decided that for naval purposes this gun shows no superiority to the 3-pr.

The ·303 Maxim gun is now being adopted for naval purposes in lieu of the ·45, which has hitherto been the naval pattern. This will make the ammunition interchangeable with that used for rifles.

The requirements of ammunition and projectiles for the Navy during the past year have been met, and considerable assistance has been given to the Army since the commencement of hostilities.

Supplies of cordite are well maintained, and the substitution of cordite for powder charges in all quick-firing guns will be nearly complete by the end of 1900–1901.

Trials of projectiles, which have been in progress for some time have resulted in the adoption of a pattern of shell suitable for the attack of the light armour now being generally adopted for the protection of secondary armaments.

A new pattern of naval cutlass and a new pattern of accoutrements have been approved, and will be gradually introduced.

The replacement of older types of torpedoes is being continued, and the adoption of gyroscopic fittings will result in a considerably increased efficiency of all torpedoes so fitted.

The results obtained from the new types of heavy gun mountings are very satisfactory, the rapidity of fire having been considerably increased beyond previous results.

Designs for new mountings of improved type are being worked out, and the manufacture of all gun mountings now on order is keeping pace with the requirements of the Fleet.

Wireless Telegraphy.

The Marconi system of Wireless Telegraphy was tried in the naval manœuvres of 1899, and proved very successful so long as only one ship was signalling. Signals were taken in successfully at a distance of sixty miles.

Negotiations have been carried on between Her Majesty's Government and the Marconi Company, but the question of terms has not yet been settled. In the meanwhile the Admiralty are endeavouring to procure, for further and more extended trials, some more installations of Wireless Telegraphy, both from the Marconi Company and from other sources.

NEW WORKS.

New Works in the Estimates.

The principal new works for which provision is made in these Estimates are—

A new building slip and workshops at Devonport.

The establishment of a coaling depôt at the Falkland Islands.

Hospital and other shore accommodation for the Ganges training ship at Harwich.

A new general hospital at Portland to replace the existing huts.

Cold meat stores at Gibraltar.

Works in Progress.

The extension of No. 5 slip at Portsmouth is finished. Good progress has been made with the new Angle Iron Smithery, and with the new coaling arrangements. The extension of the jetties will be finished by the end of the financial year 1899–1900.

The new coal sheds at Gibraltar and the extension of the boat-house at Malta will also be completed this financial year.

The new cooperage in Plymouth Victualling Yard has been finished.

Some delay has occurred in completing the plans for the new slip at Chatham, but it is expected that considerable progress will be made with this work during the year 1900–1901.

The rifle ranges at Sheerness and Wei-hai-Wei are finished, and those at Eastney and Tipnor are nearly complete.

Progress under the Naval Works Acts.

Inclosure and Defence of Harbours.

Gibraltar.—The Admiralty Mole extension was brought up to water-level by 30th September, 1898. It is now being increased to its full section. About 1800 ft. of quay wall have been completed.

The detached mole has been brought above water-level for a length of over 1400 ft.

Three dredgers are at work deepening the harbour.

On the Commercial Mole the bonded warehouses are well advanced. The extension of the Old Mole to the eastern end of the viaduct is completed, and the embankment beyond the viaduct is in progress. The dredging of the trenches for the walls of the Western Arm is finished, and rubble is being deposited.

Portland.—The whole of the new breakwater was brought up to low-water level by the middle of April, 1899, five months in advance of the contract time.

Dover.—Admiralty Pier Extension.—Block-making for this work has proceeded steadily, and there are now over 1600 blocks in stock. The temporary staging has been completed for nine bays—that is, for a length of about 450 ft.—and there are now in use on the staging one 60-ton and two 40-ton goliaths. Divers have been at work for some weeks preparing the foundations, and block-setting on the permanent work of the Admiralty Pier extension has been begun.

As a protection to the staging and shipping a lightship has been moored off the end of the staging, and a fog-signalling apparatus established on the outer end of the staging itself.

East Reclamation.—The piles for the trestle railway have been erected for a total length of 3650 ft., and the superstructure is completed for 3200 ft. The foundations of the reclamation wall are complete for a length of 3050 ft., and the wall itself has been brought up to the level of high-water of neap tides for a full length of 2700 ft.

The excavation of the chalk cliff for filling has proceeded steadily, and for a length of about 600 ft. the backing of the reclamation wall has been completed to a height of 25 ft. above low water.

East Arm.—A commencement has been made with the erection of the temporary staging for this work.

Sandwich, &c.—Block-making for the turret winding wall and the making of apron blocks for the east reclamation wall has been carried on satisfactorily.

Adapting Naval Ports to present needs of Fleet.

Deepening Harbours and Approaches.—Work at Chatham and Haulbowline is finished. Dredging is proceeding at Portsmouth and Devonport.

Keyham Dockyard Extension.—Nearly all the mud has been

removed from the site of the docks. Good progress has been made with the works generally.

Gibraltar Dockyard Extension.—The new dockyard buildings and the dam for No. 3 dock are in course of construction. Good progress has been made with the slipways for torpedo-boat destroyers.

Hong Kong Dockyard Extension.—The recently-acquired buildings are being used as stores, workshops, &c., and the necessary alterations are being made in them. The widening of the Queen's Road is proceeding. The contract for the main portion of the work has just been approved.

Pembroke Jetty.—This work has not made as much progress as was expected, and the contract time for completion will be exceeded.

Portsmouth—Widening Caisson.—Practically completed.

Haulbowline Improvements.—Work is nearly finished.

Chatham Dock.—Tenders will be invited shortly.

Malta Dockyard Extension.—The preparation of the site is proceeding rapidly.

Bermuda Dockyard Extension.—The contract will be let during the present year.

Simon's Bay Dockyard Extension.—The contract particulars are being prepared.

Naval Barracks, &c.

Progress on the new naval barracks at Chatham, Portsmouth, and Keyham has been very satisfactory.

Sheerness Naval Barracks.—As explained last year, the proposal to build new barracks for the gunnery school at Sheerness has been abandoned, and a site is being sought at Chatham. No satisfactory arrangements have yet been arrived at for the acquisition by the Admiralty of a sufficient area of land for the purpose; but this delay is of little consequence, as, owing to the difficulty of obtaining sufficient labour, it would in any case be inexpedient to start another large building work at Chatham until the new naval barracks and hospital have made further progress.

Chatham Naval Hospital.—The foundations of the main buildings are completed and the brickwork is progressing.

Britannia Royal Naval College.—Tenders have been invited. Much work has been done on foundations, preparation of the site, &c.

Magazines.—Work is progressing satisfactorily.

Haslar hospital extension is making good progress, and the infectious hospital at Haulbowline is nearly finished.

A statement of the estimated expenditure for the year ending on the 31st March, 1900, will shortly be laid before Parliament.

GEORGE J. GOSCHEN.

17*th February,* 1900.

Abstract of Navy

Votes.		Estimates,	
		Gross Estimate.	Appropriations in Aid.
	I.—NUMBERS.		
A.	Total Number of Officers, Seamen, Boys, Coast Guard, and Royal Marines	114,880	....
	II.—EFFECTIVE SERVICES.	£	£
1	Wages, &c., of Officers, Seamen and Boys, Coast Guard, and Royal Marines	5,643,016	116,016
2	Victualling and Clothing for the Navy	2,186,175	470,875
3	Medical Establishments and Services	230,175	21,375
4	Martial Law	13,320	20
5	Educational Services	120,744	28,444
6	Scientific Services	81,185	14,285
7	Royal Naval Reserves	271,213	113
8	Shipbuilding, Repairs, Maintenance, &c.:		
	Section I.—Personnel	2,524,815	12,815
	Section II.—Matériel	4,248,000	164,000
	Section III.—Contract Work	6,367,055	38,055
9	Naval Armaments	3,060,008	55,308
10	Works, Buildings, and Repairs at Home and Abroad	865,800	20,000
11	Miscellaneous Effective Services	281,912	10,712
12	Admiralty Office	276,100	9,000
	Total Effective Services . . . . £	26,169,518	961,018
	III.—NON-EFFECTIVE SERVICES.		
13	Half-Pay, Reserved, and Retired Pay	798,972	12,272
14	Naval and Marine Pensions, Gratuities, and Compassionate Allowances	1,145,550	21,950
15	Civil Pensions and Gratuities	343,882	382
	Total Non-Effective Services . . £	2,288,404	34,604
	IV.—EXTRA ESTIMATE FOR SERVICES IN CONNECTION WITH THE COLONIES.		
16	Additional Naval Force for Service in Australasian Waters—Annuity payable under	95,300	35,000
	GRAND TOTAL . . . . £	28,553,222	1,030,622

Note.—Under an Act of the Cape of Good Hope Legislature, entitled, "The Navy Contribu-
contribution towards the annual expenditure by the Imperial Government in connection with
A gift of 12,000 tons of coal for the use of Her Majesty's Ships, etc., is made annually by the

Estimates for 1900-1901.

1900-1901.	Estimates, 1899–1900.			Difference on Net Estimates.		Votes.
Net Estimate.	Gross Estimate.	Appropriations in Aid.	Net Estimate.	Increase.	Decrease.	
Total Numbers.			Total Numbers.	Numbers.	Numbers.	
114,880	....	....	110,640	4,240	....	A.
£	£	£	£	£	£	
5,527,000	5,361,017	118,317	5,242,700	284,300	....	1
1,715,300	2,051,712	445,012	1,606,700	108,600	....	2
208,800	197,890	21,290	176,600	32,200	....	3
13,300	12,232	32	12,200	1,100	....	4
92,300	119,756	29,156	90,600	1,700	....	5
66,900	82,341	12,841	69,500	....	2,600	6
271,100	271,113	113	271,000	100	....	7
						8
2,512,000	2,429,815	12,815	2,417,000	95,000	....	Sec. I.
4,084,000	3,960,000	161,000	3,799,000	285,000	....	Sec. II.
6,329,000	6,638,460	37,460	6,601,000	....	272,000	Sec. III.
3,004,700	2,755,585	44,785	2,710,800	293,900	....	9
845,800	806,830	11,730	795,100	50,700	....	10
271,200	258,645	10,445	248,200	23,000	....	11
267,100	270,600	9,000	261,600	5,500	....	12
25,208,500	25,215,996	913,996	24,302,000	1,181,100	274,600	
786,700	786,914	12,214	774,700	12,000	....	13
1,123,600	1,137,936	21,936	1,116,000	7,600	....	14
343,500	341,893	393	341,500	2,000	....	15
2,253,800	2,266,743	34,543	2,232,200	21,600	....	
60,300	95,300	35,000	60,300	....	....	16
27,522,600	27,578,039	983,539	26,594,500	1,202,700	274,600	

Net Increase £928,100

tion Act, 1898," a sum of £30,000 is paid annually out of the public revenue of that Colony as a Her Majesty's Naval Service.
Natal Government.

STATEMENT showing the Actual and Estimated EXPENDITURE for NAVAL SERVICES for the Three Years ending the 31st March 1901.

		£	s.	d.
1898–99	Estimated Expenditure (after deducting Appropriations in Aid)	£23,778,400	0	0
	Supplementary Estimate (28th February, 1899)	£350,000	0	0
		24,128,400	0	0
	Net Expenditure, as per Final Account	23,880,875	11	11
	Net (Expenditure less than Estimate)	£247,524	8	1
1899-1900.	Estimated Expenditure (after deducting Appropriations in Aid)	£26,594,500	0	0
1900–1901.	Estimated Expenditure (after deducting Appropriations in Aid)	£27,522,600	0	0

STATEMENT of the Principal Points of DIFFERENCE between the ESTIMATES of 1899–1900 and those for 1900–1901.

INCREASES.		£
Wages, &c., of Officers, Seamen, and Marines		281,300
Victualling and Clothing		108,600
Medical Establishments and Services		32,200
Martial Law		1,100
Educational Services		1,700
Wages, &c., of Men in Dockyards		85,921
Naval Stores		282,000
Hulls of Ships (Contract)		419,339
Auxiliary Machinery for Her Majesty's Ships and Vessels (Contract)		4,558
Wages of Artificers and Crews of Vessels (Naval Ordnance Establishments)		10,134
Projectiles and Ammunition		251,050
Small Arms and Miscellaneous Naval Ordnance Stores, &c.		74,864
Inspection, Proof, Experiments, and Freight (Naval Ordnance Stores)		12,800
Works, Buildings, and Repairs		50,700
Miscellaneous Effective Services (Passage Money, &c.)		23,000
Non-Effective Services		21,600
Miscellaneous Items		8,511
		1,672,377
DECREASES.	£	
Scientific Services	2,600	
Propelling Machinery (Contract)	521,515	
Gun Mountings and Air-compressing Machinery (Contract)	112,582	
Machinery for Her Majesty's Shore Establishments (Contract)	50,000	
Repairs, &c., of Ships, &c. (Contract)	5,405	
Inspection of Contract Work	4,000	
Royal Reserve of Merchant Cruisers	1,800	
Guns	18,700	
Torpedoes and Gun Cotton	27,675	
		744,277
Net Increase £		928,100

STATEMENT showing the Total Estimated **EXPENDITURE** for the **NAVAL SERVICE**, including Amounts provided in the **NAVY ESTIMATES**, as well as in the **CIVIL SERVICE** and other **ESTIMATES**, for the following Services:—

		1900–1901.	1899–1900.
NAVY ESTIMATES:			
Estimated Expenditure (after deducting Appropriations in Aid)		£ 27,522,600	£ 26,594,500
CIVIL SERVICE ESTIMATES:			
Estimated Expenditure under—			
Class I. Vote 8.—Public Buildings, Great Britain:	£		
Maintenance and Repairs, including New Works, Alterations, &c.	4,830		
Rents, Insurance, Tithes, &c.	8,045		
Fuel, Light, Water, &c.	5,025		
Furniture	2,000		
Removal into and furnishing New Block of Admiralty Extension	10,000		
		29,900	19,800
Class I. Vote 9.—Surveys of the United Kingdom		200	150
„ I. „ 12.—Rates on Government Property		(a) 88,800	84,800
„ I. „ 13.—Public Works and Buildings, Ireland:			
Coast Guard, viz.:	£		
Purchase of Sites	—		
New Works and Alterations, including Naval Reserve Stations	4,490		
Maintenance and Supplies	5,826		
Furniture, Fittings, &c.	—		
	£10,316		
Naval Reserve, viz.:			
Maintenance and Supplies	154		
		10,470	13,098
Class II. Vote 8.—Board of Trade:			
Staff and Incidental Expenses in connection with the Royal Naval Reserve Force		3,380	3,280
„ II. „ 9.—Mercantile Marine Services:			
Staff and Incidental Expenses in connection with the Royal Naval Reserve Force		2,450	—
„ II. „ 14.—Exchequer and Audit Department (Cost of Audit):	£		
Navy Cash Accounts	7,096		
Expense and Manufacturing Accounts	4,752		
Store Accounts	5,658		
		17,506	17,092
Class II. Vote 23.—Stationery and Printing		71,000	71,000
„ III. „ 1.—Law Charges, England (Net)		3,080	3,400
„ III. „ 7.—Prisons, England and the Colonies: Maintenance of Naval Prisoners		4,928	3,262
„ III. „ 13.—Prisons, Scotland		100	95
„ III. „ 20.—Prisons, Ireland		59	56
REVENUE DEPARTMENT ESTIMATES:			
Vote 1.—Customs.—Payment of Navy Wages and provision of funds for District Paymasters of the Coast Guard		142	162
Vote 1.—Customs.—Staff and Incidental Expenses in connection with the Royal Naval Reserve Force		3,261	—
Vote 2.—Inland Revenue.—Analysis of Food, &c.		100	140
Vote 3.—Post Office.—Postage of Official Correspondence (including Parcels)	£ 15,177		
Vote 5.—Post Office Telegraphs.—Official Telegrams and Expenses in connection with Telegraphs (Admiralty Wires, and Services of Clerks)	14,567		
		29,744	29,625
Total	£	27,787,720	26,840,460

Note.—In addition to the Services shown above, an annuity of £16,243 18s. is payable to the Commissioners of Woods, &c., from the Consolidated Fund, under the Public Offices Sites Act of 1882 (45 & 46 Vict. c. 32).

(a) This sum is exclusive of the Admiralty share of the amount included in the Rates Estimate for Drainage Works at Malta.

VOTE (A.)

NUMBERS of OFFICERS, SEAMEN, BOYS, and ROYAL MARINES, Borne on the Books of Her Majesty's Ships, and at the ROYAL MARINE DIVISIONS.

One Hundred and Fourteen Thousand Eight Hundred and Eighty.

I.—SEA SERVICE.

Under which Vote Provided.	RANKS, &c.	NUMBERS, ALL RANKS. 1900–1901.		1899–1900.		Numbers of all Ranks borne on 1st January, 1900.
Vote 1	FOR HER MAJESTY'S FLEET					
	Flag Officers	15		15		
	Commissioned Officers	3,628		3,482		
	Subordinate Officers	806		810		
	Warrant Officers	1,413		1,331		
	Petty Officers and Seamen	73,259		69,984		
	Boys (Service)	3,700		3,700		
			82,821		79,322	78,536
	COAST GUARD.					
	Commissioned Officers	89		89		
	Chief Officers of Stations	237		237		
	Petty Officers and Seamen	3,874		3,874		
			4,200		4,200	4,101
	ROYAL MARINES (for Service Afloat and on Shore).					
	Commissioned Officers	444		434		
	Warrant Officers	32		32		
	Staff Sergeants and Sergeants	1,361		1,307		
	Buglers and Musicians	620		604		
	Rank and File	16,133		15,913		
			18,590		18,290	18,143
	Total		105,611		101,812	100,780
	Net Increase		3,799			

II.—OTHER SERVICES.

Under which Vote Provided.	RANKS, &c.	1900–1901.		1899–1900.		Numbers borne on 1st January, 1900.
Vote 1	Naval Cadets	260		265		
	Engineer Students	178		183		
	Pensioners in Home Ships and in the Reserves	987		990		
	Boys under Training	6,200		6,200		
			7,625		7,638	7,239
Other Votes	Various Services	..	1,644	..	1,190	1,190
	Total		9,269		8,828	8,429
	Net Increase		441			

	1900–1901.		1899–1900.	
Total, Sea Service	105,611		101,812	
„ other Services	9,269		8,828	
		114,880		110,640
Net Increase		4,240		

	1900–1901.		1899–1900.
Including Officers and Seamen	1,531	—	1,407
„ Pensioners (Vote 1)	987	—	990
„ Pensioners (other Votes)	16	—	16
„ Boys (Training, Seaman Class)	6,200	—	6,200
„ Boys (Training, Artisans)	320	—	—
„ Royal Marines	215	—	215
	9,269	—	8,828

VOTE 8.

SHIPBUILDING, REPAIRS, MAINTENANCE, &c.

I.—ESTIMATE of the SUM which will be required, in the YEAR ending 31st March, 1901, to defray the EXPENSES of SHIPBUILDING, REPAIRS, MAINTENANCE, &c., including the COST of ESTABLISHMENTS of DOCKYARDS and NAVAL YARDS at HOME and ABROAD.

DOCKYARD WORK.

SECTION I.—PERSONNEL.—Two Million Five Hundred and Twelve Thousand Pounds.

(£2,512,000.)

SECTION II.—MATÉRIEL.—Four Million and Eighty-Four Thousand Pounds.

(£4,084,000.)

CONTRACT WORK.

SECTION III.—CONTRACT WORK.—Six Million Three Hundred and Twenty-Nine Thousand Pounds.

(£6,329,000.)

I.—SUB-HEADS under which SECTION I., PERSONNEL, of this VOTE will be accounted for.

——	ESTIMATES.		Increase.	Decrease.
	1900–1901.	1899–1900.		
	£	£	£	£
DOCKYARD WORK.				
SECTION I.—PERSONNEL.				
Dockyards at Home.				
A.—Salaries and Allowances . . .	(a) 178,562	161,014 (b) 14,000	17,548	..
B.—Wages, &c., of Men, and hire of Teams	1,945,392	1,885,827 (c) 14,000	45,565	..
C.—Wages, &c., of Police Force . .	41,040	41,395	..	355
D.—Contingencies	6,600	6,400	200	..
Naval Yards Abroad.				
E.—Salaries and Allowances . . .	(a) 66,442	63,000	3,442	..
F.—Wages, &c., of Men, and hire of Teams	271,167	244,811	26,356	..
G.—Wages, &c., of Police Force . .	14,412	12,268	2,144	..
H.—Contingencies	1,200	1,100	100	..
£	2,524,815	2,429,815	95,355	355
Deduct,—				
I.—Appropriations in Aid	12,815	12,815	..	..
£	2,512,000	2,417,000	95,355	355
		Net Increase .	£95,000	

(a) These amounts include the sums of £23,398 and £1,233 for pay of Inspectors of Trades at Home and Inspectors of Shipwrights Abroad respectively, which is charged direct to the cost of shipbuilding.
(b) Provided under Sub-head B. in 1899–1900. (c) Transferred to Sub-head A. in 1900–1901.

Note.—Provision has been made for New Construction in the above Vote to the extent of—

Section 1	£913,650
„ 2	1,699,386
„ 3	5,847,110
	£8,460,146

VOTE 8.—SHIPBUILDING, REPAIRS, MAINTENANCE, &C.—*continued.*

II.—SUB-HEADS under which SECTION II., MATÉRIEL, of this VOTE will be accounted for.

———	ESTIMATES.		Increase.	Decrease.
	1900–1901.	1899–1900.		
	£	£	£	£
DOCKYARD WORK—*continued.*				
SECTION II.—MATÉRIEL.				
Naval Stores.				
A.—Timber, Masts, Deals, &c.	136,000	118,000	18,000	..
B.—Metals and Metal Articles	2,043,000	1,979,500	63,500	..
C.—Coals for Yard purposes	79,500	64,000	15,500	..
D.—Hemp, Canvas, &c.	191,500	200,000	..	8,500
E.—Paint Materials, Oils, Pitch, Tar, Tallow, Boats, Furniture, and other Miscellaneous Articles	500,000	555,000	..	55,000
F.—Electrical, Torpedo, and other Apparatus	190,000	196,500	..	6,500
G.—Coals for Steam Vessels	1,000,000	750,000	250,000	..
H.—Freight	60,000	52,000	8,000	..
I.—Rents, Water, &c., Dockyards at Home, and Naval Yards Abroad	33,900	31,645	2,255	..
K.—Gas, &c., Dockyards at Home, and Naval Yards Abroad	14,100	13,355	745	..
£	4,248,000	3,960,000	358,000	70,000
Deduct,—				
L.—Appropriations in Aid	164,000	161,000	3,000	..
£	4,084,000	3,799,000	355,000	70,000
	Net Increase		£285,000	

VOTE 8.—SHIPBUILDING, REPAIRS, MAINTENANCE, &c.—*continued.*

II.—SUB-HEADS under which SECTION III., CONTRACT WORK, of this VOTE will be accounted for.

——	ESTIMATES. 1900-1901.	ESTIMATES. 1899-1900.	Increase.	Decrease.
	£	£	£	£
SECTION III.—CONTRACT WORK.				
A.—Propelling Machinery for Her Majesty's Ships and Vessels	1,999,130	2,520,645	..	521,515
B.—Auxiliary Machinery for Her Majesty's Ships and Vessels	69,059	64,501	4,558	..
C.—Hulls of Ships, &c., Building by Contract	3,381,961	2,962,622	419,339	..
D.—Purchase of Ships, Vessels, &c.	..	..	..	..
E.—Repairs and Alterations by Contract of Ships, &c., and their Machinery and Stores	93,055	98,460	..	5,405
F.—Inspection of Contract Work	48,000	52,000	..	4,000
G.—Gun Mountings and Air-Compressing Machinery	612,650	725,232	..	112,582
H.—Machinery for Her Majesty's Shore Establishments at Home and Abroad	100,000	150,000	..	50,000
I.—Royal Reserve of Merchant Cruisers	63,200	65,000	..	1,800
£	6,367,055	6,638,460	423,897	695,302
Deduct,—				
K.—Appropriations in Aid	38,055	37,460	595	..
£	6,329,000	6,601,000	423,302	695,302
	Net Decrease		£272,000	

PROGRAMME of the ESTIMATED EXPENDITURE in CASH, and in NET
REPAIRS, MAINTENANCE, &c.,
SUB-HEADS under which this ESTIMATED EXPENDITURE will be
provisions of Sec. 1 (2), ARMY

	ESTIMATED EXPENDITURE IN				
	Direct Expenditure.				
	Dockyard Work.		Contract Work, Sec. III.	Total Direct Expenditure. (A)	
	Personnel, Sec. I.	Matériel, Sec. II.			
	£	£	£	£	
NEW CONSTRUCTION:					
A.—DOCKYARD-BUILT SHIPS—					
Hulls, &c. (*c*)	783,070	1,585,281	310,471	2,678,822	1
Machinery	36,220	42,940	664,413	743,573	2
	819,290	**1,628,221**	**974,884**	**3,422,395**	3
B.—CONTRACT-BUILT SHIPS—					
Hulls, &c. (*c*)	88,210	65,115	3,615,130	3,768,455	4
Machinery	..	..	1,168,803	1,168,803	5
	82,210	**65,115**	**4,783,933**	**4,937,258**	6
C.—SMALL VESSELS (*d*)	6,150	6,050	88,293	100,493	7
TOTAL NEW CONSTRUCTION	**913,650**	**1,699,386**	**5,847,110**	**8,460,146** (*f*)	8
D.—RE-CONSTRUCTION, REPAIRS, ALTERATIONS, &c.	842,540	482,280	292,991	1,617,811	9
E.—SEA STORES, COALS, &c.	..	1,605,000	15,628	1,620,628	10
F.—ESTABLISHMENT, INCIDENTAL, AND MISCELLANEOUS CHARGES, UNAPPROPRIATED	..	..	..	..	11
TOTAL	1,756,190	3,786,666	6,155,729	11,698,585	12

(*c*) Including Hydraulic and Transferable Gun Mountings, &c.
(*d*) Including Harbour Craft, and excluding Torpedo Boats, &c., the value of which is included under other Sub-Heads.
(*f*) Exclusive of £25,300 provided under Vote 2 for new Tank Vessels and Lighters for Victualling Yard Service; also £8,950 provided under Vote 9 for new Vessels for Naval Ordnance Store Service.

VALUES OF STORES issued for SHIPBUILDING, RE-CONSTRUCTION, in the Year 1900–1901.

accounted for in the NAVY EXPENSE ACCOUNTS, under the AND NAVY AUDIT ACT, 1889.

1900–1901.		EXPENDITURE AS ESTIMATED IN NAVY ESTIMATES, 1899-1900.			Difference between Direct Expenditure, 1899–1900 (B) and 1900–1901 (A).	
Establishment, &c., Charges, apportioned.	Aggregate, 1900-1901.	Direct Expenditure. (B)	Establishment, &c., Charges, apportioned.	Aggregate, 1899–1900.	Increase.	Decrease.
£	£	£	£	£	£	£
269,791	2,948,613	2,839,047	272,703	3,111,750	..	160,225
22,320	765,893	800,768	19,197	819,965	..	57,195
292,111	**3,714,506**	**3,639,815**	**291,900**	**3,931,715**	..	**217,420**
78,829	3,847,284	3,509,000	84,459	3,593,459	259,455	..
17,529	1,186,332	1,520,896	22,813	1,543,709	..	352,093
96,358	**5,033,616**	**5,029,896**	**107,272**	**5,137,168**	..	**92,638**
1,736	102,229	185,770	2,950	188,720	..	85,277
390,205	**8,850,351**	**8,855,481**	**402,122**	**9,257,603**	..	**395,335**
182,879	1,800,690	1,463,150	144,989	1,608,139	154,661	..
52,110	1,672,738	1,396,962	44,634	1,441,596	223,666	..
1,318,361	1,318,361	..	1,299,644	1,299,644	..	..
1,943,555	**13,642,140**	**11,715,593**	**1,891,389**	**13,606,982**		

NET DECREASE ON DIRECT EXPENDITURE . £17,008.

RECAPITULATION OF ESTIMATED EXPENDITURE.

	ESTIMATED EXPENDITURE.		ESTIMATED DISTRIBUTION OF THE DIRECT AND INCIDENTAL EXPENDITURE.							
			Naval Construction.					Establishment and Incidental Charges Unapportioned to Ships, &c.		
				Re-construction, Repairs, Alterations, and Refits.						
	Charged direct as Incurred. (a.)	Establishment and Incidental Charges Apportioned. (b.)	New Construction. (c.)	Ships Re-constructing. (d.)	Ships for Reliefs, or Re-commission. (e.)	Ships in Commission and Reserve. (f.)	Stores for Ships, Coals for Steaming, &c. (g.)	Fleet, Port, National, and Unappropriated Charges. (h.)	Unappropriated Charges (Haulbowline, West India Docks and Naval Yards Abroad). (i.)	Total Amount of Estimated Expenditure. (k.)
SUB-HEADS OF EXPENDITURE.	.	.	A., B., and C.	D.			E.	F.		
	£	£	£	£	£	£	£	£	£	£
SECTION I.—DOCKYARD WORK PERSONNEL	1,756,190	949,002	1,085,893	..	435,110	505,376	34,836	423,176	220,801	2,705,192
SECTION II.—DOCKYARD WORK MATÉRIEL	3,786,666	740,146	1,829,772	..	230,723	332,859	1,622,274	306,086	298,538	4,526,812
SECTION III.—CONTRACT WORK	6,155,729	254,407	5,934,686	..	161,696	134,926	15,628	69,760	..	6,410,136
TOTAL ESTIMATED EXPENDITURE for 1900–1901 £	11,698,585	1,943,555	8,850,351	..	827,529	973,161	1,672 738	799,022	519,339	13,642,140
TOTALS OF SUB-HEADS £	13,642,140		8,850,351	1,800,690			1,672,738	1,318,361		13,642,140

DIAGRAM SHEWING THE ACTUAL EXPENDITURE UPON THE CONSTRUCTION OF NEW SHIPS DURING THE 29, YEARS BETWEEN 1872-73 & 1900-1.

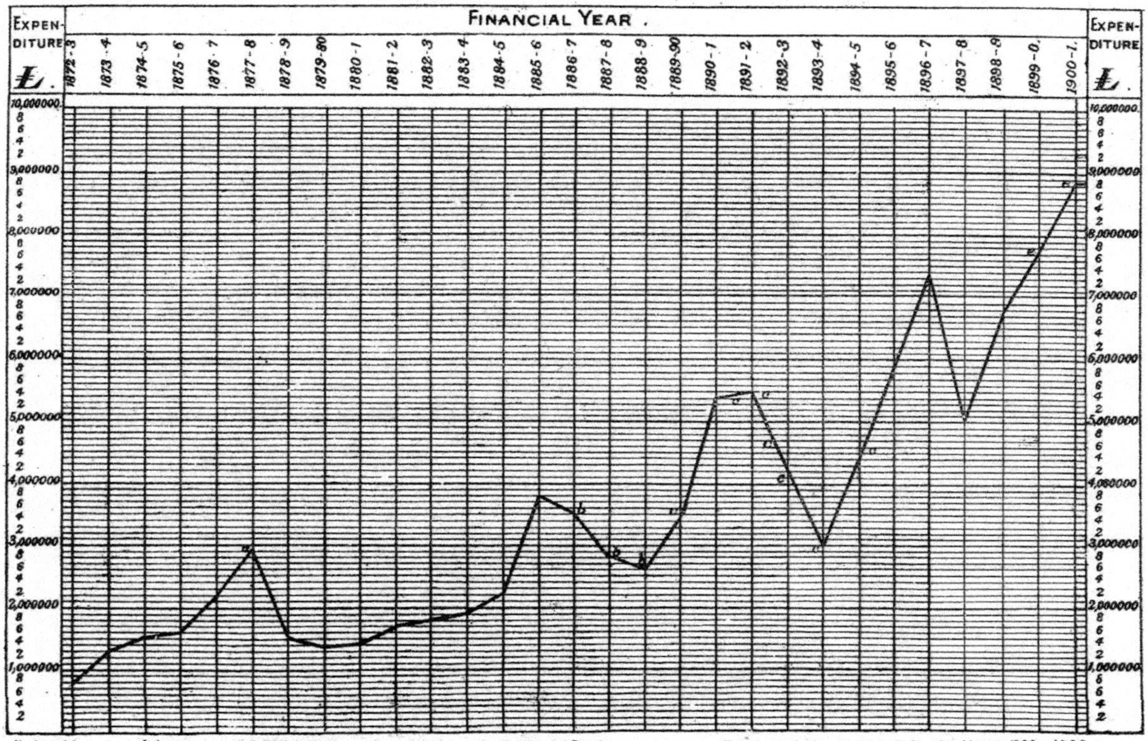

(a) Includes £1,523,000. for purchase of Ships under the Vote of Credit.
(b) Includes Expenditure under Lord Northbrook's Special Programme
(c) Includes Expenditure under the Defence Acts of 1888 and 1889.

e,e, Estimated Expenditure for the Years 1899-1900 & 1900-1 at the time when Navy Estimates for 1900-1 were prepared

LIST of NEW SHIPS and VESSELS Estimated to be Passed into the FLEET RESERVE during the Years 1900–1901 and 1899–1900.

1900–1901.

NAME OF SHIP.	Load Displacement in Tons.	Indicated Horse Power.	Number of Guns.
ARMOURED SHIPS.			
Formidable . . .	15,000	15,000	16
Implacable . . .	15,000	15,000	16
Irresistible . . .	15,000	15,000	16
Albion	12,950	13,500	16
Glory	12,950	13,500	16
Vengeance . . .	12,950	13,500	16
PROTECTED SHIPS.			
Spartiate	11,000	18,000	16
Pandora	2,200	*7,000	8
UNPROTECTED SHIPS.			
Victoria and Albert .	4,700	11,000	—
Shearwater . . .	980	1,400	6
Vestal	980	1,400	6
Mutine	980	1,400	6
Rinaldo	980	1,400	6
TORPEDO-BOAT DESTROYERS {16 No. . }	various		
TORPEDO-BOATS {2 No. }	—	—	—

1899–1900.

NAME OF SHIP.	Load Displacement in Tons.	Indicated Horse Power.	Number of Guns.
ARMOURED SHIPS.			
Canopus	12,950	13,500	16
Goliath	12,950	13,500	16
Ocean	12,950	13,500	16
PROTECTED SHIPS.			
Ariadne	11,000	18,000	16
Amphitrite . . .	11,000	18,000	16
Andromeda . . .	11,000	16,500	16
Gladiator . . .	5,750	10,000	10
Hyacinth . . .	5,600	10,000	11
Highflyer . . .	5,600	10,000	11
Hermes	5,600	10,000	11
Pioneer	2,200	*7,000	8
Perseus	2,135	*7,000	8
Prometheus . . .	2,135	*7,000	8
Pyramus	2,135	*7,000	8
Pomone	2,135	*7,000	8
UNPROTECTED SHIPS.			
Condor	980	1,400	6
Rosario	980	1,400	6
Bramble	700	*1,300	6
Britomart . . .	700	*1,300	6
Dwarf	700	*1,300	6
Thistle	700	*1,300	6
TORPEDO-BOAT DESTROYERS {1 No. . }	various		—

* Forced draught.

French Navy Estimates, 1900.

Cap. in French Esti-mates.	Heads of Expenditure.	Credits voted for the year 1900.*	Credits granted for the year 1899.
		£	£
	PERSONNEL.		
1, 2	Admiralty Office	132,986	132,916
3, 4	Navy Pay	1,878,220	1,846,268
5	Marines.	622,420	611,726
6	Gendarmerie Maritime	30,792	30,792
7	Inspection of Administrative Services	10,368	10,444
8	Construction Staff	195,440	77,016
9, 10, 11	Administrative Staff, Commissariat, etc.	268,491	265,450
12	Medical and Religious Staff	87,357	88,448
13	Fisheries and Navigation	26,872	26,875
	LABOUR.		
	Wages—		
14	Shipbuilding; new construction; fitting for sea	539,360	598,768
15	Shipbuilding; repairs	295,137	293,600
16	Armaments; construction of new guns	111,806	116,396
17	Armaments; repairs	26,582	27,680
18	Works	30,758	38,044
19	Victualling	33,545	35,076
20, 21	Master-attendants' and Storekeepers' Departments	211,883	221,720
22	Miscellaneous	14,597	14,896
	MATÉRIEL.		
	Stores and Supplies—		
23	Admiralty	10,100	9,980
24	Shipbuilding in Dockyards	1,600,000	1,472,488
25, 26	Shipbuilding by contract	1,339,000	1,696,264
27	Fitting for sea; maintenance; repairs	502,000	385,000
	Carried forward	£7,967,714	£7,999,847

* Amounts voted by Chamber of Deputies in March, 1900.

Cap. in French Estimates.	Heads of Expenditure.	Credits voted for the year 1900.*	Credits granted for the year 1899.
		£	£
	Brought Forward	7,967,714	7,999,847
	MATÉRIEL—*continued.*		
	Stores and Supplies—*continued.*		
28	Hydrographic Service	22,544	21,372
29, 30	Repairs, conversions, &c., in dockyards and by contract	445,260	441,700
31	Armaments; new guns and conversions	999,680	727,680
32	Armaments; powder, ammunition and repairs	68,400	68,400
33	Torpedoes	128,720	101,920
34	Works; new and large alterations	434,972	300,972
35, 36	Ditto, supplementary for defence of military ports	36,872	62,000
37	Works; repairs	60,920	60,920
38	Clothing	212,160	214,488
39	Barracks	34,081	44,324
40	Victualling	960,117	941,437
41	Hospitals, etc.	107,959	107,959
42 to 46	Machinery, tools, etc.	212,732	212,732
47	Fuel and lighting	33,800	29,720
48	Office furniture, printing, etc.	21,416	21,416
	MISCELLANEOUS.		
49, 50	Travelling expenses and freight Allowance for lodging, etc.	244,200	247,600
51	Charitable and subscriptions	40,934	69,816
52	Fisheries and Commerce (materials for protection, etc.)	12,948	9,008
53	Pensions	453,788	454,392
54	Secret Service	4,000	4,000
55 to 57	Miscellaneous	4,444	2,320
	Total	£12,507,661	£12,144,023

PROGRAMME OF NEW CONSTRUCTION, TO BE CONTINUED OR UNDERTAKEN IN 1900.—BUILDING IN DOCKYARDS.

Class.	Names of Ships.	Where Building.	Date of Commencement.	Proposed Date of Completion.	Estimated Cost.	Expenditure proposed for 1900.
					£	£
Battleships	Charlemagne	Brest	1894	1897	1,056,224	5,244
	Saint-Louis	Lorient	1895	1898	1,103,350	15,126
	Gaulois	Brest	1896	1898	1,049,396	5,244
	Henri IV.	Cherbourg	1897	1900	801,248	185,040
	Iéna	Brest	1898	1900	1,114,260	244,496
	Suffren	„	1898	1901	1,195,564	318,256
Armoured Cruisers, First-class	Jeanne d'Arc	Toulon	1896	1901	1,056,636	128,372
	Dupetit-Thouars	„	..	1901	819,368	238,040
	Gueydon	Lorient	1898	1902	832,288	212,900
	Condé	Cherbourg	..	1903	882,960	189,892
	Gloire	Lorient	..	1902	902,460	211,640
	La Marseillaise (ex C 9)	Brest	..	1902	902,460	189,480
	C 11	Cherbourg	..	1905	1,159,300	2,800
	C 12	Brest	..	1904	1,159,300	50,388
	C 13	Toulon	..	1905	1,159,300	46,640
	Dupleix	Rochefort	1899	1901	652,352	178,060
	Jurien de la Gravière	Lorient	1897	1900	453,496	109,036
Third-class Protected Cruiser	D'Estrées	Rochefort	1897	1899	205,568	11,004
Torpedo-gunboats and Destroyers	Dunois	Cherbourg	1896	1898	122,124	8,000
	La Hire	„	1896	1899	121,540	8,000
	Pertuisane	Rochefort	..	1902	68,908	20,400
	Escopette	„	..	1902	68,908	20,400
	Flamberge	„	..	1902	68,908	17,668
	Rapière	„	..	1902	68,908	17,668
	M 12	„	..	1903	68,908	2,826
	M 13	„	..	1903	68,908	2,826
	Decidée	Lorient	1898	1899	57,736	7,804
	Zélée	Rochefort	1899	1900	63,184	18,116
				CARRIED FORWARD	£17,283,562	2,465,366

PROGRAMME OF NEW CONSTRUCTION, TO BE CONTINUED OR UNDERTAKEN IN 1900.—BUILDING IN DOCKYARDS—*continued.*

Class.	Names of Ships.	Where Building.	Date of Commencement.	Proposed Date of Completion.	Estimated Cost.	Expenditure proposed for 1900.
					£	£
			Brought	forward .	17,283,562	2,465,366
Submarine Boats .	Narval . . .	Cherbourg	1898	1900	23,652	8,956
	Sirène	"	..	1902	24,006	7,656
	Triton	"	..	1902	24,006	7,656
	Farfadet . . .	Rochefort	..	1902	24,006	16,556
	Korrigan . . .	"	..	1902	24,006	16,556
	Gnome . . .	"	..	1902	24,006	16,556
	Lutin	"	..	1902	24,006	16,556
	Français . . .	Cherbourg	1899	1900	32,972	23,696
	Algérien . . .	"	1899	1900	32,972	23,696
	Q 13	"	..	1902	28,972	5,680
	Q 14	"	..	1902	28,972	5,680
First-class Torpedo-boats .	Six First-class Torpedo-boats Nos. 223, 224, 225, 226, 242 and 244	Various .	..	1901	116,157	32,008
		TOTAL BUILDING IN DOCKYARDS IN 1900 .			£17,691,295	2,646,618

PROGRAMME OF NEW CONSTRUCTION, TO BE CONTINUED OR UNDERTAKEN IN 1900.—BUILDING BY CONTRACT.

Class.	Names of Ships.	Places of Building and Completion.	Date of Contract.	Date of Completion.	Total Estimated Cost.	Expenditure proposed for 1900.
					£	£
Armoured Cruisers First-class	Montcalm	La Seyne—Toulon	1897	1901	891,360	197,964
	Sully	" "	1899	1903	949,336	284,184
	Amiral Aube	St. Nazaire—Brest	1899	1903	968,700	166,112
	Desaix	" "	1897	1901	711,280	166,664
	Kléber	Bordeaux—Rochefort	"	"	711,040	166,384
Fast Cruisers	Guichen	St. Nazaire—Toulon	1895–8	1898	620,252	36,688
	Châteaurenault	La Seyne—Toulon	1895–7	1899	618,962	72,768
Third-class Protected Cruiser.	Infernet	Bordeaux—Rochefort	1896–8	1899	193,848	18,800
Torpedo Cruisers	Hallebarde	Le Havre—Cherbourg	1896–7	1899	67,404	3,400
	Fauconneau	" "	1897	1900	67,640	5,736
	Espingole	" "	"	"	67,640	6,288
	Pique	" "	"	"	66,904	7,320
	Epée	" "	"	"	66,904	7,284
	Framée	Nantes—Lorient	"	"	66,584	7,212
	Yatagan	" "	"	"	66,584	13,240
	Eight boats, M 14 to M 21.	..	1900	..	539,712	53,424
River Gunboats	Argus	London—Hong Kong	1899	1900	22,424	11,352
	Vigilante	" "	"	"	22,424	11,352
Sea-going Torpedo Boats	Siroco	Le Havre—Cherbourg	1898	1900	42,584	14,088
	Mistral	" "	"	"	42,585	16,652
	Simoun	" "	"	"	40,904	11,692
	Typhon	" "	"	"	40,904	10,712
	Trombe	Nantes—Lorient	"	"	41,084	16,172
	Audacieux	" "	"	"	41,084	14,300
	Bourrasque	Le Havre—Cherbourg	1899	1901	40,984	12,972
	Rafale	" "	"	"	40,984	8,169
	Borée	Bordeaux—Rochefort	"	"	40,384	16,570
	Tramontane	" "	"	"	40,384	16,570
		Carried forward			£7,130,899	1,374,069

PROGRAMME OF NEW CONSTRUCTION, TO BE CONTINUED OR UNDERTAKEN IN 1900.—BUILDING BY CONTRACT—*continued.*

Class.	Names of Ships.	Places of Building and Completion.	Date of Contract.	Date of Completion	Total Estimated Cost.	Expenditure proposed for 1900.
					£	£
			Brought	forward	7,130,899	1,374,06
'irst-class Torpedo Boats	31 Boats:— Nos. 229, 231, 232, 236, 237, 238, 239, 240, 241, 243, 245, 246, 247, 248, 249, 250, 251, 252, 253, 254, 255, p. 75, p. 76, p. 77, p. 78, p. 79, p. 80, p. 81, p. 82, p. 83, p. 84 . . .	Various	1898–9	1899–1900	541,926	147,348
Portable Torpedo Boats . . .	Three Boats . .	Chalon—Toulon	1898	"	14,772	1,440
Corpedo Scout .	Libellule . . .	Le Havre—Cherbourg	1899	1900	13,756	6,936
				TOTAL BUILDING BY CONTRACT IN 1900	£7,701,353	1,529,793

German Navy Estimates, 1900.

(Converted at £1 = 20·43 marks.)

ORDINARY PERMANENT ESTIMATES.

—	Proposed for the financial year 1900.	Granted for the financial year 1899.
	£	£
Imperial Naval Office	63,743	62,129
Observatories	15,651	14,617
Accounts	16,222	14,762
Martial Law	3,486	1,748
Divine Service and Schools	3,909	3,842
Military Personnel	809,562	761,858
Maintenance of the Fleet	845,462	776,022
Victualling	54,932	47,793
Clothing	15,144	13,945
Barrack Administration, Cashiers and Accountants	114,673	110,745
Lodging Allowance	62,104	58,516
Medical	58,571	54,902
Travelling Expenses, Freight Charges, &c.	121,644	113,619
Training Establishments	15,591	16,261
Dockyard Expenses	1,000,628	940,514
Ordnance and Fortification	324,702	303,560
Accountant-General's Department	24,563	22,881
Pilotage and Surveying Services	25,257	24,538
Miscellaneous Expenses	41,756	37,648
Administration of Kiau-chau Protectorate	1,898	
Total . . . . . £	3,619,490	3,379,900

SPECIAL ORDINARY ESTIMATES.

Shipbuilding Programme for the Financial Year, 1900.

For the Construction of—	
1st class Battleship, Kaiser Wilhelm der Grosse (Ersatz König Wilhelm), 4th and final instalment	£ 195,790
Battleship A, 3rd instalment (Kaiser Barbarossa)	195,790
„ Kaiser Karl der Grosse (B), 3rd instalment	195,790
Large cruiser A, 3rd instalment	171,317
Small cruiser Nymphe (A), 3rd and final instalment	11,258
„ Niobe (B), „ „	11,258
Battleship C, 2nd instalment	234,949
„ D, „	234,949
„ E, „	234,949
Small cruiser C, „	88,106
„ D, „	88,106
Battleship F, 1st instalment	122,369
„ G, „	122,369
Large cruiser B, „	97,895
Small cruiser E, „	63,632
„ F, „	63,632
Gunboat A, „	24,473
One Torpedo-boat Division, 2nd and final instalment	121,096
„ „ 1st instalment	117,475
Total	£2,395,203

SUMMARY.

—	Proposed for the Financial Year 1900.	Granted for the Financial Year 1899.
	£	£
Ordinary Permanent Estimates	3,619,490	3,379,960
Shipbuilding	2,395,203	2,182,379
Armaments and Torpedo Equipments	1,008,370	590,749
Other Items	183,773	183,529
Extraordinary Expenditure	254,528	210,475
Total £	7,461,364	6,547,032

Italian Navy Estimates, 1900-1901.

FINANCIAL YEAR, 1ST JULY, 1900, TO 30TH JUNE, 1901.

Converted at £1 = 27 lire.

——	Proposed for 1900–1901.	Revised Estimates 1899–1900.
ORDINARY EXPENDITURE—GENERAL EXPENSES.	£	£
Admiralty	49,338	50,733
Pensions	190,963	184,851
Expenditure on various services connected with the Mercantile Marine	527,250	278,477
Total . . £	767,551	514,061
EXPENDITURE FOR NAVAL SERVICES.		
Ships fitting out	220,670	229,630
General Staff of the Navy	120,119	117,785
Corps of Constructors	49,074	48,326
Commissariat Service	33,333	30,585
Medical Service	24,867	24,867
Wages—Men	444,444	458,148
Gratuities	55,619	45,248
Assistants to Constructors and others	53,348	53,642
Accountants, &c.	53,518	52,440
Police	11,326	11,481
Telegraph Service	6,666	6,666
Telegraph Materials	8,518	8,518
Forts—*Personnel*	11,111	10,555
Victualling	290,370	283,333
Lighting	7,333	7,037
Hospital Services	17,828	19,241
Honorary Distinctions	444	444
Fuel and Stores	205,255	206,959
Salaries and Wages—Workshops and Fortifications	4,130	4,487
Training Establishments	15,555	15,666
Naval Academy	5,431	5,333
Scientific Services—*Personnel*	1,419	1,419
„ „ *Matériel*	9,259	9,963
Law Charges	1,185	1,185
Travelling Expenses	18,518	18,518
Transport of *Matériel*	4,629	4,629
Carried forward . . . . £	1,673,969	1,676,105

—	Proposed for 1900–1901.	Revised Estimates. 1899–1900.
	£	£
Brought forward	1,673,969	1,676,105
Materials and Labour for repair of Ships	668,518	487,037
Guns, Torpedoes and Small Arms	*81,481	292,592
Labour for construction and repairs of Armaments	82,334	82,334
Works Department—Repairs	87,171	87,171
Construction and Completion of the following Vessels, viz.: 1st Class Battleships: Benedetto Brin, at Castellamare; Regina Margherita, at Spezia. Turret Ship: Emanuele Filiberto, at Naples. Armoured Cruisers: Francesco Ferruccio, at Venice, and one unnamed at Taranto. Armoured Cruisers: Giuseppe Garibaldi, and Varese. 3rd Class Cruiser: Puglia, at Taranto. Torpedo Cruisers: Agordat, at Castellamare; and Coatit, at Naples. Torpedo-boat Destroyers. Sea-going Torpedo Boats. Small Craft	888,888	851,851
Total £	3,482,361	3,477,090

EXTRAORDINARY EXPENDITURE.

	£	£
General Expenses and Half Pay	2,832	3,185
Shipbuilding	18,518	18,518
Coast Defence and Fortifications	14,812	11,109
Torpedoes	18,518	18,518
Total £	54,680	51,330
Depreciation of Ships in Commission	129,629	111,111
Rent of Lands occupied by Government	98,532	121,427

SUMMARY.

	£	£
Effective Expenditure (Ordinary and Extraordinary)	4,304,592	4,042,481
Depreciation of Ships in Commission	129,639	111,111
Rent of Lands occupied by Government	98,532	121,427
Grand Totals £	4,532,753	4,275,019

* The charges under this head are diminished by transfers to other accounts, for Construction and Maintenance of Ships.

Russian Navy Estimates, 1900.

(Converted at £1 = 9·6 Roubles.)

Heads of Expenditure.	1900.	1899.
	£	£
Central and Ports' Administration	237,014	204,690
Salaries and Assistance	52,744	52,553
Educational	102,469	93,148
Medical Establishments and Services	108,885	101,887
Pay of Officers, Seamen, etc.	494,941	463,876
Victualling	183,441	129,979
Clothing	246,631	204,494
Expenses of Ships in Commission	1,489,351	1,224,872
Hydrographic Department	83,867	103,752
Hydrographic Survey of the Mouths of the Yenesei and Obi	5,698	
Naval Armaments and Electric Lighting	905,837	958,200
New Construction	2,402,128	3,548,181
Repairs and Refits	652,801	
Admiralty Yards and Workshops	488,463	420,722
Buildings, Rent and Repairs	421,825	373,405
Building and Maintenance of Lighthouses	28,646	—
Retired Pay	82,344	75,937
Various Expenses	149,534	129,469
Works of Port Alexander III.	333,333	541,667
Improvement of the Port of Vladivostok	312,500	
Improvement of Port Arthur	312,500	—
Expenditure on account of Estimates of 1901	26,369	25,770
Total . . . . . . £	9,121,321	8,652,602

United States Navy Estimates, 1900 and 1901.

Converted at £1 = $4 8665 (Par, as adopted by Congress).

Detailed objects of Expenditure and Appropriation.	Estimates, 1900.	Appropriations, 1900.	Estimates, 1901.
	£	£	£
Pay of the Navy	2,615,111	2,774,102	2,631,315
Pay, Miscellaneous . . .	102,743	102,743	102,743
Contingent Navy	4,110	2,055	2,055
Bureau of Navigation . .	107,906	108,796	93,366
Naval Academy	40,163	40,100	42,703
Bureau of Ordnance . . .	645,867	645,867	514,357
" Equipment. . .	537,440	568,261	666,285
" Yards and Docks .	93,177	93,177	109,590
Public Works—			
Yards and Docks . .	1,206,573	1,200,203	2,829,233
Naval Academy . . .	480,836	147,950	415,288
Naval Observatory . .	2,055	2,055	6,267
Bureau of Medicine and Surgery.	39,556	39,556	36,987
" Supplies and Accounts	661,755	661,755	661,755
" Construction and Repairs	672,630	672,630	1,293,706
" Steam Engineering .	248,207	402,588	570,060
Marine Corps	518,618	518,618	554,890
Increase of the Navy—			
Construction and Machinery	1,231,357	1,231,357	3,849,400
Armour and Armaments .	821,916	821,946	821,946
Equipment	41,097	82,194	51,372
Total . . .	£10,071,147	£10,110,953	£15,253,318

Draft of a New Bill additional to the Act concerning the German Navy of the 10th April, 1898.*

We, William by the Grace of God, German Emperor, King of Prussia, &c., decree in the name of the Empire, and with the assent of the Federal Council and the Reichstag, as follows :—

I. Establishment of Ships.

Section 1.

1. The establishment of ships fixed by the Law of the 10th April, 1898, concerning the German Fleet, shall be increased by

(*a*) *Ready for service :*

1 Flagship for the fleet.
2 Squadrons each of 8 battleships.
2 large cruisers } as scouting vessels for the fleet in home
8 small cruisers } waters.
5 large cruisers } for Foreign Service.
5 small cruisers }

(*b*) *As a Reserve :*

2 battleships
1 large cruiser } for Foreign Service;
2 small cruisers }

and shall be decreased by :

2 divisions each of 4 coast-defence ships.

2. In this increase the 8 coast-defence ships shall be accounted as battleships until they are replaced.

* This Act provided that the establishment of the German fleet should consist of one flagship, and two squadrons, each of eight battleships, two large and eight small cruisers for scouting with the home fleets, and five large and five small cruisers for foreign service, as well as a reserve of two battleships and one large and two small cruisers, the cruisers being intended for foreign service. The eight coast-defence vessels (Siegfried class) were to be reckoned as battleships until they should be replaced by new ships.

II. Ships in Commission.

Section 2.

As a consequence of this increase, the ships of the fleet in home waters shall be kept in commission upon the following principles:—

1. The first and second squadrons constitute the active fleet and the first and third squadrons the reserve fleet;

2. Of the battleships and cruisers of the active fleet, all shall be in commission; of those of the reserve fleet, half of the battleships and cruisers.

3. Individual ships of the reserve fleet not in commission may be commissioned for the manœuvres.

III. The Provision of Means.

Section 3.

The financial provision made necessary by this law shall be borne upon the annual estimates of the Empire.

Given, etc.

I.—The Necessity and Extent of the Increase of the Navy.

The German Empire requires peace at sea.

To the German Empire the security of its economical development, and more especially of its foreign trade throughout the world, is a vital question. To attain this the German Empire requires not only peace on land but peace also at sea, not peace *at any price*, but peace *with honour*; making provision for its legitimate requirements.

A naval war waged on account of economical interests, and more especially of commercial interests, will probably be of extended duration, since the aim of an enemy of superior strength will be attained all the more completely the longer the war lasts. Moreover, a naval war which, after the German naval forces had been annihilated or shut up, would be confined to the blockade of the coasts and the capture of merchant vessels on the high seas, would cost little to the enemy; on the contrary, the cost of such a war would be amply covered by the simultaneous impetus given to the enemy's own trade.

An unfortunate naval war, of even one year's duration only, would annihilate Germany's sea trade, and thereby bring about the most calamitous conditions, in the first place in economical respects, and, as an immediate consequence thereof, in social respects also.

Quite apart from the consequences which the conditions imposed in making peace after such a war might entail, the destruction of the maritime trade during the war could not be made good, even after the termination of hostilities, within a measurable period, and consequently a grave economical decline must be added to the sacrifices involved by the war.

The "Navy Act" of 1898 insufficient.

The "Navy Act" has not made provision for the possibility of a naval war against a great Sea Power, because, when it was drafted in the summer of 1897, the immediate object was to ensure the execution of the programme of naval construction of 1873, by the building of up-to-date vessels, while limiting the increase to that small number of battleships which were necessary, at least for a double squadron, to carry out the organisation imperatively suggested by tactical considerations.

The preamble to the "Navy Act" did not leave any doubt as to the military value assigned to the battle fleet. It says expressly: "If opposed to superior naval powers the fleet merely counts as a *sallying* fleet" (Ausfallflotte), that is to say, the fleet must retire into a port and await a favourable opportunity for a sortie. Even supposing it scores a success in such a sortie, it will nevertheless suffer considerable loss of vessels, as well as the opponent. The enemy possessing a superior number of vessels may replace his losses, but we cannot. In a war against a materially superior Naval Power, the fleet provided for in the Naval Construction Act will render a blockade difficult, more especially during the earlier stage of the war, but it will never be able to prevent it. It can merely be a question of time, for sooner or later the fleet will be overpowered or enclosed in its own port after being considerably weakened. As soon as this has happened, no great country can be shut off more easily than Germany from any sea traffic worth speaking of, both of German vessels and of vessels of neutral Powers. For this purpose there is no need of blockading long stretches of coast-line; it will suffice to blockade a few large seaports.

In the same way as the traffic of German ports, the German merchant vessels on all seas are at the mercy of an enemy more powerful at sea. Hostile cruisers on the chief trade routes—in the Skagerrak, in the British Channel, in the north of Scotland, in the Straits of Gibraltar, at the entrance of the Suez Canal, and at the Cape of Good Hope—will render German shipping traffic almost an impossibility.

On this point also the preamble to the Naval Construction Act expresses itself in no doubtful terms. It sets forth that: "The protection of the maritime trade on all seas is mainly contemplated in times of peace; in case of war, it will be the task of the cruisers on foreign stations to afford protection, 'as far as possible,' to our merchant vessels." That is to say, the men-of-war will do what they "possibly" can. What is "possible" in this respect will appear when we remember that the Naval Construction Act provides in all for forty-two cruisers; while, for instance, the greatest Naval Power of the day has already as many as 206 cruisers (ready or on the stocks), and, moreover, commands points of support and coaling-stations on all the principal trade routes.

The necessity for a strong navy.

Under the existing circumstances, in order to protect Germany's sea trade and colonies, there is one means only, viz., Germany must have a fleet of such strength that, even for the mightiest Naval Power, a war with her would involve such risks as to jeopardise its own supremacy.

For this purpose it is not absolutely necessary that the German fleet should be as strong as that of the greatest Sea Power, because, generally, a great Sea Power will not be in a position to concentrate all its forces against us. But, even if it should succeed in confronting us in superior force, the enemy would be so considerably weakened in overcoming the resistance of a strong German fleet that, notwithstanding a victory gained, the enemy's supremacy would not at first be secured any longer by a sufficient fleet.

In order to attain the proposed aim, viz., protection of our sea trade and our colonies by insuring Peace with Honour, Germany will require, in accordance with the proportions of strength of the great Naval Powers, and in consideration of our own tactical formations, two double squadrons of efficient line-of-battle ships, with the requisite complement of cruisers, torpedo-boats, etc. As the Naval Construction Act provided for two squadrons only, the construction of a third and fourth squadron must be provided for. Out of these four squadrons, each two will form one fleet (or double squadron). The second fleet is to be organised, as regards its tactical composition, in exactly the same way as the first fleet provided for in the Naval Construction Act.

As regards the extent to which vessels should be kept commissioned in peace time, we must be guided by the following considerations. As, even after the projected increase has been carried out, the number of vessels of the German Navy will still be more or less inferior to that of other individual Great Powers, our endeavours must be directed towards compensating this superiority by the individual training of the crews, and by tactical training by practice in large bodies.

A satisfactory personal training of individual crews, as well as sufficient tactical training by practice in large bodies, can only be guaranteed by permanent commissioning in peace time. Economy as regards commissioning of vessels in peace time means jeopardising the efficiency of the fleet in case of war. The minimum extent of commissioning in peace time would be the permanent formation of a fleet comprising the best and most modern vessels, as an active force constantly commissioned, *i.e.*, a force in which all the battleships and cruisers are in commission. This fleet will form the school for the tactical training in the double squadron, and in case of war will bear the first brunt. As regards the second fleet, which will comprise the older battleships, it will have to suffice if one-half of the number of its vessels only are in commission. Of course, for the purpose of practice in larger bodies, it will be necessary to commission certain further vessels temporarily, for manœuvres. In the event of war this second fleet, the reserve fleet, protected by the active battle fleet, will have to supplement the inferior training of its various crews and the insufficient practice in manœuvring in large bodies, by making good this deficiency after mobilization.

A summary of the intended organisation of the active and reserve fleets, showing their tactical composition, and the extent to which they are to be commissioned in peace time, will be found in annexe 1.*

If Germany possesses four squadrons of first-class battleships, it will be less important to have a coast defence squadron consisting of small ironclads.

Increase of vessels on foreign stations.

Besides the increase of the fleet at home, an increase of vessels on foreign stations will also be necessary. In consequence of the seizure of Kiau-Chau and the great increase of our transmarine interests during the last two years, it has already become necessary to send out to foreign stations, at the expense of the scouting vessels of the battle fleet, two more large vessels than were provided for under the programme of the Naval Construction Act.† For an effective representation of our interests, even still more vessels ought to have been sent out, had they but been available. In order to realise of what importance an increase of the number of vessels on foreign stations would be, we must remember that they are the representatives of German military force abroad, and that frequently it may be

* The provision is for 34 battleships, 8 large cruisers, 24 small cruisers, and 80 torpedo-boats, with a battleship for each squadron and a torpedo-boat for each division as a reserve; the active fleet to comprise 1 flagship and 4 divisions (2 squadrons), each division having 4 battleships; 4 divisions of cruisers, each consisting of 1 large and 3 small cruisers; and 4 torpedo flotillas, each in 2 divisions, and each division comprising 5 boats. The organization of the reserve fleet will be exactly the same, but in peace time only 2 battleships of each division, 2 divisions of cruisers, and 8 torpedo-boats will be in commission.

† Reference is made to an appendix to the Bill: "The growth of German maritime interests from 1896 to 1898."

incumbent on them to gather the fruits which the sea power created by the home fleet of the Empire has matured. Moreover, a sufficient representative force on the spot, supported by a strong fleet at home, will in many cases avert differences, and will thus also contribute, on its part, towards the maintenance of peace with a full preservation of German honour and German interests.

A particular explanation of the additional requirements cannot be given for an extended period, in the same way as is done in the case of the home fleets fundamentally organized.

If it is demanded that the foreign fleet shall be able—

1. To represent German interests everywhere energetically in peace time;
2. To be equal to warlike conflicts with countries across the sea not possessing any strong navy—

an increase by at least five large and five small cruisers, and one large and two small cruisers as a reserve, appears necessary. The Naval Construction Act provides for three large and ten small cruisers ready for use, and three large and four small cruisers by way of reserve.

A distribution of the foreign service fleet upon the foreign stations cannot be given, as this distribution may depend on political circumstances, and decisions on this point can only be made specially for each case.

II.—Carrying out the Increase. The Expense. Provision of the Funds.

It will be impossible to postpone the increase until after the expiration of the sexennate.

If once the necessity of such a strong fleet for Germany is recognised, it will no longer be possible to dispute that the honour and welfare of the country imperatively demand that the home fleet should be increased *as soon as possible* to the requisite strength.

With the Estimates for 1900 the increase of the Navy provided for by the Naval Construction Act will have been completed with the exception of one small cruiser. After the Budget has been passed, all ships for the *increase* of the fleet will be laid down in the course of the summer. Those remaining for the subsequent years will be *new vessels substituted for old ones.* For the next three years the construction scheme of the Naval Construction Act projected the laying down of five large and seven small vessels to replace old ones (Ersatzbauten). As regards the small vessels estimated for, these are intended to replace totally obsolete vessels, entirely unfit for active service. If we proceed, in the first place, with the construction of "substitutes" for them as being particularly urgent, scarcely any funds will remain for the laying down of *large* ships, since the sum estimated for this purpose in the shipbuilding scheme, viz., thirty-five millions of marks, is almost entirely consumed by increases of prices as regards the cost of the other ships, and by an increase, which has become requisite, of the stores of ammunition. If, therefore, we were to confine ourselves to the limit of the funds estimated for in the Naval Construction Act, we should be unable to lay down any more large vessels between 1901 and 1903.

In consequence of the urgent need of strengthening the navy on the one hand, and owing to the restriction placed on the construction of large vessels by the limits of the Naval Construction Act on the other, it became an imperative necessity that the bringing in of a Bill for the increase of the fleet should not be delayed until the end of the six years' Budget period, but should take place at once.

Explanation of the shipbuilding scheme.

The constructions for the increase of the fleet (Vermehrungsbauten) should expediently fit into the gaps left by the constructions to replace old vessels (Ersatzbauten), falling due next year, and above all, on account of their heavy cost, the construction of substitutes for large vessels must be considered.

If we disregard the limit as to funds set by the Naval Construction Act, and only consider the age of the vessels, it will be found that the following vessels require replacing:—*

1. In 1901: Seven large vessels (four of the Sachsen type and the König-Wilhelm, Kaiser, Deutschland);
2. In the twelve years from 1902 to 1913: Three large vessels (Oldenburg, Kaiserin Augusta, Siegfried);
3. In the four years from 1914 to 1917: Seventeen large vessels (seven Siegfried type, four Brandenburg type, five Hertha type, Fürst Bismarck).

Taking into consideration the vessels required to be built as substitutes for old vessels, it would therefore be necessary to carry out the requisite increase of the Navy during the years from 1902 to 1913. But even then the annual activity in shipbuilding would still be so very unequal, that it appears appropriate to distribute the total requirement of forty-six large ships equally over sixteen years, and, as a rule, to lay down three large vessels every year. Judging from the experience of the last few years (during which, likewise, three large vessels were laid down annually), there is no reason to doubt that this rate of shipbuilding can be kept up.

As regards small cruisers it would be expedient to adopt a similar

* A nominal list of the vessels is given in an appendix.

rate of building. Within the next sixteen years twenty-nine ships will require to be replaced, while the increase should be sixteen ships. Consequently a building programme calculated for sixteen years would involve laying down, as a rule, three vessels each year.

Concurrently with these, there will be the construction of torpedo-boat divisions, gun-boats, and special service vessels.

As regards the term of endurance (Lebensdauer) of the new large torpedo-boats we are as yet lacking experience. Assuming it to be sixteen years, it will be necessary to lay down one torpedo-boat division each year (viz., four additional divisions for the new squadron, and twelve divisions as substitutes for those of the existing two squadrons and the coast-defence armour-clad squadron).

The number of the gunboats and special service vessels required during the next sixteen years, by way of substitutes and additions, cannot be estimated at present.

Consecutive order of construction.

It still remains for us to decide as regards the consecutive order of the constructions. In order to attain greater naval efficiency as soon as possible, it is necessary, in the first place, to complete a third squadron, consisting of modern battleships with accompanying vessels. After the completion of this squadron it will be possible to form an active battle-fleet, consisting of seventeen line-of-battle ships of the latest construction, and a reserve battle-fleet, consisting (with the exception of the Brandenburg type) of inferior ironclads, four of the Brandenburg type, four of the Sachsen type, eight of the Siegfried type, and the Oldenburg.

It is true that this would necessitate a postponement of the replacing of the Sachsen type by new vessels until after the additional constructions for supplementing the active battle fleet. This is a material disadvantage; nevertheless it appears permissible, as the Sachsen type vessels have been subjected to reconstruction on an extensive scale during the last few years. Of course this has not converted coast-defence ironclads twenty-five years old into high-class modern line-of-battle ships; nevertheless the vessels are still seaworthy, and are but slightly inferior in utility to the Siegfried class, likewise belonging to the reserve battle-fleet. A building programme drawn up in accordance with the above considerations is contained in Annexe II., 2d.*

Cost of shipbuilding and armament.

To execute this building programme there will be required, for shipbuilding, including torpedo-boat divisions (additions and substitutes), on the basis of the unit prices as by the Budget for 1900, a total of 1306 million marks (£64,000,000), or on an average 81·6 million marks (£4,000,000) annually.

* See pp. 438 and 439.

But it will further become necessary to increase the strength of the heavy ordnance in battleships and large cruisers, and the provision of ammunition for all types of vessels. Consequently, somewhat higher unit prices have been assumed in the calculations, whereby we arrive at an average annual requirement of 87·6 millions of marks (£4,292,000). But still further expense will arise:

(1.) Through the construction of gunboats and special service vessels.

(2.) Through the necessary reconstruction of older vessels, as far as this expense cannot be met from the estimates for current expenditure.

(3.) Through increase of cost in consequence of technical improvement.

(4.) Through increase in the cost of materials and wages.

The presumable monetary needs to meet these requirements cannot be estimated. In the monetary estimate an addition of 12·4 millions of marks (£607,600) annually extra has been put on to the above average annual quota of 87·6 millions of marks. We thus arrive at a presumable average requirement, for shipbuilding and armament, of a sum of about 100 million marks (£4,900,000) annually.

Summary of the Total Expenses of Naval Constructions and Armaments.

(Additional vessels and substitutes for old ones.)

No.	Description.		Amounts.
			£
1	Remaining instalments on vessels already voted, or demanded for 1900, for procuring ammunition, &c.. .		6,630,000
2	*For the construction and armament of—*		
	28 battleships at £1,225,000 each . . .	£34,300,000	
	18 large cruisers at £882,000 each . .	£15,876,000	
	45 small cruisers at £269,500 each . .	£12,127,500	
	16 torpedo-boat divisions at £294,000 each	£4,704,000	
			67,007,500
	Total . .		73,637,500
	To be deducted—		
	The remaining instalments of votes for construction and armament of vessels, falling due after 1916 . .		4,964,000
	Balance . .		68,673,500
	Making for the years from 1901 to 1916 an annual average of		4,292,400
3	Additional amount for construction of gunboats and special service vessels, for reconstruction, increases of price, &c.		607,600
	Making a total annual average of		£4,900,000

DISTRIBUTION OVER 16 YEARS OF VESSELS TO BE BUILT AS

Bu'get Year.	Battleships.	Large Cruisers.	Small Cruisers.	Total.
1901	2 additions	1 substitute.	1 addition 2 substitutes	
1902	2 „	1 „	3 substitutes	6
1903	2 „ (1901–1905: For active battle Fleet.)	1 „	2 „	5
1904	„	1 addition (1904–1905: For active battle Fleet.)	3 additions (1904–1905: For active battle Fleet.)	6
1905	2 „	1 „	3 „	6
1906	2 substitutes.	1 addition	2 additions	5
1907	1 substitute.	2 „ (1906–1908: For foreign service.)	3 „ (1906–1908: For active service.)	6
1908	1 „	2 „	2 additions 1 addition (1908–1909: For the reserve battle Fleet.)	6
1909	1 addition for the reserve battle fleet.	...	1 addition 2 substitutes	5
1910	1 substitute.	2 substitutes.	3 substitutes.	6
1911	2 substitutes.	1 substitute.	3 „	6
1912	2 „	1 „	3 „	6
1913	2 „	1 „	2 „	5
1914	2 „	1 „	3 „	6
1915	2 „	1 „	3 „	6
1916	2 „	1 „	3 „	5
Total	28 Battleships.	18 Large Cruisers.	45 Small Cruisers.	91

"SUBSTITUTES" FOR OLD ONES, AND AS "ADDITIONS" TO THE FLEET.

Torpedo Boat Division.	Remarks.
1 substituted division 1 additional division 1 " " 1 " " 1 " " } For active battle Fleet.	1ST BUILDING PERIOD, 1901–1905. *Battleships.* 10 additions to supplement the active fleet by a squadron of 1st class battleships (1 flagship, 8 ships for active service and 1 as a reserve). *Large and Small Cruisers.* Building the small cruisers still remaining on account of the Naval Construction Act. 2 + 6 additions for the two groups of scouting vessels belonging to the new squadron. Substitutes for unserviceable vessels. *Large Cruisers.* König Wilhelm, Kaiser, Deutschland. *Small Cruisers.* Zieten, Blitz, Pfeil, Arcona, Alexandrine, Greif, Schwalbe. *Torpedo Boat Divisions.* 1 substituted division, 4 additional ones for the two torpedo boat flotillas belonging to new squadron.
1 substituted division. 1 " " 1 " " 1 " "	2ND BUILDING PERIOD, 1906–1909. *Battleships.* Substitution of ships of the "Sachsen" type. 1 addition (reserve ship for the squadron of the "Brandenburg 4" and "Sachsen" types). *Large and Small Cruisers.* 6 + 7 additions for foreign service, 2 additions as a complement of the scouting groups of the coast defence ironclad divisions. Substitutes for 2 small cruisers: "Sperber," "Bussard." *Torpedo Boat Divisions.* 4 substituted divisions.
1 substituted division. 1 " " 1 " " 1 " " 1 " " 1 " " 1 " "	3RD BUILDING PERIOD, 1910–1916. *Battleships.* 1910–1914. Substitutes for the 8 coast defence ironclads of the "Siegfried" class and the "Oldenburg." 1915–1916. Substitutes for the 4 battleships of the "Brandenburg" class. *Large and Small Cruisers.* Substitutes for the "Kaiserin Augusta," the 5 ships of the "Hertha" class, for the "Fürst Bismarck" as well as for the small cruisers. Falke, Cormoran, Comet, Seeadler, Geier, Wacht, Condor. Meteor. Jagd. Irene, Gefion, Niobe, D. Prinzess Hela, Nymph, E. Wilhelm. Gazelle. C. F. *Torpedo Boat Divisions.* 7 substituted divisions.
16 Torpedo Boat Divisions.	

Increase of personnel.

The increase in the number of vessels will necessitate an increase of the personnel by 35,551 officers and men by the year 1920 (when the vessels laid down in 1916 would be ready for sea).

	Altogether.	Annual Average.
Executive Officers	1212	60
Engineers	283	14
Medical Staff	188	9
Paymasters	122	6
Men	33,746	1687
Total . . .	35,551	1776

(Here follow particulars of other extraordinary (non-recurring) expenses; the necessary enlargement of the dockyards and harbour works; coast fortifications; barrack and hospital accommodations; depôts for ordnance, torpedo and mining material; a calculation of other non-recurring expenses; the increase of regular expenses; the increase of total expenditure for naval purposes (from 196 million marks [£9,604,000] in 1900 to 232 million marks [£11,368,000] in 1916); and the mode of providing the funds).

III. Fixing the Increase by Legislative Act.

The necessity of fixing, by law, the strength and organisation.

By the passing of the Naval Construction Act the *necessity* of placing the strength and organisation of the Navy on a legal footing by Act of Parliament has been *acknowledged*. This fact alone implies that an increase of the Navy should require to be similarly settled by Act. Although under these circumstances there is no need to give further reasons for the necessity of fixing this by law, we will nevertheless recapitulate once more, below, the reasons why the Federal Governments consider the settlement of the matter in the form of an Act indispensable.

The Federal Governments consider that an increase of the Navy will only be able to accomplish the intended purpose, viz., securing peace even if confronted with the mightiest Naval Power, provided that it is carried out to the full extent in which it is projected. Fractions of a squadron will not constitute a formation, and from the naval point of view can only be ranked as an addition to the reserve of matériel. Nor will the limitation to three squadrons suffice, because the object of the increase would not be realised thereby.

It is therefore necessary, before the execution of the programme is taken in hand, that the legislative powers should agree *as to whether the entire programme is considered in order, and is to be executed.* On this question a decision must be arrived at, and a decision of lasting validity. This can only be ensured by an Act.

Before commencing the execution the entire programme must be passed.

Apart from these considerations, fixing the projected increase by Act is requisite for the following reasons:

1. It is only by fixing the proposed increase by Act that the determination to create the fleet can be characterised. Without such determination, expressed in a manner leaving no doubt, considerable difficulties will arise as regards the carrying through of the great project, both from the point of view of personnel and the matériel.

   It is only if a positive guarantee is afforded for the carrying through of the programme that the participation of any considerable number of capable shipyards in the construction of the fleet can be depended upon, as it is only in such event that they will care to invest the capital required for the expensive plant for building men-of-war. But a healthy competition will only be rendered possible if the Navy is not limited to a very few large contractors.

   It is only if a further development of the Navy is rendered certain by an Act that there will be a sufficient supply of naval cadets, boys and volunteers—that is to say, of persons who wish to enter the Navy as a profession for life.

   It is only if the purpose is assured by an Act that the internal arrangements of the Navy, and in particular the extension of dockyards and harbour works, can be adapted from the first to the subsequent requirements.

2. *The unanimous decision by Federal Council and Reichstag,* providing by legal enactment for the increase of the Navy to twice its former strength, is of the greatest importance as regards the respect which Germany will command abroad, and hence as regards the entire political and economical development of the German Empire.

Objections raised against an enactment.

In reference to fixing the increase by enactment, the objection has been raised that the time required for carrying out so large a programme is so long that it cannot be foreseen whether the technical, political, and financial preliminary conditions of such an enactment might not be radically altered in the meantime.

The Federal Governments do not consider any such radical alterations probable. If, contrary to expectation, they should occur, the two legislative powers, between them, are able at any time to alter the Naval Construction Act, together with the proposed new Bill. Similar conditions exist with regard to any legislation.

Surely no one could believe that the Federal Governments would ever oppose any alteration of the Naval Construction Act which may become necessary in consequence of altered technical or naval conditions of warfare.

It is further objected that the very Naval Construction Act has shown that the proposed strength of a navy, that is to say, the aim of its development, is not a matter suited for being fixed by enactment, and that, therefore, the same mistake ought not to be made again.

This objection can only be accounted for by an improper conception of the Naval Construction Act. That Act consists of two parts, viz., one of permanent applicability, which in the former transactions of the Reichstag was referred to as the "Aeternat," and another part, of transient importance only, which was referred to as the "Sexennat."

The first is the essential part. It regulates the strength and organisation of the fleet (Art. I., 1), the building of vessels to replace old ones (Art. 2), the commissioning of ships (Art. 3), and the *personnel* (Arts. 4 and 5). The object of the new Bill is to enlarge, in the essential part,—the "Aeternat"—the strength (Art. I., 1), and as a necessary consequence thereof, the rules as to commissioning (Art. 3). The other enactments of the Naval Construction Act are not affected hereby.

The second part of the Naval Construction Act, the so-called "Sexennat," solely serves to fix a stated period for attaining the statutory number of vessels, but *not for carrying out the construction of vessels to replace old ones* (See Art. I., 3), as the construction of vessels for substitution will continue in an unbroken series even beyond the "sexennial" period. In the discussion of the Naval Construction Act in the Reichstag, it was the enactment of a stated period for providing the prescribed number of vessels which, from considerations of fiscal principle, met with by far the greatest objection. The consequence was, that the actual essence of the Naval Construction Act was thus thrust into the background, and that it was the legal enactment of a fixed period for procuring the proposed number of vessels (that is to say, that part which after the passing of the Budget for 1900 has been assured and thus rendered objectless), which in public opinion imparted to the Act its character, and the name of the "Sexennat."

In consequence of the difficulties then experienced, and in acknowledgment of the fact that it is not altogether unobjectionable to fix by law a stated time for the execution of a programme, the carrying out of which will take a lengthy period, the Federal Governments have considered that they should desist from such a demand, and have solely confined themselves to demanding the fixing, by legal enactment, of the purpose of the projected increase of the Navy, and of the commissioning of ships necessitated thereby. In doing so, they believe that the Reichstag, having agreed to the aim of the development, will do its utmost to further this purpose towards its realisation in proportion to the financial capabilities of the Empire.

Demand of the Federal Governments in respect of the Bill.

INDEX.

B.

C.

D.

E.

F.

G.

H.

O.

P.

Q.

R.

S.

T.

LONDON: PRINTED BY WILLIAM CLOWES AND SONS, LIMITED,
STAMFORD STREET AND CHARING CROSS.

Printed in Dunstable, United Kingdom

91221457R00332